QM 556 .P74 1997
OCLC:35025785
HUMASON'S ANIMAL TISSUE
  TECHNIQUES

H*umason's Animal Tissue Techniques*

HUMASON'S
Animal Tissue Techniques
FIFTH EDITION

JANICE K. PRESNELL
and
MARTIN P. SCHREIBMAN

The Johns Hopkins University Press
Baltimore & London

Humason's Animal Tissue Techniques is the fifth edition of *Animal Tissue Techniques* by Gretchen L. Humason, © 1962, 1967, 1972, 1979 by W. H. Freeman and Company, San Francisco.

© 1997 The Johns Hopkins University Press
All rights reserved. Published 1997
Printed in the United States of America on recycled acid-free paper
06 05 04 03 02 01 00 99 98 97   5 4 3 2 1

The Johns Hopkins University Press
2715 North Charles Street
Baltimore, Maryland 21218-4319
The Johns Hopkins Press Ltd., London

Library of Congress Cataloging-in-Publication Data will be found at the end of this book.
A catalog record for this book is available from the British Library.

ISBN 0-8018-5401-6

*In memory of Gretchen Lyon Humason,
a true pioneer in histotechnology*

Contents

Preface to the Fifth Edition *xiii*
Preface to the Fourth Edition *xvii*

Part I Introduction

1 *Historical Overview of the Development of Microtechnology* 3
2 *Practical Aspects of Laboratory Safety* 7

Part II Basic Procedures

3 *Fixation* 17
    Chemicals Commonly Used in Fixatives 19
    Maceration 24
    Fixing the Tissue 25
    Washing the Tissue 26
    Fixatives and Their Uses 27
    Fixation by Perfusion 38
    Postfixation Treatments 40
    Decalcification 40
    Other Methods of Tissue Preparation 43
4 *Dehydration: Preparation for Embedding* 45
5 *Clearing, Infiltrating, and Embedding: Paraffin Method* 48
    Clearing 48
    Dehydration and Clearing Combinations 49
    Infiltrating with Paraffin 50
    Embedding (Blocking) with Paraffin 51
    Timing Schedule for Paraffin Method 52
    Automatic Tissue Processors 54
6 *Microtomes and Microtome Knives* 55
    Microtomes 55
    Microtome Knives 56
7 *Paraffin Sectioning and Mounting* 58
    Sectioning 58
    Mounting 62
8 *The Microscope* 67
    The Compound Microscope 67
    The Operation of a Microscope 69

viii   Contents

      Measuring Devices Used on a Microscope   73
      Specialized Microscopy   74
9 Stains and Staining Action   84
      Natural Dyes   84
      Mordants   85
      Synthetic Dyes   88
      Nature of Staining Action   92
      Standardization of Stains   93
10 Mounting and Staining Procedures   95
      Mechanical Aids   95
      Processing Slides for Mounting   96
      Cover Glass Mounting   96
      Mounting Media (Mountants)   98
      Aqueous Mounting Techniques   100
11 Hematoxylin Staining   101
      Single Solutions   102
      Double Solutions   103
      Substitutes for Hematoxylin Solutions   105
      Counterstains (Plasma Stains) for Hematoxylin, Gallocyanin, and Hematein   106
      Hematoxylin Staining Procedures   108
      Hematoxylin Substitute Procedures   115
      Red Nuclear Staining   116

Part III   Specific Staining Methods

12 Staining Connective Tissue and Muscle   121
      Mallory Staining   121
      Trichrome Staining   126
      Collagen and Elastin Staining   131
      Bone Staining   136
      Muscle Staining   140
13 Silver Impregnating Reticulum   143
      Silver Impregnation   143
      Silver Impregnation for Reticulum   146
14 Silver Impregnating and Staining Neurological Elements   153
      Glia   154
      Astrocytes   157
      Nerve Cells, Processes, and Fibrils   160
      Myelin   172
      Degenerating Axons   177

15 Periodic Acid-Schiff, Feulgen Techniques, and Related Reactions 182
Schiff Reagent 182
Schiff Reactions 183

16 Staining Hematological Elements and Related Tissues 193
Blood Smears 193
Blood Tissue Elements and Inclusion Bodies 201
Hemoglobin Staining 205
Bone Marrow Staining 208
Staining for Fibrin 212

17 Staining Pigments and Minerals 214
Staining for Iron 214
Bile Pigment (Bilirubin) Staining 217
Melanin and Lipofuscin Staining 219
Staining for Calcium Deposits 222
Removal of Pigments 224

18 Staining Proteins and Nucleic Acids 228
Protein Staining 228
Nucleic Acid and Nucleoprotein Staining 235
Control Slide Techniques 240

19 Staining Lipids and Carbohydrates 246
Lipids 246
Carbohydrates (Saccharides) 253

20 Staining Cellular Elements 264
The Argentaffin Reaction 264
Enterochromaffin Cell Staining 264
Amyloid Staining 269
Metachromasia 271
Mast Cell Staining 274
Endocrine Gland Staining 276

21 Staining Golgi Apparatus, Mitochondria, and Living Cells 290
Golgi Apparatus Staining 290
Mitochondria Staining 295
Supravital Staining 299

22 Staining Microorganisms 302
Staining Bacteria 302
Staining Spirochetes 308
Staining Fungi 312
Staining of Rickettsiae and Inclusion Bodies 317

Part IV Histochemistry and Miscellaneous Special Procedures

23 *Histochemistry*   327
    Fixation   328
    Dehydrating and Embedding   331
    Cryostat Sectioning   334
    Sectioning without a Cryostat   336
    Acetone Fixation and Embedding   337
    General Suggestions   338
    Alkaline Phosphatase   340
    Acid Phosphatase   345
    Aminopeptidase (Proteolytic Enzyme)   348
    Esterases (Nonspecific) and Lipases   350
    Succinic Dehydrogenase   353
    The Oxidases   356
    Peroxidase Methods   357
    Substrate Film Methods   359
    Osmium Black Methods   360

24 *Immunohistochemistry*   361
    Immunostaining Methods   362
    Troubleshooting (or Dealing with the Unexpected)   373
    General Immunohistochemical Directions and Hints   375
    A Standard Table   378

25 *Microwave Histology*   380

26 *Special Procedures I*   386
    Freezing Techniques   386
    Fixation, Blocking, and Sectioning   386
    Mounting   388
    Rapid Staining Methods   389
    Nitrocellulose Method   391
    Ester Wax Embedding   392
    Sectioning   393
    Glycolmethacrylate Processing for Thin Sections   397
    Epoxy Resin Processing   400

27 *Special Procedures II*   404
    Exfoliative Cytology   404
    Gynecological Cytology   404
    Nongynecological Cytology   405
    Sex Chromatin   408
    Chromosomes   410
    Permanent Mounts   413
    Culture Methods   416
    Chromosome Analysis   419

28 Special Procedures III   *422*
        Preparation of Invertebrates for Whole Mounts and Sections   422
        Preparation of Chick Embryos   431
        Whole Mounts   433
        Animal Parasites   442
        Fish and Reptiles   447
        Handling Small Tissue Samples   448

29 Special Procedures IV   *449*
        Autoradiography   449
        Procedures for Autoradiographs   450

Part V   Solution Preparation and General Laboratory Aids

30 Solution Preparation   *461*
        Abbreviations and Terms   461
        Stock Solutions   462
        Stain Solubilities   477

31 General Laboratory Aids   *481*
        Labeling and Cleaning Slides   481
        Restaining Faded Slides   481
        Recovering Broken Slides   482
        Restoring Basophilic Properties   482
        Two Different Stains on One Slide   482
        Reclaiming and Storing Specimens   483
        Removing Laboratory Stains from Hands and Glassware   486
        Suppliers of Equipment and Chemicals   487

References   *491*
Index   *559*

Preface to the Fifth Edition

Gretchen Lyon Humason passed away December 2, 1991, in Oak Ridge, Tennessee. We knew her both personally and through the use of her text, *Animal Tissue Techniques*. As we now revise this text, a task that was initiated while she was alive and alert, her presence is strongly felt. Her desires, her philosophies, and what she thought we were to accomplish have formed an integral part of this book. We include here her preface as it appeared in the fourth edition (1979) because her words are so similar to our own goals and mission.

Gretchen L. Humason graduated from the University of California at Los Angeles in 1929 with a B.A. degree in zoology. She began working as an assistant in the Zoology Department at UCLA and advanced the following year to lecturer in microscopic technique. In 1933, she completed a master's degree in vertebrate paleontology. Her work toward her doctorate was interrupted by World War II. At that time she was called upon to teach a premedical crash course to army and navy personnel. She remained in the Zoology Department until 1958. After her husband Louis retired from his position as a physicist at UCLA, they moved to New Mexico. Humason began working at the Health Division of the Los Alamos Scientific Laboratories in New Mexico. She was a research assistant to Dr. C. C. Lushbaugh and, when he later went to the Associated Universities in Oak Ridge, Tennessee, in 1963, she went too. She continued her research on the spontaneous colon cancer of the marmoset monkey until her final official retirement in 1989.

Gretchen L. Humason was aware that the Johns Hopkins University Press would probably publish a revised edition of *Animal Tissue Techniques*. She truly wanted the book to continue. When she first wrote the book in 1961, she stated that she had written it for the students. She believed in her students and wrote in her first preface, "students are a blessing because there is no surer way to master a subject than to teach it." We agree. We have striven to retain her goals and aspirations in preparing this new edition. We are preparing this revision with love and dedication to the subject and with deep respect and sincere admiration for this gentle and generous person. We sincerely hope that we have retained the flavor and utility of the book and that it will reach as many of the new, as well as the old, generations of microtechnicians as did the previous editions.

This fifth edition, now known as *Humason's Animal Tissue Techniques,* is written in five parts. Part I includes a brief overview of the history of microtechnique so that the new student can develop a perspective of the field; for the experienced individual it reviews interesting developments leading to the present state of the art. Considerable attention has been given to the area of safety and the handling of hazardous laboratory materials in chapter 2—safety is an integral component to the training and practice of laboratory workers. We have also included a section devoted to special provisions for the physically challenged in the laboratory. We believe this to be a seminal inclusion for a text of this kind—essential and long overdue.

Part II covers those basic procedures and general considerations with which every tissue technician should be familiar. Part III provides detailed information about specific staining methods for most tissues. An instructor might choose a few favorite methods from this section to round out a course, while the professional technician will find here most of the specific methods required on the job. Part IV deals with special procedures, those that are special in the sense that they are not common in most laboratories, although they may be very important in some. Although the discussion of some of these procedures is brief, references have been cited extensively for the benefit of those who might wish to refer to more thorough discussions. Chapter 24 in this section is devoted to immunohistochemistry and related procedures and is all new. Part V is devoted largely to laboratory aids and the preparation of solutions—useful information in any laboratory. In this edition some methods are removed, new ones are added, and others are modified. The list of references has been carefully amended to cover the recent important publications in the field. Information on suppliers cited in the text has been consolidated for easier reference.

For those planning to take the registry exam to become certified as a histologic technician or technologist, we recommend the following supplemental references: *Histotechnology: A Self Instructional Text* (by Frieda Carson; ASCP Press, 1990), *Theory and Practise of Histotechnology* (by Denza Sheehan and Barbara Hrapchak; distributed by Shandon Lipshaw, 1980), *Theory and Practise of Histological Techniques* (by J. D. Bancroft; Churchill-Livingstone, 1990), and *Laboratory Methods in Histotechnology* (by Edna Proffitt et al.; AFIP, 1992). The current guidelines for taking the certification examination are available from the American Society of Clinical Pathology in Chicago (800-621-4142).

We express our sincere appreciation and gratitude to the following individuals, who have facilitated the preparation of this work:

—The editors and staff at Johns Hopkins University Press, present and past, including Richard O'Grady, Eric Halpern, and Robert Harrington.
—Professor Emeritus Louis G. Moriber, who introduced MPS to the world of microtechnique in an undergraduate class at Brooklyn College and who continues to this day to be his mentor, advisor, and friend.

—The countless students who have passed through our laboratories and classes providing inspiration and the platform to instruct them in our beloved subject matter.

—Our families, especially Harold and Lucia, who provided the love, understanding, and encouragement to bring our desires to fruition.

—And, finally, Gretchen Lyon Humason, who was mentor to many, including the authors, and who encouraged us to complete this work.

—Janice K. Presnell
Martin P. Schreibman

Preface to the Fourth Edition

This book of basic and standard histological procedures (and some specialized techniques) was designed to meet the diverse needs of premedical students, medical technicians, zoology majors, and research assistants. Most histological reactions follow a logical and specific sequence, and I have attempted to include simplified discussions of the basic methods that are applicable both to normal and to pathological conditions in zoology and medicine.

It is not intended that this text should be a complete reference book on histology; the experienced worker knows of numerous such tomes, as well as journals that specialize in histology and related topics. However, special methods of wide usage and exceptional merit are included, particularly those that are not overly complicated or unpredictable. It is hoped that technicians, once familiar with the material covered here, will watch the literature for modifications and improvements of standard techniques; in this way, with this book as a foundation, their work can be kept up to date and, perhaps, simplified.

Methods for fixation are fairly well established, with only occasional variations. The section on fixation presented herein is as modern as I can make it, and it includes a brief description of the chemicals employed. Old staining techniques continue to be perfected and new ones developed; I have tried to include the best of these and, for the sake of the student, to adapt them to the standard three-hour laboratory period and to the kinds of equipment most widely available. Some special methods that are more time-consuming have been included for special projects and research. They have been simplified wherever possible to serve as introductory techniques for the student who plans to proceed to more complicated techniques later.

Some instructors may not agree with the way I have organized the text, but to me it is logical. Thus fixation is treated first because it is usually the first process in tissue preparation; this is followed by embedding in some kind of medium, sectioning on a microtome, mounting sections on slides, and, finally, staining them with the help of a microscope. A logical arrangement of staining methods is hard to come by, so I have followed my own inclinations: Some sections are organized by related tissues, others by related methods. The latter was considered desirable for such processes as silver impregnation, metachromasia, and the use of Schiff reagent. The final chapters include such specialized techniques as histochemistry, chromosome preparation, autoradiography, and invertebrate mounts. Wherever possible, I have referred to my own experience with these methods to help students succeed with their first efforts, and I have included modifications that might appeal to other adventurous technicians.

This book is in four parts. Part I covers those basic procedures and general considerations with which every tissue technician should be familiar. Part II provides detailed information about specific staining methods for most tissues. An instructor might choose a few favorite methods from this section to round out a course, while the professional technician will find here most of the specific methods required on the job. Part III deals with special procedures, those that are special in the sense that they are not common in most laboratories, although they may be very important in some. Although the discussion of some of these procedures is brief, references have been cited extensively for the benefit of those who might wish to refer to more thorough discussions. Part IV is devoted largely to laboratory aids and the preparation of solutions—useful information in any laboratory. In this edition some methods are removed and new ones added, others are modified. The arrangement of chapters has been altered by moving freezing, nitrocellulose, and other techniques to the Special Procedures section. This leaves the first two parts devoted almost exclusively to the paraffin method and the use of the microscope to examine the finished slides.

In the fourth edition, *Animal Tissue Techniques* has been extensively revised and updated. Many of the changes have been to improve its usefulness for graduate and undergraduate teaching. The typography has been altered and the design improved with an eye to making the book more readable and, hence, more useful to students and technicians alike. The list of references has been carefully emended to cover recent important publications in the field. Information on all suppliers cited in the text is consolidated for easier reference in one list beginning on p. 574.

Many of us have not regarded with proper respect the potentially dangerous materials that are used in the laboratory. Accidents do happen, so a section is included on pp. 572–74 concerning the hazards of laboratory materials. This should be called to the attention of all students and technicians.

To have included everything necessary to satisfy everyone and still to have kept the price of the book within the means of the average student would have been impossible. Some topics, necessarily, have been treated only in passing. The electron microscope, for example, is much too specialized for students in beginning technique classes, and an entire book could be devoted to instructing students in its operation alone. The topic of photomicrography is equally complex. Methods for preparing plastic whole mounts have not been included; excellent leaflets on the subject are published by the companies that supply the materials necessary for their preparation. Good color photographs are helpful, but they are also, unfortunately, expensive—even a few of them can add appreciably to the cost of a book. In my teaching, I have used a demonstration set of slides to help my students recognize proper staining. The set started with a few of my own slides, and it was gradually enlarged by additions from the students in my classes. The students were happy to contribute examples of their best work, and the collection eventually increased to several hundred excellent slides. Other instructors might consider building a study collection of slides in the same way.

I have derived invaluable personal satisfaction from my association with students. I am grateful to them for helping me to develop my tolerance and patience—two qualities that are essential in my profession. I am grateful to them, too, for what they have helped me to learn, for there is no surer way to master a subject than to teach it to others. One former student in particular should receive credit for her encouragement and for prodding me toward writing this book—Marlies Natzler of the University of California at Los Angeles.

Grateful acknowledgments are also due to Marvin Linke, Jeanne Simmons, and Leta Burleson, the three artists who contributed to the four editions of this book; to Julie Langham, for help with photography; to Nellie M. Bilstad, for valuable suggestions; to the Cytogenetics Division of Oak Ridge Associated Universities, for information about late developments in chromosome preparation; to the Zoology Department of the University of California at Los Angeles, for the lessons I learned there as a student, a departmental technician, and a lecturer; and to Dr. C. C. Lushbaugh, for his continued encouragement.

—Gretchen L. Humason
September 1978

PART I
Introduction

1
Historical Overview of the Development of Microtechnology

The invention of a primitive microscope in 1590 by the Janssen brothers, two Dutch opticians, opened virgin horizons for the investigation of fundamental biological principles. In the seventeenth century such well-known scientists as Leewenhoek, Hooke, and Malphighi constructed their own "microscopes," which yielded, because of limited magnification, little more information than could be discerned with the naked eye. Nevertheless, the observations made by these investigators are astounding; consider the description of the "cell" by Robert Hooke in 1865, which formed the foundation for the cell doctrine theory and modern biology as proposed in the mid–nineteenth century by the German scientists, Schleiden and Schwann, after the microscope as we now know it had been constructed. Development of the microscope necessitated development and refinement of histological methods that would permit the analysis of tissue sections (i.e., microtechnique); together they were responsible for the remarkable advances made in the study of such basic biological principles as growth, development, differentiation, and basic microscopic anatomy.

In the years following the 1850s, the importance of viewing minute and delicate structures was realized. The last quarter of the nineteenth century witnessed tremendous improvement in the microscope. This was due, in great part, to a partnership that developed between Ernest Abbe, who designed microscopes, and Carl Zeiss, who produced them. By 1875 instruments of exceptional quality were being manufactured. Concomitant with improvements in the microscope came the refinement and expansion of microtechnique, essentially based on technology first developed in England and later expanded in Germany. Many of the names of the scientists of that period are associated with microtechnique methods that are still in common use. The ability to study the details of cell structure ushered in what many people consider to be the golden age of cytology.

The fundamental steps in microtechnique needed to prepare material for microscopic examination (i.e., fixation, embedding, sectioning, staining, clearing, and mounting) developed as the need for refinement of methodology was dictated by advances in microscopy and means of examination. In order to study tissues and cells with the microscope, it became essential to make them thin and transparent. The first attempt at this was to utilize a method of dissociation, that is, to tear apart (or tease) tissues into smaller compartments.

This method was augmented by further softening of the intercellular connective tissue elements by first soaking or macerating them in weak alcohol or salt solution. Ranvier's methods for maceration developed in 1868 are still in use today. However, it soon became apparent that a method of cutting thin sections, rather than dissociating cells, would provide considerable advantage by retaining cellular elements in their original morphological association with other cells.

Contrary to popular notion, the first devices (microtomes) to cut thin freehand sections of plant material were developed in England as early as 1770 (Cummings 1770; Adams 1798; Custace 1799). The instrument made by Cummings was a hand model that held the specimen in a cylinder and raised it for sectioning with a screw. Pritchard adapted the instrument in 1835 to a table model by fastening it to a table with a clamp and cutting across the section with a two-handled knife. These instruments were called *cutting machines* until Chevalier introduced the term *microtome* in 1839.

Freehand sectioning continued to be used in Germany until the mid-1860s, when improved staining techniques demanded the development of methods and microtomes for producing thinner and more uniform sections. The first microtomes did not appear in Germany until 1843 (Oschatz); they consisted essentially of a screw device to advance the tissue plus a guide for a knife. Procedures to support animal tissues after fixation (i.e., by infiltrating tissues with a supporting, hard substance) paralleled the development of special microtomes. The sliding microtome (see Chap. 6), invented by Rivet (1868), served well with collodion (celloidin)-supported and frozen tissues. Although paraffin preceded the use of celloidin to support tissues, it did not become popular until the *rotary microtome* was invented by Pfeifer in 1886. One year later, Minot developed the rotary microtome into an instrument similar to the ones we now use. The Spencer Lens Company manufactured the first clinical microtome in 1901. The large Spencer rotary microtome, with its increased precision, became available in 1910.

Tissue sections were originally mounted dry or with water onto glass slides. In 1832 a combination of gum and isinglass (a semitransparent whitish gelatin prepared from the air bladder of fishes and used in jellies and glues—Prichard 1832) was introduced along with Canada balsam (T. J. Cooper 1835) as a mounting medium. This was later replaced by the use of glycerine (Warrington 1847) and a mixture of glycerine and gelatin (Lawrence 1859). Canada balsam was popular as a mounting material from its inception; however, it become more practical for use with soft animal tissues only after methods were introduced for dehydrating specimens in alcohols and clearing them in turpentine (Clarke 1851).

To be able to study and distinguish between different components of biological tissues, it became necessary to increase their visibility by accentuating their differences through the use of stains. Hartwig (1854) first discovered that killed tissues were more easily stained than fresh, living ones. Müller

(1860s) provided the first step toward killing and hardening tissues by *fixing* them in potassium dichromate, washing, and following with alcohol for additional hardening. Expansion in the uses of fixing agents was made by Lang (1878, mercuric chloride), Kleinenberg (1879, picric acid), Flemming (1880, chromic, osmic, and acetic acids), Rabl (1884, formic acid), and Blum (1893, formaldehyde).

J. von Gerlach (1858) is credited with bringing staining into general use by demonstrating that dyes could be used to color parts of the tissue differentially. He used carmine, an extract from the bodies of cochineal insects, to stain nuclei and other cell components. It has been suggested that this stain had been used earlier (Hill 1770; Ehrenberg 1838; Goeppert and Cohn 1849; Corti 1851) than von Gerlach's introduction.

Hematoxylin, another natural extract derived from logwood, was first used by Boehmer in 1865 for biological staining; Boehmer discovered that its combination with alum enabled coloring nuclei sky blue. This observation was used by Kleinenberg (1876), Weigert (1884), Ehrlich (1886), Mayer (1891), Benda (1891), and Heidenhain (1892) to provide a variety of staining characteristics using hematoxylin as a base. The use of two (Schwartz 1867) and three (Flemming 1891) contrasting stains to accentuate differences between nucleus and cytoplasm followed soon after.

As brilliant and attractive aniline dyes were introduced, their value for biological staining was appreciated. The use of aniline dyes, which are derived from coal tars, was begun in 1856 in England by William Perkin. However, it was in Germany that the use of coal tar derivatives really flourished. Thousands of dyes were developed, primarily for use in the textile industry, but some began to wend their way into the field of biological staining of plant and animal material. Beginning in 1880, George Grübler, a German pharmacist, began to make and sell biological stains to scientists around the world—the name Grübler became synonymous with fine stains. Karl Holborn bought the company from Grübler and continued his work. World War I brought an end to the sale of Grübler dyes to England and America. For many years after World War II, the company functioned under the name of Chroma Gesellschaft-Schmid GmbH and Co. They are now represented by Cell Point Scientific in Rockville, Maryland.

It was H. J. Conn and his associates in the early 1920s who began the Biological Stain Commission to oversee the production of dyes in the United States as the German ones disappeared from the market. It was their mission to test and standardize all dyes. If approved, the stains were given a commission certification number, which assured the chemical content and purity of the compound.

Beginning in the early 1900s, coincident with the development of modern physiology and related sciences such as endocrinology, there developed the need to assess cell and tissue function by analyzing histological sections, that is, interpreting dynamic function by studying fixed, static structures. This interpretation of fixed processed tissues was accomplished by comparing cell number, cell and nuclear size, and staining characteristics and was soon augmented by assessing biochemical activity by using specialized staining (i.e., for chemical groupings) or by biochemical activity (i.e., localizing specific enzymes and substrates and determining their rate of disappearance). The employment of immunological methods in histo- and cytochemistry (ICC) in the mid-1970s enabled the localization of proteins within specific cells and tissues (see Chap. 24). The emergence of molecular biological approaches during the 1980s permitted the use of in situ hybridization (ISH) to visualize which cell types in a complex tissue are expressing the gene of interest. These two powerful tools, ICC and ISH, enabled the investigator to differentiate between storage and synthesis of a particular protein. At last histophysiologists had an adequate (but by no means foolproof) method of relating structure and function!

The explosion of technology that we are currently experiencing is producing bewildering and astonishing innovations in microtechnology. Four centuries after the appearance of the first compound microscope, we are using newly developed, highly sophisticated instruments that enable us to peer down to the atomic level. We are also witnessing the introduction of heretofore inconceivable tools for microscopic analysis; automated devices for processing tissues; and user-friendly, computer-assisted means for evaluating dimensions, shapes, contents, and activity of cells (see appropriate chapters). The advances in technologies have also ushered in increased rapidity and accuracy in data gathering, evaluation, and reporting and have in part contributed to the profound literature explosion that is plaguing conscientious scientists.

What the future portends can only be imagined and finally evaluated as it is experienced. It is needless to say, however, that the histotechnicians and histologists, as well as researchers (the people!), remain essential components to successful laboratories and scientific programs. (*Note:* Previous citations in this chapter do not appear in the reference section; they are included here merely for historical interest.)

Suggested Readings

For additional reading on the history of microtechnique, see Titford (1993), Bracegirdle (1987), and McCormick (1987).

2
Practical Aspects of Laboratory Safety

Potentially hazardous conditions are commonplace, and they exist as well in microtechnique laboratories. The physical safety of the personnel must be safeguarded at all times, and good laboratory habits must be established to prevent accidents. These concerns should extend to the presence of nonskid floors and adequate access ways.

To begin with, think before you act. Give thought to any chemical, especially dangerous ones, and always read labels carefully. Have respect for equipment. Inspect electrical instrumentation periodically. Replace frayed or loose electrical wiring and make sure a ground plug is used. Microtome knives should be handled with care at all times. Never leave knives in place on the microtome when you leave the instrument for an extended period. Use knife edge guards on the microtome and lock the handwheel while removing sections from the knife edge.

All laboratories are governed by the Occupational Safety and Health Administration (OSHA), and the various rules are established on a state as well as a national level. The Environmental Protection Agency (EPA) wants to make sure that all hazardous chemicals are properly disposed. All laboratories should have a safety manual and regular in-service practice of these regulations should be enforced. The National Institute for Occupational Safety and Health (NIOSH) has published a guide regarding chemicals and their hazards. OSHA also instituted a "right to know" policy, and everyone must be informed about the dangers and hazards in their laboratory. The Centers for Disease Control (CDC) and the National Committee for Clinical Laboratory Standards (NCCLS) has initiated "universal precautions" for the handling of potentially infectious human tissues and body fluids. For clinical and some research laboratories, the universal precautions must be posted and adhered to at all times.

Animal tissue should not be processed or sectioned on instruments that have been used for human specimens unless the universal precautions are part of the daily routine. Procedures regarding decontamination should be on hand in case of such contamination. The basic procedure is to use a tuberculocidal hospital disinfectant (wear gloves and protective clothing); clean with detergent and then a bleach dilution (1:10), and rinse with water. Dispose as a biohazard waste. Various manufacturers of equipment will also offer a recommended procedure.

The proper kinds of fire extinguishers must be on hand (dry chemical and CO_2 extinguishers are recommended), in addition to strategically placed sand buckets. All laboratories should be well ventilated, and an exhaust or fume hood should always be used with noxious chemicals. Make a habit of keeping all containers, including staining set-ups, covered when not in use. To decrease evaporation, staining dishes could be covered with clear plastic wrap before putting on the glass lids.

Commercial containers of chemicals now have label warnings for their safe use and antidotes for misuse. Manufacturers are required to provide material safety data sheets (MSDS), which give information about safety, precautions, and hazards, as well as the proper methods for disposal. Read them! If your laboratory does not have an MSDS for a chemical, do not use that chemical. Make certain that the labels are tightly adhered to the container. If some of the contents are transferred to smaller shelf stock containers, place a similar permanent label on that can or bottle. Poisons and carcinogens must be clearly labeled as such. Use these chemicals with care. Never allow any chemical to remain on the skin; wash off immediately with soap and warm water. Neutralize acids with mild alkalis and alkalis with weak acids. Whenever eyes are involved, consider medical attention. After flooding with water, liquid spills can be absorbed on paper toweling and allowed to evaporate in a hood or open area outdoors. Solid spills can be picked up on paper to be disposed of safely.

Some Specifics

Formaldehyde is now considered a carcinogen, and all laboratories are governed by a federal regulation on the use of this chemical. Latex gloves should be worn when handling specimens preserved in formalin solutions. Protective clothing and eye/face shields are also recommended during gross dissections. All containers of formaldehyde should be labeled with a hazard warning. Periodic monitoring of exposure should be done to ensure overall safe levels. Formaldehyde can be recycled. Check with local waste water treatment facilities before disposing in the sewer. If necessary, a licensed waste removal firm should be employed.

Acids are highly reactive with water. Never add water to acids (e.g., H_2SO_4, HNO_3) when measuring them for dilution. Add the acid *slowly* to the water, and avoid breathing the fumes. Keep concentrated acids in partially empty small bottles. Store large stock bottles in a safe place, preferably a cool one (never warm). If acids are involved in a fire, never use water; use a dry chemical or a CO_2 extinguisher. Acids have a corrosive action on the skin; wash off immediately with continuous washing with soap and warm water.

The caustic *alkalis* are corrosive and must be washed off thoroughly, preferably first with a weak acid or vinegar and then by repeated washing with water. Get medical attention for any in the eyes.

Alums can cause burns. Wash thoroughly.

Xylene and other clearing agents are very flammable and should be stored in explosion-resistant cabinets. These reagents should be used only in well-ventilated laboratories. Prevent prolonged breathing of the fumes. Xylene may be distilled and reused, or proper disposal is necessary. Do not dump in a sink. Whenever you have a doubt about the disposal of a chemical, call the local sewage treatment facility for advice. If contact is made with skin, wash off immediately and cleanse thoroughly with soap. Rewash several times.

Chromic acid is a strong oxidizer and is toxic. It can cause inflammation and ulcers on the skin. Wash off immediately and thoroughly. Avoid inhalation; it can damage the respiratory tract. Do not use the drain for disposal; contact a licensed waste removal firm.

Ethers are highly flammable; a can of ether can be ignited by a spark or open flame. Use a flame hood if available. A charge can even take place when pouring ether; ground it by keeping the two containers in contact when pouring. Quickly and tightly cover burning vapors to cut off oxygen. Ether, in contact with air and exposed to light, could contain peroxides and then become explosive. Keep tightly closed in a metal can or brown bottle, and store in an explosion-proof refrigerator, an important component of every laboratory. Ethers are not highly toxic but will cause dizziness and possible collapse. Dioxane and ethylene oxide are considerably more toxic than ether and isopropyl alcohols. They can be absorbed through the skin and may produce nausea and vomiting and even liver and kidney damage. Proper disposal is necessary. Use a licensed waste hauler.

Ethylene glycol: Avoid breathing the vapor and wash well after contact on skin. Use propylene glycol whenever possible. Ethylene glycol is very toxic and must be properly disposed of by a licensed waste removal firm.

Iodine is poisonous; avoid contact. Medical attention is required if it is swallowed. Check with the local waste water treatment facility for allowable amounts of discard and if necessary contact a licensed waste hauler.

Nitrocellulose: Avoid inhaling. To dispose of nitrocellulose, allow the solvent to evaporate in a fume hood or open outside area. It is highly flammable.

Osmium tetroxide is toxic; keep it from skin and eyes and avoid inhaling the fumes. Proper disposal is necessary.

Oxalic acid can be absorbed in the blood to form calcium oxalate. Wash it off immediately. The small amounts used in the laboratory should be no problem. Neutralize dilute solutions with sodium bicarbonate, and discard in drain (if local waste treatment facility permits).

Periodic acid is an oxidizing agent and can cause skin burns. Wash well.

Peroxides could be irritating to the skin, eyes, and respiratory tract. Wash peroxide off immediately if it contacts the skin. Keep bottles from heat. It should be stored in the refrigerator to prolong its shelf life.

Picric acid can be explosive. It must always be kept as an aqueous solution; do not allow it to dry. Handle with respect.

Silver compounds become dangerous when combined with ammonia. Aging of these solutions or exposure to air or light can form explosive silver compounds. Silver nitride and silver azide may be formed if formalin or alcohol is present in the solution. Do not store silver solutions. Prepare them only just before use and in clean glassware (none with silvered sides). Unused solutions can be inactivated by adding NaCl or dilute HCl and discarded down a drain if local waste water regulations permit. Silver compounds can also be recovered by precious metal recoverers.

Tetrahydrofuran (THF) produces eye and nose irritation, dermatitis, and, with long exposure, kidney and liver damage. Store in safety cans. If it is spilled, leave the area to allow it to evaporate, but do not spread it by dosing with water. THF is not recommended for routine use, and proper disposal by a licensed waste hauler is necessary.

Mercury is frequently used in histopathology labs. It is considered very hazardous and must be disposed of properly.

All shelves holding bottled chemicals and stains in solution should have a barrier across the front to prevent shifting and falling of bottles.

OSHA Rules and Regulations

All laboratories are governed by OSHA and must be in compliance with the Hazard Communication Standard. There are six basic principals to which all laboratories should adhere.

1. Employers must have material safety data sheets (MSDS) for any hazardous chemical used in the work place. If not available from the manufacturer of the chemical, your state OSHA office will be willing to help.
2. These MSDS have to be readily accessible to everyone during work hours.
3. There must be a written hazard communication program available.
4. All hazardous chemicals are required to be labeled with the identity of the chemical and the appropriate hazard warning. The chemical name should correspond to that on the MSDS.
5. Employees must be provided with information and training on hazardous chemicals, and this training should be updated whenever new hazards are introduced.
6. Employers must provide annual reviews of hazardous chemical information and training in their use.

Most facilities now have a safety committee that helps individual departments develop proper safety documentation.

Recommended Further Reading

Crookham, J. N., and Dapson, R. W. *Hazardous Chemicals in the Histopathology Laboratory: Regulations, Risks, Handling & Disposal*, 3rd ed. Battle Creek, MI: Anatech (1995).

State and federal guidelines (they vary from state to state) can be obtained from the appropriate offices and should be readily available for consultation.

Special Provisions for the Physically Challenged Student

With a few minor adaptations, many physically challenged students can function efficiently in a tissue technique laboratory. The first area to be addressed should be safety and the need to guard against mishaps. Safety is a concern for all students, but there must be special emphasis on safety for individuals with specific physical limitations. Fire extinguishers and fire blankets should be within easy reach. Other safety measures require readily accessible eyewash stations and emergency shower pulls. Eye protection and a heavy rubber apron to protect against spills may be necessary as special requirements.

A work station with lower (preferably adjustable) counters and benches should be available. If this is not feasible, a ramp with a platform that can accommodate wheelchairs would deal with this problem. This platform could be moved to other laboratory areas as needed.

Positioning all needed materials for a procedure should be avoided for it adds clutter to the physically challenged student's working environment. Most physically challenged people can readily adapt to less than perfect work areas and can transport the necessary items and chemicals to the work space as needed. Many physically challenged individuals can transfer to other seating and are not limited to the wheelchair. The laboratory chairs and stools with rollers can also give greater flexibility to the work area.

A decision will have to be made regarding the possible limits of the physically challenged individual. In some projects he or she may have to be an observer. This is still firsthand exposure and will enable the student to understand the procedure. Many laboratory sessions incorporate a lab partner system, and this can greatly assist the student with limitations. Good handouts are always an assistance to any special procedure. The best possible seating arrangement should be made in consideration with the student to accommodate hearing or visual disabilities.

Many of the basic histology instruments can be modified or are commercially available in models that make them easier to use. For instance, microtomes and cryostats can be operated with a foot pedal (or hand pedal) for sectioning. Touch keyboards, now available, can control all functions of the microtome, except for the actual placing of the paraffin block in the microtome or the insertion of the knife or blade. Some cryostats have a height adjustment to allow the user to sit or stand. Microscopes can be interfaced with computers to allow image enhancement or attached to video monitors for better viewing.

For more information on accessible labs and classrooms, contact:

Heath Resource Center
American Council on Education
One Dupont Circle, Suite 800
Washington, DC 20036
(202)939-9320

American Association for the Advancement of Science
1333 H Street, N.W.
Washington, DC 20005
(202)326-6400

For more information on helping students who are deaf, contact:

Alexander Graham Bell Association for the Deaf
3417 Volta Place
Washington, DC 20007
(202)337-5220

For more information on helping blind or visually challenged people, contact:

American Council of the Blind
1010 Vermont Avenue
Washington, DC 20005
(202)467-5081

For more information on helping students with mobility impairments, contact:

Spinal Cord Injury Network
American Paralysis Association c/o Kernan Hospital
2200 Kernan Drive
Baltimore, MD 21218
(410)448-2500

For more information on communication aids and technology for people who cannot speak, contact:

Artificial Language Laboratory
Michigan State University
405 Computer Center
East Lansing, MI 48824
(517)353-5399

For more information on helping students who have learning disabilities, contact:

National Center for Learning Disabilities
99 Park Avenue
New York, NY 10016
(212)545-7500

PART II

Basic Procedures

3
Fixation

As soon as a tissue ceases to be alive, its cells start to change. Multiplying bacteria begin to destroy them, and the process of autolysis (self-digestion) by intracellular enzymes begins to dissolve them. The activity of these enzymes is reversed from that in live cells; instead of synthesizing amino acids into proteins, they begin to split protein into amino acids. These amino acids diffuse out of the cells; as a result, cell proteins are no longer coagulable by chemical reagents. These cell changes are called *postmortem conditions* and must be prevented if tissue is to be examined in the laboratory.

The prevention of postmortem conditions is the primary objective of tissue preparation, but it is also necessary to treat tissue to differentiate the solid phase of the protoplasm from the aqueous phase, to change cell parts into materials that will remain insoluble during subsequent treatment, and to protect cells from distortion and shrinkage when subjected to such fluids as alcohol and hot paraffin. Other important objectives of tissue preparation are to improve the staining potential of tissue parts and to alter their refractive indices for better visibility.

The procedure used to meet these requirements is called *fixation,* and the fluids used are called *fixatives* or *fixing solutions*. An ideal fixative should:

1. Penetrate rapidly to prevent postmortem changes.
2. Coagulate cell contents into insoluble substances.
3. Protect tissue against shrinkage and distortion during dehydration, embedding, and sectioning.
4. Allow cell parts to be made selectively and clearly visible.
5. Prepare the tissues for staining.
6. Some fixatives may have a mordanting effect on the tissue—that is, they combine insolubly with it—and enhance the attachment of dyes and proteins to each other.

Suffice it to say that there is no single ideal fixative. Ensuing discussion will show that each fixing solution has limitations; that is why there are a number from which to choose.

Tissues should be placed in fixatives as soon as possible after death. If delay is unavoidable, they should be put in a refrigerator, thus reducing autolysis and putrefaction to a minimum until the fixative can be applied.

In addition, because a single chemical seldom has all the qualities of a good fixative, a fixing solution is rarely composed of only one chemical—familiar exceptions are formalin and glutaraldehyde. Most reliable fixatives contain one or more coagulant chemicals and one or more noncoagulant chemicals. Coagulants change the spongework of proteins into meshes through which paraffin can easily pass, thus forming a tissue of the proper consistency for sectioning. They also strengthen the protein linkages against breaking down during later procedures. Used alone, however, coagulants may form too coarse a network for the best cytological detail or may include the formation of artificial structures (artifacts). Noncoagulants produce fewer artifacts, but if used alone they give the tissue a poor consistency for embedding.

Thus, the most efficient fixing fluids are combinations of protein coagulants and protein noncoagulants. No fixing solution is ideal because all chemical agents cause some chemical change in the protein structure of the cells. The choice of fixative will depend on the type of investigation to be undertaken, and an effort should be made to discover what adverse effect the fixing agent may have on the cellular components under study. It is always prudent to use more than one fixative to compare the different effects they may have on the same tissue sample.

Because they contain ingredients that act upon each other, many mixtures are most efficient when made fresh. The individual ingredients can usually be made into stock solutions, which can then be mixed together immediately before use. Among the frequently used chemicals are formaldehyde, glutaraldehyde, ethyl alcohol, acetic acid, picric acid, potassium dichromate, mercuric chloride, chromic acid, and osmium tetroxide. Since every chemical has its own set of advantages and disadvantages, each component should, whenever possible, compensate for a defect in some other component. As an example let us discuss the widely used fixative, Bouin solution: (1) Formaldehyde fixes the cytoplasm but in such a manner that it retards paraffin penetration. It fixes chromatin poorly and makes cytoplasm basophilic. (2) Picric acid coagulates cytoplasm so that it admits paraffin, leaves the tissue soft, fixes chromatin, and makes the cytoplasm acidophilic. (See p. 88 for a discussion of "acidophilia" and "basophilia.") Its disadvantages are that it shrinks the tissue and that it makes chromatin acidophilic. (3) Acetic acid compensates for the defects of both formaldehyde and picric acid.

Another example: Birge and Tibbitts (1961), by adding 0.7% sodium chloride to two fixing solutions (formalin and Bouin), reduced the amount of shrinkage caused in nuclei and cytoplasm.

When the future use of a tissue is in doubt or if it is to be stored for an indefinite time, formalin is usually the conservative choice for a fixative; it permits secondary fixation (postfixation) and will not harden excessively. If the primary objective of tissue preparation is the compilation of a simple anatomical study of cell components, then routine fixatives can be used: formalin, Gomori, Susa, Zenker, Helly, or Bouin. Special fixatives of cell inclusions are Carnoy, Flemming, Champy, Helly, Schaudinn, Regaud, and others. For histochemistry the researcher is generally limited to aldehydes, acetone, or ethyl alcohol (p. 328).

Most fixing solutions are named after the person originating them (e.g., Zenker and Bouin). If the same person originated more than one combination of chemicals, additional means of designating them have been used: Flemming weak and strong solutions: Allen B3, B15 series; and so forth. A few fixatives have been named arbitrarily; the name Susa was coined by Heidenhain from the first two letters of the words *su*blimate and *sä*ure.

CHEMICALS COMMONLY USED IN FIXATIVES

Acetic Acid

Acetic acid (CH_3COOH) is one of the oldest fixatives on record: in the eighteenth century vinegar (4–10% acetic acid content) was used to preserve hydras. In modern techniques it is rarely used alone but is an important component of many fixing solutions because of its efficient fixing action on the nucleus and its rapid penetration. It fixes the nucleoproteins but not the proteins of the cytoplasm. It does not harden the tissue; actually, it prevents some of the hardening that, without it, might be induced by subsequent alcohol treatment. In some techniques, however, acetic acid must be avoided because it dissolves out certain cell inclusions, such as Golgi and mitochondria, and some metals, such as calcium. Many lipids are miscible with acetic acid or are soluble in it. It neither fixes nor destroys carbohydrates. Pure concentrated acetic acid is called glacial acetic acid because it is solid at temperatures below 17°C.

Acids in general cause swelling in tissues—in collagen in particular—by breaking down some of the cross linkages between protein molecules and by releasing lyophilic radicals that associate with water molecules. An acid's swelling action can be a desirable property because it counteracts some of the shrinkage caused by most fixing chemicals. To curtail swelling after fixation with acetic or trichloroacetic acid solutions, tissues should be transferred to an alcoholic washing solution rather than to water.

Acetone

Acetone (CH_3COCH_3) is used only for tissue enzymes, such as phosphatases and lipases. It is used cold and penetrates slowly. Only small pieces of tissue are fixed in this chemical.

Chromium Trioxide (Chromic Acid)

Crystalline chromium trioxide (CrO_3) is called *chromic acid* when it is added to water, usually in a 0.5% amount. Chromic acid is a valuable fixative but is rarely used alone. It penetrates slowly, hardens moderately, causes some shrinkage, forms vacuoles in the cytoplasm, and often leaves the nuclei in abnormal shapes. It is a fine coagulant of nucleoproteins and increases the stainability of the nuclei. It oxidizes polysaccharides and converts them into aldehydes—an action forming the basis of the Bauer histochemical test for glycogen and other polysaccharides. To fix water-soluble polysaccharides, it is better to use acetic acid and then posttreat them with chromic acid.

Chromic acid can also be used to oxidize fats partially to make them insoluble in lipid solvents. The oxidizing action may go too far, however, and potassium dichromate, which acts in a similar fashion, is safer for this purpose and is therefore more commonly used. However, even with potassium dichromate it is important not to leave tissues in the fixative beyond the recommended time.

Excess chromic acid must be washed out because it may later be reduced (undesirably, for our purposes) to green chromic oxide (Cr_2O_2). Because formalin and alcohol are reducing agents, they must not be mixed with chromic acid until immediately before use.

Alcohols

Alcohol cannot be used as a fixative for lipids because it makes them soluble. It does not fix carbohydrates, but neither does it extract mucins, glycogen, iron, and calcium. Alcohol is seldom used alone, although it is occasionally used for fixing enzymes.

Ethyl alcohol (ethanol, CH_2H_5OH) hardens tissue but causes serious shrinkage. It is a strong cytoplasmic coagulant but does not fix chromatin. When alcohol is used, nucleic acid is transformed into a soluble precipitate and is lost in subsequent solutions and during staining.

Isopropyl alcohol (isopropanol, $CH_3CHOHCH_3$) is an excellent substitute for ethanol in processing tissues for embedding in paraffin; however, it is not generally recommended for fixation. In addition, isopropanol cannot be substituted for ethanol in the preparation of staining solutions, since many stains are not soluble in it.

Methyl alcohol (methanol, CH_3OH) is also used as a fixative, principally for hematological tissues.

Aldehydes

Formaldehyde (HCHO) is a gas sold as a solution (approximately 37–40%) in water, in which form it is known as *formalin*. The use of the term *formol* is incorrect, since a terminal -ol designates an alcohol or phenol. Unless the author of a technique specifies the dilution of formalin in terms of actual formaldehyde content, dilutions must be made from the commercial product (e.g., a 10% solution would be 10 volumes of concentrated formaldehyde [37–40% formaldehyde-saturated water] to 90 volumes of water). When using formaldehyde solutions, care should always be taken to avoid the fumes and gloves should be worn to prevent skin contact. Formaldehyde is considered a carcinogen and OSHA recommends limited contact. Always use in a well-ventilated area. For these reasons, several companies have developed "formalin" substitutes. Some of them seem to be more effective than others and should be evaluated (see Bostwick et al., 1994).

Formaldehyde, if left standing for long periods, may either polymerize to form paraformaldehyde or oxidize into formic acid. A white precipitate in a stock solution indicates polymerization; the solution has been weakened. Cares (1945) suggests the following correction: Shake the solution to resuspend the sediment; pour into autoclavable containers and seal them tightly; autoclave at 15 lb for 30 minutes; and cool. This should produce a clear solution. More dilute solutions (such as 10%) tend to oxidize more readily than do stock solutions.

Formalin, when reacting with proteins, seems to form links between adjacent protein chains. The final effect, therefore, depends on the number of chemical reactions taking place between the formalin and the reactive groups of the protein molecules. Although some of these protein linkages will be broken down by washing in water, excess formalin must be removed; enough formalin-bound proteins will remain to react with future reagents when they are applied. Histologists frequently do not realize that the quantity of irreversibly bound protein drops as the pH of the solution rises above pH 10. For this reason, formalin reacts most efficiently as a buffered solution around the neutral point of pH 7.5–8.0. Never use it unbuffered on tissues for histological study.

Formalin has a moderate speed of penetration, but its action is slow. Although formalin preserves the cells adequately, it may not protect them completely unless it is given a long time to harden the tissue. Shrinkage may take place if dehydration, clearing, and infiltration are started before the hardening action is complete.

Formalin is a good fixative for lipids; it does not dissolve lipids or fats. It does not fix soluble carbohydrates, and it does dissolve some glycogen and urea.

So-called formalin pigment may appear in tissues rich in blood. This pigment is formed when hematein of the hemoglobin escapes from red blood corpuscles before or at death and reacts with the formalin. Its formation may be prevented by a short period of fixation in formalin followed by a prolonged soaking in 5% mercuric chloride. Once formed, the pigment can be removed in a solution of 1% potassium hydroxide in 80% alcohol or in picric acid dissolved in alcohol (Baker 1958). A 30-minute treatment in either of these solutions is recommended (p. 226).

Glutaraldehyde and paraformaldehyde have special uses in histochemistry and electron microscopy. See Ericsson and Biberfeld (1967) for studies on aldehyde fixation.

Mercuric Chloride (Corrosive Sublimate)

Mercuric chloride ($HgCl_2$) is usually applied as a saturated aqueous solution (approximately 7%) and is acidic in action, owing to the release of H^+ and Cl^- ions in water. It is a powerful protein precipitant and forms intermolecular mercury links between SH, carboxyl, and amino groups. It penetrates reasonably well but not as rapidly as acetic acid. It shrinks tissue less than the other protein coagulants, hardens it moderately, and distorts the cells less than other fixatives. It is excellent for mucin.

One disadvantage of mercuric chloride is that it deposits in the tissue a precipitate. The precipitate is in part crystalline (needle-shaped), which is highly conspicuous at the microscope level. It can easily be removed with iodine in the following reaction: $2HgCl + I_2 - HgCl_2 + HgI_2$; thus, any metallic mercury is converted into mercuric iodide. Mercuric iodide is soluble in alcohol, but the brown color of iodine may remain in the tissue. This can be removed by prolonged soaking in 70% alcohol or more quickly by treatment with sodium thiosulfate, 5% aqueous. A further disadvantage is that mercuric chloride crystals inhibit freezing, making it difficult to prepare good frozen sections.

Most stains react brilliantly on tissue fixed in mercuric chloride. Chromatin stains strongly with basic stains and dye lakes; cytoplasmic structures react equally well with acidic or basic stains (see Chapter 9).

Another disadvantage with the use of mercuric chloride is the Environmental Protection Agency (EPA) regulations regarding the proper disposal of the residual mercury. Fixatives containing mercuric chloride require proper disposal by precipitating it with 13% thioacetamide solution (13.0 g thioacetamide + 100 ml distilled water; if mixed and stored in a capped bottle, it will be stable for 1 year). Barszcz and Yevich (1976) suggest that zinc chloride be substituted for mercury in Zenker-type fixatives because it is less toxic and leaves no pigment artifacts in tissues.

Procedure for Extraction of Mercury

1. For each liter of fixative, add 20 ml of thioacetamide. Mix thoroughly.
2. Leave undisturbed in a *tightly capped bottle* in fume hood for 24 hours or until a precipitate (mercuric sulfide) is formed.
3. Filter through filter paper or Buchner filter (in fume hood!).
4. Discard supernatant into the sink. The mercuric sulfide residue can be stored indefinitely or turned over to a safety official.
5. Avoid contact with skin and eyes; use under a fume hood.

Osmium Tetroxide

In solution, usually 1% aqueous, osmium tetroxide (OsO_4) takes up a molecule of water and becomes H_2OsO_5, erroneously but commonly called *osmic acid*. (Osmic acid would be H_2OsO_4.) The substance is chemically neutral, is not an acid, and cannot be isolated. Baker (1958) suggested that it might be named *hydrogen per-persomate*. Osmium tetroxide should always be handled with care and used in a well-ventilated area (under a fume hood), and special care to avoid eye and nasal contact should be taken.

In solution, the ionization of osmium tetroxide is so minute that the pH is almost exactly that of the distilled water used in making the solution. The solution penetrates poorly and leaves tissue soft, the tissue later becoming so friable in paraffin that it sections badly. Osmium tetroxide fixes the cytoplasm and nuclei, but, although it increases the stainability of chromatin in basic dyes, it reduces the stainability of the cytoplasm. Fats and other lipids reduce osmium tetroxide or combine with it to form a dark compound. Thus, fat sites become black and insoluble in absolute alcohol, cedarwood oil, and paraffin, but they remain soluble in xylene, toluene, and benzene, and if left more than 5 minutes in xylene or toluene become colorless. Osmium tetroxide is not a fixative for carbohydrates.

When fixation is complete, excess osmium tetroxide must be washed out of the tissue or it will reduce to an insoluble precipitate during treatment in alcohol. Since osmium tetroxide is also easily reduced by light and heat, it must be stored in a cool, dark place.

Osmium tetroxide is a valuable chemical for electron microscopy, and many tissues are postfixed in this solution. Good preservation can be achieved with the controlled pH of a buffer solution, which can be a veronal, symocollidine, or phosphate buffer (considered a physiological solution) (Glauert 1965).

Picric Acid

Picric acid [$C_6H_2(NO_2)_3OH$] is most often used in a saturated solution, 0.9–1.2% aqueous. It is an excellent protein coagulant, forming protein picrates that have a strong affinity for acid dyes. However, it penetrates slowly, causes extreme shrinkage, and offers no protection against subsequent shrinkage. The shrinkage has been found to be close to 50% of the original tissue volume by the time the tissue has undergone paraffin infiltration. In spite of its acidity, picric acid does not cause swelling. It does not dissolve lipids, nor does it fix carbohydrates, but it is recommended as a fixative for glycogen. It is a desirable constituent of many fixatives because it does not harden tissues, but it cannot be used alone because of the shrinkage it causes. Acetic acid is frequently used with it to counteract this undesirable characteristic. Dry crystals of picric acid can be explosive; however, as purchased it is not. It contains about 10% moisture and it is safe as long as it stays moist. This can be achieved by keeping the jar tightly capped. If it is suspected that the moisture content has fallen below 10%, add distilled water so that the reagent looks like damp sand. In some laboratories more water is added to the picric acid to play safe or picric acid is kept as a supersaturated solution, since this is the form that most preparations require. If anhydrous picric acid is required, a small amount can be dried in a dessicator overnight and the unused portion can be rehydrated. Picric acid is toxic by skin absorption. Wear gloves.

Potassium Dichromate

Potassium dichromate ($K_2Cr_2O_7$) is a noncoagulant of proteins; it makes them more basic in action but dissolves nucleoproteins. Chromosomes are therefore poorly shown, if at all. It fixes the cytoplasm without forming gross precipitates. Potassium dichromate leaves tissues soft and in poor condition for paraffin sectioning. One valuable use, however, is in preparations for fixation of mitochondria; it makes the lipid components insoluble in lipid solvents. After fixation, tissues may be soaked in 3% aqueous potassium dichromate solution to ensure that lipids will be well preserved.

Potassium dichromate can be mixed with mercuric chloride, picric acid, and osmium tetroxide, but it reacts with formalin and must not be mixed with it until immediately before use. If acetic acid is added, mitochondria will be lost, but chromosomes will be more clearly shown than they are when the chromate is used alone. It is usually necessary to wash out the dichromate with water to prevent the formation of an insoluble precipitate of Cr_2O_3; this will form if the tissue is placed directly in alcohol. Potassium is considered a carcinogen, and caution with the dust includes no eye or nasal contact.

"Indifferent Salts"

Baker (1958) applied this term to a group of chemicals (sodium sulfate, sodium chloride, and others) whose action is not clearly understood. Zenker and Helly solutions often have sodium sulfate added. Sometimes sodium chloride is added to formalin or mercuric chloride fixatives, particularly if marine forms are to be fixed or if it is necessary to hold shrinkage to a minimum.

MACERATION

Some tissues are extremely dense and cannot be manipulated while fresh. When working with these it may be desirable to separate the individual fibers of a muscle or nerve, and this is simplified by maceration. Some macerating fluids are not fixatives; these must usually be followed by some form of fixation. Other maceration fluids rely on the differential effect of weak fixatives on interstitial substances to separate the main fibers of a tissue. For example, if a fixative contains a connective tissue solvent, the solution will produce dissociation along with fixation. If a small piece of Bouin-fixed muscle is shaken in water or in 70% alcohol, the block of tissue will separate into individual fibers. Sublimate-acetic acid does not perform quite as well, but the acetic acid dissolves some connective tissue. The following are the most commonly used macerating fluids (Hale 1958; McClung 1939, 1950).

1. 30% alcohol: 24 hours or longer (up to 4 days)
2. Formalin, 1 part in 10% salt (NaCl) solution, 100 parts: 24 hours or longer.
3. 1% sodium chloride: 24 hours or longer.
4. Chromic acid, 0.2% aqueous: 24 hours.
5. Nitric acid, 20% aqueous: 24 hours. Particularly recommended for smooth muscle of bladder.
6. Boric acid, saturated solution in saline (sea water for marine forms), plus 2 drops Lugol iodine solution (p. 463) for every 25 ml: 2 to 3 days.
7. Potassium hydroxide, 33% aqueous. Good for isolation of smooth and striated muscle. After 1 to 1.5 hours, tease tissue apart with dissecting needles.
8. Maceration by enzymes. Good for connective tissue, reticulum (Galigher 1934). Place frozen sections in:

pancreatic siccum	5.0 g
sodium bicarbonate	10.0 g
distilled water	100.0 ml

    Wash thoroughly and stain.
9. 0.01% osmium tetroxide dissociates and fixes muscle fibers in a few days. 1% acetic acid added to it causes quicker dissociation and almost as good fixation. Use only small pieces of tissue.

FIXING THE TISSUE

The first consideration in the choice of fixative should be the purpose to be served in preparing the tissue for future use. Is a routine all-purpose fixative adequate, or must some special part of a cell be preserved? An aqueous fixing fluid will dissolve glycogen, and an alcoholic one will remove lipids. Thought should be given to the rate of penetration of the fluid and the density of the tissue to be fixed; obviously, an extremely dense tissue will not fix well in a fixative that penetrates slowly and poorly. If fixatives of poor penetration are used, the pieces of tissue must be as small as possible. In any case, pieces should never be larger than is absolutely necessary—the smaller the bulk, the better the fixation.

The hardening effect of fixatives should be considered. A fixative that produces excessive hardening may lead to sectioning difficulties with liver and muscle, and a fixative with the desired attributes but with less hardening effect should be used. If there is any doubt concerning the future needs of a tissue, fix it in formalin; this can be followed by postfixation treatments.

Use a large volume of fixing fluid, preferably 20 times the volume of the tissue being fixed, if possible. Remove the tissues from the animal and place them in the fixative as rapidly as is feasible, thereby keeping postmortem changes to a minimum. In most cases, do not attempt to fix an entire organ; it will be too large to allow rapid and complete penetration of the fixative, particularly if it is covered by a tough membrane. An ideal piece would be 1 to 2 cm in diameter; place a larger piece in the fixative for 15 to 30 minutes, then trim it to smaller size and return it to the fixative. This sometimes is necessary when tissue is very soft or easily crushed. Trim the piece with a new razor blade or sharp scalpel; this will cause little damage to the tissue. Care should be taken not to crush or tear the tissue while removing it because this can cause mechanical damage and artifacts. Never allow tissue to become dry before placing it in the solution; keep it moist with physiological saline. Wash off accumulated blood with saline because blood retards the penetration of a fixative. After placing the tissue in the solution, shake the container gently several times to make certain that the fluid has reached all surfaces of the tissue and that pieces are not adhering to the bottom or sides.

Thin pieces of tissue that show muscular contraction or that may turn inside out (tissues of the gastrointestinal tract are particularly likely to do this) may be placed on thick filter paper, outside wall against the paper, and dropped in the fixative. Use the same technique for sheets of tissue (e.g., the integument). Tiny, easily lost specimens—biopsies, bone marrow, and so on—may be wrapped in lens wrapper or filter paper.

The length of time required for complete fixation depends on the rate of penetration and the action of the fixative. Most coagulant fixatives produce complete fixation as fast as they can penetrate the tissue. But some fixatives, such as formalin, exhibit progressive improvement in fixing action after the tissue is completely penetrated. Prolonged action in this case improves the condition of the tissue and is rarely harmful. Occasionally, some type of postfixation treatment is advisable (p. 40).

Many fixatives must be prepared just before use, although there are exceptions. However, in all cases fixatives should not be used more than once.

WASHING THE TISSUE

After fixation is accomplished, the excess of many fixatives must be washed out of the tissue to prevent interference with subsequent processes. Often washing is done with running water; sometimes the tissue may be carried directly to alcohol, 50% or higher. Some technicians maintain, for instance, that, if tissue prepared with Bouin fixative is washed in water, it loses some of its soluble picrates and should therefore be transferred from the fixative directly to 70% alcohol plus several drops of ammonium hydroxide. When freezing is planned, formalin-fixed tissue may be washed briefly in water, but if necessary it can taken from the fixing solution directly to the cryostat. Alcohol inhibits freezing, and if it is used it must be thoroughly washed out before any attempt to freeze the tissue.

When tissue has been fixed with a mercuric chloride solution, additional treatment is necessary. Later, when staining the sections, iodine-alcohol or Lugol solution (p. 463) must be included in the staining series to eliminate the remaining crystals that would otherwise persist as dark clumps and needles in the finished slides. Tissue samples may also be washed before dehydration in 70% alcohol to which iodine crystals have been added to make a dark red-brown solution.

After washing, the tissue may be stored for several weeks or months in 70% alcohol, but it is always safer to dehydrate and embed it as soon as possible. Storage in alcohol for long periods (a year or longer) tends to reduce the stainability of tissues, as does long immersion in chromate and decalcifying solutions. Avoid the use of corks or metal caps on bottles containing fixing solutions.

FIXATIVES AND THEIR USES

Routine Fixatives and Fixing Procedures for General Microanatomy

AZF Fixative

```
zinc chloride ................................................ 25.03 g
formaldehyde (37–40%) ..................................... 150 ml
acetic acid, glacial ........................................... 9.5 ml
distilled water, fill to 1000 ml
```

Dissolve zinc chloride in about half of the water; then add water and acid. Fill to 1000 ml with distilled water. Limit processing time to less than 4 hours. Good fixative for glycol methacrylate sections and immunohistochemistries. It also eliminates the need to use mercuric chloride fixatives and corrosive acid dichromates.

Bouin Fixative

Fix for at least 24 hours—several weeks causes no damage, but long periods (months) results in poor nuclear staining.

```
picric acid, saturated aqueous ................................ 75.0 ml
formaldehyde (37–40%) ..................................... 25.0 ml
glacial acetic acid ............................................ 5.0 ml
```

Because of acetic acid content, do not use for cytoplasmic inclusions. Large vacuoles often form in tissues. Wash in 70% alcohol plus several drops of ammonium hydroxide for several days until no more yellow comes out. The yellow color must disappear before sections are stained. Otherwise, treat slides in 70% alcohol plus a few drops of saturated lithium carbonate until the yellow disappears. Some yellow will be lost in the series of alcohols.

Bouin-Duboscq (Alcoholic Bouin) **Fixative** (Pantin 1946)
Fix for 1 to 3 days

80% alcohol	150.0 ml
formaldehyde (37–40%)	60.0 ml
glacial acetic acid	15.0 ml
picric acid, saturated	1.0 g

Prepare just before using. This is better than Bouin for tissues difficult to penetrate. Go directly to 95% alcohol.

Neutral Buffered Formalin (NBF)

May remain in fixative indefinitely; action is progressive. NBF is often referred to as the "universal" fixative.

formaldehyde (37–40%)	100 ml
distilled water	900 ml
sodium acid phosphate ($NaH_2PO_4 \cdot H_2O$)	4.0 g
anhydrous disodium phosphate (Na_2HPO_4)	6.5 g

Wash in water or proceed to 70–80% alcohol for routine paraffin processing.

Formalin (Baker 1958)

May remain in fixative indefinitely; action is progressive.

formaldehyde (37–40%)	10.0 ml
calcium chloride	1 g
distilled water	80.0 ml

Wash in water. This is a good fixative for the demonstration of phospholipids.

Glutaraldehyde, see p. 328

Gomori 1-2-3 Fixative

May remain in fixative 2 to 3 weeks.

formaldehyde (37–40%)	1 part
mercuric chloride, saturated aqueous	2 parts
distilled water	3 parts

Excellent quick penetration, good for general cell morphology. Wash tissue 6 to 8 hours. Posttreat for mercuric chloride. The tissue will be hard and require soaking before sectioning (p. 61) but stains well. Humason preferred Susa or Stieve fixatives if acetic acid can be used.

Helly Fixative (Zenker Formol)

Fix for 6 to 12 hours. If tissue seems to harden excessively, follow a maximum of 18 hours in fixative with 12 to 24 hours in 3% potassium dichromate.

potassium dichromate	2.5 g
mercuric chloride	4.0 to 5.0 g
sodium sulfate (may be omitted; see Indifferent Salts, p. 24)	1.0 g
distilled water	100.0 ml
formaldehyde (37–40%; add just before using)	5.0 ml

Formalin reduces dichromate and should not be left in stock solution. The above stock, without formaldehyde, can be used for Zenker fixative (p. 80). Excellent for bone marrow, blood-forming organs, and intercalated discs. Wash tissue in running water overnight; posttreat for mercuric chloride.

Hollande Bouin Fixative (Romeis 1948)

Fix for 8 hours to 3 days.

copper acetate	2.5 g
picric acid	4.0 g
formaldehyde (37–40%)	10.0 ml
distilled water	100.0 ml
glacial acetic acid	1.5 ml

Dissolve copper acetate in water without heat; add picric acid slowly while stirring. When dissolved, filter and add formaldehyde and acetic acid. Keeps indefinitely. Wash for several hours in 2 or 3 changes of distilled water. Hartz (1947) recommends this for fixation of calcified areas, as in lymph nodes or fat necroses. It is a good general fixative.

Orth Fixative (Galigher 1934; Gatenby and Beams 1950)

Fix for 12 hours at room temperature or 3 hours at 37°C.

formaldehyde (37–40%)	10.0 ml
potassium dichromate	2.5 g
sodium sulfate	1.0 g
distilled water	100.0 ml

Mix fresh. Wash in running water overnight. A good routine fixative, also for glycogen and fat. Lillie considered this fixative to be one of the best for demonstrating rickettsias and bacteria.

Stieve Fixative (Romeis 1948)

Fix for 24 hours.

```
mercuric chloride, saturated aqueous ........................... 76.0 ml
formaldehyde (37–40%) ....................................... 20.0 ml
glacial acetic acid ............................................. 4.0 ml
```

Similar in effect to Susa but simpler to prepare. Penetrates rapidly, good for large pieces. Time not critical. Go directly to 70% alcohol. Posttreat for mercuric acid.

Susa Fixative (Romeis 1948)

Fix for 12 to 24 hours.

```
mercuric chloride saturated in 0.6% NaCl ....................... 50.0 ml
trichloracetic acid ............................................. 2.0 g
glacial acetic acid ............................................. 4.0 g
formaldehyde (37–40%) ....................................... 20.0 ml
distilled water ............................................... 30.0 ml
```

Good substitute for Zenker if potassium dichromate is not required; Susa hardens tissue less. Rapid penetration. Go directly to 70% alcohol. Posttreat for mercuric chloride.

Zenker Fixative

Fix for 4 to 24 hours. If tissue seems to harden excessively, follow a maximum of 18 hours in Zenker with 12 to 24 hours in 3% potassium dichromate. Fixation time must be controlled. The nuclei shrink during the first couple of hours and then begin to swell to normal size after the first 3 to 4 hours. Treatment beyond 8 to 10 hours reduces the quality of nuclear staining, and the sections should be treated with sodium bicarbonate (p. 481).

```
potassium dichromate ......................................... 2.5 g
mercuric chloride (substitute zinc chloride) ................. 4.0 to 5.0 g
glacial acetic acid ............................................. 5.0 ml
sodium sulfate (optional) ....................................... 1.0 g
distilled water .............................................. 100.0 ml
```

Excellent general fixative, fairly rapid penetration. Wash in running water overnight. Posttreat for mercuric chloride. Russell (1941) substituted zinc chloride for the mercuric chloride.

Fixatives for Cell Inclusions and Special Techniques

Altmann Fixative (Gatenby and Beams 1950)

Fix for 2 hours; do not fix longer or material will swell.

chromium potassium sulfate (chrome alum)	3.0 g
formaldehyde (37–40%)	30.0 ml
glacial acetic acid	2.0 ml
distilled water	238.0 ml

Good for yolk-rich material and insect larvae. Does not harden and gives good cytological detail. Wash in 70% alcohol or water.

Bilstat (Modified Carnoy's)

Fix for 1 to 8 hours.

isopropyl alcohol	60 ml (6 parts)
chloroform	30 ml (3 parts)
formic acid	10 ml (1 part)

Recommended for worms and small insects. Avoid xylene and high grades of alcohol because they harden the chitins.

Carnoy Fixative (Gatenby and Beams 1950)

Fix for 3 to 6 hours.

Formula A

acetic acid	20.0 ml
absolute alcohol	60.0 ml

Formula B (the true Carnoy)

glacial acetic acid	10.0 ml
absolute alcohol	60.0 ml
chloroform	30.0 ml

Chloroform is said to make action more rapid. Important fixative for glycogen and Nissl substance but dissolves most other cytoplasmic elements. Wash 2 to 3 hours in absolute alcohol to remove chloroform, particularly if embedding in nitrocellulose.

Methacarn (Puchtler et al. 1970)

methyl alcohol	60.0 ml
chloroform	30.0 ml
glacial acetic acid	10.0 ml

32 Basic Procedures

For embedding, transfer through two changes of methyl alcohol (2 to 3 hours and 4 hours); two changes of methyl benzoate (3 hours and 1 hour); methyl benzoate-xylene (1 hour); xylene; infiltrate and embed. Do not wash tissues in water or saline before fixation and do not contaminate with water during processing; shrinkage and alterations of tissue structure will result. Do not overexpose to alcohol; it causes shrinkage and hardening. Methacarn is recommended over Carnoy because it causes less hardening.

Carnoy-Lebrun Fixative (Lillie 1965)

Fix for 3 to 6 hours.

absolute alcohol	15.0 ml
glacial acetic acid	15.0 ml
chloroform	15.0 ml
mercuric chloride	4.0 g

This is a very fine fixative, but it does not keep; mix fresh. Penetrates rapidly. Wash well in alcohol to remove chloroform.

Champy Fixative (Gatenby and Beams 1950)

Fix for 12 hours.

potassium dichromate, 3% aqueous (3 g/100 ml water)	7.0 ml
chromic acid, 1% aqueous (1 g/100 ml water)	7.0 ml
osmium tetroxide 2% aqueous (2 g/100 ml water)	4.0 ml

Prepare immediately before use. Good for cytological detail, mitochondria, lipids, and so forth. Also recommended for invertebrates and lower vertebrates. Penetrates poorly; use only for small pieces. Wash in running water overnight.

Flemming Fixatives

Fix for 12 to 24 hours.

Strong solution

chromic acid, 1% aqueous (1 g/100 ml water)	15.0 ml
osmium tetroxide, 2% aqueous (2 g/100 ml water)	4.0 ml
glacial acetic acid	1.0 ml

Mix just before using. Acts slowly: use small pieces with no blood on outside. Wash in running water for 24 hours. High acetic acid content makes it poor for cytoplasm fixation. Excellent for demonstrating fat.

Modification for cytoplasmic study
    chromic acid, 1% aqueous (1 g/100 ml water) .... 15.0 ml
    osmium tetroxide, 2% aqueous (2 g/100 ml water) .... 4.0 ml
    glacial acetic acid .... 2 to 3 drops

Weak solution
    chromic acid, 1% aqueous (1 g/100 ml water) .... 25.0 ml
    osmium tetroxide, 2% aqueous (2 g/100 ml water) .... 5 ml
    acetic acid, 1% aqueous (1 ml/99 ml water) .... 10.0 ml
    distilled water .... 60.0 ml

Lewitsky saline modification (Baker 1958)
Omit acetic acid and replace the water with 0.75% sodium chloride.

Flemming solutions are good for mitotic figures, but are not suitable for general work because they penetrate poorly, harden excessively, blacken material, and interfere with the action of many stains. The weak solution is a fine fixative for minute, delicate objects. For dense tissues, use the strong solution. Iron hematoxylin is excellent following these fixatives.

Formol-Alcohol (Tellyesniczky) (Lillie 1965)

Fix for 1 to 24 hours.

    70% alcohol .... 100.0 ml
    formaldehyde (37–40%) .... 5.0 ml
    glacial acetic acid .... 5.0 ml

Good for insects and crustaceans. Widely used by botanists. Transfer to 85% alcohol.

Gendre Fluid (Lillie 1965)

Fix 1 to 4 hours.

    85% alcohol saturated with picric acid .... 80.0 ml
    formaldehyde (37–40%) .... 15.0 ml
    glacial acetic acid .... 5.0 ml

Good for glycogen. Wash in several changes of 80% alcohol.

Gilson Fixative (Gatenby and Beams 1950)

Fix for 24 hours; may be left several days.

    nitric acid .... 15.0 ml
    glacial acetic acid .... 4.0 ml
    mercuric chloride .... 20.0 g

60% alcohol	100.0 ml
distilled water	880.0 ml

Good for invertebrate material. Does not give a good histological picture: shrinks cytoplasm badly. Wash in 50% alcohol. Posttreat for mercuric chloride. Good for beginners, easy to work with.

Johnson Fixative (Gatenby and Beams 1950)

Fix for 12 hours.

potassium dichromate, 2.5% aqueous (2.5 g/100 ml water)	70 ml
osmic acid, 2% aqueous (2 g/100 ml water)	10.0 ml
platinum chloride, 1% aqueous (1 g/100 ml water)	15.0 ml
glacial acetic acid	5.0 ml

Shrinks protoplasm less than Flemming. Wash in water.

Kolmer Fixative

Fix for 24 hours.

potassium dichromate, 5% aqueous (5 g/100 ml water)	20.0 ml
10% formaldehyde (37–40%)	20.0 ml
glacial acetic acid	5.0 ml
trichloroacetic acid, 50% aqueous (50 g/100 ml water)	5.0 ml
uranyl acetate, 10% aqueous (10 g/100 ml water)	5.0 ml

Good for entire eye (Walls 1938) or nerve tissue, due to presence of uranium salts. Wash in running water.

Lavdowsky Fixative (Gray 1954)

Fix for 12 to 24 hours.

distilled water	80.0 ml
95% alcohol	10.0 ml
chromic acid, 2% aqueous (2 g/100 ml water)	10.0 ml
glacial acetic acid	0.5 ml

Good for glycogen (Swigart et al. 1960). Transfer to 80% alcohol. Good for beginning students.

Navashin Fixative (Randolph 1935)

Solution A

chromic acid	1.0 g
glacial acetic acid	10.0 ml
distilled water	90.0 ml

Solution B

formaldehyde (37–40%)	40.0 ml
distilled water	60.0 ml

Mix equal parts of A and B just before using. At end of 6 hours change to a new solution for another 18 hours. Useful for preserving cellular detail in plant materials; as good as Flemming on root tips and less erratic. Transfer to 75% alcohol.

Perenyi Fixative (Galigher 1934)

Fix for 12 to 24 hours.

chromic acid, 1% aqueous (1 g/100 ml water)	15.0 ml
nitric acid, 10% aqueous (10 ml/90 ml water)	40.0 ml
95% alcohol	30.0 ml
distilled water	15.0 ml

Good for eyes. When fixing the entire eye, always make a small hole near the ciliary body so the fluids of both chambers can exchange with the fixing fluid. For the best fixing results, inject a little of the fixative into the chambers. Decalcifies small deposits of calcium; good fixative for calcified arteries and glands. Trichromes stain poorly; hematoxylin is satisfactory. Wash in 50% or 70% alcohol.

Regaud Fixative (Kopsch) (Romeis 1948)

Fix for 4 to 24 hours.

potassium dichromate, 3% aqueous (3 g/100 ml water)	40.0 ml
formaldehyde (37–40%)	10.0 ml

Mix immediately before use. Recommended for mitochondria and cytoplasmic granules. Tends to harden. Follow fixation by chromating several days in 3% potassium dichromate, which should be renewed every 24 hours. Wash in running water overnight.

Rossman Fixative (Lillie 1965)

Fix for 12 to 24 hours.

absolute alcohol, saturated with picric acid	90.0 ml
formaldehyde (37–40%)	10.0 ml

Good for glycogen. Wash in 95% alcohol.

Sanfelice Fixative (Baker 1958)

Fix for 4 to 6 hours.

chromic acid, 1% aqueous (1 g/100 ml water)	80.0 ml
formaldehyde (37–40%)	40.0 ml
glacial acetic acid	5.0 ml

Mix immediately before use. Good for chromosomes and mitotic spindles. Fix small pieces. Produces less final shrinkage than others of this type. Wash in running water 6 to 12 hours.

Schaudinn Fixative (Kessel 1925)

Fix smears for 10 to 20 minutes at 40°C.

mercuric chloride, saturated aqueous	66.0 ml
95% alcohol	33.0 ml
glacial acetic acid	5.0 to 10.0 ml

Recommended for protozoan fixation and for smears on slides or in bulk. Not for tissue; produces excessive shrinkage. Transfer directly to 50% or 70% alcohol. Posttreat for mercuric chloride.

Sinha Fixative

Specimens may remain in fixative 4 to 6 days.

picric acid, saturated in 90% alcohol	75.0 ml
formaldehyde (37–40%)	25.0 ml
nitric acid	8.0 ml

Sinha (1953) adds that 5% mercuric chloride may be included; he probably means 5 g per 100.0 ml of above solution. Recommended for insects; softens hard parts with no damage to internal structure. Transfer directly to 95% alcohol.

Smith Fixative (Galigher 1934)

Fix for 24 hours.

potassium dichromate	5.0 g
formaldehyde (37–40%)	10.0 ml
distilled water	87.5 ml
glacial acetic acid	2.5 ml

Mix immediately before use. Good for yolk-rich material (Laufer 1949). Wash in running water overnight.

The pH of some fixatives changes after mixing and again after tissue is added. This may be a factor worth considering when stainability or silver impregnation is important (Freeman et al. 1955).

B-5 Fixative

Fix 1 to 8 hours (no more) depending on size of specimen. Sections will need to be treated for mercuric chloride removal.

Stock solution
mercuric chloride	12 g
sodium acetate	2.5 g
distilled water	200 ml

Working solution
B-5 stock solution	20 ml

Add formaldehyde (37–40%), 2.0 ml, just before use.
Greatly increases the nuclear detail in lymph material and bone specimens.

Zinc Formalin (ZF)

Zinc formalin is becoming more popular because of the toxicity of mercuric chloride–based fixatives and postfixatives. L'Hoste and Torres (1995) have used ZF routinely for hematoxylin and eosin (H&E) staining as well as with immunostaining, and the results are definitely enhanced, especially with certain lymphoid markers and nuclear structure.

formaldehyde (37%)	10.0 ml
distilled water	90.0 ml
zinc sulfate	1.0 g

The pH is about 4.2, and this can cause some formalin pigment deposition in the tissues. To remove this pigment, treat the tissues with 10% ammonium hydroxide for 10 minutes. Taylor (1993) substitutes zinc chloride for the zinc sulfate, and the result is a fixative with a pH of 6.2. With the higher pH there is little or no formalin pigment seen. Churukian (1993a) recommends ZF for histochemical as well as for immunohistochemical procedures.

Helly's Fluid

This solution is the same as Zenker's fluid except for the last step. Add 5.0 ml of formaldehyde (37%) before use.

Much of the current literature is very positive about the results obtained with zinc chloride (Herman et al. 1988, Churukian 1993).

Fixation by Perfusion

Perfusion (forceful flooding of tissue) is advantageous only for tissue that requires rapid fixation but is not readily accessible for rapid removal. A prime example is the central nervous system. Many organs are not adequately fixed by this method because the perfusion fluid may be carried away from the cells rather than to them.

Special equipment necessary for perfusion includes a cannula that fits the specified aorta to be used and rubber tubing to connect the cannula to the perfusion bottle. Many laboratories also incorporate a peristaltic pump to facilitate quick and even fluid replacement.

When the animal is dead (practiced humanely) or under deep anesthesia, cut the large vessels in the neck and drain out as much blood as possible. Expose the pericardium by cutting the costal cartilages and elevating the sternum. Cut the pericardium and reflect it back to expose the large arteries. Free part of the aorta from the surrounding tissue and place a moistened ligature behind it. Make a small slit directed posteriorly in the wall of the aorta, and insert the moistened cannula. Bring the two ends of the ligature together and tie the cannula firmly in place. Cut open the right atrium to permit escape of blood and other fluids.

Precede fixation with an injection of a small amount (50 to 100 ml) of physiological saline (isotonic to the serum of the organism being studied; e.g., 0.9% NaCl for mammals) to wash out residual blood before it attaches to the vessel walls. Fill just the rubber tubing leading from the perfusion bottle to the cannula. (Separate the isotonic saline from the fixative with a clamp near the attachment of the rubber tubing to the bottle.) If a formalin-dichromate fixative is being used, substitute 2.5% potassium dichromate for normal saline. Fill the perfusion bottle with fixative (500 to 1000 ml depending on size of animal) warmed to body temperature.

When ready to start perfusion, hold the bottle at table level and open the clamp on the rubber tubing. Gradually raise the bottle to a height of 4 to 5 feet, at which point enough pressure will be exerted to force out most of the blood. After 5 minutes, open the abdomen and examine the organ to be perfused. If the surface vessels are still filled with blood and the organ has not begun to take on the color of the perfusing solution, it is possible that the perfusion has failed. However, stubborn cases may require 10 to 30 minutes to perfuse. Whenever the blood color is absent, perfusion is complete, but it should be continued long enough to keep shrinkage to a minimum.

Observe these suggestions: (1) The cannula used should be as large as possible to permit as rapid a flow as possible. This aids in washing out blood ahead of the fixative. (2) If the head alone is to be fixed, clamp off the thoracic duct, and, if the brain and spinal cord are to be fixed, clamp off intestinal vessels. The fixative is then directed toward brain and spinal cord. (3) When the perfusion bottle is being filled, allow some of the fluid to flow through the rubber tubing and cannula to release air bubbles. This can also be done with the saline. Air bubbles will block the perfusion. (4) Do not allow the injection pressure to exceed the blood pressure; artifacts will result.

If liver is to be perfused—particularly if by glutaraldehyde, which penetrates slowly—perfusion should be done through the portal vein followed by perfusion through the hepatic vein (Fahimi 1967).

If only a small piece of an organ or a small animal (mouse, fish) is to be fixed, a modified and easier perfusion is usually adequate. Inject the organ with a hypodermic syringe of fixative, and immediately after injection cut out a small piece of tissue close to the injection site and immerse it in the same type of fixative.

Lillie (1954b) lists two disadvantages of perfusion: (1) The blood content of the vessels is lost, and (2) perfusion is not possible if postmortem clotting is present. But he does favor perfusion as the outstanding method for brain fixation, saying that immersion of the whole brain without perfusion "can only be condemned." He suggests the following as the preferred method of fixation for topographic study, if whole brain perfusion is not possible.

1. Cut a single transverse section anterior to oculomotor roots and interior margin of anterior colliculi, separating the cerebrum from midbrain and hindbrain.
2. Make a series of transverse sections through the brain stem and cerebellum (5- to 10-mm intervals), leaving part of the meninges uncut to keep slices in position.
3. Separate two cerebral hemispheres by a sagittal section. On the sagittal surface identify points through which sections can be cut to agree with standard frontal sections. Make cuts perpendicular to the sagittal surface. Cut rest of brain at 10-mm intervals.
4. Fix in a large quantity of solution.

Less shrinkage occurs in the brain after perfusion than after immersion without perfusion. Lodin et al. (1967) recommend formalin-saline (1:1) as the best fixative, ethanol (not tertiary butyl alcohol) for dehydration, and embedding in celloidin or epoxy (or frozen cut sections). Paraffin embedding results in more shrinkage than do the above methods; if it is used, try vacuum to reduce the shrinkage. Mounting on slides with 80% ethanol, rather than with water, will recover a good portion of the shrinkage in the sections.

For additional references on perfusion, see Bensley and Bensley (1938), Emmel and Cowdry (1964), and Lillie (1965). See also Koenig et al. (1945) and Eayres (1950).

POSTFIXATION TREATMENTS

Chromatization

Chromatization improves preservation and staining, particularly of mitochondria and of the myelin of nervous tissues, and enhances nuclear detail.

Place the fixed tissue in a 2.5–3% aqueous potassium dichromate solution (2.5 to 3 g/100 ml water): overnight for small gross specimens (1 to 2 cm); 2 to 3 days for larger ones; 1 to 2 hours for sections on slides before staining (may be left overnight).

Wash thoroughly in running water: overnight for large gross tissues; 15 to 30 minutes for slides.

Deformalization

The removal of bound formalin is frequently necessary in silver impregnation methods, such as the Ramón y Cajal and del Río-Hortega methods and the Feulgen technique.

DECALCIFICATION

Calcium deposits may be so heavily concentrated in the tissue that they interfere with sectioning and result in torn sections and nicks on the knife edge. If deposits are sparse, overnight soaking of blocked tissue in water will soften the deposits sufficiently for sectioning (p. 62). Heavy deposits may be removed by any of several methods, but do not leave tissues in any of the fluids longer than necessary.

If any doubt arises about the completion of decalcification, check for calcium by the following method.

To 5 ml of the solution containing the tissue, add 1 ml of 5% sodium or ammonium oxalate. Allow to stand 5 minutes. If a precipitate forms, decalcification is not complete. A clear solution indicates that it is complete. Sticking needles in the tissue to check hardness is a sloppy technique that can damage the cells.

There are several good decalcifying solutions that may be purchased in large volumes. We have used "RDO" and "Decal" for several years and have consistently achieved excellent results. The rapid action of RDO is remarkable and the quality of staining and histological detail after its use is excellent. Old bones cut down to 1 cm in thickness, if possible, require a 6-hour treatment; small and young pieces only 1.5 to 2 hours. Teeth will require overnight and up to 18 to 24 hours. Do not overdecalcify; this detracts from the staining quality. Decalcifying may be followed by a brief washing in water, but this is not necessary. Fixation and decalcification may be combined in a mixture of 1 part undiluted formalin with 9 parts RDO. Avoid leaving any of the residue on the microtome knife edge or any metal surface. RDO is very corrosive to metals.

Acid Reagents

After the use of an acid for decalcification, prevent swelling in the tissue and impaired staining reactions by transferring into 70% alcohol for 3 to 4 hours or overnight. If acidity will interfere with staining, treat with 2% aqueous lithium carbonate solution or 5% sodium sulfate: 6 to 12 hours, then into 70% alcohol. Chromate treatment (p. 40) also improves staining.

Because calcium ions are soluble at pH 4.5, buffer solutions may be used. They are slower than other methods but cause no perceptible tissue damage.

Formic acid A
    formic acid .... 5.0 to 25.0 ml
    formaldehyde (37–40%) .... 5.0 ml
    distilled water to make 100.0 ml

With 5 ml formic acid content, 2 to 5 days. If increased to 25 ml, less time is required, but there is some loss of cellular detail.

Formic acid B
    formic acid, 50% aqueous (50 ml/50 ml water) .... 50.0 ml
    sodium citrate, 15% aqueous (15 g/100 ml water) .... 50.0 ml

Kristensen fluid (1948), pH 2.2
    8 N formic acid (see p. 462) .... 50.0 ml
    1 N sodium formate .... 50.0 ml

Treat for 24 hours. Wash in running water for 24 hours. Highly recommended.

Villanueva Fixative (1979)
    sodium chloride .... 0.99 g
    distilled water .... 27.5 ml
    absolute alcohol .... 65.0 ml
    hydrochloric acid (concentrated) .... 7.5 ml

Time depends on size of tissue—usually 12 to 24 hours. Wash well in running water. Process as usual.

Chelating Agents for Decalcification

Chelating agents offer the advantage of maintaining good fixation and sharp staining potential during decalcification. They are organic compounds that have the power of binding certain elements, much as calcium and iron. A commercial preparation of the disodium salt of ethylene diamine tetraacetic acid (EDTA) is the most commonly used agent. The method does have two disadvantages: The tissue tends to harden, and decalcification is slow; however, if enzyme staining is necessary, many prefer this method.

Hilleman and Lee (1953) recommend 200 ml of a 5.5% solution of EDTA in 10% formalin, for pieces 40 × 40 × 10 mm. Decalcification may require up to 3 weeks. Renew the solution at the end of each week. Transfer directly to 70% alcohol.

Vacek and Plackova (1959) report that a 0.5 M solution of EDTA at pH 8.2–8.5 yields better results in silver methods than does decalcification with acids.

Schajowicz and Cabrini (1955) adjust the solution to pH 7.0 with NaOH and HCl. Hematoxylin and eosin stain as usual, but glycogen is lost, and alkaline phosphate has to be reactivated after the use of chelating agents.

We have had good success in decalcifying material, including entire small animals and embryos, using a commercial preparation that combines a che-Baxter lating agent with dilute HCl (S/P decalcifying solution, VWR Scientific Products). Tissues are decalcified for 24 to 48 hours, washed for 12 hours in running water, and then processed.

Decalcification Combined with Fixation (See RDO, p. 40)

Lillie Fluid (1965)

This is used for 1 to 2 days.

>   picric acid, 1–2% aqueous (1 to 2 g/100 ml water) .... 85.0 ml
>   formaldehyde, 37–40% .... 10.0 ml
>   formic acid, 90–95% aqueous (90 to 95 ml/10 to 5 ml water) ..... 5.0 ml

To extract some of the yellow: 2 to 3 days in 70–80% alcohol

Lillie Alternate Fluid (1965)

Add 5% of 90% formic acid to Zenker fixative.

Perenyi Fluid

This is also a fixative. Good for small deposits. Little hardening effect. Excellent cytological detail. Good for calcified arteries and glands such as thyroid: 12 to 24 hours. Wash in 50–70% alcohol.

Schmidt Fluid (1956)

Schmidt uses the following for 24 to 48 hours, pH 7–9:

4% formalin (4 ml/96 ml water) plus 1 g sodium acetate	100.0 ml
EDTA	10.0 g

No washing necessary. Transfer tissue directly into 70% alcohol, dehydrate, and embed as usual.

Decalcification Combined with Dehydration

Jenkin Fluid (Culling 1957)

absolute alcohol	73.0 ml
distilled water	10.0 ml
chloroform	10.0 ml
acetic acid	3.0 ml
hydrochloric acid, concentrated	4.0 ml

The swelling action of the acid is counteracted by the inclusion of alcohol. Large amounts of solution should be used, 40 to 50 times the bulk of tissue. After decalcification, transfer to absolute alcohol for several changes, then clear and embed.

OTHER METHODS OF TISSUE PREPARATION

The major portion of this book is devoted to sectioning methods for preparing tissue for staining because of the complexity and quantity of such methods. However, the student should recognize that there are other means of examining tissues.

Exceedingly thin membranes can be examined directly by mounting them in glycerol or other aqueous medium. Considerable detail can be observed with reduced light or with phase microscopy. Sometimes a bit of stain can be added to sharpen or differentiate certain elements. More permanent preparations can be secured by fixation.

"Touch" preparations (impression films) are made by pressing the cut surface of fresh tissue against a dry slide. Cells adhere to the surface and can be examined unstained, or the slide can be immediately immersed in a fixative and then stained.

Smears are one of the commonest devices for simple slide preparation, such as blood and bone marrow (p. 193), Papanicolaou (p. 406), fecal (p. 443), and chromosomes (squash preparations, a modification of smears, p. 410).

Dried smears and touch preparations can develop artifacts by structural distortion of the cells. Rehydration destroys some of the cellular detail. More satisfactory results follow a 3-minute treatment in a mixture of glycerin and water, equal parts. Follow by washing in running water, fix, and stain.

Mikat and Mikat (1973) flatten cells for better visibility of their contents; 5-mm cubes of tissue are placed on formalin-moistened filter paper and centrifuged 16 hours at 2200 rev/minute. The nuclear detail is improved with the spreading of the chromosomes.

"Cell blocks"—concentrated clusters of individual cells or grouped cells—are described in detail on p. 52.

4
Dehydration: Preparation for Embedding

Tissues fixed in aqueous solutions maintain a high water content, which can hinder later processing. Except in special cases (freezing method, water-soluble waxes, and special cell contents), the tissue must be dehydrated (water removed) before some steps in processing can be successful.

During fixing and washing, tissues lack the ideal consistency for sectioning, or cutting thin slices a few micrometers in thickness. They may be soft, or they may contain hollow spaces (lumen) and be easily deformed by sectioning. If the cells are pierced by a knife, their fluid content may be released, causing them to collapse. To preclude these problems, it is necessary to replace the fluids in the tissue by a medium that hardens to a firm, easily sectioned material. The cells are filled intracellularly (impregnated or infiltrated) and enclosed extracellularly (embedded) with the medium and are thereby protected during handling. Universally used media for this purpose are paraffin and nitrocellulose and variations of these. Other media, less frequently used, are gelatin, water-soluble and ester waxes, and plastics.

Various conditions determine the choice of medium. Paraffin is suitable for most histological and embryological purposes when sections of 2 to 15 μm are required. Nitrocellulose and plastics can also be used, but serial sections are made more easily with paraffin-impregnated tissue, and paraffin preparation requires much less time. Impregnating with nitrocellulose has distinct advantages when it is desirable to avoid heat and when a tissue becomes hard too readily or is too large for the paraffin technique. With nitrocellulose, shrinkage is kept at a minimum, whereas with paraffin it can amount to as much as 20%. Gelatin can be used for extremely friable (fragile) tissue in the freezing technique, and water-soluble waxes are used when alcohols, hydrocarbons, and the like must be avoided. Ester waxes are recommended for hard and smooth tissues when it is necessary to avoid hardening agents, such as hydrocarbons.

Before embedding the tissue in paraffin, nitrocellulose, plastics, or ester waxes, all water must be removed from it. This dehydration is usually achieved by immersing the tissue in a series of solutions of ethyl alcohol (ethanol) in water, with gradually increasing percentages of alcohol. Changing through solutions of 30%, 50%, 70%, 80%, 95%, and absolute alcohol will reduce some of the shrinkage occurring in the tissue. If time does not permit such a series, the 30%, 50%, and 80% steps may be eliminated without great damage to the tissue. The time required for each step depends on the size of the object—30 minutes to 2 hours or maybe 3 hours for large, dense pieces (greater than 1 cm). A second change of absolute alcohol should be included to ensure complete removal of water; a total of 2 to 3 hours in both changes should be ample, even for large pieces. Too long an exposure to either 95% or absolute alcohol tends to harden the material, making it difficult to section.

There are other agents that are just as successful dehydrants as ethyl alcohol. The ideal dehydrating fluid would be one that would mix in any proportion with water, ethyl alcohol, xylene, or paraffin. An example is n-butyl alcohol (1-butanol). Absolute butyl alcohol is miscible with paraffin; after infiltrating tissues with a warm (50°C) butyl alcohol-paraffin mixture, infiltration with pure paraffin can follow. The butyl alcohols have the added disadvantage of having an odor that is disagreeable and irritating to some individuals; they should be used in a fume hood. Isopropyl (99%) alcohol (isopropanol) is an excellent substitute for ethyl alcohol and is sufficiently water-free for use as absolute alcohol. Actually, isopropyl alcohol produces less shrinkage and hardening than does ethyl alcohol and is free from Internal Revenue restrictions. There are two disadvantages to the use of isopropyl alcohol for dehydrating purposes: (1) It cannot be used on tissues that are to be embedded in nitrocellulose, since nitrocellulose is practically insoluble in it; and (2) most dyes are not soluble in it so, with a few exceptions, it cannot be used during staining procedures. Methanol can be used.

For many years, we have used a combination of ethanol and butanol for dehydrating tissues before paraffin infiltration. This is a method developed by Zirkle (1940) for use with botanical material. We have found that it is equally suitable with animal material and avoids the use of xylol, which tends to harden tissue. In essence, one uses a series of alcohols, which begins with a combination of water, n-butanol, and ethanol; in a series of steps, the water is eliminated and the ratio of ethanol (or isopropanol) to butanol is shifted so that tissues come to rest in 100% n-butyl alcohol. Tissues are now transferred to paraffin (the xylol step is avoided) for infiltration and embedding.

Stock Solutions (to make 4 liters)

85Z (Zirkle) 600 water + 1400 ml butanol + 2000 2-propanol
95Z 0 water + 2200 ml butanol + 1800 2-propanol
100Z 0 water + 3000 ml butanol + 1000 2-propanol

PROCEDURE

After fixation and washing, tissues are taken from 70% or 80% ethanol, where they may have been stored, and exposed for 3 hours in each of the "Z" solutions (85Z, 95Z, 100Z). From 100Z, tissues are transferred into 3 changes of pure n-butanol for 3 hours each and then finally placed into paraffin for the completion of the process.

For the preparation of dilutions of ethyl alcohol, it is customary to use 95% alcohol and dilute it with distilled water in the following manner: If a 70% solution is required, measure 70 parts of 95% alcohol and add 25 parts of distilled water to make 95 parts of 70% dilution. In other words, into a 100-ml graduated cylinder pour 95% alcohol to the 70 ml mark and then add distilled water up to the 95 ml mark.

Absolute alcohol is not exactly 100% but may contain as much as 1% or 2% water. If the water content is no higher than this, the absolute alcohol is considered 100% for practical purposes in microtechnique. If it is necessary to make certain that the water content is no more than 2%, add a few milliliters of the alcohol to a few milliliters of toluene or xylene. If turbidity persists, there is more than 2% water present. But if the mixture remains clear, the alcohol is satisfactory as an absolute grade.

Dilutions of isopropyl alcohol (99%) can be handled as a 100% solution; that is, for 70%, use 30 ml of water to 70 ml of alcohol, and so forth.

If distilled water is not provided in the laboratory building, there are compact systems available that can provide a sufficient amount of water for the average microtechnique laboratory. The stills are available in sizes that produce 0.5 to 10 gallons of water per hour. Deionized water is a satisfactory substitute for distilled water in most solutions.

Special Treatment for Small, Colorless Tissues

A small, colorless tissue often seems to disappear into the opaqueness of the paraffin. An easy answer to this problem is to add some eosin to the last change of 95% alcohol; the tissue can then be seen more readily and oriented with greater facility. (However, eosin cannot be added if isopropyl alcohol is being used, since eosin is not soluble in isopropyl.) The eosin will not interfere with future staining; it is lost in the hydration series after deparaffinization.

5
Clearing, Infiltrating, and Embedding: Paraffin Method

CLEARING

In most techniques that require dehydration and infiltration (or impregnation), an intermediary step is necessary between the two. Because the alcohol used for dehydration will not dissolve or mix with molten paraffin (an exception is tertiary or normal butyl alcohol), the tissue must be immersed in some fluid miscible with both alcohol and paraffin before infiltration can take place.

"Clearing" may seem a strange name for this intermediary step, but it describes a special property of the reagents that are used. They remove or *clear* opacity from dehydrated tissues, making them transparent. Blocks of tissue appear to deepen in color; they seem almost crystalline, never milky.

Two hydrocarbons—toluene and xylene—are reagents commonly used for clearing. However, if the tissue contains much cartilage or is fibrous or muscular, it is wise to avoid these because of their tendency to harden such tissue. Xylene, the most widely used clearing agent, has the tendency to harden such tissues and therefore its use should be limited to minimum treatment times.

An additional problem with these agents is their rapid rate of evaporation; unless they are handled quickly, they evaporate out of the tissue and leave air pockets that will not infiltrate with paraffin.

Caution: See Laboratory Safety, Chapter 2.

Chloroform is used in many laboratories but has outstanding disadvantages. It desiccates some tissues, connective tissue in particular, and has a boiling point of 61°C and a vapor pressure of 160 mm Hg, making it highly volatile and difficult to use. Aniline can be used with good results but is difficult to remove during infiltration. A mixture of equal parts of methyl salicylate (oil of wintergreen) and aniline followed by pure methyl salicylate, and then methyl salicylate-paraffin offers quicker and more certain results. Methyl salicylate can also be used alone. Other oils, such as bergamot, clove, creosote, terpinol, can be used. Amyl acetate and cellosolve (ethylene-glycol-monoethyl ether) do not harden excessively; however, cellosolve is highly volatile.

Oils, including those above, can be used for clearing if tissue hardening is a serious problem. Cedarwood oil is well known and is relatively safe for the beginner, but it has disadvantages. Overnight immersion is usually required to ensure complete replacement of the alcohol in the tissue. Also, as with all oils, all traces must be removed before infiltration, which is sometimes difficult since the oil may have a boiling point in the 200s and be slow moving out of the tissue. Oil will move out of the tissue more rapidly with the addition of an equal part of toluene. Furthermore, cedarwood oil and the absolute alcohol which must be used with it are both expensive.

During the use of xylene, toluene, or benzene for clearing, if the solution becomes turbid, it means that water is present and the tissue is not completely dehydrated and may shrink. The only remedy is to return the tissue to absolute alcohol to eliminate the water and then to place it in a fresh clearing solution. Embedded tissue containing water can shrink as much as 50% and cause difficulties in sectioning and in mounting sections.

Biodegradable clearing agents that yield satisfactory results are now commercially available (e.g., "Tissue Clear," Fisher Scientific). See Andre et al. (1994), Langman (1995), and Wynnchuk (1993) for further discussion.

DEHYDRATION AND CLEARING COMBINATIONS

Dioxane Method

Any procedure that shortens the preparation time of tissues to be embedded has merit and finds favor among technicians. Graupner and Weissberger in 1931 proposed the use of dioxane (diethyl dioxide), which dehydrates and clears tissue in a minimum of steps. Dioxane is miscible with water, alcohol, hydrocarbons, and paraffin. It seems to eliminate some shrinkage and hardening, and is a relatively inexpensive method because fewer solutions are required. However, there are several reasons for discontinuing its use, the most important being that it is a suspected carcinogen.

It is well to be cautious with dioxane: Use it only in a well-ventilated room; avoid inhalation and unnecessary contact with hands; and keep dioxane containers tightly closed at all times. The residual dioxane must be properly disposed and according to OSHA is very toxic and should not be considered a routinely used clearing agent.

If a fixative containing potassium dichromate is used, tissues must be washed thoroughly before using dioxane or the dichromate will crystallize.

Tetrahydrofuran (THF) for Dehydration

Haust (1958, 1959) recommends tetrahydrofuran (THF) for dehydration for several reasons. It mixes with water in all proportions; it also mixes with paraffin depending on the temperature, becoming increasingly miscible as the temperature rises. It is miscible with nearly all solvents and can be used as a solvent for mounting media. Most dyes are not soluble in THF, but iodine, mercuric chloride, acetic acid, and picric acid are soluble in it. THF has a low boiling point of 65°C, so it evaporates rapidly and must be kept in a tightly closed container at all times. It has a tendency to form peroxides, but storing it in amber bottles lessens this problem All mixtures of THF should be stored at 4°C (Salthouse 1958). It can be irritable to the eyes and, if used in an open processor, it should be kept in an exhaust-ventilated room. Avoid skin contact as well as inhaling the vapors.

Check with your local EPA to determine proper disposal methods.

Haust Method

Following fixation, proceed to step 1.

1. 1 part THF to 1 part water: 2 hours.
2. THF, 3 changes: 2 hours each.
3. 1 part THF to 1 part paraffin, 53–54°C: 2 hours.
4. Paraffin: 2 hours.
5. Embed.

INFILTRATING WITH PARAFFIN

Paraffin, the original substance used in this method, has been replaced by refined mixes of paraffin and resin, but the title *paraffin* is still with us. Paraffin is considered to be either soft or hard depending on its melting point. The melting point of soft paraffin lies in the 50–52°C or 53–55°C ranges; that of the hard, in the 56–58°C or 60–68°C ranges. The choice of melting point depends upon the thickness at which the tissue is to be sectioned or upon the type of tissue—hard paraffin for hard tissues and soft paraffin for soft ones. If relatively thick sections are to be cut, use a soft paraffin; otherwise the sections will not adhere to each other in a ribbon. If thinner sections are desired (5 to 7 μm), use a paraffin in the 56–58° grade. For extremely thin sections of less than 5 μm, the best results can sometimes be obtained with a hard paraffin of 60–68° melting point. The sections will retain their shape and size without excessive compression and will ribbon better than if the paraffin is softer. Temperature can also influence the choice of paraffin; hot weather will force the use of a harder paraffin than can be used in cool weather. If it is impractical to stock more than one kind of paraffin, that with a 56–58°C melting point is usually the best choice.

There are many excellent embedding media, rigidly controlled mixtures of paraffin and several plastic polymers of regulated molecular weights. Many require little ice for embedding, and cut well with less compression than paraffin. The important factor is to limit the time in the paraffin because the heat can cause shrinkage and to adhere to the manufacturer's temperature limits at all times. Do not overheat. Overheating can destroy some of the properties of the paraffin and reduce the quality of the tissue sections. Maintain quality control of the temperature and record.

Normally, tissues are transferred directly from clearants to paraffin. Keep the oven temperature just high enough to maintain the paraffin in a melted state, no higher. This state lessens the danger of overheating tissue, which can harden and shrink. The paraffin must be fully molten to infiltrate the tissue effectively. Melted paraffin that has been kept in a warm oven for several days or weeks is better for infiltrating and embedding than freshly melted paraffin. After 30 minutes to 1 hour in the first bath, the tissue is removed to a fresh container of paraffin for a similar length of time. Two changes of paraffin are sufficient for most requirements, but for some tissues that are difficult to infiltrate—such as horny skin, bone, or brain—a third change may be necessary, and the total time of infiltration may need to be extended to as much as 6 hours, or even overnight. Fortunately, such cases are rare.

Most tissue processors incorporate the use of vacuum throughout the entire processing procedure to remove air from the tissues (e.g., lung) and eliminate holes in the final paraffin block. Otherwise, an ordinary vacuum oven may be used to attain these ends.

EMBEDDING (BLOCKING) WITH PARAFFIN

As soon as the tissue is thoroughly infiltrated with paraffin, it is ready to be embedded; the paraffin is allowed to solidify around and within the tissue. The tissue is placed in a small container or mold already filled with melted paraffin, and the whole is cooled rapidly. Before transferring the tissue, warm the instruments that manipulate it. This warming will prevent congealing of paraffin on metal surfaces. Handle the tissue and paraffin as rapidly as possible to prevent the paraffin from solidifying before the tissue is oriented in it. Orientation is important. If the tissue is placed in a known position, it is easy to determine the proper surface for sectioning. Most laboratories incorporate a cassette system in which the properly trimmed tissue travels through the entire processing system in its cassette and the base of the cassette becomes the support for the mold. These cassettes can be easily identified with pencil and there is little chance for loss or confusion of the specimen.

There are many cassette and embedding molds systems available. Each technician eventually adopts his or her own preferred method. A few suggestions: Petri dishes, empty slide boxes can be used to accommodate larger specimens. Lightly coat the insides of glass dishes with glycerol; then the solidified paraffin block loosens readily.

When small amounts of paraffin are ready to be solidified, they can be cooled immediately on ice or on the cold plate provided on the embedding station. Make certain that the block has fully solidified before attempting to section it. The perfect block is one in which the paraffin crystals are contiguous, and the paraffin appears clear and homogeneous. Paraffin may contain 7–15% air dissolved in it and will appear clear when that air is distributed evenly through its mass, but pockets of air produce milky spots, a condition called "crystallization." Slow hardening of the block in the air or too rapid cooling may cause the crystallization effect, particularly in large blocks. Quick hardening of outer surfaces will trap the air.

If the paraffin does crystallize, difficulty may be encountered during sectioning; the only remedy is to return the block to melted paraffin, allow it to remelt, and repeat the embedding process. Experienced technicians soon learn how fast to handle the paraffin to reduce crystallization to a minimum.

The embedding centers are being used extensively by professional technicians and can be a practical addition for the classroom.

Thoroughly hardened paraffin blocks can be stored indefinitely without injury to the tissue, but they must be kept in a cool place where they cannot soften.

Embedding Cellular Contents of Body Fluid "Cell Blocks"

"Cell blocks" are clusters of individual cells that have been concentrated and embedded for sectioning. The process for embedding them is as follows:

1. Collect the material in centrifuge tubes and add fixative: 1 hour, or overnight. Agitate occasionally.
2. Concentrate by centrifuging (preferably in conical-bottom tubes); decant and add water or alcohol depending on requirement of fixative. Carefully loosen material at bottom of tube and stir with a pick or probe and wrap in lens or filter paper and process as a routine specimen.

The filter paper method for bone marrow (p. 210) can be adapted to use with many body fluid cells.

For other methods of handling clumps of cells, see Del Vecchio et al. (1959), de Witt et al. (1957), McCormick (1959b), Seal (1956), and Taft and Arizaga-Cruz (1960).

TIMING SCHEDULE FOR PARAFFIN METHOD

The timing that follows is for tissue blocks ±10 mm in size—smaller pieces will require less time; larger pieces, more time.

Schedule Using Ethyl Alcohol

1. Fix overnight or 24 hours. (Bouin, Gomori, Susa, Stieve, neutral buffered formalin, Zenker. If tissue fixed in Zenker tends to harden, do not leave it overnight in the fixative; transfer to 3–5% aqueous potassium dichromate overnight, then proceed as usual on schedule).

2. Wash in water, running if possible: 6 to 8 hours or overnight. (Exceptions: Bouin, Gomori, Susa, and Stieve-fixed tissue can be transferred directly to 50% or 70% alcohol without washing.) Bouin-fixed tissues should be placed into ammoniated 70% alcohol for 2 to 3 days.
 3. Transfer to 50% alcohol: 1 hour. (optional)
 4. Transfer to 70% alcohol: 1 hour.
 5. Transfer to 95% alcohol: 1 to 1.5 hours.
 6. Transfer to absolute alcohol #1: 0.5 to 1 hour.
 7. Transfer to absolute alcohol #2: 0.5 to 1 hour.
 8. Transfer to xylene (or xylene substitute) #1: 0.5 to 1 hour.
 9. Transfer to xylene #2 (or xylene substitute): 0.5 to 1 hour.
 10. Transfer to melted paraffin #1: 0.5 to 1 hour.
 11. Transfer to melted paraffin #2: 0.5 to 1 hour.
 12. Embed.

Alternate Schedule Using Isopropyl Alcohol

 1–4. Same as for ethyl alcohol.
 5. Transfer to isopropyl alcohol #1: 0.5 to 1 hour.
 6. Transfer to isopropyl alcohol #2: 0.5 to 1 hour.
 7. Transfer to xylene (or xylene substitute) #1: 0.5 to 1 hour.
 8. Transfer to xylene (or xylene substitute) #2: 0.5 to 1 hour.
 9. Transfer to melted paraffin #1: 0.5 to 1 hour.
 10. Transfer to melted paraffin #2: 0.5 to 1 hour.
 11. Embed.

As noted in Chapter 4 (p. 45), if the tissue is well hardened by the fixative, it is not necessary to dehydrate the tissue through a number of graduated steps, such as 50%, 60%, 70%, 80%, 95%, and absolute alcohol. The 50%, 60%, and 80% steps are frequently eliminated, except when special cytological studies are done. Also, the use of isopropyl alcohol considerably shortens the schedule.

With hand processing, pass tissues from one fluid to another, use the decantation method to avoid excessive manipulation with forceps and reduce injury to the tissue. After pouring off a solution, drain the tissue bottle briefly against a paper towel or cleansing tissue to pull off as much as possible of the discarded solution. Do not allow tissues to dry. The removal of the preceding solution reduces contamination and dilution of the new solution. Since 95% alcohol, absolute alcohol, clearers, and melted paraffin all contribute to hardening tissue, avoid leaving it in any of these fluids for longer than the maximum time recommended above (preferably only for the minimum period), and never overnight.

AUTOMATIC TISSUE PROCESSORS

Most laboratories now accomplish the foregoing processes with automation. There are two basic types of processors. The most economical version consists of a timing device, and the tissues are shifted automatically through a series of beakers. The timing device can be present to handle changes during the night so that the tissues will be ready for embedding in the morning. Small metal or plastic receptacles with snap-on lids hold the tissues and labels, and are deposited in a basket that clips into the bottom of the lid of the instrument. When the time arrives for removal of the tissues to a new solution, the lid rises and rotates to lower the basket into the next container. The two final beakers are thermostatically controlled paraffin baths. A technician can set the timing device for any interval desired—15 minutes, 30 minutes, 1 or more hours, etc.—over a period of 24 hours. These models have a clock that can control the instrument over a weekend or for several days. These open machines should always be operated in a well-ventilated area or under a fume hood.

There are several other tissue processors on the market and most of these incorporate a single station and the different fluids are pumped into this station. The systems are totally enclosed and prevent the fumes from escaping. These instruments are microprocess controlled and the programs can be easily changed and modified to meet the needs of the various processing schedules. Vacuum can be used throughout the process and temperature can be monitored.

6
Microtomes and Microtome Knives

MICROTOMES

The following types of microtomes and instruments are used in most laboratories:

1. The rotary microtome, most widely used for paraffin sections and for cryostat sectioning. Special rust-proof rotary microtomes are manufactured for use in cryostats.
2. The heavy duty, or retracting, microtome which has the capacity to cut very thin (0.5 or 1 μm). This microtome retracts on the up stroke to prevent tissue damage. Plastic as well as paraffin sections can be cut on this machine and these instruments often come equipped with various control panels and motorized versions are available. These microtomes often have the accessories to allow the user to use glass knives as well as metal (even disposable) ones.
3. The sliding microtome for large paraffin or bone specimens, nitrocellulose (celloidin) and plastic sections: Not always the most practical method; slow and expensive, but often excellent for hard and large objects such as eyes, bone, and cartilage; also for cases in which shrinkage must be kept to a minimum. These also are available in motorized as well as "heavy duty" versions. The motorized models are recommended for sectioning undecalcified bone and large plastic blocks. This instrument also is available in a large low-temperature cabinet that will allow frozen sectioning of large and undecalcified specimens.
4. The ultrathin-sectioning microtome for sections thinner than 1 μm for electron microscopy.
5. The vibrating blade microtome (Vibratome) uses an oscillating blade sectioning in a liquid bath to lubricate the blade and prevent heat buildup. This instrument is used mostly to cut fresh unfixed tissue and bones.

Microtomes should always be kept well oiled and maintained according to the manufacturer's recommendations to prevent parts from wearing unnecessarily or sticking. Most of the individual companies that produce the instruments also provide excellent technical service if repairs are necessary. The older microtomes are difficult to repair and replacement parts are often no longer manufactured. There are several repair facilities available (such as Ridge

Microtome Service in Knoxville, TN) that claim to fashion even obsolete microtome parts.

MICROTOME KNIVES

There are four common types of microtome knives:

1. The "C" profile for frozen sections and paraffin ribbons.
2. The "D" profile for hard materials. This knife can also have a tungsten carbide edge. The profile can also be a specific degree or angle, such as a 45 or 55° angle. These larger knifes are usually used on the large sliding motorized microtomes. These knives are difficult to sharpen and must often be sent to the manufacturer for sharpening. There are several reliable companies (Dorn & Hart and Sturkey) that offer a speedy and necessary service.
3. Disposable blades are great for routine paraffin sectioning and the thicker blades can be used for harder specimens also. These are available from several sources and some are better than others as determined by trial and error. Some need to be precleaned in xylene to remove surface oils and some are influenced by different brands of paraffin. Some disposable blades are coated with Teflon.
4. Glass knives are necessary for the glycol methacrylate specimens. These can be broken into triangles (like knives used for sectioning Epon for electron microscopy) or a "Ralph" knife, which has a larger cutting surface.

Because knives seem to demand hours of attention, they often become the technician's nightmare, and keeping them in optimum condition is difficult.

Theoretically, a perfect cutting edge is the juncture of two plane smooth surfaces meeting at as small an angle as is feasible—ideally 15–18°. These cutting surfaces are called the cutting facets. The cutting edge of a very sharp knife, when examined by reflected light under 100X magnification, appears as a fine discontinuous line. It may vary slightly in width, but it should show only a light reflection, a narrow, straight, bright line. At a magnification of about 500 times, the edge will have a finely serrated appearance. The fineness or coarseness of these serrations depends on the degree of success in sharpening the knife. The facets are determined by the angle at which the knife is sharpened.

The importance of taking good care of a knife cannot be overemphasized. Clean it after use; materials such as blood and water, if left on the knife edge, corrode it. Clean the knife with xylene on soft cleansing tissue and wipe it dry. Do not strike the edge with hard objects, such as a section lifter or dissecting needle; the edge can be easily damaged or dented. Store in the hard box in which the blade was received from the manufacturer.

Sharpening Knives

The glass plate has had the longest and most successful use as a sharpening instrument. The earlier models of automatic blade sharpeners incorporated a glass plate system and a version of this machine is still available (Leica, Inc.). Another model (Shandon-Lipshaw) uses metal plates and diamond abrasive to achieve appropriate cutting edges. Adjustments for angle can be made on this machine. A number of sharpening machines permit grinding off the old edge and leaving a new one (e.g., the Hacker instrument). This machine shortens considerably the time needed to get a good cutting edge. The life of the knife can be reduced considerably if the manufacturer's instructions are not carefully followed.

The clearance angle of the knife can be adjusted in the knife holder on the microtome. This angle is usually set ($10-11°$) and forgotten, but as the knives get smaller with age and with sharpening, the angle may need to be adjusted.

The art of hand stropping knives is almost obsolete. Everyone is always in a hurry, but if you can still find a leather strop, a sharp knife can be obtained (see Humason (1979), 4th ed., pp. 51–54, for discussion).

The life of the sharp edge of a good knife can be prolonged if the technician has another knife for the preliminary trimming of paraffin blocks. After undesirable parts of the tissue block have been trimmed away, the good knife can be substituted for the old one and the required sections collected.

7
Paraffin Sectioning and Mounting

SECTIONING

Before sectioning, embedded blocks should be trimmed into squares or rectangles, depending on the size of the mold and the shape of the tissue, with two edges parallel. The two short or side edges need not be parallel (sometimes with advantage, as will be indicated later). Wooden blocks or metal object discs, still used in classrooms, are covered with a layer of paraffin. A heated spatula is held between the paraffin on the wooden block or metal disc and the undersurface of the tissue block. When both surfaces are melted, the instrument is withdrawn and the tissue block pressed firmly into the paraffin on the holder. There should be at least several millimeters of paraffin between the tissue in the paraffin block and the face of the block holder. After the paraffin has cooled, the tissue block is ready to be sectioned. These steps are rarely used in research and clinical laboratories where cassette systems, designed to form paraffin blocks, can be clamped unmounted directly in the microtome.

Paraffin blocks should be kept on a cooled surface or on ice for at least 10 to 15 minutes before sectioning. (This will have little value where serial sections are needed.)

Set a rotary microtome for section thickness (5 or 6 μm) or depending on the cytological detail required for appropriate analysis. A setting of 10 μm is suggested for beginning students. Place the cooled specimen block in the specimen retaining clamp and tighten the clamp. If rectangular in shape, the greatest edge of the block should strike the blade first. The two parallel sides should also be parallel to the blade edge. Insert the microtome knife and tighten its clamps. If using a wooden block, allow it to extend about 50 mm beyond the metal clamps; this will prevent the paraffin block from breaking loose from the wooden block when the clamps are tightened. The knife must be held in the clamps at a proper angle for optimal sectioning, producing a minimal amount of compression, and allowing the sections to adhere to each other. Many suggestions can be made concerning angle determination, but in most instances the technician finds the answer after a process of trial and error. When placed in the microtome, the knife must be tilted just enough so the cutting facet next to the block clears the surface of the block. If the tilt is not sufficient, the surface of the block is pressed down with the welding effect of the facet, and no section results (Fig. 7-1a). At the next stroke of the knife, this compression increases, and finally a thick section is cut, a composite of the compressed sections. Too great a tilt of the knife makes it scrape through like a chisel, rather than slicing through the tissue (Fig. 7-1b). The use of knife guards should be practiced to protect against exposed blade edges.

Set the advancing head on the microtome back to its beginning with the handle. Turn the course advance mechanism to align the face of the tissue block close to the knife edge, or carefully move the blade holder carriage back to meet the paraffin block. With an old knife start trimming into the block until the desired area is reached. Change to a good knife (not recommended for serial sections) and readjust tissue carrier if necessary. Never touch the edge of the knife with anything hard. Fragments of paraffin can be flicked off with a camel's hair brush. Scratches appearing in sections often can be remedied by rubbing the knife edge with a fingertip—*upward*. Extreme care—this could be dangerous! Also, this motion on the back of the knife will remove bits of paraffin, which can cause scratches. Cleansing tissue dipped in xylene is also helpful. (*Warning:* Discard the first section after cleaning the knife; it probably is a thick one.) When all sectioning is completed, use xylene to clean the knife; leave no corrosive material on its edge.

Section with any easy rhythm; never rush, or the result is likely to be sections of varying thicknesses. As the sections move down on the knife they form a "ribbon," each section adhering to the one that precedes it as well as the one that follows it. During the cutting of the sections, a bit of heat is generated, enough to soften the paraffin and cause the individual sections to stick to each other. As the ribbon forms, hold it away from the knife with a camel's hair brush and ease it forward so that the sections do not remain on the knife. This is advisable to prevent the sections from bunching and stacking on each other. They can stack high enough to topple over the edge and get caught between the tissue carrier and the next stroke of the knife. The parallel edges of the paraffin block must be cut clean and parallel. If the edges have

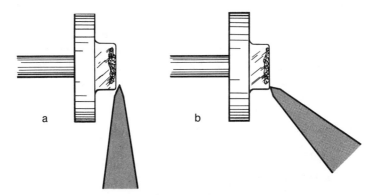

Figure 7-1. Paraffin knife tilt: (*a*) insufficient; (*b*) excessive.

not been trimmed but remain as the original sides formed by the mold, a ribbon will not form.

In dry weather, static is frequently a problem during sectioning. Woolen and nylon clothing tends to create more static than cotton clothing, particularly when in contact with plastic-covered chairs. The static causes the sections to stick to the knife or to parts of the microtome. They break apart and stay bunched up instead of lying flat. The friction of the knife as it crosses the paraffin block forms static electricity on its surface. Antistatic sprays can be used to relieve this situation. Grounding the microtome is sometimes helpful, as is increasing the humidity in the room by boiling water.

The ribbons formed during sectioning can be mounted directly on slides, floated on a water bath, or laid in order in a box. It is particularly convenient to lay the ribbons in the box when the slides cannot be made immediately or when numerous sections—serial sections, for instance—have to be cut. Shallow boxes make handy containers and, if painted black (black construction paper works well) provide an excellent dark background for the sections. If the shorter edges of the block have not been trimmed parallel, a serrated edge along the ribbon indicates the exact position of each individual section, minimizing the danger of cutting through one. The sections can be stored in boxes until all required slides have been mounted. Valuable sectioning time is conserved by this means.

Suggestion: Use a 1-inch paint brush to free the microtome of paraffin litter after sectioning. Wiping with an absorbent towel moistened with waste xylene is especially useful in cleaning up.

Difficulties Encountered in Sectioning: Suggested Remedies (Modified from Richards 1949)

1. Ribbons are crooked.
    a. Wedge-shaped sections caused by poor trimming; sides of paraffin block are not parallel or not parallel to edge of knife.
    b. Part of knife edge may be dull; try another part of it.
    c. Uneven hardness of the paraffin; one side may be softer than the other, or contain areas of crystallization; re-embed.
2. Sections fail to form ribbons (usually due to hardness of paraffin).
    a. Use softer paraffin (lower melting point).
    b. Blow on knife to warm it or dip it in warm water.
    c. Cut thinner sections.
    d. Place table lamp near knife and block to warm them both.
    e. Resharpen knife.
    f. Lessen tilt of knife and clean edge.
    g. Dip block in softer paraffin and retrim so a layer of this paraffin surrounds original block.

3. Sections are wrinkled or compressed.
   a. Resharpen knife; a dull knife compresses badly.
   b. Paraffin too soft; re-embed in harder paraffin.
   c. Cool block and knife.
   d. Increase tilt of knife.
   e. Clean edge of knife with finger or xylene; remove any paraffin collected there.
   f. Tissue is not completely infiltrated (Poor infiltration is usually caused by traces of water or alcohol). Correct by first removing paraffin that is present. Soak in xylene for 2 or 3 hours (or more); change twice, then place in absolute alcohol for 1 or more hours. This should remove all traces of water. Clear again in xylene (check against milky appearance); reinfiltrate and embed.
   g. Soak tissue block in water. When soaking in water is recommended, the cut face of the tissue is exposed to tap water for 30 to 60 minutes. This treatment is generally satisfactory; however, some technicians advocate the addition of glycerin (1 part to 9 parts water) or fabric softeners, same ratio, or several drops of liquid dish detergent per 100 ml of water, or 60% alcohol instead of water (Lendrum 1944). These fluids work in through the cut tissue surface and soften tough parts. (*Exception:* Do not soak nervous system tissue at any time and lymph nodes and fatty tissue only briefly. Paraplast will not absorb water.)
4. Ribbons are split or scratched longitudinally.
   a. Nick in knife; move to another part of edge or resharpen knife.
   b. Knife dirty or gritty along edge.
   c. Dirt or hard particles in tissue or in paraffin; crystals from fixing solution not adequately removed; filter stock paraffin or decalcify tissue.
   d. Decrease tilt of knife.
   e. Tissue too hard; soak in water or RDO decalcifier briefly (1 to 5 minutes).
5. Tissue crumbles or falls out of paraffin.
   a. Poor infiltration; reinfiltrate and re-embed.
   b. Not completely dehydrated.
   c. Not completely de-alcoholized.
   d. Too long in paraffin bath or too hot while there; soak in water.
   e. Clearing fluid made tissue too brittle; soak in water.
6. Sections cling to block instead of knife.
   a. Knife dull or dirty.
   b. Increase tilt of knife.
   c. Paraffin too soft or room too warm; try harder paraffin or cool block.
   d. Infiltrating paraffin too hot, or too long exposure to solutions that harden; soak in water.

7. Tissue makes scratching noise while sectioning.
    a. Tissue too hard; paraffin too hot or exposed too long to solutions that harden; soak in water.
    b. Crystals in tissue; fixing reagents not adequately removed by washing; calcium or silicon deposits present. Soak block briefly, 5 to 30 minutes depending on size, in full-strength RDO. Wipe block clean with tissue paper and resection.
8. Knife rings as it passes over tissue.
    a. Knife tilted too much or too little.
    b. Tissue too hard; soak in water.
    c. Knife blade too thin; try a heavier one.
9. Sections curl, fly about, or stick to things, owing to static electricity from friction during cutting, especially in weather of low humidity.
    a. Increase humidity in room by boiling water in open pan.
    b. Ground microtome to water pipe.
    c. Postpone sectioning until weather is more humid; early morning sectioning often is best.
    d. See p. 60 for suggestions concerning clothing, furniture, and static eliminator.
10. Sections are skipped or vary in thickness.
    a. Microtome in need of adjustment or new parts.
    b. Tighten all parts, including knife holder and object holder clamp. Always!
    c. Knife tilt too great or too little.

MOUNTING

Since the embedding medium usually has to be removed from the sections to allow uniform staining throughout the cells, the tissue sections must be affixed to some solid surface or they will disintegrate during removal of the embedding medium. The sections are, therefore, mounted on glass slides that support the tissue and also permit the mounting of many sections in sequence.

Slides and Cover Glasses

Most microscopic materials (sections, whole mounts, smears, touch preparations) are mounted on glass slides and are covered with microscopic cover glasses. The slides may be obtained in the regular 3- × 1-inch size or in 3- × 2-inch pieces for larger tissue sections. These slides are precleaned and are ready to use. Additionally, there is a variety of specialty slides (e.g., poly-L-lysine coated, silane [aminoalkylsilane] treated, polystyrene) available that enhance tissue section adherence or can be used for special materials such as touch preps and cell pellets.

Cover glasses are routinely manufactured in circles, squares, and rectangles, ranging in thickness from #1 to #2. Thickness #1 (about 0.15 mm) usually is adequate for oil immersion work. Most labs use the #1 coverslips. Thickness #2 (about 0.20 mm) can be used for whole mounts or when oil immersion microscope objectives are unnecessary. Circles and rectangles may be purchased in sizes from 12 to 25 mm in diameter. Circles are designed for mounts for which ringing the cover is necessary to create a seal against evaporation, as for whole mounts in glycerol jelly or any other volatile mountant. Rectangles come in 22- and 24-mm widths and in 30-, 40-, 50-, and 60-mm lengths.

Mounting Techniques

Before sections are mounted, the slide must be identified with pencil on the frosted end of the slide or a diamond or carbide tipped "pen" can etch the necessary information to prevent confusion with other slides. In addition, this mark enables the worker to distinguish on which side of the slides the sections are mounted and thus to avoid failure during the staining process. Some felt-tipped ink pens are reasonably water resistant and can be used on slides with frosted ends, but this may be removed by xylene or alcohol.

There are two general methods for adhering sections to slides. Many clinical and research laboratories use a water bath. Paraffin ribbons are floated on the surface of the water and picked up on slides. The other method used frequently in research laboratories and student teaching labs is to collect ribbons in a flat box to be later mounted onto slides.

The customary means of affixing sections to slides is with albumin (a mixture of egg white and glycerin). Elmer's glue, poly-L-lysine, or aminoalkylsilane may also be used for adhering sections. Aminoalkylsilane ("Bind-Silane," Pharmacia LKB Biotechnology Inc.) is recommended to adhere sections for in situ hybridization preparations. Gelatin, Elmer's glue, or agar can be used in a water bath. (See also subbed slides, p. 469.) It is essential that slides be absolutely clean to ensure adherence of the sections throughout the staining procedure. With one finger, smear a thin film of albumen fixative (see p. 467) on the slide and, with a second finger, wipe off excess albumen. This should keep the film thin enough. Thick albumen can pick up stain and can prevent a clear image of sharp, uncluttered tissue elements.

The water bath should have a temperature 5–10° below the melting point of the paraffin being used. Excessive heat for flattening may cause shrinkage and tearing and displacement of tissue parts. When removing the sections from the microtome knife, stretch them as flat as possible as they are placed on the water surface. After they have warmed a bit, they can be pulled to more nearly their original size with a couple of dissecting needles. Separate the required sections from the rest of the ribbon. Dip an albumenized slide under them and with a needle hold them against the slide while removing it from the bath. When the sections are spread smoothly, the excess water can be drained off; this is facilitated by touching a piece of tissue or filter paper to the edge of the slide and around the sections. Some labs use old magazines cut to size for this purpose. Additional adhesion and flattening can be achieved by placing WET filter paper over the sections that have just been drained and then rolling your thumb, with pressure, on the paper. It is imperative that both the filter paper and the section not be dry. Permit slides to dry overnight on a warm plate or in an oven of comparable temperature. If the sections must be stained on the same day that they are mounted, dry them in one of the mechanical hot-air dryers or microwave them in a wooden slide box for 5 to 10 minutes (rotating after half of the time). If glue or gelatin is added to the water bath, unalbumenized slides can be used.

Where a water bath is not used, the sections are collected (glossy side down) in a box. The sections can now be separated with a spatula, scalpel or razor and removed to an albumenized or subbed slide containing enough distilled water to spread well beyond the edge of the sections. In another approach, several drops of albumen are added to 40 to 50 ml of water. The sections are now floated on a slide flooded with this medium and placed on a slide warmer for spreading and to correct for any compressions acquired during sectioning. Keep the slide on the warmer until the sections are stretched flat. The use of dissecting needles to "stretch" the ribbon as it begins to spread, is very useful to get out the folds and compression. Dense tissues will compress less than the paraffin surrounding them and often will develop folds because the paraffin does not expand enough on the warm plate to permit the tissues to flatten. Pieces with a lumen will invariably demonstrate this fault. One of the simplest ways to correct it is to break away the paraffin from the outside edge of the tissue. Do this when the ribbon is floating on the water on the slide, but not while it is warm, or the paraffin will stick to the dissecting needles. Allow the slide to cool and then split away the paraffin with care so that the tissue is not injured (Fig. 7-2). If the paraffin tends to stick to the needles during this process, wipe off the needle tips with the fingers. Clean needles will minimize this problem. A rather severe method for straightening stubborn ribbons is the addition of some acetone, alcohol to the water (just a few milliliters).

Certain tissues frequently develop cracks through the sections while drying—a tendency common to spleen, liver, lymph nodes, and nervous tissue. To prevent this, as soon as the sections are spread, drain and blot them free of as much water as possible and dry them at a higher temperature (60–65°C). Properly mounted sections will have a smooth, almost clear appearance. If they have a creamy, opaque texture, and if they reflect light when examined from the undersurface, air is caught between the glass and the sections. Tissues so mounted will float off in aqueous solutions. The cause usually lies in poorly cleaned slides or in drying the slides at too low a temperature or not drying them long enough. If, when mounting the sections, bubbles

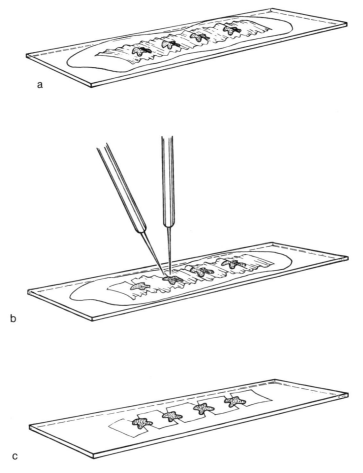

Figure 7-2. Paraffin sections: (a) compressed or folded sections; (b) breaking paraffin with dissecting needles to permit ribbon expansion without folds; (c) expanded and flattened sections.

appear under them, use distilled water that has been boiled for some time in order to release trapped air.

Several other conditions may cause tissues to loosen from slides. These include brittle sections, a large yolk or osmium tetroxide content, and the use of strong acids or alkalis during staining. In these situations, the sections will require special treatment as described below. In addition, poorly infiltrated sections and nervous tissue that have been soaked in water will usually fail to adhere but can sometimes be saved by the following method.

After drying, and when ready to stain, dissolve out the paraffin with xylene. Place the preparations for 1 to 2 minutes in absolute alcohol, followed by 1 to 2 minutes in a dilute solution (0.5%) of nitrocellulose in ether-alcohol (1:1). Drain off excess nitrocellulose, but do not allow it to dry completely; wave the slide in the air for a second and place in 70% or 80% alcohol. Continue with the planned staining method. Also try mounting this type of tissue on subbed slides (p. 469) instead of an albumenized slides. This method is much easier than coating the mounted sections, especially for nervous tissue. When water used in mounting sections does not puddle uniformly on subbed slides, but forms tight-up pools, sections are difficult to spread; to correct this, apply a thin coat of egg albumen over the subbed surface before adding the sections and water.

Serial Sections

With a few exceptions, the paraffin method is used for serial sections, and in most cases every section is mounted and in correct order. While sectioning, arrange the ribbons in a cardboard box from left to right and top to bottom (like a printed page). When the sectioning is finished, remove the sections to slides in the above order. Serially number each slide. This system affords simple sorting of slides when all processing is complete.

8
The Microscope

THE COMPOUND MICROSCOPE

The fundamental parts of a compound microscope are the objectives and the eyepieces or oculars. Both are systems of lenses; the eyepieces at the top and the objectives at the bottom of a tube. The other basic parts are the stage on which the object to be viewed is placed and a mirror or system of mirrors for directing light. The ordinary laboratory microscope uses bright-field illumination with direct light amplified by a substage condenser and mirror. The condenser is also a system of lenses, the purpose of which is to concentrate light on the object to be viewed so it is seen in a bright, well-lit field.

The objectives magnify the specimen a definite amount, forming an image, which is magnified again by the eyepiece. The lenses of a good microscope should not only magnify but also improve the visible detail. This is the resolving power, and it is a function of (1) the wavelength of the light used; (2) the lowest refractive index (refractive index is a measurement of the refraction or bending of light rays as they pass through one medium to another at an oblique angle) between the objective and the substage condenser (p. 68); and (3) the greatest angle between the central and most extreme rays of light entering the front lens of the objective. Blue light increases resolving power over white light; ultraviolet increases it even more, but it cannot be seen, and the results must be photographed.

Eyepieces are manufactured in a variety of powers of magnification (engraved on them); 10× is the customary lens for student and laboratory microscopes. Most of these microscopes are equipped with three objectives: a scanning lens or low power, usually 2.5× or 4×; an intermediate lens, 10× or 20×; and finally a high-power dry lens, 40×. The oil immersion lenses, 50× and 100×, are reserved for microbiological and hematological applications and detailed cytological examinations. All are engraved with their magnifying power. As one shifts from low power to higher powers, the magnification increases, but the size and depth of the examining field decrease. Also, the clearance of the objective above the specimen (working distance) decreases, and more light is required. The working distance is the distance between the front of the objective and the top of the coverslip when the specimen is in focus.

Achromatic objectives (producing an image essentially free of color fringes) are found on most student laboratory microscopes. These objectives are corrected chromatically for the light of two wavelengths, the C (red) and F (blue) lines of the spectrum, and spherically for the light of one color, usually in the yellow greens. Achromats give a good image in the central portion of the field but not as sharp an image at the edges as can be obtained with apochromatic (corrected also for spherical aberrations) objectives. These special objectives are essential for critical microscopy and color photography because they are corrected for three colors, including the G (violet) lines of the spectrum, and thereby reduce color fringes even further. Lenses, if not corrected in this manner, exhibit chromatic aberration, and color fringes will confuse the border of the specimen image. For best results, apochromatic objectives should be accompanied by a corrected condenser of a numerical aperture at least equal to that of the objective aperture and compensating ocular eyepieces. A correction collar on the apochromatic objectives can be adjusted to compensate for the thickness of the cover glass on the slide. The collar can be adjusted with one hand while the fine adjustment is manipulated with the other hand.

At the bottom of the student microscope, under the stage, there is a mirror that directs light through the diaphragm and into a condenser, which increases the illumination of the specimen. Usually there are two mirrored surfaces: one plane (flat) to direct the light, reflecting a moderate amount of light with parallel rays but no change in form; the other concave to converge the light rays to form a cone, concentrating a large amount of light. The concave surface replaces a condenser and can be used for low magnifications, but it should not be used when a condenser is present, nor should it be used for high magnifications in place of the plane mirror.

Immediately below the microscope stage is the substage condenser, whose lenses serve to converge the parallel beam of light from the mirror so that a cone of light passes through the aperture of the stage. This cone of light is focused on the specimen under examination and then is extended to fill the back lens of the objective. Without a condenser the back lens of high-power objectives is not filled with the proper amount of light. By opening or closing the iris diaphragm mounted below the condenser, one can control the diameter of light entering the condenser. A two-lens Abbe condenser is found on most laboratory and student microscopes. For critical research microscopy, a corrected condenser with a centering mount should be used. It should be carefully centered to the objective, and immersion oil should be used between it and the slide for finest detail.

The image seen in the compound microscope is inverted: it is upside down and turned right side to left. The movement of the slide will be reversed. For clarity, most images depend on color in the specimen or color added to it and/or differences in refractive indices between different parts of the specimen and the medium in which it is mounted.

Innumerable types and sources of light are used in laboratories, although most student microscopes today have a built-in illumination source. The surface of the bulb or of the filter is focused on the specimen by way of the substage condenser. If the lamp is the small substage type of lamp, it may be used in front of the mirror; if the mirror is removed, it is placed under the condenser. A more critical illumination (Köhler method, 1893) can be obtained with better kinds of lamps provided with a condensing lens and a coil or ribbon filament lamp.

Controlled illumination is lighted by a cone of rays whose proportions are regulated by a stop at the illuminator and by the iris diaphragm of the condenser. The whole area of the aperture of the objective is used, and none of the refracted light should fall on the lens mounting or on the body tube.

THE OPERATION OF A MICROSCOPE
Student Microscope

Without a substage condenser
1. Place slide on stage and center the field to be examined over the opening in stage.
2. Turn low-power objective over specimen a short distance above it—one-half inch will do.
3. Adjust concave mirror, moving it back and forth, until light is as uniform as possible through the black hole and onto the specimen.
4. Raise the body tube by coarse adjustment until the specimen is in focus. Check mirror, adjusting for better light.
5. Higher powers may be swung into position and light adjusted if necessary for greater magnification.

With substage condenser and lamp without lenses
1. Place slide on stage as above. Adjust objective, plane mirror, and body tube for focus, as above.
2. Remove eyepiece and, while looking down the tube, open iris diaphragm of the substage condenser until it coincides with the margin of the back lens of the objective, just enough to fill it with light, no more. Focus the condenser up and down until the light is uniform through the objective.
3. Replace eyepiece. Make similar adjustments for all objectives.

With substage condenser and lamp with condenser lenses
1. Place lamp 8 to 10 inches from microscope and direct light onto center of plane mirror.
2. Adjust lamp bulb or lamp condenser until the filament image focuses on the microscope condenser. The visibility of the filament image can be increased by closing the lamp iris and focusing the image on its bottom surface.
3. Place slide on stage and focus on it. With the lamp iris closed, raise or lower the microscope condenser until the lamp iris appears in sharp focus with the slide specimen. Open the lamp iris until it disappears at the edge of the field.

4. Remove the eyepiece and adjust the microscope iris by opening it until it coincides with the margin of the back lens of the objective.
5. Do not change diaphragm positions when proper adjustments have been made. Intense light can be reduced by the use of neutral density filters or by changing the rheostat control on the lamp, if one is available.

To adjust both parts of a binocular microscope, close the left eye and focus with the fine adjustment until the image is sharp for the right eye. Leave the fine adjustment alone and focus for the left eye with the focusing adjustment on the tube outside the eyepiece. Between the two eyepieces is an adjustment for changing the distance (interpupillary) between the eyes. Turn the adjustment slowly until the right and left images blend and a single field is seen with both eyes.

Research Microscopes

Critical illumination (Köhler method) using a lamp with condensing lenses and an iris diaphragm

1. Place a slide with fine-detailed specimen in position on microscope stage. Center the condenser in relation to the stage aperture.
2. Place the lamp so the diaphragm is about 10 inches from the microscope mirror and line it up with the microscope so that the light filament centers on the plane surface of the mirror. Insert a neutral filter.
3. Using a 10× ocular and 10× objective, adjust the mirror until the light passes through the substage condenser to the slide. Focus microscope on the specimen and center the light.
4. Rack up the substage condenser until it almost touches the slide, close the substage diaphragm about halfway, and adjust the lamp condenser until an image of the filament is focused on the closed substage condenser diaphragm. The image of the filament can be observed in the microscope mirror or with a small hand mirror. Partly open the substage iris diaphragm.
5. Focus with the substage condenser until a sharp image of the lamp diaphragm coincides with the specimen on the slide.
6. Take out the eyepiece and, looking down the tube, observe the back lens of the objective. Open the substage diaphragm until its image edge almost disappears in the back lens but remains just visible. The back lens should be filled with light—full cone illumination. Because the light source is imaged on the diaphragm of the substage condenser, the light source is in focus at the back focal plane of the objective, and the filament of the lamp can be seen.
7. Replace the eyepiece.

Certain precautions are imperative for good illumination and bear repetition. The back lens should be filled, or nearly filled, with even light. If the substage diaphragm is opened more than this, glare (random light not used in image formation) increases. Closing down the diaphragm to less than the diameter of the back lens decreases glare, increases contrast, and decreases resolution. (Resolution is the ability to resolve as separate entities two points that are close together. If they cannot be distinguished as separate entities, then the resolving power of the microscope has been exceeded.) Closing the iris diaphragm to observe some features of relatively transparent, unstained living specimens may be useful. However, closing the diaphragm to reduce light intensity is poor technique. Light intensity should be controlled by the kind of bulb, the use of filters, or a rheostat or variable transformer inserted in the lamp system.

Unless you have had previous experience with a certain microscope, when changing from one objective to another of higher power (high dry or oil-immersion), do not take it for granted that the microscope is parfocal (focal planes of all objectives lie in same position on the tube). The higher power may crush a valuable specimen. Proceed as follows: Examine slide with the naked eye for approximate part to be examined. Center the area over the stage aperture. Check with lower power and then middle power for the particular area of interest, and then change to high power. Do this while watching the bottom of the objective from the side. Recall that the objectives are parfocal and that one or two turns with the fine adjustment (*only*) knob will bring the image in focus. If not, go back and repeat the steps. Remember, when an objective of a different numerical aperture is used, the condenser diaphragm must be adjusted (step 6). The numerical aperture (NA) number (also engraved on each objective) is used to compare the resolving power of the lenses: The higher the numerical aperture, the greater the resolving power and the wider the condenser diaphragm must be opened.

Use of Oil-Immersion Objective

The working distance with this objective is smallest, and great care must be taken to prevent crushing a cover glass. One must use thin cover glasses or mount the specimen on the cover glass itself. Greater illumination is required than with the dry objectives, and a wider cone of light is needed. The space between the objective and the cover glass, also frequently between the slide and the condenser, must be filled with a suitable medium with a refractive index and dispersion like that of the glass of the lenses. This medium is provided by immersion oil.

1. Using high dry objective (40×), locate and center the area to be examined.

72 Basic Procedures

2. Rotate the nosepiece so that the stage opening is between the high dry and oil-immersion objectives.
3. Place a drop of immersion oil on the cover glass over the area to be examined. Watching from the side, swing the oil-immersion objective into place. The objective will make contact with the oil. As with all the other lenses, the immersion objective is par focal and the same directions given for the high dry objective apply here, too. (Oil may also be placed on the substage condenser lens.)
4. Make critical focus with the fine adjustment.
5. Remove eyepiece and adjust the light on the back lens of the objective; open the condenser diaphragm.

After use, clean oil-immersion objective; never leave oil on it. Wipe it with clean lens paper. If necessary, use a little xylene or chloroform on lens paper; dry immediately with dry lens paper. Do not leave xylene or chloroform on the lens for it will loosen the adhesive holding the front lens in place.

Other Hints for Efficient Microscopy

Always keep all surfaces clean and free from dust and grease. This includes the lamp condenser, filters, mirror, lenses or the substage condenser, lenses of objectives, slides, cover glasses, binocular prisms, and lenses of ocular eyepieces. Use a good grade of lens paper and a fine camel hair brush and compressed air to clean lenses and other parts of the microscope.

If glasses are worn, protect them from scratching by mounting a rubber band around the top of the ocular. There are commercially available protectors for this purpose. Microscopes may be equipped with high eye-point eyepieces, which enable the microscopists to wear glasses while using the microscope. When not wearing glasses, raise the spacer provided on such high eye-point eyepieces to keep the eyes at the appropriate distance.

If it is convenient to have a pointer in the eyepiece, unscrew and remove the top lens of the ocular eyepiece. Inside, about halfway down its tube, is a circular shelf. Place a small drop of Permount (or other quick-drying mountant) on this shelf and, with small forceps, place an eyelash in this drop. Arrange the eyelash to lie flat and project slightly short of center of the hole formed by the circular shelf. Screw lens back in place.

When carrying a microscope, carry it upright, preferably with both hands. Never tilt it; the eyepiece can fall out. When using a microscope to check staining effects or with any wet mounts, place a thin piece of glass on the stage. A lantern-slide cover glass (3.25 × 4 inches) is usually available and is an ideal size for most stages. Clean it after use to prevent contamination of future slides or material.

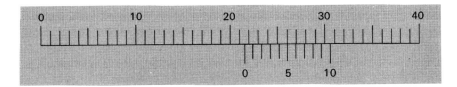

Figure 8-1. A vernier scale.

MEASURING DEVICES USED ON A MICROSCOPE

Reading a Vernier

Most mechanical stages are equipped with a vernier—a device for the purpose of relocating the same spot previously examined on a slide (Fig. 8-1). Two numerical scales run side by side, a long one constructed on a millimeter scale and a short one constructed on a scale of 9 millimeters divided into 10 equal divisions. When the area of reference on the slide is centered in the objective, read the vernier. Check the point of coincidence of the zero of the short scale against the long scale. The lower whole figure on the long scale, to the left of the zero on the short scale, is the whole number, the number to the left of the decimal point in the final reading (21 on the long scale in Fig. 8-1). *Exception:* If the zero of the short scale coincides exactly with a whole number on the long scale, then the number is recorded as a whole number with no decimals following it. The decimal is determined by the point of coincidence of the line of the short scale that perfectly coincides with any line on the long scale (0.5 of the short scale in Fig. 8-1). The reading in Figure 8-1, therefore, is 21.5.

Measuring Objects with an Ocular Micrometer

The measurement of slide specimens usually is done with a micrometer disc placed in the ocular, but first the disc's actual value with respect to the magnification at which it is being used has to be calibrated against a stage micrometer. Stage micrometers usually have a 2-mm scale divided into 0.01-mm divisions or a 0.2-inch scale divided into 0.001-inch divisions. The ocular micrometer has a 5-mm scale divided into 50 (0.1-mm) divisions or 100 (0.05-mm) divisions.

Unscrew the top lens of the ocular, place the ocular micrometer on the circular shelf inside, and replace the lens. Focus on the stage micrometer, moving it until the zero lines of both micrometers coincide. Then a definite distance on the stage micrometer will correspond to a certain number of the divisions on the ocular micrometer.

The rest is simple. For example:

The ocular micrometer has 100 divisions.

Suppose that 30 divisions (each measuring 0.01 mm) of the stage micrometer equal the 100 ocular micrometer divisions. Then 30×0.01 mm $= 0.30$ mm, the length of the 100 divisions on the ocular micrometer.

Next, $0.30/100 = 0.003$ mm (3.0 μm), the value of each single division.

Therefore, when the ocular micrometer (each division measuring 3.0 μm) is focused on a specimen and the specimen requires 9 divisions to meet its length, multiply 9 by 3.0 μm to equal 27.0 μm (0.27 mm), the total length of the specimen.

Obviously, each combination of oculars and objectives and each different tube length must be calibrated in this fashion.

SPECIALIZED MICROSCOPY

The image of a biological specimen can be formed and used in several ways. The most common form, known in all laboratories, is the use of color images, either natural color in the specimen or differential color applied to it. But unless one of a few vital stains is employed, the color image method cannot be used on live material. Most living material is transparent to light; that is, the light wave passes through it with very little loss of intensity, so other means for examining it must be used.

Dark-Field Microscopy

In dark-field microscopy, the objects themselves turn the light into the microscope by reflecting it or scattering it, and the object appears luminous on a dark background. No light from an outside source reaches the eye. A "dark" stop in the substage condenser blocks out the central part of the solid cone of light formed in the condenser, and only oblique rays in a hollow cone of light, striking from the sides, illuminate the object.

The simplest form of dark-field can be developed with a black central patch stop inserted in the carrier under the condenser. Dark-field elements in a threaded mount can be used in place of the upper lens of an Abbe condenser. This is practical with all objectives if the numerical aperture does not exceed 0.85 and immersion oil is used with the slide.

Dark-field illumination can be modified with colored stops and outer rings below the substage condenser, instead of black stops, to produce optical staining. The central stop determines the background color, and the outer ring determines the color of the object, giving optical coloring to unstained objects and helping to reveal detail and structure.

Dark-field and optical staining can be useful for examining the following: (1) bacteria, yeasts, and molds; (2) body fluids, plant or animal; (3) colloids; (4) living organisms in water; (5) foods, fibers, and pigments; (6) insects and scales; (7) crystals; (8) bone, plant, and rock sections.

Dark-field staining is a common means of studying the results of microincineration, the investigation of minerals present in different parts of tissues. Paraffin sections of tissue fixed in a solution of formalin and alcohol are mounted on slides and placed in an electrical furnace. The temperature is then slowly raised until all organic matter is burned away and only the mineral skeleton of the tissue is left. This appears white under dark-field observations.

Phase and Interference Microscopy

Although most living material is transparent to visible light, different components of tissue do alter to a different extent the phase of light waves passing through them. That is, the light velocity is altered, advanced, or retarded, and its vibration is said to be changed in phase. When the waves come together and are in phase, brightness increases, but if they are out of phase and are of equal amplitude, interference occurs and the eye sees black, not light. Waves of intermediate difference in phase produce a series of grays. It is the aim of the phase microscope to change slight phase differences into amplitude differences and to produce a variation in intensity from light to dark contrast observable in the specimen under the microscope.

Both phase and interference microscopes follow this principle of interference phenomenon—combining light waves that are out of phase with each other to produce combined waves of greater and lesser amplitude. The means of applying the principle is different.

Phase Microscopy

The cost of phase microscopy is not excessive, and the parts can be adapted to any bright-field microscope. The system requires phase contrast objectives (achromatic objectives with fixed-phase plates and a two-lens Abbe condenser with an iris diaphragm and a revolving ring below the condenser to carry four annular ring diaphragms, which are stops that produce different-sized cones of light). The correct annular ring diaphragm must be centered to the phase ring of the correct objective and must match its numerical aperture. The annular diaphragm causes the light to strike the object in the shape of a hollow cone and gives rise to two types of waves: one type passes straight through the object (undiffracted), and the other type is diffracted into a different course. Different components in the tissue diffract differently, producing the differences in intensity.

In phase microscopy, images exhibit a "halo" as a result of the diffraction of light at the phase-changing annulus (annular ring). The light that has passed straight through the specimen is made to interfere with the light diffracted sideways by it. Only the refracting structures are observed, and these edges and abrupt changes of refractive index produce the halo around the images.

Interference Microscopy

A more sensitive and accurate instrument is the interference microscope, which is better adapted to measuring the refractive indices of a specimen. This microscope produces an image of extremely high contrast with clearly defined boundaries as pseudo-3-dimensional images. It also reveals colorless or homogeneously colored structures that would be invisible or not pronounced enough to be viewed in the conventional light microscope.

In this microscope the light splitting and recombining is carried out externally to the specimen. Birefringent plates (doubly refracting) are cemented to the top lens of the condenser and to the front lens of the objective. One set of rays passes through the object, the other set passes through a clear region at one side, and they are then recombined. Any phase differences between them remain constant and can interfere to give light or dark. A refractile object in one beam causes a change in light intensity. The two rays can be arranged to bring out a black background and a bright object or vice versa and also to give high contrast between the object and inclusions in it. The lateral separation of the two rays is only a few micrometers. The Nomarski prism avoids the need for two prisms (one in the objective and one in the substage condenser) and can be used with all objectives. The prism is made of two cemented components of a uniaxial birefringent crystal mounted in an interference contrast slide in the substage condenser. It can be used with either reflected or transmitted light.

Phase versus Interference Microscopy

Both operate on the same principle—the interference phenomenon, which changes phase difference into amplitude difference. However, the special features of each should be considered in choosing between the two.

Phase
(1) Light passing through the specimen is made to interfere with light diffracted sideways by it. (2) It only shows up diffracting structures in the specimen; it produces a halo. (3) Apparatus is simple, reasonable in cost, and easy to operate and can be added to any conventional microscope. (4) It can be used to study living material, cytoplasm, cell inclusions, nucleus, and the action of physical and chemical agents on living cells. (5) It is adequate for routine examination.

Interference

(1) Light splitting and recombining is carried out on the outside of the specimen and under the control of the experimenter. (2) No halo; variations in the optical path through the object are easily interpreted. Phase change can be measured. (3) Apparatus is expensive and complicated; requires constant checking and adjusting. The Nomarski prism simplifies this. (4) It can be used on living material to determine dry mass, for example, changes in mass during cell activity; protein distribution, both in cytoplasm and nucleus.

Polarizing Microscopy

Closely related to the above types of observation is the use of polarizing attachments. These may be used on most types of microscopes, but for continued use a polarizing microscope is preferable. When a ray of plane-polarized light (vibrating in one plane) falls on the object, it is split into two rays: one obeys laws of refraction, and the other passes through the object with a different velocity. After emerging from the object, the two rays are recombined, but because their velocities are different, they will be out of phase. This phase difference is the quantity measured in a polarizing microscope.

A polarizing prism in the fork substage (instead of a condenser) or a polaroid disc in the condenser slot polarizes the light so that it vibrates in one plane only. Part of the light (ordinary ray) is reflected to the side of the prism or disc and does not illuminate the object; the other part (extraordinary ray) continues straight through the prisms to emerge as polarized light. (There is obviously a loss of light, so plenty of it must be used.) If the polarizer is a prism, it usually can be rotated 360° and the amount of rotation in the field checked.

In or above the ocular is fitted another polarizing prism, the analyzer, whose vibration direction is set at 90° to that of the polarizer ("crossed polarizers"). The extraordinary ray from the polarizer becomes the ordinary ray in the analyzer and is reflected out of the field, unless an anisotropic substance (doubly refracting when placed between the analyzer and polarizer) rotates the plane of polarization and interferes with the path of light. Such a substance divides the light from the polarizer into two beams, one of which passes through the analyzer, making the object appear to glow against a dark background. Isotropic substances (singly refracting) do not polarize and therefore do not divide the beam of light and do not glow.

Some Uses of Polarizing Microscopy

1. To determine whether an object is isotropic or anisotropic (if it rotates the plane of polarization).
2. To determine differences in physical properties in different directions, such as those occurring in the study of mitotic spindles.

3. To reveal molecular orientation of structures, chemical constitution, chemical and physical intervention in the cell, pressures, or tensions; all can produce anisotropic effects.
4. To determine refractive indices.
5. Can be used on natural and artificial fibers; cellulose fibrils; lamellar plasma differentiation, pseudopodia; spindles and asters; nerve fibers, muscle fibers, chromosomes; chemical and mineral crystals; crystallized hormones and vitamins; dust counts; starch grains; and horn, claw, and bone sections.
6. Can detect birefringent substances (e.g., amyloid)
7. Can be used on fresh unfixed material.

Fluorescence Microscopy

Fluorescence microscopy has become an invaluable tool for investigation. Fluorescent molecules absorb light at one wavelength and emit it at another, longer wavelength (so-called "primary fluorescence"). If a compound is illuminated at its absorbing wavelength and then viewed through a filter that permits only light of the emitted wavelength to pass, then the compound is seen to glow against a dark background. Secondary fluorescence can be induced by the use of fluorochromes (strongly fluorescent dyes or chemicals) applied to the specimen. Fluorochromes can induce fluorescence in objects that do not fluoresce naturally and will intensify fluorescence in those that do. The use of fluorochromes has made it possible to identify subcellular components with a high degree of specificity. Fluorochromes are stains that attach themselves to cell components with a high degree of specificity. They absorb and then reradiate light with known intensity curves of excitation and emission.

A fluorescence microscope is used to detect the fluorescent dyes that are used to stain the cells to be studied. This microscope is similar to a light microscope except that the illuminating light comes from a strong source, usually arc lamps. The most common arc lamps are the mercury and xenon burners. Compared to the mercury burners, the xenon burners have a more even intensity across the visible spectrum but lack high spectral peaks and are deficient in the ultraviolet. Infrequently, tungsten-halogen lamps are used, especially for blue or green excitation with brightly emitting specimens. More recently lasers have been used, especially the argon-ion laser. Laser sources have become especially useful in confocal microscopy (see below).

There are three classes of filters used in fluorescence microscopy; exciter filters, barrier filters, and dichromatic beam splitters (dichroic mirrors). The first, the "exciter" filter, filters the light before it reaches the specimen. It selects only those wavelengths that excite the particular fluorescent dye in use. The second, the barrier filter (filters the light from the specimen), allows only those wavelengths to pass that are emitted when the dye fluoresces. The dichroic mirror is positioned in the light path after the exciter filter but before the barrier filter; they are designed to efficiently reflect excitation wavelengths and pass emission wavelengths.

A very powerful technique now used in fluorescence microscopy is to couple fluorescent dyes that emit different colors when they fluoresce to specific antibodies. This combination permits the "labelling" of specific macromolecules that can be recognized and followed within cells and tissues during particular dynamic biological processes. Fluorescein (emits green fluorescence when excited with blue light) and rhodamine (red fluorescence when excited with green-yellow light) are two favorite combinations to detect and trace different macromolecules. These procedures are now being utilized to monitor changes in concentration and localization within living cells after intracellular microinjections of the appropriate compounds.

Electron Microscopy

Electron microscopy is considered routine in most institutions. Work on electron microscopes began in the 1930s, and by the 1940s electron microscopes had been developed in different parts of the world. Knoel and Ruska in Germany had a practical, marketable research scope by 1939. England began work in 1936, and RCA in the United States began the development of one in 1938 and had one on the market in 1941. The electron microscope has a higher resolving power than the light microscope (5 angstroms versus 2000 angstroms) and reveals submicroscopic structures that were previously invisible. It uses electrons as the illuminating beam and focuses the beam on the object by the use of magnets. As the beam passes through the object under observation, some of the electrons are scattered by the object, causing shadows of the scattered electrons on photographic film.

Electron beams must be handled in vacuum; therefore, no wet or living tissue ordinarily can be observed. Sections must be thin (0.01 to 0.05 μm) to prevent appreciable electron absorption because the image depends on the differential scattering rather than on absorption. The specimen is supported on a metal grid and is placed in the column at the electron-gun side of the object-lens aperture.

Special fixation, embedding, and sectioning are required to prepare ultrathin sections for use in the electron microscope. To facilitate the scattering of electrons, the density of the specimen must be increased. Because the usual fixing solutions do not meet this requirement, the so-called electron stains must be used in their place. These are compounds of heavy metals (tungsten and osmium), such as phosphotungstic acid and osmium tetroxide. Potassium permanganate (Luft 1956), acrolein (Luft 1959), and more recently aldehydes (Sabatini et al. 1963, 1964) have been advocated. Only plastic (methacrylates) embedding permits ultrathin sectioning. There are several companies that manufacture ultramicrotomes. Glass or diamond knives are essential for ultrathin sectioning; steel knives are rapidly dulled by cutting plastic. Crang and Komparens (1988), Dykstra (1992, 1993), and Hayat (1986) are good references.

Three-dimensional images may be obtained by metal shadowing. Metals, such as carbon, silver, platinum, and gold, are evaporated on the specimen at a small angle. A thin film of the metal covers the specimen except for a "shadow" caused by the specimen being in the way of passage of the metal. From the shape and dimension of the shadow, the height and contour of the specimen can be observed. Of the many references, important ones are Glauert (1974, 1975), Hayat (1993), Kay (1965), Mercer and Birbeck (1961), Parsons (1970), Pease (1964), Schultze (1969), and Bancroft and Stevens (1982). Also available are publications by Bozzola (1992) and Hunter (1993).

Excellent booklets about all types of microscopes can be easily obtained from microscope suppliers.

There is an exciting renaissance occurring in microscopy resulting from the integration of standard light microscopy with video methods for enhancing contrast, the use of well-known sources of energy in newly designed devices for their delivery to the sample, and new computational methods linked to computer terminals and video screens for detecting, manipulating, analyzing, and reporting information. Novel applications and uses have moved optical microscopy to the forefront of technology. The discussion that follows is intended to introduce concepts and to provide the necessary background and understanding for further reading in the new age of microscopy.

It is now possible to obtain real-space (three-dimensional) images using samples in aqueous environments with a resolution considerably better than that available with the visible light microscope. This has been achieved with the Nobel Prize–winning invention of the scanning tunneling microscope (STM) in 1981 and the atomic force microscope (AFM) in 1986 by Binnig, Rohrer, and their associates (1982). These instruments permit the observation of structures that range from large ones, such as whole cells, down to individual molecules of proteins and DNA, with subnanometer resolution. The STM and the AFM constitute the major members of the ever and rapidly increasing family of scanning probe microscopes (SPM).

With SPM systems a suitable probe is placed in contact with or close to the surface to be studied, and it measures a component. For example, it will measure a voltage or a magnetic field that reveals the distance between the probe tip and the surface of the sample in 3 spatial dimensions. The probe is scanned across the surface of the sample in small increments. Scanning is controlled with piezoelectric ceramics with 0.1 angstrom resolution and which can change shape in a reproducible way when a voltage is passed across them. At each step the distance between the probe and the surface is measured; successive scans build up a fine-grained image of the surface's highs and lows. A

3-dimensional image is constructed using the x, y, and z data to generate the surface topography or the interaction magnitude data to create a map of the probe-surface interaction. The ability to use biological samples in aqueous environments, thus avoiding sample preparation damage brought about by other methods, such as those used for EM, allows the study of the dynamics of cellular processes in biological systems, such as protein adsorption, secretory granule processing, and the structure of individual protein molecules.

The limitation of the STM is that you get measurable tunneling current only when the sample being studied is an electrical conductor. Thus, molecules like DNA cannot be imaged unless they are coated with a film of metal. These limitations were overcome by the next member of the family of probe microscopes—the AFM. It too measures the electron atmosphere of a sample's surface, but it does not require that electrons actually move from the surface to the probe. Instead of creating images by measuring changes in current, the AFM measures variations in the repulsive forces between the atoms on the microscope tip and atoms on the surface of the sample. Unlike electron microscopy, which requires special, elaborate preparation and working in a vacuum, an AFM works in air or under water and can examine a sample in its natural state. In the past 2 years variations have been designed in the AFM that are extremely sensitive to changes in magnetism, temperature, and electrochemical properties.

Near-Field Scanning Optical Microscopy (NSOM)

NSOM, conceptually the oldest form of scanning probe microscopy but only recently developed by Betzig, offers the proven contrast mechanisms of optical microscopes plus the resolving power and the easy use of scanning probe microscopes. By using a highly tapered fiber optic wire that emits light, it overcomes the physical limitation of conventional optics—the inability to resolve beyond half the wavelength of light, or about 200 nm. The NSOM can provide 12-nm resolution using 500-nm light. The NSOM has its basis in the fact that the diffraction limit to resolution in optical microscopy is not a fundamental restriction. By scanning the source of light or detector of light very close to the surface of a sample, it is possible to generate an image whose resolution is dependent only on the probe's size and the probe-to-sample separation, each of which can be made much smaller than the wavelength of light.

NSOM builds on the optical contrast of microscopes and is therefore highly useful for studying fundamental dynamic biological processes and systems in living cells, such as the study of cytoskeletons, organelles, phagocytosis and cell division. It is possible, for example, to use an optical microscope at 100× magnification and then use NSOM to get higher resolution imaging exceeding 50,000× of a much smaller domain, all in the same sample.

Confocal Microscopy

It has been suggested that confocal microscopy links optical microscopy and scanning electron microscopy for the results it is capable of delivering. It can visually probe a translucent specimen in three dimensions and yield a series of high-contrast images that can be assembled into a 3-dimensional image on a video. Central to the confocal approach in microscopy is the concept of focusing a point source of light onto (or into) a spot within the specimen and then imaging the light signal from this point through a small aperture or pinhole and onto a photodetector. The pinhole is optically conjugate to both the point source of light and the focused spot in the specimen. That is, all three planes are mutually in focus—therefore, confocal.

Confocal imaging requires that the systems for excitation and detection are aligned so that both are focussed on a single point in the specimen. For this reason epi-illumination is used in most confocal microscopes because, in this case, light for both illumination and recording passes through the same objective lens and through the same part of the specimen as well. Although this method of illumination precludes the use of those contrast techniques that depend on coherent interactions between illumination and specimen, such as phase or interference contrast, it works exceedingly well with fluorescence. The ability to produce monoclonal antibodies by newly developed technology, has made it possible to produce very specific fluorescent probes. This combination has become an important factor in the success of confocal microscopy for it is now possible to label specific nucleic acid sequences with fluorescent markers.

Optical sections can be obtained because the effective focal plane of the objective lens is very thin (0.5 μm) for an objective with a high numerical aperture. A laser beam spot is scanned across an object at the focal plane. An aperture in front of the detector blocks light reflected from out-of-focus planes. Thus, only images on the focal plane are relayed to the monitor and not those from below or above. Scanning at various levels within the specimen by raising or lowering the focal plane will deliver incredible images at each level. As with computerized tomography, the relatively new technology that has revolutionized medical diagnoses, the images collected can be assembled to reproduce a 3-dimensional image on the video monitor. The extra special advantage is that organisms do not have to be sectioned or destroyed to be examined. Images can be derived from transparent specimens. Resolution can be achieved down to 0.1 μm with useful magnification that ranges from $100\times$ to as much as $10,000\times$.

The SPM appeared a little more than 10 years ago. More than 20 scanning probe microscopes have been developed since then. There is no limit to the number of variations of the scanning probe theme. This is especially true for applications in industry, especially those devoted to the production of semiconductors and hard disks. For example, Kumar Wickramasinghe has built a laser force microscope in which a vibrating wire is brought close enough to a chip's surface to feel attractive forces, essentially surface tension from water molecules that condense on the surface. The forces change the wire's vibration rate, which is detected by a laser. This method can reveal bumps as small as 25 atoms. He has also magnetized the wire to produce magnetic force microscopy and given the wire an electric charge to deliver an electrostatic force microscope. It is anticipated that before long there will be positron and nuclear magnetic resonance microscopy, thus presenting the possibility to do NMR on single atoms. We can expect continued growth of this renaissance in microscopy for its industrial and biological use and because of the lucrative rewards they will continue to bring to the companies involved in their production.

Suggested Readings

See Betzig and Trautman (1992), Binnig et al. (1982), Pawley and Smallcomb (1992), Shuman et al. (1989), Wickramasinghe (1992), and Wilson. For details regarding microphotography, see Delly (1988) and Smith (1994).

9
Stains and Staining Action

Because most tissues do not retain enough color after processing to make their components visible under a bright-field microscope, it is expedient to add colors to tissues by staining them. Correctly chosen stains aid in identifying tissues and their elements, and in diagnosing pathological conditions. Knowledge about the structure and action of the chemicals and stains used for tissue identification must be reviewed and understood. The subject is extensive, and the student is advised to consult specialized references for detail that cannot be included here.

NATURAL DYES
Cochineal and Carmine

These are members of a group of dyes called natural dyes. Unlike other natural dyes, cochineal and carmine are derived from an animal source—a minute insect, *Coccus cacti*, which lives on spineless cacti. The dye is present as a purple sap in the females, which are harvested, dried, and pulverized to produce cochineal. This dye by itself has little affinity for tissue unless iron, aluminum, or some other metal is present. With the salt of one of these metals as a mordant (see p. 85), staining will result. Alum cochineal, a commonly used form of this dye with mordant, is an efficient nuclear stain.

The dye carmine is derived from cochineal by boiling the cochineal with a salt, usually alum, to produce a precipitate. This precipitate is insoluble in water, and before it can be used as a stain it must be converted into a soluble compound such as ammoniacal carmine or acetocarmine, in a process that will be described in the section on mordants (p. 85).

Hematoxylin

In many respects, hematoxylin can be regarded as the most important of the natural dyes. It was one of the first histological dyes and remains one of the most widely known and used dyes. Hematoxylin is extracted from the heartwood of longwood trees from South and Central America and the West Indies. The tree is *Caesalpina campechianum*, one of the legumes similar to acacia or cassia trees. The crude material is exported as logs, as chips, or as dried aqueous extract of the heartwood. It is then extracted with ether in a continuous extraction apparatus evaporated to dryness, dissolved in water, filtered, and crystallized out of solution. All of these steps are slow and difficult to handle and require costly apparatus, thus making hematoxylin one of our most expensive dyes.

In the crystallized condition it is not yet a dye, and its color must be allowed to develop by oxidation into hematein (color acid—no relation to hematin, the colored constituent of red blood cells). Oxidation may be accomplished in either of two ways: "naturally" (a slow process of exposure to air for 3 to 6 weeks, as in Heidenhain hematoxylin) or artificially by the use of mercuric oxide, hydrogen peroxide, or another oxidizing agent (a more rapid process, as in Harris hematoxylin). Used alone, hematein is only a weak and diffuse dye with little affinity for tissues. A weak acid will not combine with nuclear elements in sufficient quantity to produce efficient staining. Some form of mordanting is required to form a base from this dye, which will then stain the acidic nuclear elements. The most commonly used mordants are alum salts of aluminum, potassium, or iron.

When properly oxidized, hematoxylin is an exceedingly powerful dye with various shades of staining from purples, through blues, and into blue-blacks. The iron-mordanted form is one of the more valuable dyes for mitotic study, and it gives to the chromatin a precise black or blue-black color. This black color is the result of the presence in hematoxylin of some tannin, and the latter in combination with iron salts produces a lasting black color.

Hematoxylin staining is discussed in Chapter 11.

Other Natural Dyes

Other natural dyes are saffron from stigmas of *Crocus;* indigo from plants of the genus *Indigofera;* berberine from barberry; orcein and litmus from the lichens *Lecanora* and *Rocella;* and brazilin, chiefly from a few species of *Caesalpina,* a tropicopolitan genus of leguminous trees (Mohr 1950). Orcein, a specific dye for elastin (present in elastic fibers), is prepared by boiling the plants in water. The lecanoric acid in them splits to produce orcinol—a resorcinol with a methyl group attached to it. Orcinol with NH_3 (ammonia) and atmospheric oxygen forms orcein.

MORDANTS

Metallic mordants have been used for over 100 years. A mordant is a metallic salt or hydroxide that combines with a dye radical to form an insoluble compound. Such a compound is called a lake. For example, carmine dissolved in a solution of aluminum sulfate, becomes positively charged and acts as a highly basic dye. The term *mordant* should not be used for all substances that increase staining action, but only for salts and hydroxides of di- and trivalent metals.

The use of mordants in staining tissues has many advantages. Once the mordant-dye compound has formed in the tissue, the dye is relatively permanent, is insoluble in neutral solutions, and can be followed by many other forms of staining. Dehydration will not decolorize the tissue. Blocking and extraction methods have very little effect on mordant dyes. There is considerable flexibility in mordant-dye procedures, as the mordant can be used before the dye, together with the dye, or after the dye (rare).

Carmine and hematoxylin are two dyes often used with mordants for staining tissue. For these dyes, the mordants commonly used are aluminum, ferric and chromium salts, and alums (potassium alum, ammonium alum, iron alum, and chrome alum). A solution of ferric chloride can be used in place of iron alum, and Cole (1933) recommends the use of a phosphate-ripened hematoxylin with it. The ferric chloride causes the tissues to stain more rapidly and more intensely than does the iron alum. Iron alum solutions are quite acid; the greater the amount if iron alum present, the more acid the solution. Adding even a weak acid, such as acetic acid, to an alum hematoxylin makes the solution more selective for nuclei.

For long-lasting solutions of combined mordant and dye, mordants with little or no oxidizing action must be used: ammonium alum, potassium alum, phosphotungstic acid, phosphomolybdic acid, and iron alumferrous sulfate. If a long life is not required, it is possible to use mordants with a vigorous oxidizing action: ferric chloride, ferric acetate, and ferric alum (Cole 1943). Since the usefulness of these solutions lasts only a matter of hours, they must be prepared immediately before use. In iron-hematoxylin solutions that are premixed, a high mordant-to-dye ratio tends to produce nuclear staining, while a high dye-to-mordant ratio favors myelin staining. Though the term *iron hematoxylin* is widely used, it is actually chemically incorrect; hematoxylin cannot form a lake but first must be oxidized to hematein (Puchtler and Sweat 1964a).

Hematein forms blue-green lakes with copper, lilac with nickel, red with tin, dark brown with lead, and purple with bismuth.

When using two separate solutions, a mordant of any kind can be used if followed by a well-ripened hematoxylin solution. If the solution is unripened, the mordant should include a substance of considerable oxidizing power, a ferric or chromium salt. The value of two solutions lies in the fact that the dye can be preceded by a salt, which cannot be used in combination with the dye in a single solution. For example, ferric chloride, when added to an ammonia-ripened hematoxylin, will throw down a precipitate of ferric hydroxide. Double mordanting can be profitable; a mordant followed by a solution of hematoxylin containing a mordant gives excellent results. The mordants for separate use are ammonium or potassium alum, ferric ammonium alum (2 to 3 drops of HCl increases contrasts), and ferric chloride (HCl increases contrast). They yield the following colors with hematoxylin: ammonium alum, bright blue nuclei; potassium alum, lilac or violet; chrome alum, cold gray blue; iron alum, blue to black (Cole 1943).

When used with hematoxylin, some mordants are more effective than others for staining particular types of material. The following is a list of tissue elements and the metals effectively used on them (Mallory 1944).

—Nuclei: aluminum, iron, tungsten
—Myelin sheaths: chromium, iron, copper
—Elastic fibers: iron
—Collagen: molybdenum
—Connective tissue elements, neuroglia: tungsten
—Axis cylinders: lead
—Mucin: iron
—Fibrin: tungsten

Progressive and Regressive Staining

Lakes, formed by mordant and dye, can be used progressively, but when mordant and dye are used separately, regressive staining usually is more effective. In progressive staining the stain is added to the tissue until the correct depth of color is reached. In regressive staining, the sections are overstained, and excessive amounts are removed by one of the following methods.

Method I: Excess Mordant

With an excess of free mordant present outside the tissue, the mordant-dye complex in the tissue is broken up, and, since the amount of mordant in the tissue is smaller than that in the differentiating fluid, the dye moves out of the tissue into the fluid. (If the tissue is left long enough in the fluid, it is conceivable that most of the dye could move out of it into the excess mordant, causing the sections to become colorless.) Because the excess holds considerably more dye than the cytoplasm, most of the dye is lost from the cytoplasm, but some still remains in the nuclei. When enough dye has been extracted from the cytoplasm, and when the correct intensity is left in the nuclei, remove the slides from the mordant and wash them thoroughly, usually in running water, to remove excess mordant. Remaining traces can cause the stain to fade in time.

Method II: Acids

Acids are effective differentiators for some dyes.

Method III: Oxidizers

Oxidizers furnish a third method of regressive staining, and by this method the dye can be oxidized to a colorless condition. Oxidizers are slow in their action and as a result parts of the cell that contain little dye will be bleached before those containing more dye. Picric acid, a commonly used oxidizer, has both a moderate oxidizing and a weak acidic action.

Accentuators and Accelerators

Not to be confused with mordants are accentuators and accelerators. Accentuators are substances that, contrary to the action of mordants, do not become a part of the dye complex or lake. Instead they increase the selectivity or stainability of the dye (e.g., phenol in carbolfuchsin). Accelerators, as their name implies, accelerate the action of the dye—usually of importance in silver impregnation—e.g., chloral hydrate.

SYNTHETIC DYES

Natural dyes had no competition until the middle of the nineteenth century when William Perkin worked out the processes for making aniline or coal-tar dyes. These were the first synthetic dyes.

Synthetic dyes, like natural dyes,, can be used either progressively or regressively. An acid solution is often used to remove excess basic dye; an alkaline solution is used to remove excess acidic dye. In some cases alcohol can be used as a differentiator, particularly for basic dyes; but in general a sharper differentiation is achieved by using an acid. Part of the explanation of this observation lies in the property of proteins, the principal components of protoplasm. Proteins exhibit dipolar or amphoteric properties, that is they can dissociate as both an acid and a base depending on the pH of the solution that the protein is placed into. If placed into water, a protein will display acid properties if it has more acid groups (i.e., carboxyl and hydroxyl groups) than basic ones (amino) or basic properties if there are more basic groups. At the isoelectric point for each protein (stated as a pH) the degree of dissociation of the acid and base groups are equal—the net charge is zero and the proteins precipitate out of solution. Below its isoelectric point a protein behaves as a base (ionization of the acid groups is suppressed and thus the number of basic groups increases). Therefore, below its isoelectric point proteins will combine with acids or anions, such as the acid dyes eosin, orange G, aniline blue. Above its isoelectric point the reverse is true—the protein behaves as an acid and will combine with bases or cations such as the basic dyes methylene blue, hematoxylin, basic fuchsin. There is no dye binding at the isoelectric point. This is true for proteins in solution as well as those rendered insoluble by fixation. Thus, as the pH of the solution becomes more acidic, increasing amounts of acid dye are bound; as it becomes more basic, increasing amounts of basic dyes are bound. As a result, by manipulating the pH of a solution, you may enhance or remove acid and base dyes.

The real importance of synthetic dyes lies in their use for double and triple staining—the use of two or more stains on the same slide. The dyes, if properly chosen, will stain selectively; that is, each dye, because of a known specificity, will stain only specific parts of the cells. This type of staining has obvious advantages. Owing to their chemical nature, synthetic dyes make this kind of staining possible. If the dyes are synthesized, their formula can be controlled, and the significant part of the dye is either anionic (acid) or cationic (basic) in action. Actually, dye powder as purchased is a salt, but the salts of the so-called basic dyes give up OH^- ions and act as cations, and the acidic dyes give up H^+ ions and act as anions. Therefore, an acid, or anionic, dye is the salt of a color acid, usually a sodium salt; a basic, or cationic, dye is the salt of a color base, usually a chloride. (See structure of synthetic dyes, p. 91) The terms anionic and cationic are more appropriate than acidic and basic, respectively, and are becoming more widely used. Basic dyes have an affinity for nucleic acids (i.e., nuclei and cytoplasmic RNA), which are basophilic (readily stained by basic dyes), and acidic dyes have an affinity for the general cytoplasm, which is acidophilic (readily stained by acidic dyes).

When the sodium salt of an acid dye and the chloride of a basic dye are mixed, there is a tendency for ions to interchange. The so-called neutral dyes are formed and give results differing from those obtained with ordinary double staining using separate acid and base dyes. Polychroming is a process in which a dye forms other dyes spontaneously—the basis of the development of modern blood staining (Romanovsky stains). Methylene blue in solution is oxidized into one or more compounds of lower methylation. Therefore, all methylene blue solutions, on staining, contain lower homologs, primarily azure A, B, and C, and the methylene blue is now called polychrome methylene blue. The azures are more violet in color and more selective in their action than unpolychromed methylene blue and account for the differential action of blood stains in the differentiation of white blood cells. When eosin is added, it enters into chemical combination with the above basic dyes to form an insoluble precipitate, which when redissolved, is the basis of Wright, Leishman, and May-Grunwald stains. The pure azures may now be purchased. Pure azure A-eosinate is equal to a traditional Giemsa stain. Azure C is not as good. Azure B is best and, as an eosinate, can be substituted for May-Grunwald staining.

The dyes used in polychroming are not stable in aqueous solutions and precipitate out; the dye powder must therefore be dissolved in alcohol. Since alcoholic solutions do not stain well, different methods of using them have been employed in the Romanovsky stains. Sometimes they are used immediately after mixing (Giemsa), or after applying the alcoholic solution, the stain is diluted with water (Wright). It appears that certain parts of the cell have an affinity for the neutral stain, others have an affinity for the basic dyes and break up the stain to acquire the basic part; other cell parts do the same with the acidic portion. These actions by the cell parts are termed, respectively, neutrophilic, basophilic, and oxyphilic. The cells are stained in a solution of the proper pH, or they are overstained in a neutral solution and differentiated to the proper coloration of their parts. (The Maximow method uses alcohol; the Giemsa method, dilute acetic acid; and the Wolbach method, resin.)

Polychroming must not be confused with another type of staining, metachromasia, in which certain substances are stained in one color and others in another color by the same dye. The definition of metachromasia given by Paddy (1970) is "a characteristic reversible color change that any dye may undergo by virtue of a change in its environment not involving chemical reaction of the dye."

The explanation for the reaction to the dye thionine may be as follows: Thionine stains chromatin blue; it stains mucus, ground substance of cartilage, and granules of mast cells red. The dye seems to exist in aqueous solution in two forms; (1) the normal color, blue; and (2) the metachromatic color, red. Both forms are always present, but the red is in a polymerized form of the blue, and color change may be due to some interaction between the chromophores present. The blue form is favored by an increase of temperature, a lowering of pH, a decrease of dye concentration, or the addition of salts, alcohol, or acetone. The red form is favored by a decrease of temperature, a raising of pH, or an increase in concentration of dye (Bergeron and Singer 1958).

Tissue components showing metachromasia are made up of large anionic molecules containing sulfate, phosphate, or carboxylic acid radicals in abundance, and the groups are close enough to permit secondary bonding between the dye molecules, thus permitting polymerization of the bound dye. Mucin, the ground substance of cartilage, and the granules of mast cells contain substances of this nature and therefore take up the metachromatic form.

Some other metachromatic stains are methyl violet, brilliant cresyl blue, azure A, B, and C, toluidine blue, new methylene blue, methylene blue, crystal violet, safranin, bismarck brown, and basic fuchsin.

The majority of dyes do not stain metachromatically, but are orthochromatic in action. This means that their action is direct and predictable under normal conditions; if it is a blue dye, it stains blue; if it is a green dye, it stains green; and so forth.

Some dyes are not stable in solution and will gradually give rise to other colored agents, which should be considered impurities. The presence of such agents is called allochromasia. Orthochromatic dyes may give rise to metachromatic impurities by allochromasy. Nile blue solutions, an example of allochromasia, contain cation of the dye (blue), anion of the dye (sulfate), Nile red (red or rose), and amino base of Nile blue (orange yellow).

A chromotrope is a substance that can alter the color of a metachromatic dye. Examples of chromotropic tissue substances are the matrix of cartilage, granules of mast cells, and secretions of some mucus glands. Bathochromic (Baker 1958) means a shift in the peak of the absorption curve of a dye in solution toward longer wavelengths. Hypsochromic means a shift toward the shorter wavelengths.

Structure of Synthetic Dyes

The synthetic (coal tar or aniline) dyes are derivatives of benzene, all built on the benzene ring:

There are an infinite number of ways in which this ring can combine with other radicals to produce compounds of various degrees of complexity. Certain chemical groups called chromophores (C=C, C=O, C=S, C=N, N=N, N=O, and NO_2) are associated with color. When these groups are attached to a benzene derivative, the compound acquires the property of color and is known as a chromogen. A chromogen, however, is not yet a dye. It has no affinity for tissues, will coat them only mechanically, and can be easily removed by mechanical means. The compound must also contain a group that gives it the ability to form a union with some tissue end group, either directly or through chelate action of a mordant metal. The most common type of group is one which permits electrolytic dissociation (the formation of anions and cations). This auxiliary group, known as an auxochrome, furnishes combining properties to the compound.

A simple demonstration of the principle of synthetic dye formation is picric acid. The nitro group ($-NO_2$) is a chromophore, and three of these can displace three hydrogen atoms on the benzene ring to produce trinitrobenzene:

$$O_2N-\underset{}{\underset{NO_2}{\bigcirc}}-NO_2$$

This is yellow, a chromogen but not yet a dye. Replace one or more H^+ with an OH^-, the auxochrome, and produce dissociable picric acid:

$$O_2N-\underset{NO_2}{\overset{NO_2}{\bigcirc}}-OH$$

Since the auxochromes are the slat-forming groups, they usually determine the action of the dye; the more basic or acidic groups present, the more basic or acidic is the action of the dye. If both a basic and an acidic group are present, the basic group predominates but is weakened in action by the acidic group.

Most commercial dyes are sold as salt. An acid dye is the salt of a color acid, usually sodium salt; the basic dye is the salt of a color base, usually a chloride. The basic chromophores are: azo group (—N=N—), azin group and indamine group (—N=). The acid chromophores are: nitro group (—NO$_2$) and quinoid benzene ring:

Simpler compounds can be converted into more complex ones by substituting radicals for the H$^+$ atoms present. As more H$^+$ atoms are replaced by radicals, the deeper the color becomes. The simplest dyes tend to be yellow. As methyls or other groups are added, the colors pass through reds, violets, blues, and greens. Using ethyl groups instead of methyl groups deepens the color.

Nomenclature of stains has no absolute conformity. For the names of some stains, the color is used (orange G, Martius yellow); for others, a chemical term (methyl green). If the term is followed by a letter or numeral (Sudan III, IV; Ponceau 2R, 4R), one dye is being distinguished from a related one. B indicates a more bluish color, Y or G, a more yellowish color; WS means water-soluble; A, B, and C distinguish among azures.

NATURE OF STAINING ACTION

Biologists, chemists and physicists disagree about the nature of staining mechanisms—is it chemical, physical, or a combination of both? If chemical, this can mean that some parts of the cells are acid and others alkaline, the acid parts tending to combine with cations and the alkaline with anions. Absorption and diffusion of the dye occur, penetrating the cellular elements and combining with them. This action can be combined with physical action, where there is an absorption of the dye, an attraction of plus and minus charges for each other, and a condensation of the dye on the surface of the cell parts. Minute particles of the dye are deposited on the surface of the tissue by selective adsorption and then enter into combination with the tissue. The proteins, nucleic acids, and other components of the protoplasm proceed along lines of chemical laws by exchanging ions. Stearn and Stearn (1929, 1930) maintain that the confusion lies in the term adsorption, that it may be either a chemical or a physical force, and that the adsorbent can form ions and then proceed along chemical lines. The development of physical chemistry has helped to resolve the question. There appears to be no doubt that both physical and chemical factors take part in the mechanism of salt formation—the basic action of stains.

In any case, the staining properties depend on three factors:

1. Strength of dye.
2. Rate of ionization of tissue proteins and dyes.
3. pH value of dye solution and tissue proteins.

In addition, staining can be affected by other conditions:

1. Alcoholic or aqueous solution of dye.
2. Low or high temperature during reaction.
3. Simple or multiple combinations of dyes.
4. Strong or weak concentration of dye in solution.
5. Permeability of tissues and dyes.

STANDARDIZATION OF STAINS

In the early days of tissue staining, it was difficult to secure reliable dyes. The textile dye industry was the sole source of dyes, and the products received were often unsatisfactory. There was no adequate standardization of color, and no standardization of chemical content was undertaken. There were variations in color and quality and many impurities in the dyes. Grübler in Germany, the first to try to standardize dyes, built a highly specialized business in this field. He did not actually manufacture dyes but brought up batches from other firms and tested them for technical use (see p. 5).

The lack of German dyes after the two world wars led to a new form of standardization in the United States. A body called the Commission on Standardization of Biological Stains was organized, and it later became the Biological Stain Commission. The object of this commission is to work with the manufacturers, showing them what the biologists require, testing their products, and permitting approved batches to be put on the market with the stamp of the commission on them. Specifications of the most important dyes have been drawn up by the commission. The specifications are partly chemical and partly spectrophotometric; they contain detailed instrumentations for testing the dyes and the results to be expected from the tests.

Batches of dyes approved by the commission bear a label furnished by the commission and known as a certification label. On it is a CI (Color Index) number, indicating the certification number of that particular batch. This number means that (1) a sample of that batch was submitted to the commission for testing, and a portion of it is on file; (2) the sample proved true to type in spectrophotometric tests; (3) its dye content met specifications and is correctly indicated on the label; (4) it was tested by experts in the procedure named on the label and found satisfactory; and (5) no other batch can be sold under the same certification number. Any description of the use of the stain should be followed by its CI number or the letters CC, indicating that it is Commission Certified.

The actual dye content of commercial dyes may vary from 32% to 99%, and some batches may contain colored compounds with staining properties different from those of the actual dyes. Even different batches of a dye can very. Rosenthal et al. (1965) developed a simple paper chromatography method for testing dyes that requires only simple laboratory equipment.

Suggested Readings

For further detail concerning stains and their actions, read Baker (1945, 1958), Carson (1990), Clark and Allard (1983), Cole (1933, 1943), Green (1990), Gurr (1969), Holmes (1929), Lillie (1977), Lillie and Fulmer (1976), Pearse (1985), Singer (1952), and Stearn and Stearn (1929, 1930).

10
Mounting and Staining Procedures

Certain standard principles usually apply to the staining of tissue on slides, but there are many exceptions and variations. Paraffin sections first must be deparaffinized, because most stains are applied in either aqueous or alcoholic solutions and will not penetrate paraffin-infiltrated tissues. The customary solvent for paraffin is xylene. This is followed by the removal of the xylene with absolute alcohol because stains rarely can be applied successfully in a xylene medium. After the removal of the xylene, the general rule is to transfer the slides to a medium comparable to the solvent of the dye being used. That is, if the dye is a water solution, the slides are hydrated through a series of decreasing alcoholic and increasing aqueous dilutions, such as 95%, 80%, 70%, and 50% alcohol, and finally they go into water. If the dye is dissolved in a 50% alcoholic solution, then the slides are carried only to 50% alcohol before going into the staining solution. There are a number of "new" xylene substitutes that work very well, and they are recommended for use. However, these products must also be disposed of properly for they are not miscible with water.

During the hydration process, undesirable pigments or other materials (mercuric chloride crystals, formalin pigment, etc.) are removed and the slides are washed well to remove the reagent. Counterstains (background color) or other special treatments are applied in their proper sequence, to allow each dye or chemical to maintain its specific effect on the tissue elements. Improper sequence of staining, decolorizing, or other solutions can result in a poorly stained slide. Hematoxylin-eosin staining (p. 108) is a simple example: if the eosin (counterstain) is applied before the hematoxylin (nuclear stain), the eosin stain will be completely removed by the action of the hematoxylin stain.

It is important to use quality control procedures when staining to confirm the veracity of the color response, especially in clinical diagnoses. Thus sections of tissue that contain most of the basic histological elements or special cytological components (e.g., mucus) are usually processed for staining along with the specimen samples.

MECHANICAL AIDS

If slides are to be transported in quantity, rather than individually, several types of holders on the market are useful. Most of these hold several slides.

Some are baskets; others are clips that fit into special staining dishes. The automatic tissue stainers are equipped with slide carriers to fit the instrument.

PROCESSING SLIDES FOR MOUNTING

A final processing of slides is necessary to make permanent preparations for examination and storage without deterioration. All alcohol and water must be extracted (with certain exceptions), and a medium applied that maintains the tissues in a clear and transparent condition, does not alter the color intensity of the stains, and holds a cover glass in place. The water is removed through increasing concentration of alcohol until absolute alcohol is reached. The final reagent is xylene (or a similar solvent), which removes the alcohol and makes the sections lose their opacity and become clear. Finally, a mounting medium is applied and the cover glass lowered into place, completely covering the sections. (The solvent of the mounting medium is usually either toluene or xylene.)

COVER GLASS MOUNTING

Although there are automatic coverslippers available, most laboratories still prefer the standard coverslipping methods. The tidiest method for mounting a cover glass is shown in Figure 10-1: Apply a thin streak of medium on the cover glass; turn cover glass over and rest it on edge on the slide, beside the sections; ease cover glass into place slowly to allow air to escape from under cover glass; press firmly from center outward to distribute the medium evenly. An alternate method is to apply mounting medium along one edge of sections on slide; rest one edge of cover glass adjacent to mounting medium; lower gradually to ease out air without bubble formation; press gently in place.

If, during microscopic examination of stained and mounted slides, dull black spots replace nuclear detail (Fig. 10-2b), the clearing solution was partly evaporated out of the sections before the cover glass was in place. Return such slides to xylene, dissolve all mounting medium to allow the air to leave the sections, and remount. If the slides appear dull, almost milky, instead of crystal clear, water is present. Remove all mounting medium in xylene and return slides to absolute alcohol—preferably a fresh solution—clear, and remount.

In addition, there are commercially prepared liquid coverslip media that dry to a hard, clear, and scratch-free consistency, obviating the need for additional coverslipping.

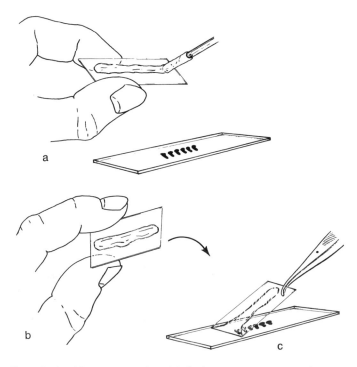

Figure 10-1. Mount a cover glass: (a) placing mountant on cover glass; (b) turning cover over; and (c) lowering it onto slide.

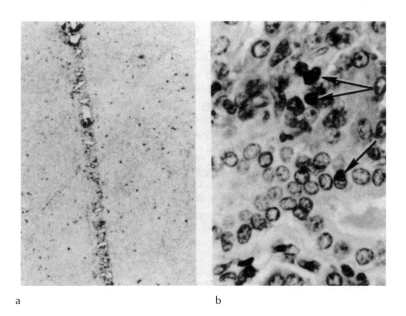

Figure 10-2. Common imperfections in sections: (a) knife scratch in paraffin section as seen under microscope before deparaffinization; (b) black areas caused by drying of sections before cover glass was applied.

MOUNTING MEDIA (MOUNTANTS)
Natural Resins

Formerly, natural resins were used as mounting media: Canada balsam, composed of terpenes, carboxylic acid, and esters; gum damar, composed of unsaturated resin acids and a little ester; and gum sandarac, an unsaturated acid resin. These dried slowly and were variable in composition and unpredictable in behavior. Some developed acidity, faded stains, and would turn yellow and crack after a few years.

Synthetic Resins

The synthetic resins now available are superior to natural resins in most respects; their composition can be controlled; they are stable and inert; they dissolve readily in xylene or toluene, do not require long drying, and adhere tightly to glass; and they have the correct refractive indices, are pale in color, and do not yellow with age. The resins should have refractive indices of 1.53 to 1.54 or better (Lillie et al. 1953).

The most widely used synthetics are the β-pinene polymers (terpene resins). There is no reason to recommend one product over the others; all are equally efficient. They are soluble in xylene, toluene, aromatic hydrocarbon solvents, and in chlorinated hydrocarbons such as chloroform, but not in dioxane.

Most synthetic mountants are allowed to air dry, but leaving synthetic resin mounts on a slide warmer overnight will make them relatively firm. Slides can be cleaned, marked with ink, and stored without dislodging cover glasses.

Aqueous Mounting Media

Aqueous mounting media are indispensable for the preservation of tissue elements that are soluble in alcohol or hydrocarbons or are demonstrated by the use of dyes soluble in those fluids. Nearly all media that have been proposed use (1) gelatin and gum arabic as solidifying agents with water; (2) sugars and salts for increasing the refractive index; and (3) glycerols and glycine as plasticizing agents. Gum arabic slowly hardens by drying, gelatin sets by cooling, and glycerin keeps the tissues from cracking or overdying. The addition of phenol or thymol prevents mold growth.

Kaiser Glycerin Jelly

> water .... 52.0 ml
> gelatin .... 8.0 g
> glycerin .... 50.0 ml
> Add as a preservative either carbolic acid (phenol), 0.1 g or a thymol crystal.

Allow gelatin to soak for 1 to 2 hours in water, and add glycerin and preservative. Warm for 10 to 15 minutes (not above 75°C) and stir until mixture is homogeneous. This keeps well in covered jar in refrigerator. If heated above 75°C, the gelatin may be transformed into metagelatin, which will not harden at room temperature.

Von Apathy Gum Syrup

gum arabic (acacia)	50.0 g
sucrose (cane sugar)	50.0 g
distilled water	50.0 ml
formaldehyde (37–40%)	1.0 ml

Dissolve lumps of gum arabic in warm water; add sucrose. When dissolved, filter and allow to cool. Add formaldehyde. Do not use powdered form of gum; it makes a milky solution, whereas the lump form produces a clear medium. The cover glass does not have to be sealed as it does with Kaiser glycerin jelly above.

The following solutions are recommended for small whole mounts, insects, worms, tiny invertebrates, and so forth.

Lactophenol Mounting Medium

melted phenol (carbolic acid)	3 parts
lactic acid	1 part
glycerin	2 parts
distilled water	1 part

Yetwin Mounting Medium for Nematodes and Ova (1944)

10% bacto-gelatin, granular, aqueous	150.0 ml
glycerin	50.0 ml
1% chromium potassium sulfate aqueous (chrome alum)	100.0 ml
phenol (carbolic acid), melted	1.0 ml

Dissolve gelatin in boiling water; add glycerin. After mixing, add chrome alum solution and phenol. The medium liquefies in 15 minutes at 65°C. Organisms may be transferred from glycerin or formalin directly into medium. The gelatin hardens to form a permanent mount.

Fluorescent Mounting Medium (Rodriquez and Deinhardt 1960)

Elvanol 51-05 (Du Pont)	20.0 g
0.14 sodium chloride buffered with 0.01 M KH_2PH_4—$Na_2HPO_4 \cdot 12H_2O$ (pH 7.2)	80.0 ml
glycerin	40.0 ml

Agitate frequently during a period of 16 hours. Remove undissolved particles of Elvanol by centrifuging at 12,000 rpm for 15 minutes. The pH should be between 6 and 7.

AQUEOUS MOUNTING TECHNIQUES

Aqueous mounts—sections or whole mounts—are removed from the water, placed on a slide, and covered with a drop of mounting medium. The cover glass, held in a horizontal position, is placed directly on the medium. Do not drop it from a slanted position as this may push the object to one side. In many cases it is not necessary to press the cover glass into place; its own weight is sufficient. Since the sections or objects are not attached to the slides, too much pressure may result in disarranged and broken material. Commercial water-soluble mounting agents are available (e.g., Aquamount, Shandon Lipshaw Scientific).

Using Two Cover Glasses

The material is mounted out of glycerin into glycerin jelly between two cover glasses, one of which is smaller than the other. Clean away excess jelly and air dry overnight. Invert the pair of cover glasses on a drop of resinous medium on a slide. This will permanently seal the mount, and it can be treated like resin mount.

Ringing Cover Glasses

If the mountant contains a volatile substance like water and the slides are to be relatively permanent, the cover glass must be sealed with a ringing material. Ringing cements or colorless nail polish that can be dissolved in xylene can be used.

Conger (1960) recommends dental wax. It is solid and slightly tacky at room temperature, but flows easily when melted. It adheres well and does not leak, but will crack off cleanly if frozen with dry ice or liquid nitrogen. Good for acetocarmine and aceto-orcein preparations.

For temporary mounts, melted paraffin can be used to ring a cover glass. Because it is susceptible to temperature damage and will crack away from the cover, it is not recommended for slides subject to hard usage or for a long time.

Commercial "wet" water-soluble mounting agents are available.

11
Hematoxylin Staining

The strength of hematoxylin as a staining agent depends on its proper oxidation and the use of a mordant, as discussed in Chapter 9. A number of methods have been developed for oxidizing hematoxylin.

Baker (1958) recommends oxidation with sodium iodate for preparing hematoxylin solutions. Start with a wholly unoxidized hematoxylin dye powder, not hematein powder. Solutions started with the hematein powder tend to lose their strength by occulation (sedimentation) of the products of oxidation. When in solution, the hematoxylin dye should be only partly oxidized by the sodium iodate. Use less chemical than would be required for complete oxidation; about one-fourth to one-half of the full amount of oxidizer is adequate. The solution will then continue to combine with atmospheric oxygen and thereby maintain its strength. Such solutions, allowed to ripen slowly for six weeks or more, will produce brilliant staining for many months. Mayer's hematoxylin is an example of this type.

The choice of oxidants can be one of convenience; all give equivalent results. Iodine is used as a 1% alcoholic solution, and the following as 0.1% aqueous: sodium iodate ($NaIO_3$), a sodium metaperiodate ($NaIO_4$), and potassium permanganate ($KMnO_4$).

The rate of oxidation is also affected by the type of solvent used. A neutral aqueous solution forms hematein in a few hours; an acid solution does this more slowly, and an alkaline solution, more rapidly. Alcoholic solutions are slow, and the addition of glycerin retards them even more. The function of the additives is not always clear. Some, thymol and salicylic acid, may be preservatives; alcohol can prevent molds; and glycerin can stabilize against overoxidation. The latter action probably forms the precipitate coating on the inside walls of the storage bottle. Overoxidation also can be prevented by a minimum of air space in the bottle. Color changes that take place in a stock solution indicate its efficiency. The changes, with no mordant present, are from water white through lilac, bright purple, deep purple red, orange red, orange brown, to brown. At the purple stage the solution is most vigorous; at the red stages, less so; at the brown stage it is no longer useful. The lifetime of alcoholic solutions is five times greater than that of aqueous ones (Cole 1943).

SINGLE SOLUTIONS

Delafield Hematoxylin (Carleton and Leach 1947)

Dissolve 4 g hematoxylin in 25 ml absolute alcohol. Mix gradually into 400 ml ammonia alum ($NH_4Al(SO_4)_2 \cdot 12H_2O$), saturated aqueous (approximately 1 part alum to 11 parts distilled water). Leave exposed to light in a flask with a cotton plug for 3 to 5 days. Filter. To the filtrate add 100 ml glycerin and 100 ml methyl alcohol. Allow to ripen for at least 6 weeks. The ripened solution will keep for years in a air-tight bottle.

Hance and Green (1961) bubbled pure oxygen for 4 hours into 3 liters of freshly prepared Delafield solution. This reduced the oxidation time from weeks to hours. The solution thus oxidized stained more rapidly than solutions oxidized by atmospheric oxygen and could be diluted to half strength for use.

Ehrlich Hematoxylin (Gurr 1956)

hematoxylin	2.0 g
ammonia alum ($NH_4Al(SO_4)_2 \cdot 12H_2O$)	3.0 g
alcohol, methyl or ethyl	100.0 ml
glycerin	100.0 ml
distilled water	100.0 ml

Ripens in 6 to 8 weeks or may be ripened for immediate use with 0.24 g sodium iodate.

Add 100 ml glacial acetic acid. Keeps for years. pH should be maintained at 5.0. Good for bone staining.

Harris Hematoxylin (Mallory 1944)

Dissolve 1.0 g hematoxylin in 10 ml ethyl alcohol. Dissolve 20 g potassium alum ($KAl(SO_4)_2 \cdot 12H_2O$) or ammonia alum ($NA_4Al(SO_4)_2 \cdot 12H_2O$) in 200 ml water and boil. Add hematoxylin and boil 1/2 minute. Add 0.5 g mercuric oxide. Cool rapidly. Add a few drops of acetic acid to keep away metallic luster and brighten nuclear structure. Does not keep longer than a month or two. (*Warning:* Mix with care; the reaction can be explosive.) Harris hematoxylin is available ready-to-use commercially.

Mayer Hematoxylin (Mallory 1944)

Add 1 g hematoxylin to 1 liter distilled water. Heat gently and add 0.2 g sodium iodate and 50 g potassium alum ($KAl(SO_4)_2 \cdot 12H_2O$). Heat until dissolved; add 1 g citric acid and 50 g chloral hydrate. Allow to ripen, probably for 6 to 8 weeks, although it can be used within 1 to 2 weeks.

Lillie's modification: hematoxylin, 5 g in 700 ml distilled water. Add

300 ml glycerin, 50 g ammonia alum, 0.2 to 0.4 g sodium iodate, and 20 ml glacial acetic acid.

Gill's Hematoxylin (1974)

hematoxylin	4.0 g
sodium iodate	0.4 g
aluminum sulfate	35.2 g
distilled water	710.0 ml
ethylene glycol	250.0 ml
glacial acetic acid	40.0 ml

The ready-for-use solution can be purchased from many sources, and a variety of strengths are available. Staining time: 2 to 4 minutes. This is one of the more popular hematoxylins in clinical histology and cytology laboratories.

DOUBLE SOLUTIONS

Weigert Iron Hematoxylin (Lillie and Henderson Modification 1968)

Solution A

ferric chloride ($FeCl_3 \cdot 6H_2O$)	2.5 g
ferrous sulfate ($FeSO_4 \cdot 7H_2O$)	4.5 g
hydrochloric acid	2.0 ml
distilled water	298.0 ml

Solution B

hematoxylin	1.0 g
95% alcohol	100.0 ml

Mix A and B, equal parts. The solution turns black at once and is usable for 2 to 3 weeks or until it turns brown. Stain progressively 3 to 5 minutes.

Groat Variation of Weigert Hematoxylin (1949)

In this variation a single solution is used.

distilled water	50.0 ml
sulfuric acid (spgr 1.84, 94% H_2SO_4)	0.8 ml
ferric alum ($FeNH_4(SO_4)_2 \cdot 12H_2O$)	1.0 g
95% alcohol	50.0 ml
hematoxylin	0.5 g

Mix in order given, at room temperature. Filter. Stain progressively 9 to 10 minutes.

Krutsay Iron Hematoxylin (1962b)

hematoxylin	1.0 g
aluminum potassium sulfate ($KAl(SO_4)_2 \cdot 12H_2O$)	50.0 g
potassium iodate (KIO_3)	0.2 g
hydrochloric acid	5.0 ml
distilled water	1000.0 ml

This can be used like Mayer's hematoxylin or converted to an iron hematoxylin like Weigert's by adding 8 volumes of the solution below to 100 of above hematoxylin just before use.

ferric alum ($FeNH_4(SO_4)_2 \cdot 12H_2O$)	2.0 g
hydrochloric acid	0.5 ml
distilled water	100.0 ml

A fresh solution is brownish violet; if it turns orange, it is no longer usable. Staining time: 5 minutes.

Slidders et al. Iron Hematoxylin (1969)

Solution A

hematoxylin	1.0 g
95% alcohol	100.0 ml
aluminum chloridate hydrate ($AlCl_3 \cdot 6H_2O$)	10.0 g

Solution B

ferrous sulfate hydrate ($FeSo_4 \cdot H_2O$)	10.0 g
distilled water	100.0 ml

Combine A and B and add:

hydrochloric acid, concentrated	2.0 ml
sodium iodate, 9% aqueous	2.0 ml

Allow to stand 48 hours, ready for use. Stain 5 to 10 minutes, wash and differentiate with 0.5% HCl in 70% alcohol, rinse and neutralize with 2% aqueous potassium acetate or other mildly alkaline solution.

Cole Hematoxylin (Clark et al. 1973 modification)

Stock Solutions
iodine, 1% in 95% alcohol
aluminum potassium sulfate ($KAl(SO_4)_2 \cdot 12H_2O$), 1.2% aqueous
hematoxylin, 10% in absolute alcohol

Working Solution

1% iodine	2.0 ml
10% hematoxylin	1.0 ml
1.2% alum	100.0 ml

The stock solutions are stable for years. Place the freshly mixed working solution in a paraffin oven (55–60°C) overnight. Cool, ready for use. Staining time: 5 minutes.

Double Solutions That Are Not Mixed before Use

Heidenhain Iron Hematoxylin

Solution A

ferric alum (FeNH$_4$(SO$_4$)$_2$ · 12H$_2$O)	4.0 ml
distilled water	100.0 ml

Keep in refrigerator to prevent precipitation on sides of bottle.

Solution B

hematoxylin	10.0 g
95% alcohol	100.0 ml

Let stand until a deep wine-red color; 4 to 5 months is not too long. Add 4 to 5 ml of this stock solution to 100 ml distilled water; this gives a practically aqueous solution and is already ripe. Saturated aqueous lithium carbonate—3 drops—added to the working solution improves color.

A and B are not usually mixed. A is a mordant solution and precedes B. If they are mixed for use, the solution deteriorates rapidly.

Testing Hematoxylin Solutions

Add several drops of the solution to tap (not distilled) water. If it turns bluish purple immediately, it is still satisfactory; if it changes slowly or stays reddish or brownish, it has weakened or broken down and should be discarded.

SUBSTITUTES FOR HEMATOXYLIN SOLUTIONS

Gallocyanin (Berube et al. 1966)

gallocyanin	0.15 g
chrome alum (CrK(SO$_4$)$_2$ · 12H$_2$O), 15% aqueous	100.0 ml

Boil 10 to 20 minutes. Cool. Filter. Restore volume to 100 ml by washing the precipitate. Keeps about a week, then deteriorates slowly. An iron lake may be prepared by substituting 5% aqueous iron alum for the chrome alum. (Proescher and Arkush 1928; also see Terner and Clark 1960a,b).

Hematein (Hemalum) (Kornhauser 1930)

> hematein .... 0.5 g
> 95% alcohol .... 10.0 ml
> aluminum potassium sulfate ($KAl(SO_4)_2 \cdot 12H_2O$), 5% aqueous .. 500.0 ml

Grind hematin with alcohol in glass mortar and add to the aqueous potassium alum. Immediately ready for use.

Iron Alizarine Blue S (Meloan and Puchtler 1974)

> alizarine blue S, 2% aqueous .... 100.0 ml
> ferric chloride, 40% aqueous .... 5.0 ml

Mix well and let stand 15 minutes. Filter. Stain for 15 minutes. Wash 10 to 15 minutes and proceed to counterstain. Do not boil the stain; it is good for 2 to 3 weeks.

Celestin Blue B

> celestin blue B .... 2.5 g
> ferric ammonium sulfate, 5% aqueous .... 500.0 ml

Bring to a boil for 3 minutes. Cool and filter. Add 70 ml glycerin. Good for 6 to 12 months.

This is an excellent stain and can be followed by acidic stains of the Mallory or Masson types. It can precede hematoxylin; the iron alum in the solution acts as a mordant for the hematoxylin. Use a Mayer-type hematoxylin for 3 to 6 minutes. Many prefer this combination over a Weigert iron hematoxylin. Lillie et al. (1973) used iron gallein for hematoxylin.

COUNTERSTAINS (PLASMA STAINS) FOR HEMATOXYLIN, GALLOCYANIN, AND HEMATEIN

Eosin

> eosin Y .... 1.0 g
> 70% alcohol .... 1000.0 ml
> glacial acetic acid .... 5.0 ml

Dilute with equal volume of 70% alcohol for use and add 2 to 3 drops of acetic acid.

Eosin (Putt 1948)

eosin Y	1.0 g
potassium dichromate	0.5 g
saturated aqueous picric acid	10.0 ml
absolute alcohol	10.0 ml
distilled water	80.0 ml
acetic acid (optional)	1 drop

Eosin-Orange G

1% eosin Y in 95% alcohol	10.0 ml
orange G, saturated solution in 95% alcohol (approximately 0.5 g per 100 ml)	5.0 ml
95% alcohol	45.0 ml

Orange G

orange G	1.0 g
phosphotungstic acid	5.0 g
90% alcohol	100.0 ml

Other Acceptable Counterstains

1. Acid fuchsin, 5% aqueous (slight acidity improves stain). If overstained, rinse with tap water.
2. Orange G, saturated in 95% alcohol.
3. Van Gieson or substitute (p. xxx).
4. Bordeaux red, 1% aqueous.
5. Biebrich scarlet, 1% aqueous; a good counterstain.
6. Additional "eosins."
   a. Eosin Y or eosin B, 0.1–0.5% in 95% alcohol.
   b. Erythrosin B, 0.1–0.5% in 95% alcohol.
   c. Phloxine B, 0.5% aqueous, plus a few drops of acetic acid.
7. Congo red, 0.5% aqueous.
8. Light green SF, yellowish, 0.2–0.3% in 95% ethyl alcohol.
9. Aniline blue WS.
10. Fast green FCF, similar to light green or aniline blue.
11. Wool green, 0.5% aqueous.
12. Methylene blue, 1% aqueous.

HEMATOXYLIN STAINING PROCEDURES

Delafield (or Harris) Hematoxylin, I (Progressive Method)

FIXATION
Any general fixative or one specific for nuclear detail.

SOLUTIONS
Hematoxylin, see p. 102.
Counterstain, see p. 106.

PROCEDURE
The slides are passed through a "down" series of jars, a process often termed *running down slides to water* (or hydration) because a series of alcohols of decreasing strength is used (from left to right, top row of Fig. 11-1). Never at any time during this procedure allow the slides to dry.

1. Xylene (or xylene substitute): 2 to 3 minutes or longer.

Two changes may be necessary to ensure removal of paraffin.

2. Absolute alcohol: 10 dips or longer.
3. 95% alcohol: 10 dips or longer.
4. 70% alcohol: 10 dips or longer.

If mercuric chloride was absent from fixative, skip steps 5 through 7 and proceed to step 8.

5. Lugol solution (p. 463): 3 minutes.
6. Running water: 2 minutes or longer.
7. 5% solution thiosulfate, $Na_2S_2O_3$: 2 to 3 minutes.

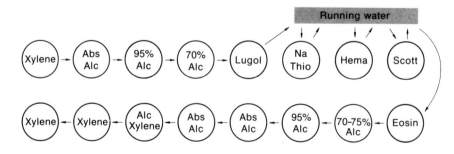

Figure 11-1. A suggested arrangement of staining jars. An alternate arrangement can include two jars of xylene in the "down series" (*left to right*) when many slides are being stained. An absolute alcohol-xylene in "up series" (*right to left*) is optional.

8. Running water: 3 to 5 minutes.
9. Delafield hematoxylin (or Gills or Harris): 2 to 5 minutes; check after 1 minute for stain intensity. Fresh solutions stain faster than old ones. If not dark enough, return slides to stain. Rinse off stain in tap water before checking under microscope. If slide becomes too dark, convert to regressive type of stain.
10. Running water: 3 to 5 minutes
11. Scott solution (p. 464) or saturated lithium carbonate: 2 minutes.
12. Running water: 3 to 5 minutes.
13. Counterstain: 1 or more minutes, depending on stain used.

Transfer slides through "up" series, termed running up slides (or dehydration), using a series of alcohols of increasing strength. Steps 14 and 15 can control intensity of the counterstain. Watch timing carefully in these solutions (from right to left, bottom row, Fig. 11-1).

14. 70% alcohol: 1 or more dips.
15. 95% alcohol: few dips.
16. Absolute alcohol: 2 minutes.
17. Absolute alcohol: 2 minutes or longer.
18. Absolute alcohol-xylene (1:1), optional: 2 minutes or longer.
19. Xylene: 2 minutes or longer.
20. Xylene: 2 minutes or longer.
21. Mounting medium: keep sections moist with xylene during this process; they must not dry. Add cover glass.

RESULTS
nuclei—deep blue
cytoplasmic structure—pink, rose, etc., depending on counterstain

(*Note:* In some cells rich in cytoplasmic RNA [e.g., those involved in synthesizing large amounts of protein as, for example, salivary gland cells], the cytoplasm will show varying degrees of blue staining.)

COMMENTS
1. Slides may be left in stronger alcohols and xylene longer than scheduled (indicated by "or longer"), but do not leave them much longer in any solutions weaker than 80% or 95% alcohol. Weaker alcohols and water can loosen the sections from the albumen.
2. Isopropyl alcohol can be substituted for ethyl alcohol in this schedule, but cannot be used as a stain solvent.
3. Two to 5 minutes for Delafield's (step 9) is an approximate time only. This may have to be varied according to the tissue used. Tissue sections are individualistic, often due to the type of fixative employed on them,

and a certain amount of trial and error may be required to develop the exact timing schedule.
4. Poor staining can result from improper fixation, leaving gross tissues too long in alcohol iodine, faulty processing during preparation for embedding, or careless handling of the slides in preparation for staining. In many instances no amount of trial and error will produce a perfect stain. The only correctable faults are those made during deparaffinization and hydration of the slides. The slides should be transferred in a reversed process through the solutions, back to xylene. Change to fresh solutions and try again.
5. Sometimes too many slides have been taken through the fluids and the fluids become contaminated. If more than 20 slides are being stained, use two changes of xylene in step 1 and follow step 20 with a third change. Prevent further contamination by draining slides properly. On removing them from a solution, touch the edges against the inner surface of the staining jar and then against paper towels. But do not be too thorough and let the slides dry; merely drain off excess liquid.
6. Hematoxylin staining must be left in a "blued" condition. The original pinkish color of the hematoxylin, like Delafield's, must be converted to a blue color by alkalinity. In the progressive method this is accomplished in running water to remove excess dye and start the bluing action. Then use Scott solution or lithium carbonate (step 11), a couple of drops of ammonia in a final wash, or a weak solution of sodium bicarbonate to ensure the alkaline condition. In the regressive method (below) the same applies, and, as a precaution against neutralization caused by carrying one solution over into the other (step 5), after several exchanges of slides add a drop or two of acid and base to them. If there is any pinkness left in the nuclei, they are not adequately blued.

Delafield (or Harris) **Hematoxylin, II** (Regressive Method)

PROCEDURE
1. Deparaffinize and run slides down (hydrate) to water, removing $HgCl_2$ if present.
2. Stain in Delafield hematoxylin or Harris (without acetic acid), until slides are well overstained: 15 to 20 minutes.
3. Wash in running water: 3 to 5 minutes or longer.
4. Transfer to 70% alcohol.
5. Destain (in acid alcohol) for a few seconds (5 to 10 quick dips). Wash in running water 1 minute or longer. Dip in weak ammonia water (1 to 2 drops ammonium hydroxide) or saturated lithium carbonate until sections turn blue. Examine under microscope. If nuclei are too dark, repeat above procedure. When nuclei stand out as a sharp blue against colorless background, proceed with next step.

6. Counterstain.
7. Dehydrate as for progressive staining.
8. Clear and mount.

RESULTS

nuclei—deep blue, sharper than in the progressive method
other elements—colors of counterstain

Mayer Hematoxylin

FIXATION

Any general fixative or one specific for nuclear detail.

SOLUTIONS

Hematoxylin, see p. 102.
Counterstains, see p. 106.

PROCEDURE
1. Deparaffinize and run slides down (hydrate) to water, removing $HgCl_2$ if present.
2. Stain in Mayer hematoxylin: 11 minutes.
3. Wash in running water: 3 minutes.
4. Blue in Scott solution (or lithium carbonate): 3 minutes.
5. Wash in running water: 3 to 5 minutes.
6. Counterstain.
7. Dehydrate quickly through 70% and 95% alcohols.
8. Complete dehydration, absolute alcohol.
9. Clear and mount.

RESULTS

nuclei—deep blue
other elements—colors of counterstain

COMMENTS

Mayer hematoxylin was preferred by Humason. It keeps well, and a single solution will stain several hundred slides over a period of two weeks without losing its intensity. To mordant with celestin blue B see p. 106.

Heidenhain Iron Hematoxylin

FIXATION

Any general fixative, preferably one containing $HgCl_2$. Excellent for kidney and salivary glands. Good stain for nuclei and to demonstrate basal rodding (i.e., large concentration of mitochondria in basal region of cell).

Solutions

Hematoxylin and iron alum, see p. 105.
Counterstains, see p. 106.

Procedure

1. Deparaffinize and run slides down (hydrate) to water, removing $HgCl_2$.
2. Mordant in 4% iron alum: 15 to 30 minutes (or overnight, see comments below).
3. Wash in running water: 5 minutes.
4. Stain in hematoxylin: 15 to 30 minutes (or overnight).
5. Wash in running water: 5 minutes.
6. Destain in 2% aqueous iron alum or saturated aqueous picric acid (see comments below) until nuclei stand out sharp and clear against colorless background. Before examining the slides, dip them in water containing a couple of drops of ammonia. *Warning:* too much ammonia will loosen sections.
7. Wash in running water: 15 to 30 minutes or longer.
8. Counterstain, if desired.
9. Dehydrate, clear, and mount.

Results

nuclei—blue black to black
muscle striations—blue gray
mitochondria—blue gray
other elements—colors of counterstain, or very light gray to colorless if no counterstain used.

Comments

A full stain response can be achieved by carrying out steps 2 and 4 on a slide warmer (52°C) for 20 minutes. Slides are placed on a 4- × 4-inch glass plate and covered with the appropriate solution (personal observation).

The shorter staining schedule produces blue-black nuclei and the overnight staining a truer black. The overnight staining is recommended for mitochondria. Picric acid usually perfects a more sharply differentiated nucleus than destaining with iron alum. Prolonged destaining with iron alum can leave a tan or yellow tinge in the cytoplasm. Yellow color left in the tissue after picric acid is due to insufficient washing after the destaining. Striations of muscle and some protozoan structures are better differentiated by iron alum. Thorough washing is essential after either method or the sections continue to destain slowly and fade (picric acid destaining, Tuan 1930).

Destaining agents act as both an acid and an oxidizer, reacting in two ways: (1) As an acid they extract stain faster from the cytoplasm than from the nuclei; (2) as an oxidizer they bleach the stain uniformly. If an acid alone is used, the staining of the nuclei is favored, and if an oxidizer alone is used, the staining of the cytoplasm is enhanced.

Iron alum mordanting can precede a hematoxylin such as Mayer's and produce a blacker (iron-type) nuclear staining. Destain, if necessary, in the same manner as for Heidenhain's.

Hutner (1934) recommendation for nuclei
1. Mordant in 4% iron alum: 1 hour.
2. Stain hematoxylin: 1 hour.
3. Destain in saturated aqueous picric acid.
4. Wash and blue by adding 1 to 2 drops ammonia to 70% alcohol.

Hutner (1934) recommendation for cytoplasm
1. Mordant in iron alum: 30 minutes.
2. Stain in hematoxylin: 30 minutes.
3. Destain in freshly made 95% alcohol, 2 parts to Merk's superoxol (30% hydrogen peroxide), 1 part.
4. Rinse in 1 change 70% alcohol: 10 minutes.

For brevity, intermediate steps have been omitted from the above procedures. Oxidizer and acid can be combined to approximate iron alum destaining effects.

> 0.25% HCl in 95% alcohol, 2 parts
> superoxol, 1 part.

Wash in 70% alcohol and blue.

Cell membranes and various cytoplasmic structures stain well with Heidenhain hematoxylin, probably because of the use of iron alum. The metal is first bound to the tissues and then combines with the hematoxylin. It is likely that proteins are most frequently the bearers of iron-bending residues and are responsible for the precise staining action of Heidenhain hematoxylin (Puchtler 1958).

Acidified Hematein (Puchtler and Sweat 1964a)

Dissolve 0.1 g hematoxylin in 2 ml 95% ethanol. Add 18 ml distilled water and 2 ml of a 1% aqueous solution of sodium iodate. Ripen for 15 minutes or longer. Add 60 ml of 10% acetic acid. This solution is not stable and should be prepared shortly before use.

Acidified hematein may be used in step 4 of Heidenhain method. Mordant (step 2) for 1 hour, wash thoroughly, and stain with acidified hematein for 1 hour. Wash thoroughly. Counterstain if desired. Dehydrate, clear, and mount. No differentiation is required.

For double iron hematoxylin solutions and their use, see p. 103.

Phosphotungstic Acid-Hematoxylin (Puchtler et al. 1963)

FIXATION

Any general fixative, preferably one containing $HgCl_2$; if tissue is formalin-fixed, mordant sections before staining in saturated aqueous mercuric chloride, rinse, and treat with Lugol's.

SOLUTIONS

Hematoxylin solution

hematoxylin	1.0 g
phosphotungstic acid	20.0 g
distilled water	1000.0 ml

Dissolve hematoxylin and phosphotungstic acid in separate portions of water. Heat hematoxylin solution until no more color change is noted. Cool to room temperature and combine with phosphotungstic acid solution. Add 0.177 g potassium permanganate. Let stand 1 month.

Langeron iodine solution

iodine	1.0 g
potassium iodide	2.0 g
distilled water	200.0 ml

PROCEDURE
1. Deparaffinize and run slides down (hydrate) to water.
2. Treat with Langeron iodine solution: 10 minutes.
3. Rinse in distilled water and transfer to 5% sodium thiosulfate: 2 minutes.
4. Wash in running water: 5 minutes.
5. Transfer to hematoxylin solution (room temperature) and place in paraffin oven (approximately 60°C): 1 hour.
6. Rinse briefly in two changes of 95% alcohol.
7. Dehydrate, clear, and mount.

RESULTS

fibrin, nuclei, muscle fibers, keratin, terminal bars—dark blue
collagen, reticulin, basement membranes, amyloid, cartilage, thrombocytes—shades of red

COMMENTS

A freshly prepared solution must stand 1 month to be sufficiently ripened; otherwise the staining will be of low contrast. The staining properties improve with age, and the solution keeps for many months, even up to a year. Do not preheat the dye solution, and store it at room temperature. If kept in the oven, it will deteriorate within 1 to 2 days. This is an excellent stain for muscle and can be used with PAS and Luxol fast blue.

HEMATOXYLIN SUBSTITUTE PROCEDURES

Gallocyanin (Berube et al. 1966; Einarson 1951)

Gallocyanin is an excellent substitute for hematoxylin and can be used in most procedures designed for hematoxylin. The solution is made in a few minutes and ready for immediate use; also an iron lake can be prepared. No differentiation is required, and it is better than hematoxylin for tissues of the central nervous system. It can replace thionine for ganglia and glia cells, but not for myelin. It is good for Negri bodies. The preferred fixatives are Zenker (acetic or formol), or formalin—fixatives that preserve chromophilic substance.

SOLUTION
Gallocyanin, see p. 105.

PROCEDURE
1. Deparaffinize and hydrate to water.
2. Leave in stain overnight, or 3 hours at 56°C.
3. Wash thoroughly and proceed to counterstain.
4. Dehydrate, clear, and mount.

RESULT
nuclei—blue

Hematein (Hemalum) (Kornhauser 1930)

FIXATION
Any general fixative, preferably one containing $HgCl_2$.

SOLUTIONS
Hemalum, see p. 106.
Counterstains, see p. 106.

PROCEDURE
1. Deparaffinize and hydrate slides down to water, remove $HgCl_2$.
2. Stain hemalum, progressively: about 5 minutes.
3. Wash in running water: 5 to 10 minutes.
4. Counterstain.
5. Dehydrate, clear, and mount.

RESULTS
nuclei—blue
other elements—color of counterstain

Eriochrome Cyanine RC (Chapman 1977)

FIXATION
A good general fixative.

SOLUTION
eriochrome cyanin R	0.2 g
ferric chloride ($FeCl_3$), anhydrous	0.9 g
HCl	5.0 ml
distilled water	100.0 ml

PROCEDURE
1. Deparaffinize and hydrate sections to water; remove $HgCl_2$ if present.
2. Rinse in 0.5 N HCl: few seconds.
3. Stain: 1 to 2 minutes.
4. Rinse in 0.5 N HCl: 2 seconds
5. Wash in running water: 30 seconds.
6. Blue in 0.1% sodium bicarbonate: 1 minute.
7. Wash in running water: 5 minutes.
8. Counterstain (if desired), but stain lightly.
9. Dehydrate, clear, and mount.

RESULTS
nuclei—blue
matrix of hyaline cartilage, some mucus, and granules—pale blue
other structures—shades of pink and red

RED NUCLEAR STAINING

Darrow Red (Powers and Clark 1963; Powers et al. 1960)

FIXATION
Any general fixative.

Solution
Darrow red	50.0 mg
0.2 M glacial acetic acid (pH 2.7)	200.0 ml

Boil briefly. Cool and filter. Stable for 1 month.

Procedure
1. Use frozen sections in water, or deparaffinize paraffin sections and hydrate to water.
2. Stain in Darrow red: 20 to 30 minutes.
3. Rinse in distilled water.
4. Differentiate and dehydrate 50%, 70%, and 95% alcohol, not so slowly as to remove too much dye from nuclei but slowly enough to decolorize background.
5. Completely dehydrate in *n*-butyl alcohol, clear, and mount.

Results
nuclei—red
other elements—depending on other stains applied

Comments
This is a good nuclear stain for contrast with blue cytoplasmic stains—after Luxol fast blue for example. It is a good stain for Nissl substance at pH 2.7. For glial nuclei, use acetic acid—acetate buffer at pH 3.5 to prepare the staining solution. Stain 10 to 15 min. For celloidin sections, use 25 mg of dye in 200 ml of acetic acid—acetate buffer, 20 minutes. Dehydrate overnight in terpineol-xylene, 1:3.

Scarba Red (Slidders et al. 1958)

Fixation
Any general fixative.

Solutions
Scarba red

melted phenol (carbolic acid)	2.0 g
neutral red	1.0 g

Mix thoroughly, allow to cool, dissolve in:

95% alcohol	15.0 ml

Add:

2% aniline in water	85.0 ml
glacial acetic acid	1.0 to 3.0 ml

Mix well and filter. Keeps for at least 6 months.

Differentiator

70% alcohol	85.0 ml
formalin	15.0 ml
glacial acetic acid	15.0 drops

PROCEDURE
1. Deparaffinize and hydrate slides to water (remove $HgCl_2$ if present).
2. Stain in Scarba red: 5 minutes.
3. Rinse in distilled water.
4. Transfer to 70% alcohol: 2 to 3 minutes.
5. Differentiate until nuclei are clear and sharp against colorless background.
6. Dehydrate, clear, and mount.

RESULTS
chromatin and calcium—red

For restoration of basophilic properties, see p. 482.

PART III
Specific Staining Methods

12
Staining Connective Tissue and Muscle

Stains can be used to demonstrate various components of connective tissue and muscle, such as collagen, reticulum, elastin, hyaline, fibrin, and striations in muscle cells.

MALLORY STAINING

Mallory Triple Connective Tissue Stain

Innumerable modifications of this method have appeared in the literature, particularly those using phosphomolybdic and/or phosphotungstic acid. Baker (1958) suggests that phosphomolybdic acid acts as a "colorless acid dye" in tissues, chiefly on collagen. It acts as a dye excluder with acid dyes such as acid fuchsin, excluding them from collagen. Then the aniline blue stains the collagen selectively but is excluded from other tissues.

It is true that various dyes (acid and basic triphenylmethane) form lakes with phosphomolybdic acid; such compounds are used in textile dying (Puchtler and Sweat 1964b). Basic dyes and some amphoteric dyes are bound by interaction between the acid groups of phosphomolybdic acid and accessible basic groups of the dye (Smith et al. 1966). If no phosphomolybdic acid is present, aniline blue stains more strongly and with very little selectivity. The action of phosphomolybdic and phosphotungstic acid has been called a mordant action, but this is disputed. Everett and Miller (1974), in a series of experiments, showed that most histological structures stain deeply by aniline blue; after phosphotungstic acid treatment only the connective tissue stains. Their preliminary conclusions are that the poly acids act not as mordants but as selective blocking against all tissues except connective tissue. Quintarelli et al. (1971) and Scott (1976) believe phosphotungstic acid combines by electrostatic bonds with cationic groups. This reaction is directly related to the pH, and a salt-type linkage is involved between the acid and organic cations.

In regard to the use of phosphotungstic acid, it may be of interest to note that certain tissue elements react more specifically to certain metals than to others: collagen to molybdenum and connective tissue and neuroglia elements to tungsten. Lillie (1952) writes that phosphotungstic acid intensifies plasma staining and that phosphomolybdic acid intensifies fiber staining. The acids may be used separately or together, depending on the desired final effect.

Oxalic acid is often used. Mallory (1944) claims that it makes the aniline blue stain more rapidly and intensely. Baker (1958) seems to agree when he suggests that oxalic acid lowers the pH, seeming thereby to aid staining with aniline blue and also with orange G.

No polychroming or metachroming takes place in the combination of dyes; they react orthochromatically in varying intensities of their own colors. In Mallory's combination of stains, aniline blue (acidic) stains connective tissue and cartilage; orange G (acidic) stains blood cells, myelin, and muscle; acid fuchsin (acidic) stains the rest of the tissue (including the nuclei) in shades of pink and red. This procedure is not, however, an efficient method for staining nuclei, and they will fade in a few years. Modifications of Mallory's using basic fuchsin, carmine, or azocarmine result in more permanent nuclear staining, since the basic stain reacts more reliably with the nuclei than does the acid one. This method may also be preceded by hematoxylin staining for a brilliant permanent nuclear stain.

The final staining may be followed by an acetic acid wash (0.5–1%): 3 to 5 minutes or longer, which produces more transparent sections without altering the color.

Suggestion: Try Sirius supra blue in place of aniline blue (Sweat et al. 1968). Aniline blue WS was formerly a mixture of water blue 1 and methyl blue; that mixture is now obsolete.

Treat sections for 5 minutes with 1% phosphomolybdic acid; wash with 3 changes of distilled water: 10 seconds each. Stain for 5 minutes in the following solution:

```
aniline blue ................................................. 2.0 g
distilled water ............................................. 100.0 ml
glacial acetic acid ......................................... 2.0 ml
```

Wash in 2 to 3 changes of distilled water: 10 seconds each. Dehydrate, clear, and mount.

Pantin Method (1946)

FIXATION
Any general fixative, preferably one containing $HgCl_2$.

SOLUTIONS
Mallory I
```
acid fuchsin ................................................ 1.0 g
distilled water ............................................. 100.0 ml
```

Phosphomolybdic acid
```
phosphomolybdic acid ........................................ 1.0 g
distilled water ............................................. 100.0 ml
```

Mallory II
    aniline blue .... 0.5 g
    orange G .... 2.0 g
    distilled water .... 100.0 ml

CONTROL
    small intestine, tongue

PROCEDURE
1. Deparaffinize and hydrate slides to water; remove $HgCl_2$. If $HgCl_2$ is absent from fixative, mordant in saturated aqueous $HgCl_2$, plus 5% acetic acid: 10 minutes. Wash, treat with Lugol's and sodium thiosulfate; wash and rinse in distilled water.
2. Stain in Mallory I: 15 seconds.
3. Rinse in distilled water to differentiate reds: 10 or more seconds.
4. Treat with phosphomolybdic acid: 1 to 5 minutes.
5. Rinse briefly in distilled water.
6. Stain in Mallory II: 2 minutes.
7. Rinse in distilled water.
8. Differentiate aniline blue in 90% alcohol
9. Dehydrate, clear, and mount.

RESULTS
    nuclei—red
    muscle and some cytoplasmic elements—red to orange
    nervous system—lilac
    collagen—dark blue
    mucus, connective tissue, and hyaline substance—blue
    chitin—red
    yolk—yellow to orange
    myelin and red blood cells—yellow and orange
    dense cellular tissue (liver)—pink with red nuclei
    bone matrix—red

COMMENTS
From distilled water (step 7) the slides can be run through the "up" series of alcohols to control the blue color. Aqueous alcoholic solutions differentiate the amount of Mallory II left in the various parts of the tissue. After the slides have remained in the absolute alcohol for a couple of minutes, they should have changed from a muddy purple to a clear blue and red. Any acetic acid rinse after step 7 contributes to the transparency of the sections.

For better nuclear detail, Mallory I can be preceded by alum hematoxylin staining (results: nuclei—blue). Lendrum and McFarlane (1940) recommend the addition of celestin blue. Krichesky (1931) proposes the following for Mallory II.

Solution A

aniline blue	2.0 g
distilled water	100.0 ml

Solution B

orange G	1.0 g
distilled water	100.0 ml

Solution C

phosphomolybdic acid	1.0 g
distilled water	100.0 ml

Keep solutions A, B, and C in separate bottles because the mixture deteriorates rapidly when combined. When ready to use, mix equal parts of each.

This modification is useful for instructors of microtechnique classes because if large stocks of Mallory II components are kept as individual solutions and freshly mixed before use, a more brilliant stain is achieved than from a single solution that has been stocked with all the components already combined.

Heidenhain Azan Stain (azan—from first syllables of *az*ocarmine and *an*iline blue), Mallory-Heidenhain (Knoeff Modification 1938)

FIXATION

Zenker formol; other general fixatives fair, but fixation is improved by mordanting slides overnight in 3% potassium dichromate.

SOLUTIONS

Azocarmine

azocarmine G	0.2 to 1.0 g
distilled water	100.0 ml
glacial acetic acid	1.0 ml

Boil azocarmine in water 5 minutes, cool, filter, and add acetic acid.

Aniline alcohol

aniline	1.0 ml
85–90% alcohol	1000.0 ml

Acid alcohol

glacial acetic acid	1.0 ml
90–95% alcohol	100.0 ml

Phosphotungstic acid
```
    phosphotungstic acid .............................................. 5.0 g
    distilled water .................................................. 100.0 g
```

Aniline blue stain
```
    aniline blue WS .................................................. 0.5 g
    orange G ......................................................... 2.0 g
    oxalic acid ...................................................... 2.0 g
    distilled water ................................................ 100.0 ml
    5% phosphotungstic acid (above) ................................... 1.0 ml
```

Acidified water
```
    glacial acetic acid .............................................. 1.0 ml
    distilled water ................................................ 100.0 ml
```

CONTROL
small intestine, tongue

PROCEDURE
For strikingly beautiful results, particularly for pituitary, Knoeff's longer method is recommended. Extra steps and changes in timing are indicated in parentheses.

1. Deparaffinize and hydrate slides down to water; remove $HgCl_2$.
2. Treat with aniline alcohol: 45 minutes.
3. Treat with acid alcohol: 1 to 2 minutes.
4. Stain in azocarmine, 56°C: 1 hour (or 2 hours). Check temperature carefully; if too high, azocarmine differentiates poorly.
5. Rinse in distilled water.
6. Differentiate in aniline alcohol. Check under microscope for brilliant red nuclei and very light red cytoplasm.
7. Treat with acid alcohol: 1 to 2 minutes.
8. Transfer to phosphotungstic acid: 2 to 3 hours (4 hours).
9. Rinse in distilled water.
10. Stain in aniline blue solution: 1 to 2 hours (4 hours).
11. Rinse in distilled water.
12. Treat with phosphotungstic acid: 3 to 5 minutes.
13. Rinse in distilled water.
14. Rinse in acidulated water: 1 to 2 minutes or longer.
15. Rinse in 70% alcohol: dip briefly.
16. Dehydrate, clear, and mount.

RESULTS
nuclei—brilliant red
collagen and reticulum—blue
muscle—red and yellow
basophilic cytoplasm—light blue
acidophilic cytoplasm—orange red
chromophobes—colorless or light gray

COMMENTS

The azan method is particularly recommended for pituitary and pancreas but is also outstanding for connective tissue. If a tissue does not "take" the azocarmine, it may be because of formalin in the fixative. To correct this condition, mordant the sections overnight in 3% aqueous potassium dichromate.

TRICHROME STAINING

Masson Trichrome Stain (Gurr Modification 1956)

When Masson (1928) developed this method, he called it a trichrome stain, although it includes four dyes. Since then the trichrome name has been applied to many modifications of Masson's method that use many combinations of different dyes.

FIXATION
Any general fixative.

SOLUTIONS
Iron alum
  ferric ammonium sulfate .... 4.0 g
  distilled water .... 100.0 ml

Hematoxylins, see p. 101.

Acid fuchsin
  acid fuchsin .... 1.0 g
  distilled water .... 100.0 ml
  glacial acetic acid .... 1.0 ml

Ponceau de xylidine
  Ponceau de xylidine .... 1.0 g
  distilled water .... 100.0 ml
  glacial acetic acid .... 1.0 ml

Lillie (1940) suggests alternate stains for Ponceau de xylidine: Ponceau 2R; nitazine yellow; Biebrich scarlet; azofuchsin 3B, G, and 4G; Bordeaux red; chromotrope 2R; chrysoidin; eosin Y; orange G; and crocein orange G.

Fast green
>   fast green FCF .... 2.0 g
>   distilled water .... 100.0 ml
>   glacial acetic acid .... 2.0 ml

Phosphomolybdic acid
>   phosphomolybdic acid .... 1.0 g
>   distilled water .... 100.0 ml

Acidified water
>   glacial acetic acid .... 1.0 ml
>   distilled water .... 100.0 ml

CONTROL
>   small intestine, tongue

PROCEDURE
1. Deparaffinize and hydrate slides to water, remove $HgCl_2$. Mordant formalin-fixed tissue in saturated aqueous $HgCl_2$: 5 minutes. Wash in running water: 5 minutes. Treat with Lugol's and sodium thiosulfate.
2. Mordant in iron alum: ½ hour.
3. Wash in running water: 5 minutes.
4. Stain in hematoxylin (Delafield or like): ½ hour.
5. Wash in running water: 5 minutes.
6. Differentiate in saturated aqueous picric acid.
7. Wash thoroughly in running water: 10 minutes or longer.
8. Stain in acid fuchsin: 5 minutes.
9. Rinse in distilled water until excess stain is removed. Check under microscope.
10. Stain in Ponceau de xylidine: 1 to 5 minutes.
11. Rinse in tap water, control under microscope for proper intensity of acid fuchsin and Ponceau de xylidine.
12. Differentiate in phosphomolybdic acid: 5 minutes.
13. No rinse; transfer directly into fast green: 2 minutes.
14. Differentiate fast green in acidified water and dehydrating alcohols.
15. Dehydrate, clear, and mount.

RESULTS
>   nuclei—deep mauve blue to black
>   cytoplasmic elements—varying shades of red and mauve

muscle—red
collagen, mucus—green

COMMENTS

Among the many modifications of Masson stain, this is one of the best because it offers good control of both red dyes by the use of two separate solutions. In most cases the full 5 minutes in the acid fuchsin is advisable; time in the Ponceau de xylidine is not quite so critical.

Pollak Rapid Method (1944)

FIXATION

Any general fixative. Formalin-fixed tissues are improved by treating slides overnight with saturated aqueous $HgCl_2$.

SOLUTIONS

Hematoxylins, see p. 101.

Trichrome stain

acid fuchsin	0.5 g
Ponceau 2R	1.0 g
light green SF, yellowish	0.45 g
orange G	0.75 g
phosphotungstic acid	1.5 g
phosphomolybdic acid	1.5 g
glacial acetic acid	3.0 ml
50% alcohol, up to	300.0 ml

Add acetic acid to alcohol; split solution in 4 parts. Dissolve acid fuchsin and Ponceau in one part, light green in the second part, orange G and phosphotungstic acid in the third part, and phosphomolybdic acid in the fourth part. Mix four solutions and filter. Keeps well.

Acidified water

glacial acetic acid	0.2 ml
distilled water	100.0 ml

CONTROL

small intestine, tongue

PROCEDURE

1. Deparaffinize and hydrate slides to water; remove $HgCl_2$.
2. Stain in Mayer hematoxylin: 10 minutes. (Pollak uses Weigert: 10 to 20 minutes.)
3. Wash in running water: 10 minutes.

4. Stain in trichrome stain: 7 minutes.
5. Rinse briefly in distilled water.
6. Differentiate in acidified water: only a few seconds. Check under microscope if necessary.
7. Dip a few times in 70% alcohol.
8. Dehydrate, clear, and mount.

RESULTS
nuclei—dark blue
muscle, elastin—red
fibrin, calcium—purple
hyalin—pale blue
collagen, mucus—blue green

Toren (1963) combines this stain with Giemsa (p. 203) for mast cell staining.

Gomori's One-Step Trichrome (1950b)

FIXATION
Neutral buffered formalin

SOLUTIONS
Bouin's solution, p. 27
Weigert's hematoxylin working solution, p. 103

Gomori's trichrome stain
chromotrope 2R	0.6 g
light green (or aniline blue)	0.3 g
glacial acetic acid	1 ml
phosphotungstic acid	0.8 g
distilled water	100 ml

1% glacial acid

CONTROL
tongue

PROCEDURE
1. Deparaffinize and hydrate to water.
2. Place in Bouin's solution for 1 hour in oven at 56°C (or overnight at room temperature).
3. Wash in running water 3 to 5 minutes.
4. Stain in Weigert's iron hematoxylin for 10 minutes.
5. Rinse in tap water.
6. Stain in Gomori's trichrome stain for 15 to 20 minutes.

7. Rinse in 1% acetic acid.
8. Rinse in distilled water.
9. Dehydrate, clear and mount.

RESULTS
muscle fibers—red
collagen—green
nuclei—blue to black

Gomori Method (1950b) (Modification of Sweat et al. 1968)

FIXATION
Any general fixative. Formalin-fixed tissues are improved by treating slides overnight with saturated aqueous $HgCl_2$.

SOLUTIONS
Bouin fixative

picric acid, saturated aqueous	1 part
formaldehyde (37–40%)	5 parts
glacial acetic acid	1 part

Mix shortly before use. Picric acid in solution is not dangerous, but it is explosive when dry. Remove used Bouin from the oven and discard.

Trichrome solution

chromotrope 2R	0.6 g
light green	0.3 g
phosphomolybdic acid	1.0 g
dissolve in distilled water	100.0 ml
add hydrochloric acid	1.0 ml

Allow to stand 24 hours in refrigerator before use. Store in refrigerator and use cold. Do not filter. The solution is stable until the reds begin to fade.

Acidified water

glacial acetic acid	0.2 ml
distilled water	100.0 ml

CONTROL
small intestine, tongue

PROCEDURE
1. Deparaffinize and hydrate slides to water; remove $HgCl_2$.
2. Treat with Bouin solution, 56°C: 1 hour.

3. Wash in running water; 5 minutes, or until decolorized.
4. Stain in trichrome mixture: 1 minute.
5. Rinse briefly in distilled water to remove some of excess stain.
6. Rinse in acidified water: 30 seconds.
7. Dehydrate in 95% alcohol, 2 changes. Be careful not to remove too much green.
8. Complete, clear, and mount.

RESULTS AND COMMENTS

There have been a number of attempts to combine the azan and Masson methods into a quicker method. Gomori considered Pollak's to be a disappointingly dull color without red shades. Almost any combination of an acid triphenylmethane dye with a sulfonated azo dye in the presence of phosphomolybdic or phosphotungstic acid will give good results. The phosphomolybdic acid favors the green and blue shades; phosphotungstic acid, the reds. A short staining time produces more red; prolonged staining, more green and blue. Rinsing in tap water weakens the reds.

The Sweat et al. modification improves the colors of the fine connective tissue fibers, so that they closely resemble those obtained with the Mallory and Mallory-Heidenhain procedures. Pretreatment with Bouin improves the affinity for staining, offers more uniform results, and reduces variations due to different fixatives. Staining with hematoxylin has been eliminated; very little of its color remains after staining in the acid trichrome mixture. If an iron hematoxylin is used, it turns the clear red of the chromotrope to a murky color.

The staining of connective tissue fibers is affected by the pH of the solution, and is optimal at around pH 1.3. The Gomori solution has a pH of 2.5 to 2.7; by replacing the acetic acid with hydrochloric acid, the pH is lowered to approximately 1.3.

The rinse in acetic acid does not change the colors, but makes them more delicate and transparent.

COLLAGEN AND ELASTIN STAINING

Picro-Ponceau with Hematoxylin (Gurr 1956)
(Van Gieson Substitute, Nonfading)

FIXATION
   Any general fixative.

SOLUTIONS
   Hematoxylins, see p. 101.

Picro-Ponceau
>Ponceau S 1% aqueous .... 10.0 ml
>picric acid, saturated aqueous .... 86.0 ml
>acetic acid, 1% aqueous .... 4.0 ml

CONTROL
large blood vessel (e.g., aorta)

PROCEDURE
1. Deparaffinize and hydrate slides down to water, remove $HgCl_2$.
2. Overstain in hematoxylin: 5 to 15 minutes (Delafield or Harris types).
3. Wash thoroughly in running water until slides are deep blue: 10 minutes or longer.
4. Stain in picro-Ponceau: 3 to 5 minutes. (This may be too strong for some tissues; in addition to staining, the stain acts to remove the hematoxylin.) Rinse for a few seconds in distilled water and check under microscope. Continue to stain and destain or differentiate in water until nuclei are sharp.
5. Dip several times in 70% alcohol.
6. Dehydrate in 95% alcohol, 2 changes, to ensure complete removal of excess picric acid. Only that which has acted as a dye must be left in the tissue.
7. Continue dehydration, clear, and mount.

RESULTS
nuclei—brown to brownish or bluish black
collagenous and reticular fibers—red
elastic fibers, muscle fibers, erythrocytes, epithelia—yellow

COMMENTS
A Weigert type of hematoxylin is excellent for this method. Delafield (or similar) hematoxylin may be preceded by mordanting. After step 1: mordant with iron alum: 10 minutes; follow by washing in running water. Then proceed to step 2.

This method is so superior to so-called Van Gieson stain, which uses acid fuchsin instead of Ponceau S, that the Van Gieson method has been omitted. The colors are identical, but Van Gieson is unsatisfactory because it fades rapidly.

Katline (1962) recommends the following procedure if the picric solution tends to fade the hematoxylin: treat the sections before the picric staining with a solution of equal parts of 2.5% aqueous phosphotungstic and phosphomolybdic acid, 0.5 to 1 minutes; wash briefly; and proceed to the picro-Ponceau solution.

Sweat, Puchtler, and Rosenthal (1964) substitute 0.1% Sirius red F3BA in aqueous picric acid of the acid fuchsin in Van Gieson method.

Verhoeff Elastin Stain (Mallory 1944)

FIXATION
Any general fixative.

SOLUTIONS
Verhoeff stain
On electric hot plate, dissolve 3 g hematoxylin in 66 ml absolute ethyl alcohol. Cool immediately, filter, and add 24 ml of 10% aqueous ferric chloride and 24 ml Verhoeff iodine solution. Usefulness is limited to 1 to 2 weeks.

Verhoeff iodine solution
- potassium iodide (KI) .... 4.0 g
- distilled water .... 100.0 ml
- Dissolve and add:
- iodine .... 2.0 g

Ferric chloride solution, 10%
- ferric chloride ($FeCl_2$) .... 10.0 g
- distilled water .... 100.0 ml

Ferric chloride solution, 2%
- 10% ferric chloride .... 20.0 ml
- distilled water .... 100.0 ml

Acid fuchsin solution
- acid fuchsin 1% aqueous .... 10.0 ml
- picric acid, saturated aqueous .... 86.0 ml
- acetic acid, 1% aqueous .... 4.0 ml

CONTROL
large blood vessel (e.g., aorta)

PROCEDURE
1. Deparaffinize and run slides down to 70% alcohol. Removal of $HgCl_2$ not necessary.
2. Stain in Verhoeff stain: 15 minutes.
3. Rinse in distilled water.
4. Differentiate in 2% ferric chloride: a few minutes. Elastic fibers should be sharp black; nuclei brown., If destained too far, return slides to Verhoeff for another 5 to 10 minutes.
5. Transfer to sodium thiosulfate, 5% aqueous: 1 minute.
6. Wash in running water: 5 to 10 minutes.
7. Counterstain in Van Gieson solution: 1 minute.
8. Differentiate in 95% ethyl alcohol, 2 changes: a few seconds each.
9. Dehydrate, clear, and mount.

Results

elastic fibers—brilliant blue black
nuclei—blue to brownish black
collagen—red
other tissue elements—yellow

Comments

If elastic fibers stain unevenly or not at all, Verhoeff solution is too old.

Verhoeff can be combined with Perl staining (p. 214) to show both iron and elastin (Pickett and Klavins 1961).

Miller's (1971) method using Victoria blue 4R, new fuchsin, and crystal violet also stains elastin black. The solution is stable.

Iron Gallein Elastin Stain (Churukian and Schenk 1976)

Solution A

gallein .... 1.0 g
ethylene glycol .... 20.0 ml

Add 80 ml absolute alcohol and mix.

Solution B

ferric chloride ($FeCl_3 \cdot 6H_2O$), 20% aqueous .... 4.0 ml
distilled water .... 95.0 ml
HCl, concentrated .... 1.0 ml

Mix equal parts of A and B just before use.

Control

large blood vessel (e.g., aorta)

Procedure

1. Deparaffinize and hydrate to water removing $HgCl_2$.
2. Stain: 30 minutes.
3. Wash in running water: 5 minutes.
4. Differentiate in 2% $FeCl_3$: 2 minutes.
5. Wash in running water: 15 minutes.
6. Proceed with counterstain of choice.

Results

elastin—black

Basic Fuchsin Stain (Horobin et al. 1974)

Solution

2.0 g basic fuchsin, bring to boil in 200 ml distilled water. Add 25 ml 30% aqueous ferric chloride and boil for 3 minutes. Filter and dry precipitate in incubator. Dissolve precipitate in 200 ml 95% alcohol using low heat until dissolved. Add 4 ml concentrated HCl and make up volume to 200 ml with 95% alcohol. Stored at 4°C this solution is good for at least 18+ months. The dry precipitate also is stable.

CONTROL
   large blood vessel (e.g., aorta)

PROCEDURE
   1. Hydrate to water as usual.
   2. Stain: 30 minutes.
   3. Rinse off excess stain in 70% alcohol.
   4. Differentiate in 1% HCl in 70% alcohol up to 5 minutes if required.
   5. Dehydrate, clear, and mount.

RESULTS
   elastin—deep purple
   strongly basophilic substances (mast cells, cartilage)—purple

Orcein (Romeis 1948)

FIXATION
   Any general fixative.

SOLUTION
   orcein .... 1.0 g
   70% alcohol .... 100.0 ml
   hydrochloric acid (concentrated) .... 1.0 ml

CONTROL
   large blood vessel (e.g., aorta)

PROCEDURE
   1. Deparaffinize and hydrate slides to water; remove $HgCl_2$.
   2. Stain in orcein: 30 to 60 minutes.
   3. Wash briefly in distilled water.
   4. Dehydrate in 95% alcohol: 2 minutes.
   5. Differentiate in absolute alcohol until background is almost colorless and elastin fibers are isolated.
   6. Dehydrate, clear, and mount.

RESULT
   elastin—red

Comments

Lillie et al. (1968) stain in orcein, 4 hours, and differentiate in ferric chloride hexahydrate (0.1 ml of 10% $FeCl_3 \cdot 6H_2O$ in 50 ml of 70% alcohol). Dehydrate and clear. This changes the elastin color to black or reddish black.

Roman et al. (1967) combine orcein, hematoxylin, ferric chloride, and iodine in a single solution and stain for two hours. Differentiation is controlled in a ferric chloride solution, and the elastic fibers are stained a deep purple.

Humason and Lushbaugh (1969) combine silver nitrate, orcein, and Sirius supra blue stains to show reticulum, elastin, and collagen in the same section.

Some disagreement appears in the literature concerning the pH at which orcein is more effective. Weiss (1954) writes that orcein stains only from an acid alcoholic solution between pH 3 and 8 and that this puts orcein in a category between basic and acidic dyes, which generally operate at the extremes of alkaline and acid pH. He suggests that there is a formation of hydrogen bond between orcein and elastin. Since this reaction takes place in acid alcohol, it is probably due to a uniquely low positive charge of elastin in such solutions. The use of alcoholic solutions is necessary to stabilize the positively charged orcein fractions. Dempsey and Lansing (1954) agree with Weiss.

Darrow (1952), experimenting with orcein, found that pH 1 to 2.4 is best for elastin staining, and that above pH 2.6 collagen stains as well. A dye content of 0.4% is adequate for elastin staining; a higher concentration adds to collagen staining. For a specific elastin reaction when collagen is present, therefore, it is advisable to check the pH of the acid alcohol used for the solution and to reduce the dye content to 0.4 g per 100 ml.

BONE STAINING

Decalcified Sections (Romeis 1948)

Fixation
Neutral buffered formalin, followed by decalcification (p. 40)

Solutions
Thionine

thionine saturated in 50% alcohol	10.0 ml
distilled water	100.0 ml

If the color does not develop properly, add 1 to 2 drops of ammonia to staining solution.

Carbol-xylol, see p. 463.

Control
decalcified bone

PROCEDURE
1. Deparaffinize and hydrate slides to water. If sections tend to loosen, try subbed slides (p. 469).
2. Stain in thionine solution: 10 minutes.
3. Wash, distilled water: 20 minutes, change several times.
4. Treat with picric acid, saturated aqueous: 1 minute.
5. Rinse in distilled water.
6. Differentiate in 70% alcohol: 5 to 10 minutes or more, until no more color comes off.
7. Dehydrate, clear and mount.

RESULTS
lacunae and canaliculi—bordered with bluish black
background—yellow

Hand-ground, Undecalcified Sections (Enlow 1954)

PROCEDURE
1. If it is necessary to remove organic materials and fat, treat in following manner:
   a. Boil in soap solution: 3 to 4 hours; and wash in running water: 3 to 4 hours.
   b. Suspend over ether or chloroform: 36 to 48 hours.
   c. Allow to dry thoroughly.
2. Cut slices as thin as possible without cracking them. A jeweler's saw is recommended.
3. Grind on sharpening hones with finely powdered carborundum or household cleansing powder. Keep surface wet with water. Optical or metallurgical grinding and polishing equipment, if available, can be used. The grinding can be done between two stones, or the section can be held on a wet rubber or cork stopper and ground against the hone (or frosted plates). If this becomes too difficult to manage, glue the section on a slide with cyanoacrylate adhesives (Krazy Glue), and continue to grind. When the section is almost thin enough, loosen it from the slide with acetone, turn it over, glue it down, and grind the opposite side.
4. Polish off both sides, with a fine leather strop.
5. Place in 95% alcohol: 5 minutes.
6. Air dry.
7. Place in plastic solution; agitate to liberate air bubbles.

   parloidin . . . . 28.0 g
   butyl (or amyl) acetate . . . . 250.0 ml

   Let stand until parloidin is completely dissolved. Stir thoroughly.

8. Transfer to slide with a drop of solution.
9. Dry thoroughly; do not add more solution.
10. When completely dry, add mounting medium and cover glass.

COMMENTS

To bring out the density and distribution of the mineral in undecalcified bone, Frost (1959) recommends methods with basic fuchsin, silver nitrate, alizarine red S, and others.

Dowding (1959) uses methyl methacrylate as a plastic embedding medium and grinds the bone down as thin as 30 μm.

Hause (1959) describes a block for holding the bone while sawing it and recommends a razor saw blade as a good cutting instrument.

Kropp (1954) describes a plastic-embedding method using heat and pressure.

Yaeger (1958) uses freeze drying and vacuum infiltration with butyl methacrylate–ethyl methacrylate.

Norris and Jenkins (1960) describe a method using epoxy resin for preparation of bone for radioautography. The resolution is good; there are no chemicals in resin to produce artifacts in nuclear emulsions; the medium does not warp or chip when machined or abraded. (Polyesters and methacrylate do distort when machined.) They include a design for a metal and Lucite mold for embedding. Sections can be made with a microtome (6 to 10 μm), with a circular saw (50 to 100 μm), or ground to give sections as thin as 10 to 50 μm. Kwan (1970) uses dental wax and a circular diamond blade. Enlow (1961) describes staining of ground and decalcified sections. Villyaneuva et al. (1964) stain ground sections with fast green FCF, orange G, celestin blue, and basic fuchsin.

Exakt Medical Instruments makes an apparatus that will grind the undecalcified bone specimens embedded in plastics to the desired thickness. The specimen size can be 100 mm × 50 mm. This system will allow the user to prepare sections of undecalcified bone with metal implants as well as teeth sections. Sanderson and Bloebaum (1993) explain the process.

Alizarine Red S Method for Whole Embryos and Small Vertebrates
(Bone Formation)

FIXATION

Vertebrate organisms should preferably be free of hair, scales, or feathers. Hollister (1934) suggests fixing in 70% alcohol for several days for fish. Hood and Neill (1948) use 95% alcohol for 3 days. Richmond and Bennett (1938) use 95% alcohol for 2 weeks. Some specimens require decolorization. The best method is to lay specimen in 95% alcohol on a white tray. Place in direct sunlight for 24 hours each side. A sunlamp was used by Hollister on sunless days. Youngpeter (1964) bleaches specimens in 30% hydrogen peroxide for 1 to 2 hours, after step 2 in procedure below. Then he immerses in KOH until specimen is transparent.

Solutions

Potassium hydroxide, 2%
potassium hydroxide, white sticks	20.0 g
distilled water	1000.0 ml

Alizarine stock solution (Hollister 1934)
alizarine red S, saturated in 50% acetic acid	5.0 ml
glycerin	10.0 ml
chloral hydrate, 1% aqueous	60.0 ml

Alizarine working solution
alizarine stock solution	1.0 ml
1–2% potassium hydroxide in distilled water	1000.0 ml

Make up at least 500 ml; use at room temperature.

Clearing Solutions (Youngpeter 1967)

Solution A
glycerin	20 parts
4% potassium hydroxide	3 parts
distilled water	77 parts

Solution B
glycerin	50 parts
4% potassium hydroxide	3 parts
distilled water	47 parts

Solution C
glycerin	75 parts
distilled water	25 parts

Procedure

1. From fixing solution, rinse a few minutes in distilled water.
2. Leave in 2% potassium hydroxide until skeleton shows through musculature: 2 to 4 hours for small embryos, 48 hours or longer for larger forms. St. Amand and St. Amand (1951) warmed solution to 38°C for quicker action.
3. When clear, transfer to alizarine working solution: 6 to 12 hours or longer for larger specimen. Skeleton should be deep red. Large specimens may require fresh changes of dye solution.

4. Transfer directly to 2% potassium hydroxide: 1 day or longer until soft tissues are destained. The KOH may be 1.0–0.5% for small specimens. Sunlight or lamp speeds up process.
5. Transfer through the three clearing solutions: 24 hours each.
6. Transfer to pure glycerin with thymol added as preservative.
7. Store in sealed tubes or bottles. Seal in museum jars or embed in plastic.

COMMENTS

Cumley et al. (1939) gradually replace the glycerin with 95% alcohol, absolute alcohol, and finally toluene. Then the specimens are transferred to toluene saturated with naphthalene and stored in anise oil saturated with naphthalene. This method is supposed to produce grater clarity than the glycerin storage.

Crary (1962) adds alizarine red S to the KOH solution and clears in a mixture of glycerin, 70% ethanol, and benzyl alcohol.

Alizarine is the most specific stain for calcium. The red coloring depends on the presence of calcium base; the only other substance staining red is strontium.

Ojeda et al. (1970) stain the cartilage of whole chick embryos with Alcian blue.

Plastic mounts can be made of these preparations and directions are generally provided by the manufacturers. For other methods, also for cartilage staining, see Wasserung (1976), cartilage and bone; Ojeda et al. (1970), bone; McCann (1971) and Love and Vickers (1972), cartilage.

MUSCLE STAINING

Milligan Trichrome Stain (1946)

FIXATION

10% formalin in normal saline. Humason preferred Gomori 1-2-3 fixative.

SOLUTIONS

Mordant
Solution A

potassium dichromate	3.0 g
distilled water	100.0 ml

Solution B

hydrochloric acid, concentrated	10.0 ml
95% alcohol	100.0 ml

Mix 3 parts A with 1 part B; use within 4 hours.

Acid fuchsin

acid fuchsin	0.1 g
distilled water	100.0 ml

Phosphomolybdic acid

phosphomolybdic acid	2.0 g
distilled water	200.0 ml

Use half of solution for orange G solution below.

Orange G

orange G	2.0 g
1% phosphomolybdic acid	100.0 ml

Fast green
Stock solution

fast green FCF	10.0 g
2% acetic acid (2 ml/98 ml distilled water)	100.0 ml

Working solution

fast green stock solution	10.0 ml
distilled water	90.0 ml

Aniline blue may be substituted for fast green.

CONTROL
    intestine or blood vessels

PROCEDURE
1. Deparaffinize and transfer slides through absolute alcohol into 95% alcohol (remove $HgCl_2$ if present).
2. Mordant in potassium dichromate–hydrochloric acid solution: 5 to 7 minutes.
3. Rinse in distilled water.
4. Stain in acid fuchsin: 5 to 8 minutes.
5. Rinse in distilled water.
6. Fix stain in phosphomolybdic acid solution: 1 to 5 minutes.
7. Stain in orange G: 5 to 10 minutes.
8. Rinse in distilled water.
9. Treat with 1% aqueous acid (1 ml/99 ml distilled water): 2 minutes.
10. Stain in fast green: 5 to 10 minutes.
11. Treat with 1% acetic acid: 3 minutes.
12. Rinse in 95% alcohol, transfer to second 95% alcohol: 5 minutes.
13. Finish dehydration, clear, and mount.

Results

nuclei, muscle—magenta
collagen—green (blue with aniline blue)
red blood cells—orange to orange red

Comments

This is a beautiful, precise stain showing strong contrast between muscle and connective tissue. Smooth muscle cells stand out sharply and clearly.

Milligan reports that this stain's weak point is the nuclear stain, which is improved by using Gomori 1-2-3 fixative.

Galigher found phosphotungstic acid-hematoxylin to be a useful stain for muscle.

Iron hematoxylin staining (pp. 103 and 111) can produce beautiful striation staining if it is followed by careful destaining with iron alum instead of picric acid. Wash thoroughly (at least 30 minutes) in running water after destaining; do not counterstain.

Most of the general fixatives are satisfactory for muscle fixation; however, Zenker is best only for the preservation of intercalated discs.

Suggested Readings

See Poley and Forbes (1964) for staining to show muscle infarction; and Puchtler et al. (1969) for a PAS-myofibril stain that is excellent for muscle fiber bands. See Churukian (1993) for discussion of the basics of staining and Shapiro and Sohns (1994) for use of PTH staining with paraffin as well as with glycol methacrylate. Gowali (1995) uses a picro-hibiscin stain to demonstrate degenerated muscle fibers.

13
Silver Impregnating Reticulum

SILVER IMPREGNATION

According to Baker (1958) methods under this heading depend on the local formation within tissues of a colored substance that is not a dye. Impregnation applies to the condition that develops when an unreduced metal (silver, etc.) is taken up from a solution of salt or other compound and deposited in a colloidal state on a tissue element. After impregnation, the tissue is removed to a reducing solution of a photographic type and the metal is reduced to the elementary state, probably in the form of a black deposit. Thus the tissue itself does not reduce the metal, but some extraneous reducer is required to perform the reaction.

Silver staining started in 1843 when Krause tried small pieces of fresh tissue in silver nitrate. Golgi, an Italian, in 1872 fixed nerve tissue for a long period in a dichromate solution. Then he tried soaking the same tissue in silver nitrate solution and found that silver dichromate was deposited selectively, leaving the impregnated components in short relief against an almost colorless background. Ramón y Cajal (1903, 1910) saw the possibilities of the silver method and experimented with it further, including ways of reducing it. In 1906, he and Golgi were jointly awarded the Nobel Prize in physiology and medicine. By continuing to improve the use of silver and gold impregnations, Ramón y Cajal and del Río-Hortega, one of his pupils, were able systematically to investigate the histology of the nervous system. In the early 1900s Protargol (Bodian method) and other organic silver compounds were introduced as substitutes for silver nitrate. The albumen fraction in the organic compound is considered to act as a protective colloid that prevents too rapid reduction by formaldehyde, thereby inducing the formation of finer-grained deposits of silver. Subsequent experimentation with silver on tissue sections led to many modifications of the methods of Ramón y Cajal, Bielschowsky (1902), and others. Silver impregnation can be very complicated because of the many factors involved to make it specific for various tissue elements.

Ammoniacal silver is the familiar complex and, when reduced by formaldehyde to metallic silver, forms a colloidal solution containing negatively charged particles. These may be precipitated out by oppositely charged surfaces, which can be changed to repel or to absorb the silver. Thus the negatively charged silver (formed by reduction) is deposited on positively charged surfaces and allows selective impregnation of neurofibrils, reticulum, Golgi, and so on. It is probable that fixation and dehydration coagulate the proteins and leave them positively or negatively charged; this would account for the difference in charges of different tissue elements. The pH of the silver solution is a strong factor in determining charges, and it depends on whether an ammonium or sodium hydroxide or sodium or lithium carbonate is used (Ramón y Cajal and del Río-Hortega methods).

As already mentioned, protective colloids, such as gum arabic and mastic, and the use of Protargol slow down the reduction of the silver to produce a finer grain (Liesegang method). Dilute reducing reagents, combined with the above, can have the same effect (Ramón y Cajal and Bielschowsky methods). Temperature is a factor because it increases the kinetic energy of the particles and permits a greater number of collisions of the particles against the tissue surfaces. In some methods, copper is added to the silver solution, supposedly to speed up impregnation by initiating the reduction to metallic silver. Too heavy a deposit is prevented by removal of some of the silver from the solution. Thus various applications of the principles can be used to control the impregnation of different kinds of tissue elements (Silver 1942). Chemical properties of tissues and their responses to these conditions all help to determine the place and amount of deposition.

The types of silver impregnation (all ammoniacal) can be classified in the following manner: (1) ammoniacal silver nitrate, (2) ammoniacal silver hydroxide, and (3) ammoniacal silver carbonate. In all of these solutions, the silver is present largely in the form of a complex silver ammonia cation $[Ag(NH_3)_2]^+$. In (1), ammonia alone is used to form the precipitate, the chief product in solution being silver diamino nitrate. In (2), the Bielschowsky method (also modified by Ramón y Cajal), sodium hydroxide is used to form the precipitate and ammonia to redissolve it, the chief product being silver diamino hydroxide. The difference between the Bielschowsky and the Ramón y Cajal methods lies in the way the silver is applied and the reduction performed. The Ramón y Cajal method uses a single ammoniated silver impregnation followed by reduction in Pyrogallol, hydroquinone, or one of the aminophenols. The Bielschowsky method uses double impregnation (silver nitrate followed by ammoniated silver) and reduction in formalin. In (3), the del Río-Hortega method, sodium carbonate (sometimes lithium carbonate) is used to form the precipitate and is followed by ammonia, the chief product being silver diamino carbonate (Kubie and Davidson 1928).

The reactions of these solutions to various conditions should be understood:

1. The ammoniacal silver nitrate is most stable, least sensitive to light, and least readily reduced, and it combines least easily with tissues. During formalin reduction, a cloud of finely divided gray dust slowly develops, and the staining is slow. This method is rarely used in preference to others.
2. Ammoniacal silver hydroxide is the least stable, the most sensitive to light, and the most readily reduced, and it combines most easily with

tissues. Almost instantly a heavy black cloud appears. It combines almost at once with the tissue, the solution darkens quickly, and a precipitate beings to form. Since silver nitrate is not reduced in an acid solution but reduces readily in an alkaline solution, it stands to reason that the ammoniacal silver hydroxide solution is the most sensitive of the three.

3. The properties of the ammoniacal silver carbonate solution are between those of the above two. Its precipitate forms more promptly than that of ammoniacal silver nitrate (1), but not as fast as that of ammoniacal silver hydroxide (2). Its precipitate is darker than that of (1), but not as dark as that of (2). The solution's reaction begins within 5 to 10 minutes and reaches optimum color before the solution begins to darken. Although solution (2) and this one are used interchangeably, this solution has the advantage in that its hydroxide $(OH)^-$ ion concentration is not high enough to render it as unstable and oversensitive as solution (2), and the presence of buffer salts makes the reduction proceed steadily and evenly. As the acid (HNO_3) is formed during the reduction, the buffer absorbs it and blocks its effect; thus the reduction is not lessened. In addition, the presence of CO_3 ion buffers the formalin and prevents formation of formic acid, also an effective stop to further reduction. Foot (1929) buffered his formalin and prolonged the reducing action, making his results darker and more intense. He warned, however, that the buffer must be kept to a minimum so the reaction would not become too intense. Equimolar silver solutions produce the most uniform results. In most cases, it is wise not to dissolve completely the precipitate that is first formed since this can cause inferior results.

The use of pyridine, a fat solvent, precedes some methods; it removes cephalin, lecithin, myelin, mitochondria, galactolipids, and so forth, and makes subsequent penetration of the silver easier. (This is particularly true for connective tissue.) Toning with gold chloride is optional in many methods, but it may yield a more desirable color and improve contrasts. The timing in gold chloride apparently is variable. It need be only long enough to make the desired change in color, usually a few seconds. If the reaction is slow, the solution has weakened. The final fixing in sodium thiosulfate (hypo) is necessary to remove all unreduced silver.

Corked bottles should be avoided, since cork extractives may disrupt selective impregnation of tissue elements (Deck and Desouza 1959). In all silver methods, take care that metal instruments do not come in contact with the silver solution; a black precipitate may dribble down the surface of slides handled in such a manner. Coat forceps with paraffin or use plastic or wooden instruments. All glassware for metallic stains (silver, gold, etc.) must be acid-clean. Soak in cleaning solution (p. 487), wash thoroughly in running water to remove cleaning solution, and rinse 4 or 5 times with distilled water. Keep one set of glassware for silver stains only. If aqueous solutions of silver nitrate only (no other chemical present) are milky, the distilled water is at fault; glass distilled is recommended. Store silver solutions in the dark and protect from dust.

The loosening of sections from slides is a common problem during silver impregnation. Davies and Harmon (1949) use a rinse (0.5 ml acetic acid, 2% aqueous, in 50.0 ml of water) before reduction in sulfite and hydroquinone in the Bodian method (p. 165). Many technicians find that Masson gelatin fixative (p. 468) is superior to Mayer albumen (p. 467) for affixing sections for silver processes. Transferring unmounted paraffin sections through all solutions and mounting and deparaffinizing at the conclusion of the method are feasible. We recommend subbed slides (p. 469).

Prepare ammoniacal silver solutions just before use. Smith (1943) warns that ammoniacal silver hydroxide solutions may become explosive if they stand for too long as the result of the formation of explosive silver amide in ammoniacal silver hydroxide solutions. (See Laboratory Safety, p. 7, and Titford [1996].)

For complete discussions concerning silver impregnation, see Baker (1958), Beech and Davenport (1933), Foot (1929), Grizzle (1996), Kubie and Davidson (1928), Long (1948), and Silver (1942).

SILVER IMPREGNATION FOR RETICULUM

For reticulum, the silver method probably depends on the local reduction and selection precipitation of the silver by the aldehyde groups of the carbohydrates in the reticulum. The silver is reduced to a dark-brown lower oxide and precipitated on the fibers. The formalin (sodium sulfate, hydroquinone) then reduces the precipitate to black metallic silver.

Among the numerous impregnation methods for reticulum, the following, when properly handled, have never failed to produce precise results and usually with a minimal loss of sections. Subbed slides (p. 469) can be used to reduce loss of sections.

Naoumenko and Feigin Method (1974)

FIXATION

Any good general fixative.

SOLUTIONS

Silver solution

To 35 ml distilled water add, in order, 7 ml 8% aqueous ammonium nitrate, 8 ml 4% aqueous sodium hydroxide, and 3.8 ml 10% aqueous silver nitrate.

Potassium permanganate
   0.25% potassium permanganate, aqueous .... 45.0 ml
   0.67% acetic acid, aqueous .... 0.5 ml

Mix fresh; not stable.

Oxalic acid
   oxalic acid .... 1.0 g
   distilled water .... 100.0 ml

Formalin solution
   formaldehyde (37–40%) .... 0.2 ml
   distilled water .... 100.0 ml

Gold chloride
   gold chloride .... 1.0 g
   distilled water .... 100.0 ml

CONTROL
   liver

PROCEDURE
   1. Deparaffinize and hydrate slides to water, remove $HgCl_2$.
   2. Treat with potassium permanganate: 2 minutes.
   3. Wash in distilled water: 15 seconds.
   4. Bleach in oxalic acid: 2 minutes.
   5. Wash in 2 changes, distilled water: 1 minute each.
   6. Impregnate with silver solution: 6 minutes or longer, depending on following steps.

From step 6 process slides individually.

   7. Dip, one slow dip, in 70% alcohol.
   8. Reduce in formalin solution, 2 changes, over a period of 2 minutes. Briefly with agitation in first, and remaining time in second.
   9. Wash in distilled water: 1 minute.
   10. Gold tone in gold water: 1 minute.
   11. Rinse in distilled water: 1 minute.
   12. Fix in 5% sodium thiosulfate: 1 minute.
   13. Rinse in distilled water: 1 minute.
   14. Counterstain if desired.
   15. Dehydrate, clear, and mount.

Results
Reticulum—black

Gridley Method (1951)

Fixation
Any good general fixative

Solution
Ammoniacal silver hydroxide

To 20.0 ml of 5% silver nitrate (1 g/20 ml distilled water) add 20 drops of 10% sodium hydroxide. Add fresh 28% (reagent) ammonia drop by drop until precipitate that forms is almost redissolved. Add distilled water up to 60.0 ml.

Control
liver

Procedure
1. Deparaffinize and hydrate slides to water. Remove $HgCl_2$ if present.
2. Treat with 0.5% periodic acid (0.5 g/100 ml water): 15 minutes.
3. Rinse in distilled water.
4. Treat with 2% silver nitrate (2 g/100 ml water): 30 minutes, room temperature.
5. Rinse in 2 changes of distilled water.
6. Impregnate in ammoniacal silver solution: 15 minutes, room temperature.
7. Rinse rapidly in distilled water.
8. Reduce in 30% formalin (30 ml/70 ml water): 3 minutes. Agitate gently.
9. Rinse in 3 or 4 changes of distilled water.
10. Tone in gold chloride (10 ml 1% stock solution/40 ml water) until yellow-brown color has changed to lavender gray.
11. Rinse in distilled water.
12. Fix in 5% sodium thiosulfate (5 g/100 ml water): 3 minutes.
13. Wash in running water: 5 minutes.
14. Counterstain if desired.
15. Dehydrate, clear, and mount.

Results
reticulum fibers—black
other tissue elements—depends on counterstain

Comments
Gridley uses periodic acid for oxidation in preference to potassium permanganate and oxalic acid because the latter combination frequently causes the sections to detach from the slides.

Gomori Method (Mallory 1944)

FIXATION
10% neutral buffered formalin.

SOLUTIONS
Ammoniacal silver solution
To 20 ml of 10% silver solution (3 g/30 ml distilled water) and 4 to 5 ml of a 10% potassium hydroxide solution (0.5 g/5 ml distilled water), add 28% ammonia water, drop by drop, shaking the flask continuously, until the precipitate that forms is completely dissolved. Carefully add silver nitrate solution, drop by drop, until the precipitate that forms disappears when the solution is shaken. Make up the solution with distilled water to twice its volume. Always use acid-clean glassware.

Potassium permanganate
potassium permanganate	0.5 g
distilled water	100.0 ml

Potassium metabisulfite
potassium metabisulfite	2.0 g
distilled water	100.0 ml

Ferric ammonium sulfate
ferric ammonium sulfate	2.0 g
distilled water	100.0 ml

Formalin
formaldehyde, (37–40%)	20.0 ml
distilled water	80.0 ml

Gold chloride solution
gold chloride stock solution (1 g/100 ml water)	20.0 ml
distilled water	80.0 ml

Sodium thiosulfate
sodium thiosulfate	2.0 g
distilled water	100.0 ml

CONTROL
liver

PROCEDURE
1. Deparaffinize and hydrate slides to water.
2. Oxidize in potassium permanganate solution: 1 minute.

3. Wash in tap water: 2 minutes.
4. Decolorize in potassium metabisulfite: 1 minute.
5. Wash in tap water: 2 minutes.
6. Sensitize in ferric ammonium sulfate: 1 minute.
7. Wash in tap water: 2 minutes. Rinse in 2 changes of distilled water: 30 seconds each.
8. Impregnate in silver solution: 1 minute.
9. Rinse in distilled water: 20 seconds.
10. Reduce in formalin solution: 3 minutes.
11. Wash in tap water: 3 minutes.
12. Tone in gold chloride: 10 minutes. Sections turn purplish gray.
13. Rinse in distilled water.
14. Reduce in potassium metabisulfite: 1 minute.
15. Fix in sodium thiosulfate: 1 minute.
16. Wash in tap water: 2 minutes.
17. Counterstain, if desired; dehydrate, clear, and mount.

RESULT

reticulum fibers—black

COMMENTS

This is a good reliable method; it is quick, and sections do not loosen with this procedure.

Kernechtrot is recommended as a counterstain because of the excellent contrast that it provides for photomicrography (McGavin, personal communication).

Also see the Nassar and Shanklin method (1961) and Staples and Clark (1990).

Wilder Method (1935)

FIXATION

Any good general fixative.

SOLUTIONS

Phosphomolybdic acid, 10%

phosphomolybdic acid .... 10.0 g
distilled water .... 100.0 ml

Uranium nitrate

uranium nitrate .... 1.0 g
distilled water .... 100.0 ml

Ammoniacal silver nitrate

Add ammonia (28% reagent), drop by drop, to 5 ml of 10% silver nitrate (0.5 g/5 ml water) until precipitate that forms is almost dissolved. Add 5 ml of 3.1% sodium hydroxide (3.1 g/100 ml water). Barely dissolve the resulting precipitate with a few drops of ammonia. Make the solution up to 500 ml with distilled water. Use immediately. Glassware must be acid-clean.

Reducing solution

distilled water	50.0 ml
formaldehyde (37–40%)	0.5 ml
uranium nitrate, 1% (above)	1.5 ml

Make up fresh each time.

Gold chloride

gold chloride stock solution (1 g/100 ml water)	10.0 ml
distilled water	40.0 to 80.0 ml

CONTROL
 liver

PROCEDURE
1. Deparaffinize and hydrate slides to water; remove $HgCl_2$.
2. Wash thoroughly in distilled water.
3. Treat with phosphomolybdic acid: 1 minute.
4. Wash in running water: 5 minutes.
5. Treat with uranium nitrate: 5 seconds or less.
6. Rinse in distilled water.
7. Impregnate with ammoniacal silver nitrate solution: 1 minute.
8. Dip quickly in 95% alcohol and immediately into reducing solution: 1 minute.
9. Wash in distilled water: 2 to 3 minutes.
10. Tone in gold chloride until yellow colors turn purplish gray.
11. Brief rinse in distilled water.
12. Fix in sodium thiosulfate, 5% (5 g/100 ml water): 3 to 5 minutes.
13. Wash in running water: 5 minutes.
14. Counterstain if desired.
15. Dehydrate, clear, and mount.

RESULTS
 reticulum fibers—black
 other tissue elements—depends on counterstain

COMMENTS
 Phosphomolybdic acid replaces potassium permanganate as an oxidizer; the phosphomolybdic acid shows less tendency to loosen sections. Sensitization with uranium nitrate reduces the time and eliminates the heat required

by some reticulum methods. Lillie (1946) disagrees that uranium nitrate is a sensitizer and claims that it is an oxidizer. (For a combination reticulum, collagen, and elastin stain, see Humason and Lushbaugh [1969].) Fitzgerald and Pohlmann (1969) adapted a method for reticulum using silver proteinate reduced by hydroquinone. Staples and associates (1986, 1990) present methods based on the argyrophil reaction for reticulum and argentaffin granules.

14
Silver Impregnating and Staining Neurological Elements

For nervous tissue, frozen sections are best; alcohol and xylene embedding may remove lipids, which can be an essential part of the tissue. The loss of lipids may result in no impregnation of oligodendroglia and weakened microglia, since their impregnation depends on lipid complexes. Periodic and chromic acid oxidation also weaken the reaction of oligodendroglia and microglia. (The reverse is true for connective tissue. Then pyridine, periodic, or chromic acid treatment is desirable to suppress the impregnation of nervous tissue elements.)

Myelin, because of its high lipid content (cholesterol, cerebroxide, and phospholipids) is soluble in fat solvents, and most of it dissolves away in dehydrating and clearing solutions. Empty spaces remain at the former sites of myelin. However, special fixatives can be used to preserve myelin through dehydration, clearing, and paraffin embedding; osmium tetroxide fixes it efficiently and at the same time colors it black. Overnight chromation of tissue blocks in 3% aqueous potassium dichromate preserves formalin-fixed myelin for paraffin processing.

If tissues from the central nervous system are to be embedded in paraffin, they require long periods of fixation to harden them; these tissues also require extended treatment with alcohol, clearing solutions, and paraffin. If the tissue is not sufficiently fixed and hardened, it will compress during sectioning, and the sections may not adhere to the slides during staining. The softness and almost jellylike consistency of nervous tissue is due to the absence of supporting tissues seen in ordinary connective tissue, which has tough intercellular substances, such as elastin and collagen. Only a delicate and ectodermally derived cellular substance—neuroglial cells and processes—supports the nervous tissue elements by lying between and binding together the nerve fibers and blood vessels. (Connective tissue wrappings—meninges—do cover the brain and spinal column.)

Ordinary stains, such as hematoxylin and eosin, do not demonstrate the innumerable processes of neuroglial cells because only the nuclei stain. Silver nitrate methods impregnate the fibers and aid in classifying them. When impregnating tissues from the central nervous system, bear in mind these facts: If the hydroxide method is used, astrocytes and microglia are not specifically impregnated; if the carbonate method is used, the opposite effect takes place. For impregnation, do not heat the solutions above normal body temperature.

Neurological techniques, as has perhaps become evident, so often necessitate highly specialized methods that many technicians prefer to avoid them. Precise attention to all details, however, can produce beautiful and exciting slides. Carefully follow directions for fixation, such as the solution composition, the duration of fixation, and whether fixation is or is not followed by washing. Always wash in distilled water unless tap water is specified. For making silver and other special solutions, use double (glass) distilled water if possible and glassware cleaned in cleaning solution (p. 487), washed well in running water, and rinsed 4 or 5 times in distilled water. All chemicals should be at least reagent grade. Use no corks in containers and no metal instruments in silver solutions. If in doubt about the age of solutions, make fresh ones. If, while being toned with gold chloride, the tissue retains a yellow or brownish hue, the gold chloride is weakened. Prepare a new solution.

Artifact precipitates are difficult to avoid, but strict adherence to procedural details will reduce them to a minimum.

The method of embedding and sectioning will depend on the impregnating or staining technique to follow. Paraffin and nitrocellulose methods are used at times, but, as mentioned above, frozen sections are usually more satisfactory. Sections of neurological tissue are frequently thicker than are sections of other tissues—7 to 20 μm, or even more. Mount sections on subbed slides (p. 469).

When fixing an entire brain, do not allow it to rest on the bottom of the container. Carefully insert a cord under the circle of Willis on the underside of the brain and support the two ends of the cord on the sides of the container. The brain, hanging upside down, should be free of the bottom of the vessel but should remain completely submerged in fixative. The spinal cord can be supported in a graduated cylinder filled with fixative. Run a thread through one end of the spinal cord and tie the thread around an applicator stick supported on the edges of the cylinder. For a perfusion method for the brain, see p. 38.

GLIA

Holzer (1921) Method for Glia Fibers

FIXATION
    10% neutral buffered formalin.

SOLUTIONS
Phosphomolybdic acid
    phosphomolybdic acid, 0.5% aqueous, freshly mixed .... 10.0 ml
    95% alcohol .... 20.0 ml

Alcohol-chloroform
    absolute alcohol .... 20.0 ml
    chloroform .... 80.0 ml

Crystal violet
    crystal violet .... 5.0 g
    absolute alcohol .... 20.0 ml
    chloroform .... 80.0 ml

Potassium bromide
    potassium bromide .... 10.0 g
    distilled water .... 100.0 ml

Differentiating solution
    aniline oil .... 30.0 ml
    chloroform .... 45.0 ml
    ammonia .... 5 drops

CONTROL
    cerebral cortex

PROCEDURE
1. Hydrate sections to water.
2. Treat with phosphomolybdic acid: 3 minutes.
3. Drain and flush sections with alcohol-chloroform.
4. Drain off alcohol-chloroform but keep sections moist.
5. Cover with crystal violet: 30 seconds.
6. Drain and add potassium bromide solution: 1 minute.
7. Drain and blot dry.
8. Add differentiating solution: 30 seconds.
9. Drain, clear, and mount.

RESULTS
    glia fibers—blue

COMMENTS
This is a rapid and beautiful stain. If it is preceded with a PAS stain, the attachment of astrocytes to capillary walls can be shown.

Penfield Modification of del Río-Hortega Silver Carbonate Method
(McClung 1950)—S

FIXATION
10% neutral buffered formalin at least 1 week (longer fixation also will give excellent results) or fix in:

    formaldehyde (37% to 40%) .... 14.0 ml
    ammonium bromide .... 2.0 g
    distilled water .... 86.0 ml

Sectioning
Frozen sections, 20 μm

Globus hydrobromic acid

40% hydrobromic acid	5.0 ml
distilled water	95.0 ml

Silver carbonate solution

Combine 5.0 ml of 10% aqueous silver nitrate and 20.0 ml of 5% aqueous sodium carbonate. Add ammonium hydroxide, drop by drop, until precipitate is just dissolved. Add distilled water up to 75 ml. Filter. The solution keeps for long periods if stored in a dark bottle.

Control
cerebellum

Procedure
1. Place sections in 1% formalin or distilled water.
2. Transfer to distilled water plus 1% ammonia. Cover to prevent escape of ammonia and leave overnight.
3. Transfer to hydrobromic acid, 38°C: 1 hour.
4. Wash in 3 changes of distilled water, 2 minutes in each.
5. Transfer to 5% aqueous sodium carbonate: 1 hour. Tissue may remain in this solution 5 to 6 hours with no ill effect.
6. Impregnate in silver solution: 3 to 5 minutes or until sections turn a smooth gray when transferred to reducer. Try single sections at 3 minutes, 5 minutes, or longer.
7. Reduce in 1% formaldehyde (37–40%). Agitate during reduction: 2 minutes.
8. Wash in distilled water: 1 minute.
9. Tone in gold chloride solution (1 g/500 ml water) until bluish gray.
10. Fix in 5% sodium thiosulfate: 3 minutes.
11. Wash in running water: 5 minutes. Dehydrate, clear, and mount.

Result
oligodendroglia and microglia—dark gray to black

Comments
If only oligodendroglia are to be shown, shorten fixation time to 2 days. Long fixation tends to increase the staining of the microglia and astrocytes.

Procedure for oligodendroglia
1. Follow fixation by treatment with 95% alcohol: 36 to 48 hours.
2. Wash, freeze, and cut sections.
3. Stain in a stronger silver solution by diluting the silver carbonate solution to only 45 ml: 15 minutes or more, until the sections begin to turn brown. Check microscopically every 5 minutes for desired contrast.
4. Wash for a few seconds in distilled water.
5. Reduce with agitation, wash, and tone as above.
6. Dehydrate, clear, and mount.

The oligodendroglia will stain black.

ASTROCYTES

Phosphotungstic Acid Hematoxylin (Mitchell 1975)

FIXATION
10% neutral buffered formalin.

SOLUTIONS
Susa fixative, see p. 30.

Potassium permanganate

potassium permanganate	0.25 g
distilled water	100.0 ml

Oxalic acid

oxalic acid	1.0 g
distilled water	100.0 ml

Phosphotungstic acid-hematoxylin, see p. 114.

CONTROL
cerebellum

PROCEDURE
1. Hydrate sections to water.
2. Mordants in Susa fixative (p. 30): 60 minutes.
3. Wash in distilled water: 5 minutes.
4. Oxidize in potassium permanganate: 5 minutes.

158   Specific Staining Methods

5. Rinse off excess permanganate in distilled water.
6. Bleach in oxalic acid until colorless.
7. Wash in several changes of distilled water: 5 minutes.
8. Stain in phosphotungstic acid hematoxylin: 2 to 4 hours.
9. Dehydrate briefly through 2 changes 95% alcohol.
10. Complete dehydration, clear, and mount.

RESULTS
astrocyte fibers—blue
nuclei—blue
collagen—red
myelin—blue

NISSL SUBSTANCE

Cresyl Violet (Powers and Clark 1955)

Nissl substance is characteristic of fixed nerve cells. It is found in a granular form distributed throughout the cytoplasm and stains brilliantly with basic aniline dyes.

FIXATION
Bouin recommended; others satisfactory. Zenker not recommended.

SECTIONING
Paraffin method, 10 μm; thinner sections are of no advantage.

SOLUTIONS
Cresyl violet
    cresyl violet .... 0.2 g
    distilled water .... 150.0 ml

Buffer solution pH 3.5
    0.1 M (approx.) acetic acid (6 ml/1000 ml water) .... 94.0 ml
    0.1 M (approx.) sodium acetate (13.6 g/1000 ml water) .... 6.0 ml

Working solution
    buffer solution .... 100.0 ml
    cresyl violet solution .... 6.0 to 12.0 ml

CONTROL
cerebellum

PROCEDURE
1. Deparaffinize and hydrate sections to distilled water.

2. Stain 20 minutes in working solution. Use solution only once.
3. Rinse quickly in 70% alcohol and 95% alcohol.
4. Dehydrate, clear, and mount.

RESULT
Nissl substance—purple

COMMENTS
Cresyl violet for Nissl staining need not be the same from each manufacturer; therefore, staining comparisons should be made.

Manns (1960) stains with lithium hematoxylin for myelin and with cresyl fast violet for Nissl substance.

Thionine (Clark 1945)

FIXATION
Bouin recommended; others satisfactory; Zenker not recommended.

SECTIONING
10 μm: thinner sections of no advantage.

SOLUTIONS
Lithium carbonate, 0.55%
lithium carbonate	5.5 g
distilled water	1000.0 ml

Thionine
thionine	0.25 g
0.55% lithium carbonate	100.0 ml

CONTROL
cerebellum

PROCEDURE
1. Deparaffinize and hydrate slides to water; remove $HgCl_2$ if necessary.
2. Treat with lithium carbonate solution: 5 minutes.
3. Overstain in thionine: 5 to 10 minutes.
4. Rinse in distilled water.
5. Dip in 70% alcohol: few seconds.
6. Dehydrate in butyl alcohol, 2 changes: 2 to 3 minutes in each.
7. Clear and mount.

RESULT
Nissl substance—bright blue

Comments

If differentiation is necessary, briefly rinse slides in 95% alcohol (after step 5) and place in aniline, then in lithium carbonate saturated in 95% alcohol. Proceed to step 6.

Gallocyanin

Fixation
   Any general fixative.

Sectioning
   10 μm

Control
   cerebellum

Solution and Procedure
   See p. 105.

Result
   Nissl substance—blue

NERVE CELLS, PROCESSES, AND FIBRILS

Ramón y Cajal Method for Blocks (Favorsky Modification 1930)

Fixation
Cut slices perpendicular to organ surface, about 5 cm thick. Place in 70% alcohol, plus 0.5% glacial acetic acid: 6 hours.

Control
   thick section 15 μm cerebral cortex

Procedure
1. Transfer to 80% alcohol: 6 hours.
2. Treat with ammoniacal alcohol: 24 to 36 hours.
   a. For cerebrum, cerebellum, spinal cord, or ganglia, add 4 drops of ammonia to 50 ml of 95% alcohol.
   b. For medulla, add 9 drops of ammonia to 50 ml of 95% alcohol.
3. Wash in distilled water, several changes, until pieces sink.
4. Treat with pyridine: 1 to 2 days.
5. Wash in running water: overnight; then wash in distilled water, several changes.

6. Blot on filter paper and place in relatively large volume of 1.5% aqueous silver nitrate: 5 days in dark, 38°C.
7. Rinse in distilled water and place in following fluid: 24 hours. Ramón y Cajal's reducing fluid:

   pyrogallic acid or hydroquinone .... 1.0 g
   distilled water .... 100.0 ml
   neutral formalin .... 15.0 ml

8. Rinse in distilled water, several changes over a period of at least 1 hour.
9. Dehydrate and embed in paraffin or celloidin or double embed.
10. Make sections perpendicular to surface of organ and about 15 μm or more thick.
11. Affix to slides, dry, remove paraffin with xylene, and mount. Lay celloidin sections on slides and mount in resin.

RESULTS
neurofibrils—black
background—brownish yellow

Silver Impregnation for Nerve Tissue (Garvey et al. 1987; Staples 1991)

SOLUTIONS
1% silver nitrate
   silver nitrate .... 1.0 g
   distilled water .... 100 ml

3% gum mastic
   gum mastic .... 3.0 g
   absolute alcohol .... 100 ml

2% hydroquinone (prepare just before use)
   hydroquinone crystals .... 0.25 g
   distilled water .... 25 ml

Developing solution
   2% hydroquinone solution .... 25.0 ml
   3% gum mastic .... 15.0 ml
   1% silver nitrate .... 1.2 ml

Mix hydroquinone and gum mastic solutions. Filter through #4 Watman filter paper. Add silver solution before use.

0.25% gold chloride
   10% gold chloride solution .... 25.0 ml
   distilled water .... 75.0 ml

1% sodium thiosulfate
   sodium thiosulfate .... 1.0 g
   distilled water .... 100.0 ml

FIXATION
10% neutral buffered formalin (NBF)

CONTROL
cerebellum

PROCEDURE
1. Deparaffinize and hydrate slides to water.
2. 1% silver nitrate for 30 minutes at 4°C.
3. Rinse in distilled water, 3 changes.
4. Rinse in 95% alcohol, 2 changes.
5. Rinse in absolute alcohol, 2 changes.
6. Drain slides and place in gum mastic solution for 5 minutes.
7. Rinse slides in absolute alcohol, 3 changes.
8. Drain slides and place in developing solution for 2 minutes. Use a 60° water bath.
9. Rinse in distilled water, 3 changes.
10. Place in gold chloride solution for 5 minutes.
11. Rinse in distilled water.
12. Place slides in sodium thiosulfate for 2 minutes.
13. Rinse in running water.
14. Dehydrate, clear, and mount.

RESULTS
axons, dendrites, neurofibrils—black
background—gray

COMMENTS
Garvey (1987) states that gum mastic prevents precipitation of silver on the sections while they are developing. Avoid artifact deposition of silver by using glassware cleaned with nitric acid and deionized water. Distilled water may be used but if not sufficiently pure it may lead to precipitate formation. In addition, prolonged storage of tissues in formalin may diminish silver deposition. The use of *d*-limonene-based xylene substitutes can leach silver content even after adequate impregnation.

Ramón y Cajal Pyridine-Silver Method for Blocks (Davenport et al. Modification 1934)

FIXATION

As soon as possible in:

> absolute alcohol .... 98.0 ml
> ammonia, concentrated .... 2.0 ml

Fix for 1 to 6 days, preferably no longer.

PROCEDURE

1. Treat tissue blocks in 5% aqueous pyridine: 24 hours. This is recommended, but Davenport et al. say it can be optional.
2. Wash in distilled water; 2 to 6 hours; change every half hour.
3. Impregnate in 1.5–2.0% aqueous silver nitrate, 37°C: 2 to 3 days or longer, depending on size of tissue blocks. A minimum time yields the best differentiated tissue. The longer the tissue is left in silver nitrate, the greater the tendency for everything to stain. The time will have to be determined by trial and error.
4. Wash in distilled water: 20 minutes to 1 hour. This is a critical step. The amount of silver washed from the tissue depends on whether the water is changed often or whether the tissue is shaken in the water; change every 10 minutes or use a large volume of water and shake every 10 minutes. Equal staining of the periphery as well as central parts of the tissue determines correct washing. A light periphery indicates too much washing.
5. Reduce in 4% aqueous Pyrogallol: 4 hours.
6. Dehydrate, clear, embed, and section.
7. Mount on slides, deparaffinize, and clear. Add mountant and cover glass.

RESULTS

nerve—yellow to brown
neurofibrils—brown to black
axis cylinders of myelinated fibers—yellow to brown
axis cylinders of nonmyelinated fibers—black

COMMENTS

If the preparation is too light, reduce the time in pyridine and washing. If the preparation is too dark, omit the pyridine and wash 48 hours after fixation or reduce the concentration of the silver nitrate solution and wash longer between impregnation and reduction.

Bielschowsky Method for Blocks (Davenport et al. Modification 1934)

FIXATION

10% formalin: 2 days. For embryos add 0.5% trichloroacetic acid to 10% formalin.

SOLUTION

Add 5.0 ml of concentrated ammonia to 40.0 ml of 2% aqueous sodium hydroxide. Mix well. Add slowly from a burette or pipette 8.5% aqueous silver nitrate until opalescence remains in the solution (about 40.0 ml). Shake the hydroxide solution while adding in the silver nitrate. Add 0.5 to 1.0 ml of ammonia. Dilute with about 5 parts of distilled water.

CONTROL

cerebellum

PROCEDURE
1. After fixation, wash: 1 hour.
2. Transfer to 50% aqueous pyridine: 1 to 2 days
3. Wash in distilled water: 2 to 6 hours, depending on size of sample. Change every half hour.
4. Impregnate with 1–1.5% aqueous silver nitrate: 3 days, 37°C.
5. Wash in distilled water: 20 minutes to 1 hour, depending on size. Change every 10 minutes or use a large volume of water and shake every 10 minutes. Periphery and central portions must be equally stained.
6. Impregnate in ammoniated silver solution: 6 to 24 hours.
7. Wash in distilled water: 15 minutes or 1 to 2 hours, depending on size.
8. Reduce in 1% formalin: 6 to 12 hours.
9. Wash in running water: 10 to 15 minutes.
10. Dehydrate, clear, and embed. Section and mount.
11. Deparaffinize, clear, and cover.

Alternate method (if gold toning is desired):
   a. After step 9, rinse in distilled water.
   b. Gold tone and fix.
   c. Return to step 10.

RESULTS

nerve fibers, neurofibrils—brown to black (no gold toning)
　　　　　　　　　　　　　—gray to black (with gold toning)

Fluorescent Method (Zeiger et al. 1951).

FIXATION

95% alcohol.

SECTIONING
   Paraffin method.

CONTROL
   cerebellum

PROCEDURE
   1. Deparaffinize and hydrate slides to water.
   2. Stain in 0.1% aqueous acridine orange: 6 minutes.
   3. Differentiate in 95% alcohol: 2 seconds.
   4. Blot with filter paper and mount in fluorescent mountant.

RESULTS
   nonmyelinated fibers—bluish gray
   myelinated fibers—brownish orange

COMMENTS
Fresh tissue can be frozen, cut, and stained in acridine orange made up in physiological saline or in Ringer solution.

Bodian Method (Russell Modification 1973)

FIXATION
   Williams (1962) recommends:

formaldehyde	40.0 ml
glacial acetic acid	10.0 ml
80% alcohol	100.0 ml
picric acid	2.0 g

Not suitable are chromic and osmic acid or mercuric chloride. 10% formalin causes excessive staining of connective tissue.

SECTIONING
   Paraffin method.

SOLUTIONS
   Protargol solution

Protargol	1.0 g
distilled water	100.0 ml

Sprinkle Protargol on surface of water in a wide dish or beaker. Do not stir—this is critical. When the granules are dissolved, pour the solution into a Coplin jar containing 6 g of copper shot. This prevents the surrounding tissue from becoming impregnated with silver and obliterating some cellular detail.

Reducing solution

hydroquinone	1.0 g
formaldehyde (37–40%)	5.0 g
distilled water	100.0 ml

Make up fresh.

Gold chloride

gold chloride	1.0 g
distilled water	100.0 ml

Oxalic acid

oxalic acid	2.0 g
distilled water	100.0 ml

PROCEDURE
1. Deparaffinize and hydrate slides to water.
2. Impregnate in Protargol solution, 37°C: 12 to 24 hours.
3. Wash in distilled water, several changes.
4. Reduce: 15 minutes.
5. Rinse in distilled water, 6 changes: 1 minute total.
6. Tone in gold chloride: 4 minutes.
7. Rinse in distilled water, 6 changes: 1 minute total.
8. Develop in 2% aqueous oxalic acid. Check under microscope until background is gray and fibers are sharply defined: approximately 3 minutes.
9. Wash in distilled water, 6 changes: 1 minute each.
10. Fix in 5% aqueous sodium thiosulfate: 5 minutes.
11. Wash in running water: 5 to 10 minutes.
12. Rinse in distilled water.
13. Counterstain, if desired.
14. Treat with acidified water: 5 minutes.
15. Dehydrate, clear, and mount.

RESULTS
nerve fibers—black
background colors—depending on counterstain

COMMENTS
Foley (1943) considers counterstaining essential, claiming that it serves as contrast between nervous and nonnervous tissue and adds to transparency of the section. To counterstain by his method, follow step 11 above with the following:

12. Stain in gallocyanin (p. 105): overnight.
13. Wash thoroughly in running water: 5 to 10 minutes.

14. Mordant in 5% aqueous phosphotungstic acid: 30 minutes.
15. Transfer directly to dilution (20 ml/30 ml water) of following stock solution: 1 hour.

aniline blue WS	0.1 g
fast green FCF	0.5 g
orange G	2.0 g
distilled water	92.0 ml
glacial acetic acid	8.0 ml

16. Differentiate through 70% and 95% alcohols.
17. Dehydrate, clear, and mount.

Consistent selective results with silver impregnation have plagued technicians. Loots et al. (1977) attacked this problem for nervous tissue and concluded that Portargol yields the best results. To prevent darkening of the solution during incubation (probably due to reduction of the silver), a small amount of oxidizing agent has been added. Such a solution can be used more than once and is not excessively light sensitive. Instead of metallic copper, add copper nitrate.

Dissolve 1 g Protargol (or Merck's albumin-silver) in 100 ml distilled water. When dissolved, add in order with agitation:

1% copper nitrate aqueous	2.0 ml
1% silver nitrate aqueous	2.0 ml
30% hydrogen peroxide (undiluted)	2 to 4 drops

Substitute for step 2 above, 3 to 5 days in dark, 37°C, and proceed to step 3 as usual.

See Herr et al. (1976) for a Bodian method for frozen sections.

Nauta and Gygax Method (1951)

FIXATION
   10% neutral buffered formalin: 2 weeks to 6 months.

SECTIONING
   Frozen sections, 15 to 20 μm

SOLUTIONS
   Silver solution A

silver nitrate	1.5 g
distilled water	100.0 ml
pyridine	5.0 ml

Silver solution B
Dissolve 0.45 g silver nitrate in 20.0 ml distilled water. Add 10.0 ml of 95% alcohol. Add 2.0 ml ammonia (concentrated) and 2.2 ml of 2.5% aqueous sodium hydroxide. Mix thoroughly and keep the container covered to prevent escape of ammonia.

Reducing solution

10% alcohol	45.0 ml
10% neutral buffered formalin	2.0 ml
1% citric acid	1.5 ml

CONTROL
    cerebral cortex

PROCEDURE
1. Demyelinate sections in 50% alcohol plus 1.0 ml ammonia per 100 ml: 6 to 12 hours. A longer time has no ill effect.
2. Wash in distilled water, 3 changes: few seconds each.
3. Impregnate in silver solution A: 12 to 24 hours.
4. With no washing, transfer into silver solution B: 2 to 5 minutes.
5. Transfer directly into reducing solution until the sections turn gold in color.
6. Transfer to 2.5% aqueous sodium thiosulfate: 1 to 2 minutes.
7. Wash in distilled water, at least 3 changes.
8. Dehydrate rapidly, clear, and mount.

RESULTS
    nerve fibers and endings—black
    cells—pale yellowish brown

COMMENTS
This method is nonselective and stains normal as well as degenerating axons. For degenerating axons see p. 177.

Nauta and Gygax sections may be counterstained with cresyl violet. Follow step 7 with the cresyl violet stain: 6 minutes in preheated solution, 57°C, just before use.

cresyl violet acetate	1.0 g
distilled water	100.0 ml

Just before using add 15 drops of 10% glacial acetic acid.
Differentiate in 90% alcohol, dehydrate, clear, and mount.

Glees Method (Novotney Modification 1974, 1977)

FIXATION
Perfuse with neutral buffered formalin and store at least 1 week. Embed and section. Use subbed slides.

Solutions

Silver nitrate

silver nitrate	20.0 g
distilled water	100.0 ml

Reducing solution 1

distilled water	400.0 ml
95% alcohol	45.0 ml
10% formalin	13.5 ml
1% acetic acid	13.5 ml

Ammoniacal silver solution

silver nitrate	5.0 g
80% alcohol	100.0 ml

Add drop by drop 25% NH_4OH until precipitate formed just redissolves. A few grains can remain. Add 3 more drops. This silver solution must be freshly prepared. Keep tightly covered during use and use clean glassware. Avoid use of metal instruments.

Reducing solution 2

10% formalin	400.0 ml
95% alcohol	50.0 ml
1% acetic acid	20.0 ml

The reducing solutions are stable and store well.
Luxol fast blue, see p. 172.
Cresyl violet, see p. 158.

Control
cerebral cortex

Procedure
1. Deparaffinize and hydrate sections to distilled water.
2. Place in silver solution: 2 hours at room temperature. Sections should appear light brown.
3. Treat with reducing solution 1: 10 minutes.
4. Place in ammoniacal silver solution: 15 minutes. Sections should appear orange brown.
5. Rinse thoroughly in absolute alcohol.
6. Rinse in reducing solution 2.
7. Treat with fresh reducing solution 2: 10 minutes.
8. Wash in running water: 10 minutes.
9. Fix in 5% sodium thiosulfate; 3 minutes.

10. Wash in running water: 10 minutes.
11. Dehydrate to 95% alcohol.
12. Stain in Luxol fast blue: overnight, 56°C.
13. Allow to cool and rinse slides in absolute alcohol.
14. Rinse in distilled water.
15. Treat with 0.05% lithium carbonate: 3 to 5 minutes.
16. Differentiate in 70% alcohol until neurophil is yellow; nerve cells, orange brown; and myelin, still blue.
17. Rinse in 1% acetic acid to stop differentiation.
18. Stain with cresyl violet acetate, see above, p. 158.
19. Differentiate in 95% alcohol, clear, and mount.

RESULTS

axons—black or dark brown
myelin—blue
Nissl bodies and glial nuclei—violet
neutrophil—yellow

COMMENTS

The method may be halted at end of step 11 and finished by dehydrating, clearing, and mounting. Step 2 is not critical; the duration may be overnight. The reducing solutions are not critical, but step 5 (absolute alcohol) is—it helps prevent formation of precipitate on slides—but do not leave longer than the recommended time. Impregnation can be lost.

Cole Method (1946) (Modified: Whole Mounts)

FIXATION

None; carry fresh tissue directly into step 1.

PROCEDURE

1. Tease striated costal muscle into strips 1 mm in diameter and a few millimeters in length, and place in either of following solutions (Zinn and Morin 1962): (1) 1 part commercial lemon juice and 1 part distilled water; or (2) 0.01 M citric acid. Minimum time is 10 minutes; maximum time, 30 minutes.

Use a separate clean glass container for each step that follows.

2. Wash in several changes of distilled water; 5 minutes.
3. Transfer to 1% aqueous gold chloride. Make up this solution the day before it is to be used. Keep in dark: 60 minutes, or until tissue turns dark yellow. (If pieces are wider than 4 mm, a longer time required.)
4. Wash in several changes of distilled water: 5 minutes.

5. Transfer to 20% formic acid (20 ml/80 ml water): 10 to 20 hours in dark. Do not use metal forceps; use Teflon forceps or paraffin-coated forceps.
6. Rinse in tap water.
7. Transfer to 95% methyl alcohol-glycerin (1:1): several hours. Then remove top of container and allow alcohol to evaporate.
8. Transfer to pure glycerin.
9. To mount: Place a piece of muscle in a very small drop of glycerin (usually amount carried over by the piece is sufficient) on a round cover glass. Lay a smaller size cover glass over it. Spread the muscle to single fiber thickness by using gentle pressure and strokes at right angles to the fibers. Turn cover glasses over and mount on a glass slide. (Double cover glass mounting, p. 100.)

RESULTS
muscle fibers—red blue
motor end plates and medullated axons—black
muscle fiber nuclei—unstained

COMMENTS
Carey (1941) uses the undiluted fresh filtered lemon juice in place of the citric acid. He claims that if the tissue requires more than 12 hours in the formic acid (step 5), the gold chloride technique is faulty. The color of the tissue should be gold, not brown.

Boyd (1962) fixes in fresh filtered lemon juice and formic acid, 3:1:2 to 10 minutes.

Cole and Mielcarek (1962) outline a fluorescent method. Cavanagh et al. (1964) stain with Sudan black B.

Pyridine-Silver Method (Gladden 1970)

PROCEDURE
1. Fix muscle in absolute alcohol, 4.5 ml; distilled water, 5 ml; and concentrated nitric acid, 0.1 ml: 24 hours.
2. Transfer into absolute alcohol, 10 ml, and ammonia, 0.1 ml: 24 hours.
3. Wash in distilled water: 30 minutes.
4. Treat in pyridine: 2 days.
5. Wash in distilled water, 5 to 8 changes: 24 hours.
6. Place in 2% aqueous silver nitrate in dark, 25°C: 3 days.
7. Reduce in 5% aqueous formic acid, 10.0 ml, and Pyrogallol, 0.4 g: 6 to 24 hours.
8. Wash in distilled water and store in glycerin.
9. Tease muscle to show nerve endings and mount in glycerin.

MYELIN

Luxol fast blue B was first used by Klüver and Barrera (1953) for staining myelin sheaths. Margolis and Pickett (1956) combined Luxol fast blue MBSN with other methods to differentiate various neurological elements. Salthouse (1962) stained myelin sheaths a deep blue with Luxol fast blue ARN, which he said had a greater affinity for phospholipids than did Luxol fast blue MBSN. In 1964 he reported Luxol fast blue G, with which he had stained myelin a blue-black color, to be superior to ARN. If dissolved in isopropyl alcohol, the Luxol dyes bind to more phospholipids than if dissolved in ethyl alcohol. The Margolis and Pickett and the Salthouse solutions are listed below; they can be used interchangeably. All Luxol fast blue solutions keep indefinitely.

Luxol Fast Blue

FIXATION

Salthouse (1962) recommends 10% neutral buffered formalin or calcium-formalin, but the following solution is best:

    calcium chloride .... 10.0 g
    distilled water .... 900.0 ml

Dissolve and add:

    cetyltrimethylammonium bromide .... 5.0 g

Dissolve and add:

    formaldehyde (37–40%) .... 100.0 ml

EMBEDDING

Paraffin method.

SOLUTIONS

Luxol fast blue MBSN (Margolis and Pickett 1956)
    Luxol fast blue .... 0.1 g
    95% alcohol .... 100.0 ml
    acetic acid, 10% aqueous .... 0.5 ml

Luxol fast blue ARN (Salthouse 1962)
    Luxol fast blue G .... 1.0 g
    95% alcohol .... 1000.0 ml
    glacial acetic acid .... 0.2 ml

Silver Impregnating Neurological Elements 173

Luxol fast blue G (Salthouse 1964)
    Luxol fast blue G .... 1.0 g
    95% isopropyl alcohol .... 1000.0 ml
    glacial acetic acid .... 0.2 ml

Periodic acid, 0.5%
    periodic acid .... 0.5 g
    distilled water .... 100.0 ml

Lithium carbonate, 0.5%
    lithium carbonate .... 0.5 g
    distilled water .... 1000.0 ml

Schiff reagent, see p. 182.

Sulfurous acid
    sodium metabisulfite, 10% aqueous .... 6.0 ml
    distilled water .... 100.0 ml
    1 N HCl .... 5.0 ml

Cresyl violet
    cresyl violet .... 0.1 g
    distilled water .... 100.0 ml

Just before use, add 15 drops 10% acetic acid. Filter and preheat solution to 57°C.

PROCEDURE
1. Deparaffinize and hydrate slides to water. (Remove $HgCl_2$ if present.) Transfer to 95% alcohol: 2 minutes.
2. Stain in Luxol fast blue MBSN, 60°C: overnight; or Luxol fast blue ARN or G, 35–40°C: 2 to 3 hours.
3. Rinse off excess stain in 95% alcohol.
4. Rinse in distilled water.
5. Dip in lithium carbonate: 15 seconds.
6. Differentiate in 70% alcohol: 20 to 30 seconds.
7. Rinse in distilled water.
8. Dip in lithium carbonate, second solution: 20 to 30 seconds.
9. Differentiate in 70% alcohol: 20 to 30 seconds.
10. Rinse in distilled water. If differentiation is not complete, repeat steps 8 and 9.
11. Oxidize with periodic acid: 5 minutes.
12. Wash in distilled water, 2 changes: 5 minutes total.
13. Treat with Schiff reagent: 15 to 30 minutes.
14. Treat with sulfurous acid, 3 changes: 2 minutes each.

15. Running water: 5 minutes.
16. Stain in cresyl violet: 6 minutes, preheated.
17. Differentiate in 95% alcohol.
18. Dehydrate, clear, and mount.

RESULTS
myelin—blue green
PAS positive elements—rose to red
nuclei—blue
Nissl granules—deep blue purple

COMMENTS

The Luxol fast blue used by Klüver and Barrera was the amine salt of a sulfonated copper phthalocyanin. Phthalocyanin-amine dyes may have one to four sulfonyl groups on each molecule and one of several kinds of bases. Because of the copper phthalocyanin, Luxol fast blue methods are sometimes called copper phthalocyanin methods. Luxol fast blue ARN, however, is a diaryl quanidine salt of a sulfonated azo dye with no copper present, and it has a greater affinity for phospholipids than does the MBS salt. Klüver and Barrera thought the reaction was due to the affinity of the copper phthalocyanin for the porphyrin present in myelin, but the superior reaction of the ARN salt proves that it does not depend on the copper. Pearse (1968) suggests that Luxol fast blue staining of fixed tissues is due to the presence of lipoproteins rather than lipids, that it is an acid-base reaction to form a salt, with the lipoprotein base replacing the phthalocyanin. Lycette et al. (1970) consider the phosphate group of phospholipids to be essential for the reaction. Whatever the explanation, the sensitivity of the dye for myelin is excellent. Also, stores of phospholipids can be clearly shown by this method. Salthouse (1962) demonstrates that the ARN salt forms complexes with many phospholipids.

Hale et al. (1960) use Luxol fast blue MBSN to differentiate abnormal from normal collagen; abnormal collagen does not stain, but normal collagen stains an intense blue.

Snodgrass and Lacey (1961) modify the Luxol fast blue method to differentiate degenerating myelinated fibers.

Lockard and Reers (1962) use it with neutral red to include Nissl staining.

Margolis and Pickett (1956) also describe methods for following Luxol fast blue with phosphotungstic acid-hematoxylin to differentiate neuroglia from myelin and with oil red O to distinguish degenerating myelin from normal myelin.

Dziabis (1958) describes a method for staining gross brain sections.

Luxol Fast Blue–Holmes Silver Nitrate (Margolis and Pickett 1956)

FIXATION
10% neutral buffered formalin preferred.

Silver Impregnating Neurological Elements 175

EMBEDDING
Paraffin method.

SOLUTIONS
Silver nitrate, 20%
    silver nitrate .... 20.0 g
    distilled water .... 100.0 ml

Boric acid
    boric acid .... 12.4 g
    distilled water .... 1000.0 ml

Borax
    borax .... 19.0 g
    distilled water .... 1000.0 ml

Impregnating fluid
In a 500-ml cylinder, mix 55 ml boric acid solution and 45 ml borax solution. Dilute to 494 ml with distilled water. With pipette add 1 ml of 1% silver nitrate (1 g/100 ml water). With another pipette add 5 ml of 10% pyridine (10 ml/100 ml water). Mix thoroughly.

Reducer
    hydroquinone .... 1.0 g
    sodium sulfite (crystals) .... 10.0 g
    distilled water .... 100.0 ml

Can be used repeatedly, but only for a few days.
Luxol fast blue, see p. 172.

CONTROL
spinal cord

PROCEDURE
1. Deparaffinize and hydrate sections to water.
2. Treat with silver nitrate in dark, room temperature: 1 hour. Prepare impregnating fluid.
3. Wash in distilled water, 3 changes: 10 minutes.
4. Impregnate, 37°C: overnight.
5. Shake off superfluous fluid and place in reducer: 2 minutes.
6. Wash in running water: 3 minutes, rinse in distilled water.
7. Tone in 0.2% aqueous gold chloride: 3 minutes.
8. Rinse in distilled water.
9. Treat with 2% aqueous oxalic acid: 3 to 10 minutes; when axons are thoroughly black, remove.

10. Rinse in distilled water.
11. Fix in 5% aqueous sodium thiosulfate: 3 minutes.
12. Wash in running water: 10 minutes.
13. Rinse briefly in 95% alcohol.
14. Stain in Luxol fast blue: overnight.
15. Rinse in 95% alcohol; rinse in distilled water.
16. Treat with 0.05% aqueous lithium carbonate: 15 seconds.
17. Differentiate in 70% alcohol: 20 to 30 seconds.
18. Rinse in distilled water.
19. Repeat steps 16 and 17 if necessary. (Three repeats are usually necessary for sharp differentiation.)
20. Dehydrate, clear, and mount.

RESULTS
axis cylinders—black
myelin sheaths—green blue

COMMENTS
Margolis and Pickett write that slides may be left in the distilled water (step 6) until it is time to prepare them for Luxol fast blue staining. However, leaving the sections in water tends to detach them from the slides. Humason preferred to transfer slides after step 13 into absolute alcohol and to coat them with nitrocellulose. Harden the nitrocellulose and store the slides in 70–80% alcohol until you are ready to stain them. Slides can be left in this condition for a weekend with completely satisfactory results. The nitrocellulose protection also permits agitation of slides in the differentiating fluids. Subbed slides can be used.

Chromic Acid–Hematoxylin (Lillie and Henderson 1968)

FIXATION
10% formalin.

SOLUTIONS
Chromic acid

chromic acid	0.2 g
distilled water	100.0 ml

Acetic hematoxylin

1% stock alcohol hematoxylin	10.0 ml
distilled water	89.0 ml
glacial acetic acid	1.0 ml

or

1% aqueous acetic acid	100.0 ml
hematoxylin	100.0 mg

Borax-ferricyanide
                borax .... 1.0 g
                potassium ferricyanide .... 1.0 g
                distilled water .... 100.0 ml

PROCEDURE
1. Cut frozen sections, 10 to 15 μm, transfer to 1% aqueous gelatin, and pick up on slides. (Can also transfer directly onto subbed slides.)
2. Drain and blot firmly with filter paper. Immerse in 1% aqueous gelatin: 1 to 2 minutes. Drain.
3. Fix gelatin in formalin vapor: 30 minutes. Transfer to 5% formalin until ready to use.
4. Wash in running water: 10 minutes.
5. Mordant in chromic acid: 4 hours.
6. Wash in running water: 10 minutes.
7. Stain in acetic hematoxylin: 2 hours.
8. Rinse in distilled water.
9. Differentiate in borax-ferricyanide: approximately 1 hour. Check under microscope.
10. Dehydrate, clear, and mount.

RESULTS
myelin—blue black
cells and gray substance—yellow to colorless

COMMENTS
This method is recommended over the Ora or Mahon method because the myelin sheaths remain a more normal size in relation to other structures in this method, whereas paraffin embedding may cause shrinkage in thin sheaths.

For a Mahon method for celloidin-embedded tissue, see Metz (1976). Angulus and Sepinwall (1971) use gallocyanin as a myelin stain.

DEGENERATING AXONS

Nauta and Gygax Method (Modified by Powell and Brown 1975)

FIXATIONS
Perfusion method seems to be preferred.

1. Perfuse with 0.9% sodium chloride (intracardiac and intra-aortic cannulation) until escaping fluid is clear.
2. Perfuse with 500 ml of 5% aqueous potassium dichromate and 2.5% aqueous potassium chlorate. (This step is peculiar to Anderson (1959); others perfuse with a formalin solution.)

178 Specific Staining Methods

3. Remove tissue and fix in 10% formalin neutralized with carbonate: 1 week or longer. Nauta and Gygax recommend 1 to 3 months as best.

Embedding and Sectioning
1. If embedding is necessary, use gelatin. Wash slices (5 to 10 mm thick) in running water: 24 hours. Incubate in 25% gelatin, 37°C (cover tightly): 12 to 18 hours. Wipe off excess gelatin and immerse in cool formalin: 6 hours or longer.
2. Frozen sections: cut sections 15 to 25 μm and place in 10% neutral buffered formalin, room temperature or cooler. Can be stored in formalin, preferably in refrigerator. Process only a few sections (6 to 10) at a time.

Adey et al. (1958) prefer a dual-freezing block method, using dry ice evaporated in 70% alcohol, over CO_2 freezing. This method produces more even freezing. Rapid freezing and thawing with CO_2 may rupture some of the fibers; the use of quick freeze spray (environmentally safe) will sometimes prevent this.

Solutions
Bleach

1% hydroquinone, aqueous; mix fresh	1 part
1% oxalic acid, aqueous	1 part

Silver solution A

0.5% uranyl nitrate, aqueous	30.0 ml
2.5% silver nitrate, aqueous	36.0 ml
distilled water	56.0 ml

Silver solution B

0.5% uranyl nitrate	40.0 ml
2.5% silver nitrate	60.0 ml

Silver solution C

2.5% silver nitrate	60.0 ml
ammonium hydroxide, fresh, concentrated	4.0 to 5.0 ml
2.5% sodium hydroxide, aqueous	2.8 to 3.0 ml

Mix in above order and use immediately.

Reducing solution

distilled water	405.0 ml
95% alcohol	45.0 ml
1% citric acid, aqueous	13.5 ml
10% formalin, aqueous	13.5 ml

Gelatin-alcohol
Dissolve 6.0 g gelatin in 400.0 ml hot, distilled water. Cool and add 400.0 ml 80% alcohol.

PROCEDURE
1. Rinse sections thoroughly in 4 changes of distilled water: 2 to 5 minutes each. Agitate.
2. Soak in 0.05% potassium permanganate, aqueous: 2 to 6 minutes.
3. Rinse briefly in distilled water.
4. Soak with agitation in freshly mixed solution: 30 seconds to 2 minutes.
5. Rinse thoroughly in distilled water. Sections can remain here overnight.
6. Place in silver solution A, freshly mixed: 30 minutes.
7. Transfer directly to silver solution B, also freshly mixed: 30 minutes.
8. Rinse thoroughly in distilled water.
9. Transfer into silver solution C, freshly mixed: 2 minutes. Do not crowd the sections or overuse the solution. Change every 8 to 12 sections. Two minutes is usually adequate, but check the intensity under the microscope after the first 30 seconds.
10. Transfer directly into reducing solution: 2 minutes in each of 2 changes.
11. Fix in 0.5% sodium sulfate: 1 minute or more.
12. Rinse thoroughly in distilled water.
13. Place in gelatin-alcohol: 5 minutes or overnight.
14. Mount on slide; blot gently but do not dry.
15. Dehydrate in 95% absolute alcohol (2 changes): 3 minutes each.
16. Clear in 2 changes xylene: 5 minutes each, and cover.

RESULTS
degenerating axons—black
normal axons—various shades of brown

COMMENTS
See also Nauta and Gygax (1951), Nauta and Ryan (1952), Wall (1950), White (1960), and additional references included therein.

The Nauta-Gygax impregnation of degenerating fibers is related to unsaturated lipids. Since the Nauta reaction is abolished by prior bromination, this suggests that the basic mechanism of the impregnation may involve unsaturated bonds, the ethylene bonds of cholesterol esters. The presence of these esters is attributed principally to the breakdown of the myelin. It is possible that degeneration of the axis cylinders is also part of the reaction (Giolli 1965).

Hamlyn (1957) and Guillery et al. (1961) describe methods for paraffin-embedded sections that are stained after mounting.

Marchi Method—B

FIXATION

10% formalin plus 1% potassium chlorate: 24 to 48 hours, no longer (Swank and Davenport 1934a).

SOLUTION

Marchi fluid (Poirier et al. 1954)

osmium tetroxide, 0.5% aqueous	11.0 ml
potassium chlorate, 1% aqueous	16.0 ml
formaldehyde (37–40%)	3.0 ml
acetic acid, 10% aqueous	3.0 ml
distilled water	67.0 ml

PROCEDURE

1. Cut tissue into thin slices, about 3 mm for easier impregnation.
2. Chromate in 2.5% aqueous potassium dichromate in a dark enclosure: 7 to 14 days, change twice.
3. Transfer directly to Marchi fluid, a volume 15 to 20 times that of tissue: 1 to 2 weeks depending on size of tissue. Turn tissue over every day to improve penetration.
4. Wash in running water: 24 hours.
5. Dehydrate and embed. If using celloidin, keep embedding steps to a minimum. If using paraffin, avoid xylene and its solvent action on osmium tetroxide.
6. Deparaffinize slides with chloroform and mount in chloroform-resin.

RESULT

degenerating myelin—black
background—brownish yellow
neutral fats—black

COMMENTS

The principle behind the Marchi method is that the myelin of medullated nerves oxidizes more easily than does degenerating myelin. Normal myelin is oxidized by chromatin and will not react with osmic acid. Degenerating myelin contains oleic acid, which does not oxidize during chromating and therefore reduces the osmic acid and stains black.

Suggested Readings

For additional reading about neurological staining, see Culling (1957), Luna (1960, 1964, 1968), Lillie (1954b), Mettler (1932), Margolis and

Pickett (1956), Mettler and Hanada (1942), Poirier et al. (1954), Swank and Davenport (1934a,b; 1935a,b), Garvey et al. (1987, 1990, 1991), Koski and Reyes (1986), Lloyd et al. (1985), Murdock and Fratkin (1988), and Vacca (1985). Also note that the *Journal of Histotechnology* 19:3 (1966) is devoted to silver stains.

15
Periodic Acid-Schiff, Feulgen Techniques, and Related Reactions

Feulgen and periodic acid-Schiff (PAS) techniques involve two chemical reactions: (1) the oxidation of α-amino alcohol and/or 1.2 glycol groups to aldehydes and (2) the reaction of the resulting aldehydes with Schiff reagent to form a purple-red color. The Schiff reaction sequence for detecting the aldehydes is complicated, and two interpretations have been suggested: (1) the formation of an amino sulfinic acid followed by an additional reaction of aldehyde (Wieland and Scheuing 1921); (2) the formation of an amino alkylsulfonic acid (Hörmann et al. 1958; Rumpf 1935). Hardonk and van Duijan (1964a,b,c) and Nauman et al. (1960) confirm the second interpretation. The two oxidizers most commonly used are chromic and periodic acids. Periodic acid breaks the carbon chains of the polysaccharides containing the 1,2 glycol groupings and oxidizes the broken ends into aldehyde groups. Chromic acid is a weaker oxidizer whose action is limited almost exclusively to glycogen and mucin (the principle of the Bauer method). If necessary, glycogen and starch can be demonstrated to the exclusion of other reactants by iodine or Best carmine (p. 254), and mucin can be demonstrated by alcian blue (p. 256) or metachromatic methods (p. 271). If it is desirable to prevent the reaction of glycogen or starch, the saliva or diastase treatment is simple and effective (p. 185).

Among the nucleic acids, oxidation will not form aldehydes, and acid hydrolysis is required (see pp. 188 and 235).

SCHIFF REAGENT

A century ago Hugo Schiff carried out extensive research on the reactions of amines with aldehydes and reported that the addition of a few drops of aldehyde would restore a red-violet color to a rosanilin (fuchsin, magenta) dye solution that had been decolorized by SO_2. His name has been applied to the colorless derivative formed by the action of SO_2 on basic fuchsin or any of its component dye moieties.

Feulgen (1914) discovered that hydrolysis of fixed tissues exposed the deoxypentose of the nucleus in an aldehyde form. Then he and Rossenbeck (1924) described the Feulgen reaction for DNA, that mild acid hydrolysis followed by Schiff reagent gave a reddish purple color to DNA-containing structures.

In an acid solution and with an excess of SO_2 present, basic fuchsin (a mixture of several related phenol methane dyes, rosanilins, and pararosanilins built on the quinoid ring) is reduced to form a colorless N-sulfinic acid (fuchsin sulfurous acid). This reagent, with the addition of aldehydes, forms a new phenyl methane dye, slightly different from basic fuchsin, since the color is more purple red than pure red. The chemical reaction is not wholly understood. Thus areas rich in DNA show deep coloration after hydrolysis. Schiff is stable as long as an excess of SO_2 with high acidity is present. Anything removing these conditions restores the original dye and produces a pseudoreaction. But when regenerated by an aldehyde, the dye becomes extremely resistant to such agents. Schiff reagent is not a dye; it lacks a chromophore and is, therefore, colorless. Baker (1958) considers this an example of localized synthesis of a dye; when Schiff reagent comes in contact with an aldehyde, the chromophore of the triarylmethane (quinoid ring structure) dye is reconstituted. The additive compound of the aldehyde with the Schiff reagent could be called a dye.

Schiff reagent deteriorates rapidly at temperatures above 40°C; but at 0–5°C, if kept tightly capped, the deterioration is slow. Always store in the refrigerator when not in use. Under these conditions it will keep as long as 6 months.

Chung and Chen (1970) restore exhausted Schiff reagent by adding 0.5 g $NaHSO_3$ to 100 ml of the reagent. This can be repeated 2 or 3 times as the solution becomes exhausted and is advisable even when the solution is kept in the refrigerator. If the solution shows a red tinge, it is partially exhausted; if it shows a purple tinge, it is completely exhausted. In either case it can be recovered by this method. "Ready-to-use" Schiff reagent is commercially available.

For references concerning the Schiff reaction, see Atkinson (1952), Baker (1958), Bensley (1959), Garvey et al. (1992), Glegg et al. (1952), Gomori (1952), Jaspers (1987), Kasten (1960, 1985), Lhotka and Davenport (1947, 1949, 1951), Lillie (1951a,b, 1954a,b), Lodin et al. (1967), McManus (1961), Meloan and Puchtler (1986), and Stowell (1945, 1946).

SCHIFF REACTIONS

PAS Technique, Aqueous

FIXATION

Any general fixative but, if glycogen or other soluble polysaccharides are to be demonstrated, fixation and washing should be done in alcoholic fluids of no less than 70% alcoholic content.

SOLUTIONS

Periodic acid

periodic acid (HIO_4) .... 0.6 g
distilled water .... 100.0 ml
nitric acid, concentrated .... 0.3 ml

Schiff reagent (Lillie 1951b)

    basic fuchsin .... 0.5 to 1.0 g
    distilled water .... 85.0 ml
    sodium metabisulfite ($Na_2S_2O_2$) .... 1.9 g
    N HCl .... 15.0 ml

Place Schiff reagent components in a bottle with approximately 50 to 60 ml of free air space. Shake at intervals for at least 2 hours or overnight. Add 2.0 g activated charcoal: 1 minute; shake occasionally. Filter. If solution is not water clear, the charcoal is old. Try a fresh batch and refilter. Store Schiff reagent in a bottle with a minimum of air space above the solution, and keep in a refrigerator. This will decrease loss of SO_2 (Elftman 1959b). It can be stored frozen but never at room temperature.

Kasten (1960) recommends the use of less than 0.5 g of charcoal per 100 ml of solution, claiming that too much charcoal reduces dye sensitivity. Do not leave the charcoal in the solution too long, only about 45 seconds, and filter.

Sodium bisulfite

    sodium metabisulfite ($Na_2S_2O_2$) .... 0.5 g
    distilled water .... 100.0 ml

CONTROL
    small intestine or known fungus-infested tissue

PROCEDURE
1. Deparaffinize and hydrate slides to water; remove $HgCl_2$.
2. Treat with periodic acid, aqueous: 5 minutes.
3. Wash in running water: 5 minutes.
4. Treat with Schiff reagent: 10 minutes.
5. Transfer through sulfite solutions, 3 changes: 1.5 to 2 minutes each (see comment 2).
6. Wash in running water: 5 minutes.
7. Counterstain, if desired (see comment 4).
8. Dehydrate, clear, and mount.

RESULTS
    fungi—red
    nuclei and other tissue elements—color of counterstain

Many tissues give positive PAS reactions: glycogen, starch, cellulose, mucins, colloid of thyroid, cartilage matrix, reticula, fibrin, collagen—rose to purplish red (see comment 6).

COMMENTS
1. When preparing slides, avoid excessive use of egg albumen, which contains sufficient carbohydrate to react with the Schiff.
2. Some technicians eliminate the sulfite rinses, but high chlorination of the water makes this a questionable practice. The reagent must be removed, and the sulfite solutions ensure this. Renew them often.
3. *Control slides:* To remove glycogen, run the slides down to water and subject them to the saliva test. Saliva contains a diastatic enzyme that dissolves glycogen and starch. Human saliva can be negative for diastase and should be tested on a known starch before its use on control slides. A reliable substitute is a 1% solution of diastase of malt in a phosphate buffer (p. 473), pH 6.0, 37°C: 1 hour.

   To remove mucin, treat slides with lysozyme, 0.1 mg to 10.0 ml of Sorensen $M/15$ phosphate buffer (p. 473), pH 6.24, room temperature: 40 to 60 minutes.

   To remove RNA, treat slides with 0.01 mg RNAse in 10.0 ml 0.2 M acetate buffer (p. 470), pH 5.0 at room temperature: 10 to 15 minutes (or tris-HCl buffer, 0.5 M, pH 7.5, 56°C: 2 hours).
4. *Counterstains:* Nuclei—hematoxylin; for glycogen—fast green, FCF, as in Feulgen method; mucin or acid polysaccharides—an acid dye (fast green); other polysaccharides—a basic dye (malachite green). Light green or methylene blue can also be used.
5. If it is necessary to remove the PAS from the tissue, treat it with potassium permanganate until all the color is removed; then bleach the permanganate with oxalic acid.
6. A number of carbohydrate and carbohydrate-protein substances give a positive PAS reaction: acid mucopolysaccharides, glycolipids, mucoproteins, glycoproteins, pituitary gonadotropins and thyrotropins, neutral mucopolysaccharides, unsaturated lipids, phospholipids, and others. Among the polysaccharides that give a positive PAS reaction are glycogen, starch, and cellulose; all of these have 1,2 glycol groups. Cartilage has a polysaccharide compound that makes this tissue react positively. Among the mucoproteins, the mucins are carbohydrates, and they give a PAS reaction. Other chemical structures that contain polysaccharides and display a positive PAS reaction include striated and brush borders, reticulin fibers, and basement membranes. Fats may become colored with Schiff's after application of periodic acid if they are glycolipids. Unsaturated fats color, probably because of the oxidation of C=C sites to 1,2 glycols. Simple sugars are always lost, no matter what technique is used. To retain complex sugars, avoid use of water solutions (McManus 1961).
7. The PAS technique can be combined with many other procedures, such as silver, aldehyde-fuchsin, Luxol fast blue, Sudan black, and alkaline

phosphatase (Elftman 1963; Himes and Moriber 1956; Lazarus 1958; Moffat 1958). For black periodic and black Bauer methods, see Lillie et al. (1961). An example of a combination stain using PAS is the triple stain for DNA, polysaccharides, and protein (Himes and Moriber 1956). Schreibman frequently uses this in the classroom (see below).

PAS Technique, Alcoholic (Bedi and Horobin 1976)

FIXATION
Same as for aqueous technique.

SOLUTIONS
Periodic acid
periodic acid (HIO_4) .... 1.0 g
90% alcohol .... 100.0 ml

Keep in dark; solution is unsatisfactory if it turns brown.

Schiff reagent
Prepare a fresh solution of de Tomasi's Schiff. Dissolve 1 g basic fuchsin in 200 ml boiling distilled water. Shake for 5 minutes. Cool to exactly 50°C. Filter and add to filtrate 20 ml N HCl. Cool to 25°C and add 1 g sodium (or potassium) metabisulfite. Keep in dark 14 to 24 hours. Add 2 g activated charcoal. Shake for 1 minute. Filter and add an equal volume of 1% phosphotungstic acid, aqueous. Shake. A white precipitate of phosphotungstic acid-Schiff complex is formed. Centrifuge and discard supernatant fluid. Suspend the precipitate in same volume of absolute alcohol as was used to prepare the complex. The precipitate can be dried and frozen at 0°C and stored for several months.

CONTROL
small intestine

PROCEDURE
1. Deparaffinize sections and remove $HgCl_2$ if present.
2. Hydrate sections to 70% alcohol and briefly rinse in distilled water.
3. Treat with periodic acid: 60 minutes.
4. Wash in absolute alcohol, 2 to 3 changes: 3 minutes each.
5. Treat with Schiff complex: 30 minutes.
6. Wash in several changes absolute alcohol: 20 minutes.
7. Clear and mount.

COMMENTS
Some polysaccharides, mucosubstances, and fatty acids are extremely water soluble and demonstrating them by aqueous solutions becomes difficult. Al-

coholic solutions are therefore recommended as more reliable for such substances.

Feulgen Reaction

FIXATION
A fixative containing $HgCl_2$ is preferred.

SOLUTIONS
5 N hydrochloric acid, see p. 462.
Schiff reagent, see p. 182.

Bleaching solution (sulfurous acid)
1 N HCl	5.0 ml
potassium bisulfite ($K_2S_2O_5$) or sodium bisulfite ($Na_2S_2O_5$), 10% aqueous	5.0 ml
distilled water	100.0 ml

or

HCl, concentrated	1.0 ml
potassium or sodium bisulfite	0.4 g
distilled water	100.0 ml

For best results, make up bleach fresh each time.

Fast green
fast green FCF	0.5 g
95% alcohol	100.0 ml

CONTROL
small intestine

PROCEDURE
1. Deparaffinize and hydrate slides to water; remove $HgCl_2$. (Leaving slides in 95% alcohol overnight will remove lipids that might cause a plasma reaction.)
2. Rinse at room temperature in 5 N HCl: 2 minutes.
3. Hydrolyze in 5 N HCl, room temperature: 1 hour (see comment 1).
4. Rinse at room temperature in 5 N HCl; rinse in distilled water.
5. Stain in Schiff reagent: 2 hours in dark.
6. Drain and transfer quickly into bleaching solution, 3 changes: 1.5 to 2 minutes each.
7. Wash in running water: 10 to 15 minutes.
8. Rinse in distilled water.

9. Counterstain in fast green: 10 seconds.
10. Dehydrate, clear, and mount.

RESULTS

DNA-containing substance—red violet
other tissue elements—shades of green

COMMENTS

1. The specificity of the Feulgen reaction has been attacked periodically, but it seems evident that properly applied the reaction can be reasonably specific. Because the results of the reaction can vary with conditions, optimum conditions are essential. For instance, excessive hydrolysis will allow the degraded nucleic acids to diffuse from the tissue. Depending on the fixative used, the following times of hydrolysis are recommended (Pearse 1968):

Carnoy	6 minutes	Formalin sublimate	8 minutes
Carnoy, formula B	8 minutes	Helly	8 minutes
Carnoy-Lebrun	6 minutes	Regaud	14 minutes
Champy	25 minutes	Susa	18 minutes
Chrome acetic	14 minutes	Zenker	5 minutes
Flemming	16 minutes	Zenker-formol	5 minutes
Formalin	8 minutes		

The conventional hydrolysis temperature has been 60°C, but Deitch et al. (1968) and Decosse and Aiello (1966) agree that 5 N HCl at room temperature is preferable. The Feulgen reaction time is less critical at room temperature, and Feulgen values may be 5–30% higher than when hydrolysis is done at 60°C.

2. Kasten and Burton (1959) make the following Schiff reagent. It is quickly prepared and colorless. It does not stain the hands and does not require refrigeration. It can be made more sensitive by boiling it for 1 minute.

basic fuchsin	0.05 g
distilled water	300.0 ml
sodium hyposulfite ($Na_2S_2O_4$)	6.0 g

Solution should decolorize immediately. Filter if necessary. Ready for immediate use.

3. Pink solutions of Schiff reagent may have lost their potency. Test by pouring a few drops into 10 ml of 40% formalin. A good solution changes rapidly to reddish purple but, if the color changes slowly and becomes blue purple, the solution is breaking down. Older batches of reagent tend to fade more than do new batches (de la Torre and Salisbury 1962).

4. Chen (1944b) uses the following weak Flemming fixative for avian parasites. It gives beautiful results on any kind of smear preparation and on very small pieces of tissue, which should be fixed for 1 to 4 hours.

> chromic acid, 1% in normal saline .... 25.0 ml
> acetic acid, 1% in normal saline .... 10.0 ml
> osmic acid, 2% in normal saline .... 5.0 ml
> normal saline .... 60.0 ml

Wash smears for 1 hour in running water; proceed to step 3, hydrolysis. Wash pieces of tissue overnight before embedding for sectioning.

5. Kasten and Lala (1975) observed false positives after glutaraldehyde fixation and blocked free aldehydes by reducing them with a fresh solution of 0.5% sodium borohydride ($NaBH_4$) in 1% sodium phosphate, monobasic ($NaH_2PO_4 \cdot H_2O$) aqueous solutions for 1 hour, room temperature. Hydrolyze, wash, and treat with Schiff.

It is always advisable to wash well after an aldehyde fixative like formalin, but it is more difficult with glutaraldehyde. The latter presents more opportunities for the introduction of free aldehydes into fixed cells.

6. The first of the series of three bleaching solutions (step 6) will begin to accumulate Schiff reagent and turn pink. Then it is advisable to remove that solution, shift steps 2 and 3 to steps 1 and 2, and add a new step 3 solution.

7. Block and Godman (1955) hydrolyzed in 1 N trichloroacetic acid and stained with a trichloroacetic acid-Feulgen. The trichloroacetic acid was substituted for hydrochloric acid, mole for mole in the hydrolyzing solution and the Feulgen reagent. In this way the histone proteins were left intact.

Deoxyribonucleic Acid (DNA) Fluorescent Technique
(Culling and Vasser 1961)

FIXATION

10% formalin for sections; methyl alcohol for smears. (Other fixatives may require a different time for hydrolysis.)

SOLUTIONS

Fluorescent Schiff reagent

> acriflavine dihydrochloride .... 1.0 g
> potassium metabisulfite .... 2.0 g
> distilled water .... 200.0 ml
> N hydrochloric acid .... 20.0 ml

Dissolve acriflavine and potassium metabisulfite in distilled water; add hydrochloric acid. Keep overnight before using.

Periodic acid

periodic acid	1.0 g
$M/15$ sodium acetate	10.0 ml
absolute alcohol	90.0 ml

Acid alcohol

70% alcohol	99.0 ml
hydrochloric acid, concentrated	1.0 ml

CONTROL
small intestine, lymph node

PROCEDURE
1. Deparaffinize and hydrate sections to water. Smears can be carried directly into next step.
2. Transfer to preheated N hydrochloric acid, 60°C; sections, 10 minutes; smears, 3 to 4 minutes (or 1% periodic acid: 10 minutes for PAS).
3. Wash briefly in distilled water.
4. Transfer to fluorescent Schiff reagent: 20 minutes.
5. Wash in acid alcohol: 5 minutes.
6. Wash in fresh acid alcohol: 10 minutes.
7. Dehydrate, clear, and mount.

RESULTS
DNA—bright golden fluorescent
other elements—dark green to black

COMMENTS

The advantage of this method over the conventional Feulgen reaction is that smaller amounts of dye molecules are more easily observed.

Culling and Vasser warn that previously heated hydrochloric acid is important; cold hydrochloric acid can produce negative results. Also, timing is important; the reaction may weaken if slides are left in hydrolysis too long.

See Betts (1961) and Kasten et al. (1959) for other fluorescent Schiff-type reagents. Also see Armstrong et al. (1957), Keeble and Jay (1962), Metcalf and Paton (1944), and Nash and Plaut (1964) for fluorescent method for DNA and RNA.

Triple Stain for DNA, Polysaccharides, and Proteins
(Himes and Moriber 1956)

FIXATION
Neutral buffered formalin.

SOLUTIONS
   Schiff reagent, see p. 184

   Bleaching solution
1 N hydrochloric acid	15 ml
potassium bisulfide, 10% aqueous	15 ml
distilled water	270 ml

   Periodic acid reagent
periodic acid	0.4 g
M/50 sodium acetate	50 ml

   Azure A-Schiff reagent
azure A	0.25 g
bleaching solution	50 ml

   Naphthol yellow S reagent
naphthol yellow S	0.2% in 1% acetic acid

CONTROL
   small intestine

PROCEDURE
   1. Hydrate slides to water.
   2. Treat slides with 1 N HCl for 3 to 12 minutes.
   3. Rinse in distilled water.
   4. Place in azure-A-Schiff reagent for 5 minutes.
   5. Rinse in distilled water.
   6. Rinse in 2 changes of bleach.
   7. Rinse in running water 30 seconds.
   8. Transfer to periodic acid solution for 2 to 5 minutes.
   9. Rinse in distilled water.
   10. Place in Schiff reagent for 2 to 5 minutes.
   11. Rinse in bleach 2 times briefly.
   12. Rinse in running water 1 minute.
   13. Stain in naphthol yellow S reagent for 30 seconds to 2 minutes (check for desired color intensity).
   14. Rinse in distilled water.
   15. Transfer to tertiary butyl alcohol, 2 changes for 5 minutes each.
   16. Clear and coverslip.

RESULTS
   DNA—blue
   polysaccharides—purplish red
   proteins—yellow
   chromatin—greenish
   mucin, glycogen, connective tissue—red
   zymogen granules, myofibrils, mitochondria—yellow

16
Staining Hematological Elements and Related Tissues

Hemopoietic (blood-forming) tissue is connective tissue specialized to produce blood cells and to remove worn blood cells from the bloodstream. There are two varieties of hemopoietic tissue: (1) myeloid, which produces erythrocytes, granular leukocytes, and platelets (thrombocytes); and (2) lymphatic, which produces most of the nongranular leukocytes.

Blood cells and fluid may become parasitized in various ways: for example, with malaria, trypanosomes, inclusion bodies of various diseases, such as rickettsia and psittacosis. Most of these foreign elements, as well as the blood cells, are best demonstrated with a Romanovsky type of stain.

See Chapter 23 for techniques for demonstrating enzymes found in hemopoietic tissue.

BLOOD SMEARS
Preparation for Thin Smears

Slides must be clean for a uniform smear. Handle slides at the edges, keeping fingers off the clean surface. Prick the finger (using sterile and safety precautions) and, when a small drop of blood appears, wipe it away. Touch the next drop of blood to the clean surface of the right end of the slide. Place the narrow edge of another slide at a 20–30° angle on the first slide and to the left of the drop of blood. Pull to the right until the slide touches the blood. As soon as the blood has spread along the line of contact in the acute angle formed by the two slides, push the right hand toward the left. Push steadily until all the blood disappears or the other end of the slide is reached. Move the hand rapidly; if the smear is spread too slowly, the leukocytes concentrate along the edges and in the tail of the smear (Christophers 1956). This method drags the blood cells but does not run over and crush them. The hand can be kept from shaking by resting it on the table. Also, do not use a slide with a rough edge, for this produces streaks in the smear. If the blood seems thick, reduce the 20° angle to feed it out at a a slower rate. For thin blood, increase the angle (Fig. 16-1).

Dry the slides rapidly in the air; waving them facilitates drying and prevents crenation (notching or scalloping of edges) of the red cells.

194 Specific Staining Methods

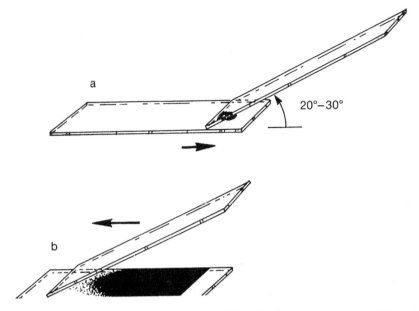

Figure 16-1. Preparation of thin smears: (a) Drop of blood is placed near one end of slide and a second slide pulled against it; (b) the second slide pulls the blood across first slide.

For cover-glass preparations (Fig. 16-2), place a small drop of blood in center of cover glass; place a second cover glass directly over the top of the drop at a slight angle to the bottom glass. Allow the blood to spread to edges of cover glasses. Quickly slide cover glasses apart. Air dry. See p. 343 for a similar method of spreading blood on slides.

Preferably, blood smears should be stained immediately or within 24 hours. If they must be stored, place them in a tight box away from dust.

Blood smears are commonly stained with a Romanovsky type of stain, or neutral stain (p. 89). Neutrality is essential, and therefore dilution is usually made with a buffer solution of a known pH. Distilled water is often too acid, and tap water, too alkaline. The smears require no fixation in the usual sense, since Wright stain (1902) includes both fixing agent (methyl alcohol) and stain.

A common practice is to leave stained blood smears uncovered, because some mounting media tend to fade Romanovsky stains after a period of time. The fading takes place slowly and is an inconvenience only if slides are to be stored for years. Plastic coatings have been recommended, but they tend to be uneven in thickness and are easily scratched. See Steinman (1955) for preparation of a plastic coating. Hollander (1963) adds 1% of 2,6-di-*ert*-butyl-*p*-cresol to the synthetic mountant to prevent the fading of blood smears. Try synthetic mountants (see p. 98).

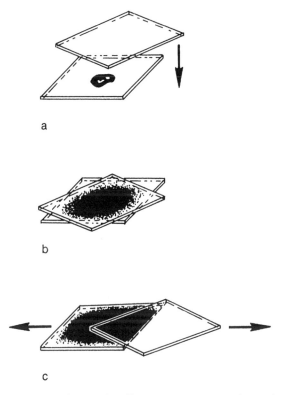

Figure 16-2. Preparation of cover-glass films: (a) A square cover glass is placed on top of blood on another square cover glass; (b) blood is allowed to spread; (c) top cover glass is pulled across and off bottom cover glass.

Thin Smear Method

SOLUTIONS

Wright stain

The stain may be purchased in two forms: (1) dry powder, (2) in solution. There are also quick stain kits available. The powder is ground up with methyl alcohol (0.1 g to 60.0 ml). The alcohol must be labeled *neutral* and *acetone-free*. Grind thoroughly in a glass mortar, and pour off supernatant (surface-floating) liquid. If undissolved powder remains, pour back the liquid and grind again. For consistency and ease of preparation, the commercially already-prepared solutions are recommended.

A slower but perhaps easier method is satisfactory. Add 0.3 g stain to 100.0 ml methyl alcohol and 3 ml glycerin in a stoppered bottle. Shake occasionally. After 24 hours, the solution is usually ready for use. To check whether a staining solution is satisfactory, with a pipette release a small drop of the solution onto filter paper. A good stain forms a deep blue spot, but a poor stain spreads into a broad pink halo around the central blue spot.

There are also quick staining kits available commercially that work quite well in class room situations.

For longer life store blood stains made up in methanol in the refrigerator.

Buffer solution, pH 6.8
    sodium phosphate, monobasic .... 7.0 g
    sodium phosphate, dibasic .... 3.0 g
    distilled water .... 1000.0 ml

Bohorfoush diluent (1963)
    sodium thiosulfate .... 0.1 to 0.2 g
    distilled water .... 1000.0 ml

Allow to stand 1 hour.

PROCEDURE
1. With wax pencil draw 2 marks across slide delineating region to be stained (length of 40-mm cover glass). (Optional step.)
2. Cover with 10 to 12 drops of Wright stain: 1 to 2 minutes.
3. Add an equal amount of buffer or diluent: 2 to 4 minutes. Blow gently on slide to mix and agitate solutions.
4. Rinse in distilled water or buffer solution, 1 or 2 dips. Precipitate deposit on the slide can be avoided by flushing the slide with a pipette, or carry the stain with the slide into the wash water; do not drain it off first.
5. Blot with bibulous paper; press but do not rub.
6. Allow slide to thoroughly dry before applying cover glass. (Optional.)

Note: For bone marrow, allow stain to remain on slide for 5 minutes.

ALTERNATE PROCEDURE
1. Fix dried film in methyl alcohol: 1 to 2 minutes. This procedure reduces red cell distortion and artifacts.
2. Dilute Wright stain with an equal amount of buffer. About 10 drops of each is sufficient for one slide. Use immediately.
3. Cover slide with diluted stain: 4 minutes.
4. Rinse and dry as in above procedure.

RESULTS
    erythrocytes—pink
    nuclei—deep blue or purple
    basophilic granules—deep purple

eosinophilic granules—red to red orange, bluish cytoplasm
neutrophilic granules—reddish brown to lilac, pale pink cytoplasm
granules of monocytes—azure
lymphocyte granules—larger and more reddish than monocyte granules, sky-blue cytoplasm
platelets—violet to purple

COMMENTS

Precipitate formation can be troublesome; the dark granules obscure the blood cells and are confusing in malarial smears. In addition to poor washing in step 4, another cause of precipitate may be evaporation during exposure to undiluted stain. Methyl alcohol is highly volatile and readily lost by evaporation. In dry warm weather use more Wright stain or shorten the time a bit, or cover the slides with a Petri dish. Rapid evaporation is easily detected, and more stain can always be added.

Proper rinsing in step 4 is one of the most important steps in this procedure. Never use tap water; it is too alkaline. Distilled water is usually satisfactory, but for best results rinse in the buffer, pH 6.4.

The longer the washing with water or buffer, the more stain is removed from the white cells. Only a dip or two is usually sufficient, but if the white cells are overstained, differentiate them by longer washing.

If the slides are overstained, the erythrocytes are too red, the white cells too pale, or the stain has precipitated, the slides can be recovered: (1) Cover the entire slide almost to excess with additional Wright stain: 15 to 30 minutes; (2) rinse with distilled water or buffer; (3) dry (Morrison and Samwick 1940).

Wright Stain for Cold-blooded Vertebrates (Heady and Rogers 1962)

PROCEDURE
1. Cover smears with Wright stain: 2 minutes.
2. Add equal portion of phosphate buffer, pH 6.7: 3 minutes (p. 473).
3. Wash in distilled water or buffer solution and blot as above.

Wright Stain for Birds (Santamarina 1964)

SOLUTIONS
Wright stain
    Wright stain powder .... 3.3 g
    absolute methyl alcohol .... 500.0 ml

This solution is more concentrated than that used for human blood.

Formalin solution
    formaldehyde (37–40%) .... 0.25 ml
    distilled water .... 500.0 ml

Adjust to pH 6.8 by adding either 0.25% sodium carbonate or 0.25% hydrochloric acid as needed.

Procedure
1. Cover dried smears with Wright solution: 8 minutes.
2. Slowly add formalin solution with a medicine dropper. Do not allow solution to overflow.
3. When a metallic sheen covers entire surface of fluid, pour off and flush immediately with distilled water adjusted to pH 6.8 as above.
4. Blot.
5. Differentiate in ether-absolute methyl alcohol (1:1). Dip slides up and down 6 to 10 times. Examine them under microscope; if they are too bluish, differentiate further.
6. Clear and mount.

Results
erythrocytes—yellow to pinkish cytoplasm, purple chromatin
thrombocytes—gray-blue cytoplasm, purple chromatin
lymphocytes—blue granules, purplish red chromatin
monocytes—light-blue granular cytoplasm, purple chromatin
heterophils—yellowish to brownish red rods, light-purple chromatin
eosinophils—similar to heterophils
basophils—dark-purple granules

Preparation of Thick Blood Films

With this type of film, a relatively large quantity of blood is concentrated in a small area, thereby increasing the possibilities of finding parasites. The concentration and timing of staining are adjusted so the action is stopped at the point when the leukocytes have stained, some hemoglobin has been dissolved, and the red cell membranes have not yet begun to stain. At this point the leukocytes, platelets, and protozoa are stained and lie on an unstained or very lightly stained background, which is yellowish from the remaining hemoglobin. Freshly prepared films stain better than films one or more days old.

Puncture the skin deeply enough to form a large drop of blood using sterile and safety precautions. On a slide, cover a space the size of a dime with enough blood (about 3 to 4 average drops) to spread easily. Too much blood will crack and peel off when dry. Smear it by circling the slide under the finger without making actual contact. Some find it easier to swirl the blood with a dissecting needle or the corner of a slide. Burton (1958) taps the slide against the table top, so the drop of blood flows transversely across the slide for 1 to 2 cm. Practice will determine the best method and how much blood to use.

An ideal film is several cell layers thick in the center, tapering off to one cell thickness at the periphery.

Allow the slide to dry in a horizontal position; if tilted, the blood will ooze to one edge of the film. Protect from dust and flies, and do *not* fix by flaming or alcohol.

Field-Wartman Thick Film Method (Field 1941; Wartman 1943)

SOLUTIONS
Solution A

methylene blue	0.8 g
azure A	0.5 g
dibasic sodium phosphate, anhydrous	5.0 g
monobasic potassium phosphate	6.25 g
distilled water	500.0 ml

Solution B

eosin Y	1.0 g
dibasic sodium phosphate, anhydrous	5.0 g
monobasic potassium phosphate	6.25 g
distilled water	500.0 ml

Dissolve phosphate salts first, then add stains. (The azure will go into solution by grinding in glass mortar with a small amount of phosphate solvent.) Set aside for 24 hours. Filter. Store in refrigerator.

PROCEDURE
1. Dip slides in solution A: 1 second.
2. Rinse gently in clean distilled water: few seconds, or until stain ceases to flow from film.
3. Dip in solution B: 1 second. Use this solution with care; it tends to decolorize the leukocytes stained with methylene blue-azure, and accelerates the dissolving of hemoglobin.
4. Rinse gently in distilled water: 2 to 3 drops.
5. Dry, do not blot. Stand slides on end and allow to air dry.
6. When completely dry, add mounting medium and cover glass.

RESULTS
There will be a thicker, central area of partially laked blood. This may not be well suited for examination. But the surrounding area, and especially at the edge toward which the hemoglobin has drained, will be creamy, sometimes mottled with pale blue. This is the best area for study.

Leukocytes
cytoplasm—pale blue, poorly defined
nuclei—dark blue, well defined
eosinophilic granules—bright red, large, well defined

neutrophilic granules—pale purple pink, small, indistinct
basophilic granules—deep blue with reddish cast.

Platelets
platelets—pale purple or lavender

Parasites
cytoplasm—blue
chromatin—purplish red or deep ruby red
pigment—unstained yellow granules of varying intensity

Wilcox Thick Film Method (1943)

SOLUTIONS
Wright-Giemsa stock solution

Giemsa	2.0 g
glycerin	100.0 ml

Heat in water bath, 55–60°C: 2 hours, stirring at intervals. Avoid absorption of moisture by covering mouth of flask with laboratory film (Parafilm). Add:

aged Wright staining solution (2.0 g/1000.0 ml methyl alcohol)	100.0 ml

Let stand overnight; add in addition:

Wright staining solution	800.0 ml

Filter; ready for use.

Working solution

Wright-Giemsa stock solution	1 part
neutral distilled water (buffer)	9 parts

Buffer, pH 7.0
Solution A

dibasic sodium phosphate, anhydrous	9.5 g
distilled water	1000.0 ml

Solution B

monobasic potassium phosphate	9.7 g
distilled water	1000.0 ml

Working Solution

> Solution A .... 61.1 ml
> Solution B .... 38.9 ml
> distilled water .... 900.0 ml

PROCEDURE
Short method

1. Stain: 10 minutes.
2. Flush scum from top of solution with buffer to avoid picking up precipitate on slides.
3. Remove slides to neutral distilled water: 1 minute.
4. Dry slides standing on end. Do not blot. (Mount if desired.)

Long method (This method produces a more brilliant stain.)

1. Stain: 45 minutes in dilution of 1 to 50 of buffer.
2. Flush off scum and transfer slides to buffer: 3 to 5 minutes.
3. Dry as above.

Thin smear and thick film can be stained on same slide. Fix thin portion in methyl alcohol for 1 to 2 minutes, taking care that thick film does not come in contact with alcohol. Stain as above (short method), but shorten washing time to 2 to 3 dips so thin smear is not too light; this shorter washing time will leave a deeper background in the thick film.

COMMENTS
Modifications of this method have been made by Fenton and Innes (1945), Manwell (1945), and Steil (1936).
Old smears will stain more brilliantly if treated 5 to 10 minutes in alcohol-acetic solution—10 drops glacial acetic acid to 60 ml absolute alcohol.

BLOOD TISSUE ELEMENTS AND INCLUSION BODIES

Giemsa Stain (Cramer et al. 1973)

FIXATION
Any good general fixative, but Zenker preferred.

SOLUTIONS
Azure-eosinate stock solution

> azure II-eosin .... 2.0 g
> azure II .... 1.0 g

azure B-eosin	1.0 g
azure A-eosin	0.5 g

Dissolve in equal parts of glycerin and methanol, 250 ml each, and allow to stand at room temperature overnight. Shake well 5 to 10 minutes and store unfiltered in a brown, well-capped bottle at room temperature. Stable for months.

Working solution A (2-hour stain)
stock solution	5.0 ml
distilled water	65.0 ml

pH should be 4.8 to 5.2. If greater than 5.2, add 1 to 2 drops of 1% acetic acid.

Working solution B (overnight stain)
stock solution	3 to 5 drops
distilled water	65.0 ml

pH should be 6.5 to 6.8; adjust with 1% acetic acid if needed.

PROCEDURE
1. Deparaffinize and hydrate sections to distilled water.
2. Stain in working solution A: 2 hours; in B: overnight.
3. Dip quickly in 1% acetic acid.
4. Blot immediately and rinse several times in absolute alcohol until only slightly bluish tint appears in alcohol running off slides.
5. Clear and mount.
6. If too blue, return through xylene and absolute alcohol into several very quick dips in 0.5–1% acetic acid. Blot, and return through absolute alcohol to xylene.

RESULTS
nuclear chromatin—dark blue
eosinophil granules—red orange
mast cell granules—dark purple
erythrocytes—pink to red
connective tissue—pink to light purple
lymphocytes—blue cytoplasm

COMMENTS
Acetic acid removes the blue colors rapidly; this has to be carefully controlled. The red tones are removed rapidly by alcohol. Too red a color, therefore, means overdifferentiation. Try restaining.

Jenner-Giemsa Stain for Malarial Parasites (McClung 1939)

FIXATION

Smears may be fixed during staining process. Fix tissue sections in any good general fixative, preferably one containing mercuric chloride and alcohol.

SOLUTIONS

Jenner solution

Jenner stain	0.2 g
methyl alcohol (neutral, acetone-free)	100.0 ml

Giemsa stock solution

Giemsa powder	3.8 g
methyl alcohol	75.0 ml
glycerin	25.0 ml

Work stain into glycerin, warm for 2 hours, 60°C oven. Add methyl alcohol.

Working solution

Giemsa stock solution	10.0 ml
distilled water	100.0 ml

PROCEDURE

1. Deparaffinize sections and run down to 50% alcohol. Remove $HgCl_2$ if present.
2. a. Sections—flood with ample amount of Jenner solution and add an equal amount of distilled water, or mix equal amounts of Jenner solution and water and flood slides or stain slide in Coplin jar: 4 minutes.
   b. Smears—flood with Jenner solution: 3 minutes. Add equal amount of distilled water: 1 minute.
3. Pour off Jenner; do not rinse.
4. Flood with diluted Giemsa solution or place in Coplin jar of solution: 15 to 20 minutes.
5. Rinse off stain with distilled water and continue to differentiate with distilled water. If too blue, eosin color can be brought out by rinsing in 0.5–1% acetic acid in water.
6. Dehydrate sections in following sequence of acetone and xylene.
   acetone 95 parts: xylene 5 parts
   acetone 70 parts: xylene 30 parts
   acetone 50 parts: xylene 50 parts
   acetone 30 parts: xylene 70 parts
   acetone 5 parts: xylene 95 parts
   Smears may be blotted and air-dried before mounting.

Suggestion: In place of the above solutions, use 3 changes isopropyl alcohol, xylene-isopropyl alcohol (1:1). Do not use ethyl or methyl alcohol, which will extract the stain.
7. Clear in xylene and mount.

RESULTS

chromatin of parasite—red or purplish red
cytoplasm of parasite—blue (other elements—color similar to Wright stain)
pigment—yellow brown to black
Schüffner's stippling—red

COMMENTS

A beautiful clear stain for any blood picture, particularly blood parasites.

For a fluorescent antibody technique, see Ingram et al. (1961) and Sodeman and Jeffery (1966). For a fluorescent stain see Hansen et al. (1970). An excellent lab manual on malaria techniques is Shute and Maryon (1966).

Giemsa Staining for Cold-blooded Vertebrates (Pienaar 1962)

1. Fix dried smears in undiluted Wright (p. 195) or Jenner (p. 203) stock stains: 3 to 5 minutes. (Smears may be previously fixed in absolute methyl alcohol.)
2. Add equal volume of distilled water buffered to pH 6.2 to 6.8: 5 to 6 minutes (pH is important).
3. Pour off solution, but do not rinse. Stain in diluted Giemsa stock (p. 203) (1 ml/10 ml distilled water): 20 to 30 minutes.
4. Flush with distilled water to prevent stain precipitate on slide.
5. Differentiate, if necessary, in distilled water, pH 5.2.
6. Rinse well in distilled water. Blot, dry, and mount.

Pienaar recommends the following modification for snake eosinophils, which are distorted in the above method.

1. Jenner solution: 1 to 3 minutes.
2. Buffered water: dilution: 10 minutes.
3. Dilute Giemsa: 10 to 15 minutes.
4. Rinse briefly in distilled water. Dehydrate rapidly through acetone, xylene-absolute alcohol (1:1), xylene. Mount.

Giemsa Staining for Birds (Lucas and Jamroz 1961)

1. Fix dried smears in Wright stain: 4 minutes.
2. Add distilled water: 4 to 5 minutes.
3. Wash in running water: 2 minutes.

4. Stain in dilute Giemsa (15 drops/10 ml water): 15 to 20 minutes.
5. Wash in distilled water. If precipitate is present, run Wright stain over the slide. Wash off with running water.
6. Stand slides on edge to dry. Do not blot.

Reticulocyte Staining (Brecher 1949)

SOLUTION

new methylene blue N	0.5 g
potassium oxalate	1.6 g
distilled water	100.0 ml

PROCEDURE
1. Mix equal amounts of stain and fresh or heparinized blood on a slide. Draw mixture into a capillary pipette and allow to stand: 10 minutes.
2. Expel mixture on a slide and mix again.
3. Make thin smears. Air dry.

RESULTS
reticulum—deep blue
red cells—light greenish blue

COMMENTS

If slides fade, fix in methanol and restain with Wright stain. The slides can be stained with Wright immediately after the air drying in step 3 if desired.

Heinz bodies are demonstrated by this stain, pale to deep blue against the pale green background of the red cells (Thompson 1961). Simpson et al. (1970) recommend Rhodanile blue as a selective stain for Heinz bodies.

Lucas and Jamroz (1961) use the following for birds: Mix a drop of blood with a drop of 1% brilliant cresyl blue in 0.85% NaCl or Ringer avian solution (p. 466). Let stand 2 minutes and smear. Dry, fix in methyl alcohol, and stain with Wright.

For cold bloods (Pienaar 1962), mix a few drops of blood and stain together in equal quantity in a small covered dish, and leave 5 minutes. Make up the stain, 1% in Ringer or Locke solution for cold blooded organisms (pp. 465–66). Make smears on slightly warm slides. Dry, fix in methyl alcohol, and stain with Wright.

HEMOGLOBIN STAINING

Buffalo Black for Hemoglobin (Puchtler et al. 1964)

FIXATION
Any general fixative.

Solutions

Tannic acid

tannic acid	5.0 g
distilled water	100.0 ml

Phosphomolybdic acid

phosphomolybdic acid	1.0 g
distilled water	100.0 ml

Buffalo black

Saturate a solution of buffalo black NBR (naphthol blue black) in methanol-glacial acetic acid (9:1). Let stand 48 hours. Do not filter. Keeps several months. Filter the used portions back into the storage bottle.

Procedure

1. Deparaffinize and hydrate to water.
2. Treat with tannic acid: 10 minutes.
3. Transfer through 3 changes of distilled water.
4. Treat with phosphomolybdic acid: 10 minutes.
5. Transfer through 3 changes of distilled water.
6. Stain in buffalo black: 5 minutes.
7. Differentiate in methanol-acetic acid (9:1), 2 changes: 5 minutes total.
8. Place in absolute ethanol or propanol, clear, and mount.

Results

hemoglobin—dark blue to purplish black
other tissues—yellow

Comments

Do not use freshly prepared tannic acid. The tannic acid should be more than 48 hours old; it keeps indefinitely.

Puchtler and Sweat (1963a) combine hemoglobin staining with a ferrocyanide staining for hemosiderin.

Cyanol Reaction (Dunn 1946)

Fixation

Formalin, preferably buffered to pH 7; fix no longer than 48 hours. Dried smears may be methyl alcohol fixed and then taken to step 2 below.

Solutions

Cyanol reagent stock solution

cyanol	1.0 g
distilled water	100.0 ml

zinc powder .... 10.0 g
glacial acetic acid .... 2.0 ml

Bring to boiling point. In a short time the solution is decolorized. Stable for several weeks.

Working reagent
Filter 10 ml of stock reagent, and add:

glacial acetic acid .... 2.0 ml
hydrogen peroxide, commercial (30%) .... 1.0 ml

PROCEDURE
1. Deparaffinize and hydrate slides down to water.
2. Treat with working reagent: 3 to 5 minutes.
3. Rinse in distilled water.
4. Counterstain in red nuclear stain.
5. Dehydrate, clear, and mount.

RESULTS
hemoglobin—dark blue to bluish gray

COMMENTS
Gomori (1952) uses 1.0 g acid fuchsin instead of cyanol and produces a red hemoglobin. Counterstain with hematoxylin or other blue nuclear stain.

Fetal Hemoglobin (Modified from Kiossoglou et al. 1963)

FIXATION
Air dry blood smears for 1 hour. Fix in absolute methanol: 5 minutes; follow by 80% ethanol: 5 minutes.

SOLUTIONS
Buffer
McIlvaine citric acid phosphate, pH 3.2 to 3.4, p. 474.

Stain
eosin Y .... 0.5 g
distilled water .... 100.0 ml

or

Ponceau S .... 0.5 g
1% aqueous acetic acid .... 100.0 ml

208 Specific Staining Methods

PROCEDURE
1. Rinse fixed slides in tap water. Follow by a rinse in distilled water.
2. Treat in buffer, 37°C: 5 minutes. Slides must be in vertical position.
3. Rinse in tap water.
4. Stain for 3 minutes.
5. Rinse in tap water: few seconds. Dry standing in vertical position.

RESULTS
cells containing fetal hemoglobin—deep pink to red
adult cells—colorless or pale-staining ghosts

COMMENTS
The smears can be stained with Mayer hematoxylin for 5 minutes before staining with eosin or Ponceau S. Also they may be cleared and mounted. Count 500 cells and express the number of F-containing cells as a percentage of the total. Smears must be fresh; if left unstained overnight they can give false determination.

BONE MARROW STAINING

Maximow Eosin-Azure Stain (Block et al. 1953)

FIXATION
Zenker-neutral formalin: 30 minutes. Neutralize by adding 2 g lithium or magnesium carbonate to 500 ml of formalin. Excess of carbonate should be present. Wash in running water 1 to 24 hours before preparation for embedding (see comment 1).

SOLUTIONS
Solution A
- eosin Y or eosin B .... 0.1 g
- Wright buffer .... 100.0 ml

Solution B
- azure II .... 0.1 g
- Wright buffer .... 100.0 ml

Working solution
- Wright buffer .... 2.0 ml
- distilled water .... 40.0 ml
- solution A .... 8.0 ml

Stirring vigorously, add gradually

- solution B .... 4.0 ml

Fresh solutions are best; their action deteriorates after 3 or 4 weeks. Working solution should appear deep violet in color, and a precipitate should not form in it for an hour or more. If a precipitate forms on the slides, the stain mixture was improperly made; solution B was added to solution A too rapidly or without stirring. If the eosin loses its brilliance, solutions are old. Store solutions A and B in the refrigerator.

PROCEDURE
1. Deparaffinize and hydrate slides to water: remove $HgCl_2$. If tissue was fixed in a fixative without potassium dichromate, chromate slides overnight in 2.5–3% aqueous potassium dichromate. Wash thoroughly in running water: 15 minutes. Proceed to step 2.
2. Stain in Mayer hematoxylin: 30 to 45 seconds, no more.
3. Wash in running water: 5 to 10 minutes.
4. Wash in distilled water: 5 or more minutes.
5. Stain in eosin-azure overnight.
6. Differentiate in 95% alcohol. In stubborn cases of differentiation (old solution) a brief immersion in colophonium alcohol (p. 480) may help to sharpen the colors. Differentiation must be done under the microscope.
7. Dehydrate in absolute alcohol, clear, and mount. If, after mounting, the stain still appears undifferentiated, remove the cover glass and mountant with xylene. Transfer to absolute alcohol: 1 to 2 minutes. Transfer for a brief period to 95% alcohol. Examine the slide under a microscope immediately, since the tissue destains rapidly after this treatment. Dehydrate, clear, and mount.

RESULTS
nuclei—dark purple blue
erythrocytes—light pink
eosinophilic granules—brilliant red
cytoplasm—pale blue

COMMENTS
1. After 30-minute fixation, cells in center are not as well fixed as those at periphery, but longer fixation produces granular cytoplasm in eosinophilic erythrocytes and erythroblasts. If 30-minute fixation is not adequate for fixing bone marrow contained in bone, follow Zenker-formol fixation by overnight treatment with 3% aqueous potassium dichromate. Also crack the bone and break away a section of the bone to permit penetration of the fixative.
2. Block (personal communication) embeds in methacrylate. This is recommended for thin sections of 1 to 3 μm with a minimum of shrinkage in the tissue and with precise cellular detail (see p. 453).

3. Lillie (1965) stains 1 hour in the following eosin-azure solution:

```
azure A, 0.1% .................................................. 4.0 ml
eosin B, 0.1% .................................................. 4.0 ml
0.2 M acetic acid .............................................. 1.25 ml
0.2 M sodium acetate ........................................... 0.75 ml
acetone ........................................................ 5.0 ml
distilled water ................................................ 25.0 ml
```

The solution should be made up immediately before use. Transfer to acetone, 2 changes; acetone-xylene (1:1), and xylene. Lillie alters the amount of acetic acid and sodium acetate in the solution. The above mixture is for material fixed in Zenker. The proportions of acetic acid and sodium acetate should be adjusted following other fixatives (Lillie 1965).

4. Bone marrow when collected should have an anticoagulant added to it. Gardner (1958) and many other workers mix the bone marrow in a tube containing 0.5 mg heparin powder. Kinsely (personal communication) wets his syringe with d-potassium EDTA as an anticoagulant. Smears can be made with some of the aspirated (drawn out by suction) marrow. The remaining material is poured into a small funnel lined with filter paper or lens paper. Rinse marrow, while in funnel, several times with saline, or until the marrow has lost most of its color. This also helps to wash the material into one mass at the apex of the funnel. Partially clotted marrow will be broken up and washed free of blood, leaving excellent clear particles of marrow. Fold in the filter paper, place in tissue receptacle and carry through fixation, washing, dehydration and infiltration without removing from filter paper. No marrow will be lost. Minute amounts of marrow may prove difficult to recover from the filter paper without loss of material. Humason allowed the paraffin around the marrow (on the filter paper) to congeal slightly, then carefully scraped paraffin and marrow together, and moved paraffin with marrow spicules intact to the mold. The congealed paraffin containing the marrow will sink to the bottom of the mold, or it can be pressed down into desired position. This method keeps the marrow concentrated reasonably well. For other methods for handling bone marrow, see Berthrong and Barhite (1964), Conrad and Crosby (1961), Leach (1960), and Raman (1955).
5. Gude and Odell (1955) recommend for dilution a 3.5% solution of polyvinylpyrrolidone (PVP) in normal saline. Endicott (1945) used plasma serum as a diluting fluid to thin smears. The serum can be kept on hand for several months if stored in the refrigerator.
6. For a fluorescent method, see Werth (1953).

Phloxine-Methylene Blue (Thomas 1953)

FIXATION

Any general fixative, Zenker preferred; or mordant slides in Zenker after hydration.

SOLUTIONS

Phloxine solution

phloxine B	0.5 g
distilled water	100.0 ml
glacial acetic acid	0.2 ml

A slight precipitate will form. Filter before use. (This is relatively unimportant because the precipitate settles at the bottom of the bottle and does not collect on the tissue.)

Methylene blue-azure solution

methylene blue	0.25 g
azure B	0.25 g
borax	0.25 g
distilled water	100.0 ml

PROCEDURE

1. Deparaffinize and hydrate slides to water; remove $HgCl_2$.
2. Stain in phloxine: 1 to 2 minutes.
3. Rinse well in distilled water.
4. Stain methylene blue-azure: 0.5 to 1 minute.
5. Partially destain in 0.2% aqueous acetic acid.
6. Complete differentiation in 95% alcohol, 3 changes.
7. Dehydrate, clear and mount.

RESULTS

nuclei—blue
plasma cell cytoplasm—blue
other tissue elements—shades of rose and red

COMMENTS

Thomas substitutes azure B for azure II of other methods because of the uncertain composition of the latter. Humason recommended Thomas' method over others. It is practically foolproof. See Greenstein's (1957) modification. He uses an alcoholic phloxine solution and mixes equal parts of separate solutions of methylene blue and azure B.

Delez and Davis (1950) make up their phloxine with oxalic acid; 1% phloxine in 0.05% aqueous oxalic acid.

This staining method is frequently used for general staining in place of a hematoxylin-eosin method.

STAINING FOR FIBRIN

Ledrum Acid Picro-Mallory Method (Culling 1957)

FIXATION
Any good general fixative.

SOLUTIONS
Celestin blue
celestin blue B	0.25 g
ferric alum, 5% (5 g/100 ml water)	50.0 ml

Boil 3 minutes. Cool and filter. Add:

glycerin	7.0 ml

Keeps for several months.
Mayer hematoxylin, see p. 102.

Picro-orange
picric acid saturated in 80% alcohol	100.0 ml
orange G	0.2 g

Acid fuchsin
acid fuchsin	1.0 g
distilled water	100.0 ml
trichloroacetic acid	3.0 g

Phosphotungstic acid
phosphotungstic acid	1.0 g
distilled water	100.0 ml

Aniline blue
aniline blue	2.0 g
distilled water	100.0 ml
glacial acetic acid	2.0 ml

PROCEDURE
1. Deparaffinize and hydrate slides to water; remove $HgCl_2$.
2. Stain in celestine blue: 3 to 5 minutes.
3. Rinse in tap water.

4. Stain in Mayer hematoxylin: 5 minutes.
5. Wash in running water: 5 minutes.
6. Rinse in 95% alcohol.
7. Stain in picro-orange: 2 minutes.
8. Stain in acid fuchsin: 5 minutes.
9. Rinse in distilled water.
10. Dip in equal parts of picro-orange and 80% alcohol: few seconds.
11. Differentiate in phosphotungstic acid: 5 to 10 minutes, until colors are clear.
12. Rinse in distilled water.
13. Stain in aniline blue: 2 to 10 minutes.
14. Rinse in distilled water: 2 to 3 minutes.
15. Dehydrate, clear, and mount.

Results
fibrin—clear red
erythrocytes—orange
collagen—blue
nuclei—blue black

Comments
This method is specific for fibrin, setting it off sharply from other tissue elements. See also phosphotungstic acid-hematoxylin (p. 114), the Gram-Weigert method (p. 303), and the Rosindole method (Puchtler et al. 1961). The latter is also considered specific for fibrin.

Several of the enzyme histochemical procedures are very useful with blood smears. These will be covered in Chapter 23.

17
Staining Pigments and Minerals

Pigments are a heterogeneous group of substances containing enough natural color to be visible without staining. Sometimes, however, color is added to give them more intense differential staining. Some pigments are artifacts, such as formalin pigment (p. 21). Others, exogenous pigments, are foreign pigments that have been taken into the tissue in some manner. Carbon is a common pigment found in the lungs of city dwellers, particularly of people from coal-burning cities. Endogenous pigments are found within the organism and arise from nonpigmented materials. Iron-containing hemoglobin can be broken down into iron-containing pigment, hemosiderin, and a brown pigment containing no iron, called hematoidin, or bilirubin, which can be oxidized to biliverdin (green). Normal hemoglobin, when not broken down into hemosiderin, will not show a positive Prussian blue reaction (p. 215). The iron is masked or occult, and the organic part of the hemoglobin molecule must be destroyed for the iron to be unmasked.

Melanin (brown or black pigment) is found normally in the skin, hair, and eye, but may occur pathologically anywhere in the body. The pigment is formed from tyrosine by the enzyme, tyrosinase. Lipofuscin, sometimes known as the "wear and tear" pigment, can be found in the heart muscle, adrenals, ganglion cells, and liver. The lipofuscins are derived from lipid or lipoprotein sources, at least partly by oxidation. Several enzymes are found in them, nonspecific esterases and acid phosphatase in particular. Melanin and lipofuscin are brownish pigments stainable by fat dyes and some basic aniline dyes such as fuchsin. They are metachromatic (p. 271) with methyl green and give a positive Schiff reaction after periodic acid treatment (p. 183).

Calcium may be present in tissues in a number of different forms, but the methods here are applicable to the insoluble forms—deposits of calcium phosphate and calcium carbonate (Ham 1957).

STAINING FOR IRON

According to Baker (1958), the iron reaction is an example of local formation of a colored substance which is not a dye. It is a type of histochemical test wherein a tissue is soaked in a colorless substance. Certain tissue elements react with the substance and become colored. This well-known test is called the Berlin blue, Prussian blue, or Perl reaction: The iron is dissolved from hemosiderin by hydrochloric acid, and then reacts with potassium ferrocyanide to form the Berlin blue precipitate, ferric ferrocyanide.

Sometimes fading occurs in slides due to the reduction of the Berlin blue to colorless ferrocyanide. The mounting resin probably takes up oxygen while drying and deprives the sections of oxygen, thereby reducing them to the colorless condition. If this takes place and it is essential to recover the slides, treat them with hydrogen peroxide. The newer synthetics are not as prone to reduce Berlin blue as former resins.

Iron Reaction (Hutchinson 1953)

FIXATION

Hutchinson recommends:

> sodium sulfate .... 12.0 g
> glacial acetic acid .... 33.0 ml
> formaldehyde (37–40%) .... 40.0 ml
> distilled water to make .... 200.0 ml

Go directly into 70% alcohol from fixative.

SOLUTIONS

Ferrocyanide-hydrochloric acid solution
Solution A

> potassium ferrocyanide .... 2.0 g
> distilled water .... 50.0 ml

Solution B

> hydrochloric acid .... 2.0 ml
> distilled water .... 48.0 ml

Prepare the two parts of the solution separately, mix them together, warm slightly, and filter. Place in paraffin oven, 56°C, a short time before using.

Safranin
> safranin O .... 0.2 g
> distilled water .... 100.0 ml
> glacial acetic acid .... 1.0 ml

Kernechtrot
> Kernechtrot (nuclear fast red) .... 0.1 g
> 5% aluminum sulfate (5 g/100 ml water) .... 100.0 ml

Dissolve with heat. Cool and filter.
Add a crystal of thymol as preservative.

216 Specific Staining Methods

Control
known positive

Procedure
1. Deparaffinize and hydrate slides to water.
2. Wash in distilled water: 3 minutes.
3. Treat with ferrocyanide-hydrochloric acid, 56°C: 10 minutes.
4. Rinse, several changes distilled water: 5 minutes.
5. Counterstain with safranin: 2 to 5 minutes, or Kernechtrot: 5 to 8 minutes.
6. Rinse in 70% alcohol.
7. Dehydrate, clear, and mount.

Results
iron pigment—brilliant greenish blue
nuclei—red
other tissue elements—shades of red and rose

Comments
Hutchinson claims that warming the solution is the most important step in this method. Do not leave the slides in the solution longer than 10 minutes. If they have been well washed in distilled water, precipitate seldom forms on the sections. This method produces deep, brilliant colors.

For masked or occult iron (Glick 1949), pretreat slides with 30% hydrogen peroxide alkalized with dilute ammonia (1 drop/100 ml peroxide): 10 to 15 minutes. Wash well and proceed to step 3. If the hydrogen peroxide tends to form bubbles under the sections and causes detachment, keep the solution and slides cool in the refrigerator during the treatment.

Humason preferred 3% nitric acid in 95% alcohol overnight. Wash in 90% alcohol and proceed as usual. This method does not unmask iron in hemoglobin.

Past (1961) decalcifies osseous tissue in 5% EDTA (ethylenediaminetetraacetic acid) adjusted to pH 12 with N NaOH before staining for iron.

Turnbull Blue Method for Ferrous Iron (Pearse 1968)

Fixation
10% (neutral) buffered formalin (NBF); other general fixatives satisfactory if acid is absent.

Solutions
Ammonium sulfide
A saturated solution of $(NH_4)_2S$, 20–30% content, analytical reagent

Potassium ferricyanide
potassium ferricyanide ($K_3Fe(CN)_6$) .... 20.0 g
distilled water .... 100.0 ml

Hydrochloric acid

hydrochloric acid (concentrated)	1.0 ml
distilled water	99.0 ml

Safranin or Kernechtrot, see p. 215

CONTROL
known positive

PROCEDURE
1. Deparaffinize and hydrate slides to water; remove $HgCl_2$.
2. Wash in distilled water.
3. Treat with yellow ammonium sulfide: 1 to 3 hours.
4. Rinse in distilled water.
5. Treat with equal parts of potassium ferricyanide and hydrochloric acid, freshly mixed: 10 to 20 minutes.
6. Rinse in distilled water.
7. Counterstain with safranin: 2 to 5 minutes, or Kernechtrot: 5 to 8 minutes.
8. Rinse in 70% alcohol.
9. Dehydrate, clear, and mount.

RESULTS
ferrous iron and ferric iron are converted to ferrous iron—deep blue
other tissue elements—shades of rose and red

COMMENTS
Kutlík (1970) stains for 2 hours at room temperature in chlorate hematoxylin (1% hematoxylin in 7.3% aqueous potassium chlorate, filtered). Wash well and counterstain. This forms a black iron hematin lake where iron compounds are present. The solution will keep for 6 months.

BILE PIGMENT (BILIRUBIN) STAINING

Glenner Method (1957)

FIXATION
10% neutral buffered formalin

SOLUTIONS

potassium dichromate, 3% aqueous	25.0 ml
buffer, pH 2.2 (0.1 N HCl, 8.0 ml; 0.1 N potassium dihydrogen phosphate, 17.0 ml)	25.0 ml

218   Specific Staining Methods

CONTROL
   tissue containing bile

PROCEDURE
1. Treat sections with dichromate solution, room temperature: 15 minutes.
2. Wash in running water: 5 minutes.
3. Counterstain with hematoxylin, safranin, or Kernechtrot.
4. Dehydrate rapidly through 95% alcohol: 5 to 10 seconds.
5. Finish dehydration, clear and mount.

RESULT
   bilirubin—emerald green

COMMENTS

The pH of 2.2 is crucial; it must be low enough to result in complete oxidation of all bilirubin, but not so acid that it causes loss of the pigment from the sections. Glenner also includes a method for the demonstration of bilirubin, hemosiderin, and lipofuscin in the same section.

This reaction is recommended in preference to Stein's iodine method (Glick 1949), which cannot be considered reliable at all times because the reactants are diffusible and the final color can spread from the original site. The test is based on oxidation of the bile pigment (bilirubin) to green biliverdin by the iodine solution.

Kutlík (1968) and Lillie and Pizzolato (1968) describe argentaffin methods (p. 264). For definite blackening of the bile pigment, Lillie and Pizzolato recommend 5 to 16 hours exposure to diammine silver at 25°C. The solution is made as follows: to 2.0 ml of 28% NH_4OH add 35 to 40 ml of 5% $AgNO_3$, rapidly at first, with swirling to dissolve the precipitate that forms. Then add smaller amounts of $AgNO_3$ until only a faint turbidity persists.

See Lillie and Pizzolato (1968) for a more complete discussion of bile pigments.

Azo-Coupling Method

FIXATION
   10% neutral buffered formalin

SOLUTIONS
   Buffer solution
   Solution A

   sodium acetate .... 1.17 g
   sodium barbital .... 2.94 g
   distilled water .... 100.0 ml

Solution B

 hydrochloric acid (concentrated) .... 0.84 ml
 distilled water .... 100.0 ml

Working solution, pH 9.2

 solution A .... 5.0 ml
 solution B .... 0.25 ml
 distilled water .... 17.75 ml

Staining solution
 fast red B .... 50.0 mg
 buffer, pH 9.2 .... 50.0 ml

CONTROL
known positive

PROCEDURE
1. Deparaffinize and hydrate sections to water or cut frozen sections.
2. Stain: 30 seconds.
3. Wash off excess stain in tap water.
4. Stain in hematoxylin: 5 minutes.
5. Wash in tap water: 5 minutes.
6. Dehydrate, clear, and mount.

RESULTS
bile pigments—orange red
nuclei—blue

MELANIN AND LIPOFUSCIN STAINING

Lillie (1956a,b) **Nile Blue Method**

FIXATION
Any good general fixative.

SOLUTION
 Nile blue A .... 0.05 g
 sulfuric acid, 1% (1 ml conc. H_2SO_4/99 ml distilled water) ...... 100.0 ml

CONTROL
skin (melanin), liver (lipofuscins)

220   Specific Staining Methods

Procedure

1. Deparaffinize and hydrate slides to water; remove $HgCl_2$.
2. Stain in Nile blue A solution: 20 minutes.
3. Wash in running water: 10 to 20 minutes.
4. Mount in glycerin jelly.

Results

lipofuscins—dark blue or blue green
melanin—pale green
erythrocytes—greenish yellow to greenish blue
myelin—green to deep blue
nuclei—poorly stained

Lillie Alternate Nile Blue Method

Control

skin (melanin), liver (lipofuscins)

Procedure

Steps 1 and 2 as above.

3. Do not wash in water. Rinse quickly in 1% sulfuric acid to remove excess dye.
4. Dehydrate at once in acetone, 4 changes: 15 seconds each.
5. Clear in xylene and mount.

Results

cutaneous, ocular, meningeal melanins—dark green
mast cells—purple red
lipofuscins—unstained but appear yellow to brownish
muscle, myelin, erythrocytes—unstained
nuclei—greenish to unstained

Comments

Nile blue sulfate stains basophilic tissue and acidic lipids, such as free fatty acids, phospholipids, and certain lipid components of some pigment deposits of a lipogenic character.

Lipofuscins stain with Nile blue by two mechanisms: a fat solubility mechanism operating at pH below 1.0; and an acid-base mechanism operating at levels above pH 3.0. When stained by the second mechanism, they retain a green stain after acetone or brief alcoholic extraction, but when the first mechanism is used, they are promptly decolorized by acetone or alcohol (Lillie 1956b, see also Lillie 1965).

Melanins stain with basic dyes at pH levels below 1.0 and retain the stain when dehydrated and mounted (Lillie 1955).

Ferro-Ferricyanide Method (Lillie 1965)

Fixation

Avoid chromate fixatives; others are satisfactory.

Solutions

Ferrous sulfate

ferrous sulfate (FeSO$_4$ · 7H$_2$O)	2.5 g
distilled water	100.0 ml

Potassium ferricyanide

potassium ferricyanide (K$_3$Fe(CN)$_6$)	1.0 g
distilled water	99.0 ml
glacial acetic acid	1.0 ml

Control

skin

Procedure

1. Deparaffinize and hydrate slides to water.
2. Treat with ferrous sulfate: 1 hour.
3. Wash in distilled water, 4 changes: total 20 minutes.
4. Treat with potassium ferricyanide: 30 minutes.
5. Wash in 1% aqueous acetic acid: 1 to 2 minutes.
6. Counterstain if desired: picro-Ponceau is satisfactory; do not use hematoxylin.
7. Dehydrate, clear, and mount.

Results

melanins—dark green
background—faint greenish or unstained; with picro-Ponceau, collagen stains red, and muscle and cytoplasm, yellow and brown

Comments

Lillie says this method is highly selective. No other pigments react in this procedure, except occasionally hemosiderin.

Indophenol Method (Alpert et al. 1960)

Fixation

Any general fixative, or use fresh-frozen sections.

Solution

sodium-2,6-dichlorobenzenone-indophenol	0.1 g
50% alcohol	100.0 ml

Prepare just before use. Filter and add 1% aqueous hydrochloric acid until color of solution is red (approximately 1:5) or until pH is 2.0.

CONTROL
liver

PROCEDURE
1. Deparaffinize and hydrate slides to water.
2. Treat with HCl-indophenol: 5 minutes.
3. Dip in tap water several times until background loses color.
4. Mount in glycerin jelly or Von Apathy (p. 99).

RESULTS
lipofuscins—red
erythrocytes—blue
other tissue elements—colorless to blue or slightly pink

COMMENTS
Avoid alcohol; it removes color immediately. Stain fades in about 1 day.

STAINING FOR CALCIUM DEPOSITS

Von Kossa Method (1901, Mallory Modification 1944)

FIXATION
10% neutral buffered formalin or alcohol, although other fixatives give reasonably good results.

SOLUTIONS
Silver nitrate
silver nitrate ($AgNO_3$) .... 5.0 g
distilled water .... 100.0 ml

Fresh solutions are always best; never use a solution more than 1 week old. *Safranin or Kernechtrot,* see p. 215.

CONTROL
tissue with calcium (e.g., calcified lesion)

PROCEDURE
1. Deparaffinize and hydrate slides to distilled water; remove $HgCl_2$.
2. Treat with silver nitrate in dark: 30 minutes.
3. Rinse thoroughly in distilled water.
4. Expose slides (in distilled water) to bright light (75 to 100W bulb

satisfactory): 1 hour. Lay them over a white background to expedite reaction.
5. Wash thoroughly in distilled water.
6. Counterstain in safranin: 2 to 3 minutes, or Kernechtrot: 5 to 8 minutes.
7. Rinse in 70% alcohol.
8. Dehydrate, clear, and mount.

RESULTS
calcium deposits—dark brown to black
nuclei and other tissue elements—shades of red and rose

COMMENTS
If the exposure to light in step 4 produces too much brown or yellow background, try developing in a photographic developer: 5 minutes (step 4).

Step 5. Rinse in distilled water and fix in hypo: 5 minutes. Wash well and continue to step 6 above.

If iron blocks out the calcium, treat 10 to 15 minutes with 0.005 M sodium EDTA in normal saline. Wash in distilled water and proceed to step 2 (McGee-Russell 1958). Rungby et al. (1993) prefer 0.05% silver lactate instead of silver nitrate because background staining is reduced.

This reaction actually demonstrates phosphates and carbonates rather than calcium itself, but since soluble phosphates and carbonates are washed out, the calcium phosphates and calcium carbonates remain to react with the silver (Lillie 1954b, 1965), and the test can be regarded as sufficiently specific.

Renaud (1959) demonstrates that a high percentage of alcohol (at least 80% content) in the fixing fluid is necessary in order to detect calcium in some tissues (heart and coronary vessels). Water and even buffered formalin can dissolve out some small deposits of calcium salts.

Pizzolato and McCrory (1962) demonstrate that neither chemical reduction nor exposure to illumination was necessary in the Von Kossa technique.

Pickett and Klavins (1961) follow step 5 with an elastin stain (preferably aldehyde-fuchsin) to demonstrate calcium elastin simultaneously.

Carr et al. (1961) describe a method using chloranilic acid (red brown) to differentiate calcium deposits from carbon deposits.

Calcium can be present in areas of calcification before it can be demonstrated by the Von Kossa metal substitution method. But there are reagents that are considered capable of complexing with calcium, such as antraquinone dyes. See Chaplin and Grace (1976) and the alizarine red S method below.

Alizarine Red S (Puchtler et al. 1969)

FIXATION
Alcohol or Carnoy fixative should be used for best preservation of calcium. Formalin (even buffered) or Zenker-formol remove calcium deposits.

SOLUTION

alizarine red S	0.5 g
phosphate or barbital buffer, pH 9	100.0 ml

CONTROL
calcified lesion or tissue with calcium

PROCEDURE
1. Deparaffinize and hydrate slides to water.
2. Stain in alizarine red: 30 to 60 seconds.
3. Rinse in buffer, pH 9: 5 seconds
4. Dehydrate, clear and mount.

RESULT
calcium—orange red

COMMENTS

The recommendations made by Puchtler et al. (1969) must be observed to obtain reliable results with this method. Proper fixation is important; calcium staining is most intense after alcohol and Carnoy fixation. Formalin and Zenker produce weakly or moderately colored deposits and staining decreases after storage in formalin, disappearing completely after 2 to 3 weeks. Alizarine red S provides more reliable staining red in a buffer solution of pH 9 than at other pH values. It stains red without a yellow tinge within 5 minutes, but intensity of color increases with staining times up to an hour. The alizarine dyes in aqueous solution of pH 7 or below become only slightly dissociated with little tendency to form calcium salts. If the solutions are made at pH 9, then many of the 2-OH groups of alizarine become dissociated and the formation of monoalizarate yields the red color for calcium. At a higher pH of 12, almost all of the 2-OH and about half of the 1-OH groups are dissociated and a blue alizarate forms. The sulfonic acid group of alizarine red S also forms as salt with calcium. Alizarine, however, cannot be considered specific for calcium alone; other metals also react with the stain. It is sensitive toward ions of uranium, titanium, bismuth, thallium, zirconium, halfnium, and thorium. The Puchtler article should be read for a better understanding of this stain and its reaction.

REMOVAL OF PIGMENTS

Melanin Pigment

This will appear as brown, grayish, or almost black granules.

Permanganate Method of Removal (Lillie 1965)

1. Hydrate slides to water.
2. Immerse slides in 0.5% aqueous potassium permanganate: 12 to 24 hours.
3. Wash in running water: 5 minutes.
4. Immerse slides in 1% aqueous oxalic acid: 1 minute.
5. Wash in running water: 10 minutes, and proceed to stain.

Performic or Peracetic Acid Method of Removal

Lillie (1954b, 1965) claims that the permanganate method an be unpredictable and that the best bleaching is done with performic or peracetic acid: 1 to 2 hours.

Performic acid
90% formic acid, aqueous	8.0 ml
hydrogen peroxide (30%)	31.0 ml
sulfuric acid, concentrated	0.22 ml

Keep at or below 25°C. About 4.7% performic acid is formed within 2 hours, but it deteriorates after a few more hours.

Peracetic acid
glacial acetic acid	95.6 ml
hydrogen peroxide (30%)	259.0 ml
sulfuric acid, concentrated	2.2 ml

Let stand 1 to 3 days. Add 40 mg disodium phosphate as stabilizer. Store at 0.5°C. Keeps for months in refrigerator.

Chlorate Method of Removal

Treat sections 24 hours in 50% alcohol plus small amount of potassium chlorate and a few drops of HCl (concentrated). Wash 10 minutes before staining.

Bromine Method of Removal

1% bromine in water: 12 to 24 hours. Wash well before staining.

Chromic Acid Method of Removal

Mixture of equal parts of 1% chromic acid and 5% calcium chloride, aqueous: 8 to 12 hours. Wash well.

Peroxide Method of Removal

10% hydrogen peroxide: 24 to 48 hours. Wash well before staining.

The peroxide method is considered by some to be the best; it is specific for melanin; other pigments resist longer than 48 hours. The permanganate and chromic acid methods allow no differentiation between pigments.

Formalin Pigment

Brown and black crystalline granules and artifacts, produced by formalin, are found in and around blood, and are considered a hematein derivative. The crystals are formed when the tissue is fixed with formalin at an acid pH. They usually will not form when the pH of the formalin fixative is above 6.0.

Barrett Method of Removal (1944)

Saturated solution of picric acid in alcohol: 10 minutes to 2 hours.

Malarial Pigment

This appears as amorphous brown-black granules. Its histochemical properties apparently are the same as those of formalin pigment.

Gridley Method of Removal (1957)

Method 1
1. Hydrate slides down to water.
2. Bleach for 5 minutes in:

    acetone .... 50.0 ml
    3% hydrogen peroxide .... 50.0 ml
    28–29% ammonia, concentrated .... 1.0 ml

or overnight in 5% aqueous ammonium sulfide (diluting 20% analytical reagent 4:1).

3. Wash thoroughly in running water: 15 minutes or longer.

Method 2
Bleach in 3% hydrogen peroxide: 2 hours.

Carbon

Carbon (opaque black) usually appears in the lungs and adjacent lymph nodes, sometimes in the spleen and liver. It is possible to distinguish it from malarial pigment, iron, or some other pigment or precipitate, since the carbon is black and is insoluble in concentrated sulfuric acid in which other pigments will dissolve.

Hemosiderin

This yellowish, brownish, or greenish brown pigment resists bleaching, does not dissolve in alkalis or acids, and is not argentaffin positive. It can be identified by Perl's test (p. 383).

Bile Pigments

These are yellowish green, resist bleaching, and are argentaffin positive. They can be converted by hydrogen peroxide, Lugol solution, nitrous acid, or dichromates into emerald green biliverdin.

18
Staining Proteins and Nucleic Acids

PROTEIN STAINING

The proteins found in cells may be classified as either (1) simple proteins, which on hydrolysis yield α-amino acids and their derivatives, or (2) conjugated proteins, which yield, in addition to amino acids, nonprotein materials. The simple proteins may be fibrous or globular proteins. Collagens, reticulins, keratin, and fibrin are simple fibrous proteins and insoluble in most aqueous media. The globular proteins—albumins, globulins, globins, and histones—are soluble in aqueous media. Among the conjugated proteins are nucleoproteins, mucoproteins, lipoproteins, and glycoproteins.

Because of their omnipresence in cells, proteins can be disappointing subjects for precise and brilliant localization. Sometimes, however, it does become necessary to ascertain the protein or nonprotein nature of the cell's contents. "Protein" techniques exist, but few are specific for proteins as such. Most of these techniques demonstrate a component of the protein (such as one of the amino acid groups), or they are specific for the aromatic nuclei, the phenolic functions, or the guanidine grouping (Gomori 1952; Pearse 1968).

The Millon reaction, one of the classic methods, is specific not for proteins, but for aromatic groups found in the amino acid tyrosine. The method, therefore, is specific for tyrosine, which is in most proteins. Two Millon variations follow.

Millon Reaction (Gomori 1952)

FIXATION
Any good general fixative.

SOLUTIONS
Solution A

mercuric acetate	5.0 g
distilled water	100.0 ml
trichloroacetic acid	15.0 g

Solution B

sodium nitrite, 1% aqueous	10.0 ml

CONTROL
small intestine

PROCEDURE
1. Hydrate sections. (Cut sections at least 10 μm thick.).
2. Incubate sections in solution A, 30–37°C: 5 to 10 minutes.
3. Add solution B; incubate 25 minutes.
4. Rinse in 70% alcohol.
5. Dehydrate, clear, and mount.

RESULT
tyrosine—pink to brick red

Millon Reaction (Romeis 1948)

SOLUTION
Dilute 400 ml of concentrated nitric acid with distilled water to make 1 liter. Let stand 48 hours. Dilute again, 1 part of this solution with 9 parts of distilled water. Add an excess of mercuric nitrate and allow to stand several days with frequent shaking to complete saturation. To 400 ml of filtrate, add 3 ml of the 40% original solution and 1.4 g sodium nitrate.

CONTROL
small intestine

PROCEDURE
1. Hydrate to water as in the procedure for the first Millon reaction above.
2. Immerse sections in reagent: 3 hours.
3. Give the slides a quick dip in 1% nitric acid.
4. Transfer quickly through 70%, 95%, and absolute alcohol, and xylene; mount.

RESULT
tyrosine—rose or red

Mercuric Bromphenol Blue Method (Chapman 1975)

FIXATION
Any general fixative.

SOLUTION
mercuric chloride	1.0 g
sodium bromphenol blue	0.05 g
2% acetic acid aqueous	100.0 ml

PROCEDURE
1. Deparaffinize sections (5 μm or less) and hydrate to water.
2. Transfer to bromphenol blue solution: 15 minutes, room temperature.

3. Rinse in 2 changes 0.5% acetic acid, aqueous, 20 minutes total.
4. Blot and dehydrate rapidly in 2 changes absolute alcohol; agitate several times.
5. Transfer to xylene: 2 minutes.
6. Treat with xylene plus 0.5 ml n-butylamine per 100 ml until section turns blue.
7. Clear in 2 changes xylene and mount.

RESULT

The response seems to be limited to proteins and peptides that are not removed by washing; they are blue.

COMMENTS

This method may not be histochemically valid, but bromphenol blue continues to be used for proteins.

See Kaniwar (1960) about the uncertain specificity of this method.

The identification of the individual amino acids in tissue sections involves a number of procedures, some of which are controversial and unreliable as to specificity and some of which are complex. Available techniques include methods for amino groups, carboxyl groups, tyrosine, tryptophan, histidine, the SS and SH groups, and arginine. Only a small number of these can be included here.

Ninhydrin Reaction (Serra 1946)

Ninhydrin (triketo-hydrindene-hydrate) reacts with the α-amino groups, oxidizing them to carbon dioxide, ammonia, and an aldehyde. The reaction is not stable and its products are diffuse, but the aldehydes that are formed will react with Schiff reagent to produce a red color. Glenner (1963) questions the value of this method, since it reacts with both free and bound α-amino acids, but is not limited to them. See also Kasten (1962) and Puchtler and Sweat (1962).

FIXATION

10% neutral buffered formalin.

SOLUTION

phosphate buffer, pH 6.98 (p. 473) .... 50.0 ml
ninhydrin, 0.4% aqueous (triketo-hydrindene-hydrate) .... 50.0 ml

PROCEDURE

1. Deparaffinize and hydrate slides to water or use frozen sections.

2. Place slides on a rack over boiling water, cover with stain, and steam for 1 to 2 minutes.
3. Drain stain off the slide and mount in glycerin jelly.

Result

Blue or violet color indicates the presence of amino acids, peptides, and proteins.

Comments

Slides should be examined at once because the color diffuses readily and begins to fade within a day or two.

Ninhydrin-Schiff Reaction (Yasuma and Ichikawa 1953)

Fixation

Any general fixative, but 10% neutral buffered formalin, Zenker, or absolute alcohol is recommended.

Solutions
Ninhydrin

ninhydrin	5.9 ml
absolute alcohol	100.0 ml

Schiff reagent, see p. 183.

Control
small intestine

Procedure
1. Deparaffinize sections. (Use 10-μm sections.)
2. Transfer to absolute ethyl alcohol, 2 changes: 2 minutes each.
3. Incubate in ninhydrin solution, 37°C: 5 to 24 hours.
4. Wash in tap water: 3 to 5 minutes.
5. Proceed as in Feulgen reaction (p. 187).

Result
protein or peptide—red or purple

Comments

See Van Duijn (1961) for acrolein-Schiff reaction, which is based on the reaction of acrolein with tissue compounds. All proteins except arginine-rich protamines are positive. Alloxan also may be used (Yasuma and Ichikawa 1953).

Staining of Histones

Histones are the basic protein of chromatin.

Alfert and Geschwind Method (1953)

FIXATION

10% neutral buffered formalin: 3 to 6 hours. Zenker, Susa, and Carnoy do not permit specific staining. Wash overnight in running water.

SOLUTIONS

Trichloroacetic acid (TCA)

trichloroacetic acid	5.0 ml
distilled water	100.0 ml

Fast green

fast green FCF	0.1 g
distilled water	100.0 ml

Adjust pH to 8.0 to 8.1 with a minimum of NaOH. Prepare fresh.

PROCEDURE
1. Deparaffinize and hydrate to water.
2. Immerse in TCA solution in boiling water bath to remove nucleic acids: 15 minutes.
3. Wash out TCA with 2 changes of 70% alcohol: 10 minutes each. Then rinse in distilled water.
4. Stain in fast green: 30 minutes.
5. Wash in distilled water: 5 minutes.
6. Transfer directly to 95% alcohol, absolute alcohol, and xylene; mount.

RESULT

histone—green

COMMENTS

This is a specific chromosomal stain, and all experiments indicate that histones or protamines are responsible for the histological stain. The stain is not retained unless nucleic acids are removed. The histones occur in a constant quantitative ratio to the DNA in cell nuclei, and neither histones nor DNA vary in relative amounts when cells undergo physiological changes.

Ammoniacal Silver Method (Black and Ansley 1964)

FIXATION

10% neutral buffered formalin.

SOLUTION

Add 10% aqueous silver nitrate, drop by drop, to 3 to 4 ml concentrated ammonium hydroxide until the first turbidity occurs. Stir constantly. Approximately 38 ml $AgNO_3$ is required for 4 ml of NH_4OH. Prepare fresh.

PROCEDURE
1. Deparaffinize and hydrate to water.
2. Treat with buffered formalin: 15 minutes.
3. Wash in 5 changes distilled water: several seconds each.
4. Stain in silver solution with gentle agitation: 5 to 10 seconds.
5. Wash in 5 changes of distilled water: several seconds each.
6. Treat with 3% neutral formalin with agitation: 2 minutes.
7. Wash in 3 changes of distilled water: several seconds each.
8. Dehydrate, clear, and mount.

RESULT
histone—black

COMMENTS

Black et al. (1964) have written an excellent article concerning the cell specificity of histones. Also see Vidal et al. (1971) for a silver stain.

Arginine Reaction

The arginine reaction uses the reaction of naphthol with guanidine derivatives in the presence of hypochlorite or hypobromite to produce an orange-red color. The reaction takes place in guanidine derivatives when one hydrogen atom of one of the amino groups is replaced by alkyl, fatty acid, or cyano radical. Proteins containing arginine give the color. In tissue fixed by ordinary methods the reaction is specific for protein-bound arginine and other guanidine derivatives such as glycocyamine or agmatine. Guanidine, urea, creatine, creatinine, and amino acids other than arginine do not give the color.

Sakaguchi Reaction (Modified by Deitch 1961)

FIXATION

Absolute alcohol-glacial acetic acid (3:1). Formalin fixation may block amino guanidyl groups and reduce the reaction. Smears may be air-dried and fixed in methyl alcohol.

SOLUTIONS
Barium hydroxide
    barium hydroxide .... 4.0 g
    distilled water .... 100.0 ml

Filter just before use.

Hypochlorite solution

Chlorox, freshly obtained (5% solution)	1.0 ml
distilled water	4.0 ml

Naphthol solution

2,4-dichloro-α-naphthol	0.15 g
tertiary butyl alcohol	10.0 ml

CONTROL
known positive tissue

PROCEDURE
1. Deparaffinize and hydrate sections to water. Use sections of 4 μm or thinner.
2. Blot and place slides in an empty Coplin jar.
3. Pour 5 parts of barium hydroxide solution in a flask. Add 1 part hypochlorite solution and then 1 part dichloronaphthol, shaking flask during each addition. Pour immediately on slides in jar. Room temperature: 10 minutes.
4. Transfer slides rapidly through 3 changes of tertiary butyl alcohol containing 5% of tri-N-butylamine. Move each slides vigorously in first change for about 5 seconds, transfer through next 2 changes: 30 to 60 seconds each. Change first solution frequently.
5. Transfer rapidly through 2 changes of xylene containing 5% tri-N-butylamine: 30 to 60 seconds each.
6. Drain rapidly and mount in Cargille's (formerly Schillaber's) oil containing 10% tri-N-butylamine.

RESULT
arginine—orange to red

COMMENTS
Always keep slides moist. Drying causes a crust of barium carbonate to form. The precipitate on the underside of the slide can be wiped off with dilute acetic or hydrochloric acid.

Deitch has substituted barium for sodium or potassium hydroxide for better tissue morphology, and the sections do not tend to fall off. The addition of an organic amine to the solutions and mountants reduces the fading problem of former methods. See also Baker (1947), Carver et al. (1953), Liebman (1951), and Thomas (1950).

The following references are only a few of the many excellent papers concerning proteins and amino acids. For proteins: Deitch (1955) and Spicer and Lillie (1961). For tyrosine: Glenner and Lillie (1959) and Lillie (1957d). For histidine: Landing and Hall (1956), Pauly (1964), and Reaven and Cox (1963). For histamine: Enerbäck (1969) and Lagunoff et al. (1961). For tryptophane: Adams (1957), Bruemmer et al. (1957), and Glenner and Lillie (1957b). For SH groups: Barnett and Seligman (1952), Bennett (1951), Bennett and Watts (1958), and Chèvremont and Fréderic (1943). For SS groups: Adams (1956) and Adams and Sloper (1955, 1956). For arginine: Notenboom et al. (1967). For fluorescent labeling of proteins: Rinderknecht (1960).

NUCLEIC ACID AND NUCLEOPROTEIN STAINING

The two types of nucleic acids are (1) DNA, principally in the nuclei and (2) RNA, which is found in the nucleolus and ribosomes of the cytoplasm. Nucleoproteins are combinations of basic proteins and nucleic acids in various highly polymerized forms.

On hydrolysis, the nucleic acids yield: (1) purine and pyrimidine bases, (2) carbohydrates, and (3) phosphoric acid. DNA contains deoxyribose (deoxypentose) sugar, and RNA contains ribose (pentose) sugar.

Because the nucleic acids exist in many states of polymerization, fixation will cause some redistribution. Organic solvents, formalin, and most acid fixatives, however, will not cause significant losses from the cell. Pearse (1968) prefers alcohol and acetic acid to formalin for nucleic acids and nucleoproteins, but Kurnick (1955) says neutral (pH 7.0) 10% formalin appears to be satisfactory. Pearse warns against long exposure if Carnoy fixative is used, since it causes the extraction of first RNA and then DNA. He recommends Lillie's acetic-alcohol-formalin, 4°C: 24 hours.

> formaldehyde (37–40%) .... 10.0 ml
> glacial acetic acid .... 5.0 ml
> absolute ethyl alcohol .... 85.0 ml

Pure nucleic acids can be stained with basic dyes, but nucleic acids in living cells are bound with protein and will not react with basic dyes. Inactivation of the protein-bound amino groups by formalin or deamination (p. 242) will increase the basophilia of the nucleic acids by releasing increasing numbers of nucleic acid phosphoryl groups to bind the basic dye. By using different exposures to these reagents, Walsh and Love (1963) demonstrated several nucleoproteins. (See also Love [1957, 1962] and Love and Liles [1959]).

For staining of nucleic acids, the Feulgen technique is generally considered to be specific for DNA (p. 187). It is thought to depend on the uncovering of aldehyde groups of deoxyribose by the hydrolysis of purine-deoxyribose bonds by hydrochloric acid. The released aldehyde gives the Schiff reaction. Proper hydrolysis is critical. Two actions occur almost simultaneously: (1) the purine bases are rapidly removed and the aldehyde groups uncovered; (2) the histones and nucleic acids are progressively removed. At the beginning, the first action predominates; as hydrolysis proceeds, the second action begins to take over and staining with Schiff will decrease. Overhydrolysis can produce a negative Feulgen reaction. The optimum time is in the area of 8 minutes; the range is 6 to 12 minutes. More recently it has been noted that the use of 5 N hydrochloric acid at room temperature extends hydrolysis time up to one hour without risk of overhydrolysis.

In methyl green-pyronine staining for nucleic acids, the two dyes are used to distinguish between polymerization of the nucleic acids, not between the acids as such. Methyl green is readily bound by highly polymerized DNA; the binding is acquired at two sites, two amino groups of the dye with two phosphoric acid groups of the DNA. Depolymerizing treatment, such as heat and certain fixatives (Zenker, picric acid), must be avoided. If formalin is used, it must be accurately adjusted to pH 7.0 and fixation must be brief, so that the solution does not become acid. Pyronine, for RNA staining, stains low polymers of both nucleic acids but it does permit methyl green to compete effectively for the highly polymerized DNA to the exclusion of the pyronine. Nuclei will stain red with pyronine alone, but they will stain green when the two dyes are used together. Control slides should be used with all methyl green-pyronine staining. If one slide is exposed to ribonuclease and one is not, and then both are stained with methyl green-pyronine, the material stained red with pyronine and removable with ribonuclease is considered RNA, and that which is not removable is not RNA. (See Kasten et al. [1962] and Tepper and Gifford [1962].)

References to some of many techniques, theories, controversies, and rationalizations for the demonstration of the nucleic acids and related components of the cell are listed below.

Spicer (1961) reports that Bouin fixes RNA as well as or better than any of a number of fixatives and yields DNA without acid hydrolysis. By staining at controlled pH (RNA stains more strongly at low pH) with Schiff reagent for DNA and methylene blue for RNA, the chromatin stains red and the cytoplasmic RNA stains blue.

Kurnick (1952) uses 1% toluidine blue in 95% alcohol as a rough screen test to demonstrate areas of high concentration of nucleic acids. By comparing an unextracted slide with a preparation made after extraction of the nucleic acids, one may differentiate the basophilia caused by nucleic acids from that caused by other components of the cell.

Daoust (1964) uses 0.1% toluidine blue in Veronal buffer, pH 5.0 (5 minutes); dehydrates in 3 changes of tertiary butyl alcohol (1 minute each); clears; and mounts.

Other stains used for nucleic acids and nucleoproteins include: azure A (Anderson 1961; Flax and Pollister 1949); Biebrich scarlet (Spicer 1962); cresyl violet (Ritter et al. 1961); methenamine silver (Korson 1964); methylene blue (Deitch 1964); fluorescent methods (Armstrong 1956; Armstrong et al. 1957; deBruyn et al. 1953); Feulgen oxidized tannin-azo technique (Malinin 1961); trimethylthionin basic fuchsin (Menzies 1963).

Aceto-orcein and acetocarmine are popular chromosome stains (pp. 410 and 412); the orcein probably depends on DNA and the carmine on protein. Hematoxylin used with an alum mordant probably stains chromatin specifically but may depend on the state of polymerization of DNA.

Tagging with ^{14}C and ^{32}P can be used for autoradiography.

Nucleoli can be "stained" with silver (Tandler 1955).

For general discussion about nucleic acids and nucleoproteins, see Gomori (1952), Kurnick (1955), and Pearse (1968).

Pyronine-Methyl Green for Nucleic Acids (Kurnick 1952)

FIXATION
Absolute alcohol, Carnoy, or cold acetone. If formalin is used, it must be adjusted to pH 7.0 and used only briefly before the solution turns acid and produces a faint green staining of cytoplasm. Picric acid depolymerizes DNA and shows no green chromatin.

CONTROL
most tissues

SOLUTIONS
Methyl green
    methyl green .... 0.2 g
    0.1 M acetate buffer (pH 4.2) or distilled water .... 100.0 ml

Before making the solution, purify the methyl green by extraction with chloroform. Add approximately 10 g methyl green to 200 ml of chloroform in a 500-ml Erlenmeyer flask and shake. Filter with suction, and repeat with smaller amounts of chloroform until the solution comes off blue green instead of lavender (usually requires at least 3 extractions). Dry and store in capped bottle. Purified dye is stable.

To detect crystal violet contaminant, place a drop of methyl green solution on a piece of filter paper. Hold the stained area over ammonia solution. The green color disappears, and violet color, if present, will remain (Kasten and Sandritter 1962).

Pyronine
Pyronine Y is saturated in acetone or, for lighter color, in 10% acetone.

PROCEDURE
1. Deparaffinize and hydrate slides to water.
2. Stain in methyl green solution: 6 minutes.

3. Blot and immerse in *n*-butyl alcohol: several minutes in each of 2 changes.
4. Stain in pyronine: 30 to 90 seconds. Shorten to a few dips if stain is too dark.
5. Clear in cedarwood oil, xylene, and mount.

Results

RNA containing cytoplasm—red
nucleoli—red
chromatin—bright green
erythrocytes—brown
eosinophilic granules—red
cartilage matrix—green
osseous matrix—pink with trace of violet

Comments

These two dyes distinguish between states of polymerization of nucleic acids, not between the acids themselves. Highly polymerized nucleic acid stains with methyl green; low polymers of both DNA and RNA stain with the pyronine. The usual techniques result in too pale a methyl green, or, if excess staining is tried, water rinses removes the methyl green and acetone removes the pyronine. Dehydrating with ethyl alcohol removes most of the color; isopropyl and tertiary butyl are no improvement. Kurnick developed the above pyronine-methyl green method when he found that *n*-butyl alcohol differentiated the methyl green but removed the pyronine. Since the pyronine was lost anyway, he therefore decided to leave it out of the first solution and tried following the methyl green with pyronine saturated in acetone. (Because the acetone must be free of water, always use a fresh solution.)

Pyronine-Methyl Green for Nucleic Acids (Elias 1969).

Fixation
Carnoy.

Solution

methyl green	0.5 g
Walpole acetate buffer, pH 4.1	100.0 ml
pyronine GS or Y	0.2 g

Procedure

1. Hydrate slides to water.
2. Stain, 37°C: 1 hour.
3. Rinse in ice-cold distilled water: 1 to 2 seconds.

4. Blot.
5. Rinse, agitating vigorously in tertiary butanol.
6. Dehydrate in 2 changes of tertiary butanol: 5 minutes each.
7. Clear and mount.

COMMENTS

The Elias method gives a consistent, qualitative, nucleic acid differentiation. The ice water differentiation rinse is critical, and tertiary butanol is superior to the *n*-butanol.

The critical factors in this method are the use of Carnoy fixative, a staining solution of low pH, and ice-cold distilled water rinse, followed by tertiary butanol dehydration.

Ahlqvist and Anderson (1972) recommend a low pH (3.8), formalin fixation, and purify both the methyl green and pyronine. Their stain solution is made up in a Walpole buffer at pH 3.8.

See also d'Ablaing et al. (1970)

Thionine-Methyl Green for Nucleic Acids (Roque et al. 1965)

FIXATION

Fix thin (2-mm) pieces in 4% formaldehyde in 1% sodium acetate: 3 hours. Dry smears fixed in methyl alcohol can be used.

SECTIONING

Cut 2- to 4-μm paraffin sections.

SOLUTION

methyl green	0.1 g
thionine	0.0165 g
citrate buffer, 0.01 M, pH 5.8 (0.01 M HCl, 42 ml; 0.01 M sodium citrate, 58 ml)	100.0 ml

Dissolve the thionine in a little water and add the buffer and methyl green. Shake well and filter. This solution is best when freshly mixed. The methyl green should be chloroform-extracted (p. 237).

PROCEDURE

1. Deparaffinize and hydrate sections to distilled water. Smears can go directly into water.
2. Stain in methyl green-thionine, 40°C: 30 minutes.
3. Rinse in distilled water.
4. Dehydrate in 3 changes of alcohol mixture; 3 butyl alcohol, 80.0 ml; absolute alcohol, 20 ml; 30 seconds in first change, 3 minutes each in second and third changes.

5. Rinse in absolute alcohol, clear and mount.

RESULTS

chromatin—green or blue green
nuclear and cytoplasmic basophilic substance—red or reddish purple

COMMENTS

Objection to the specificity of the pyronine-methyl green method centers around the unsuitable staining properties of pyronine. Roque et al. tried various basic dyes in an effort to find a suitable substitute for pyronine. Thionine gave the best results; it did not obscure the green coloring of chromatin and was less readily extracted by aqueous and alcoholic rinses. Also it stained more selectively than pyronine.

Use a short fixation time, no more than 3 hours. Zenker, Helly, and Carnoy can be used, but the staining is less intense, although the color contrasts are satisfactory.

A range of pH 6 to pH 4 is satisfactory, but the color varies with the types of buffer, becoming blue when acetate or phosphate buffers are used, and green when a citrate buffer is used.

Thin sections, 2 to 4 μm, should be cut for best color contrasts; if the RNA content is very low, sections of 4 to 6 μm may be necessary.

Nucleic acids are stained deep blue with gallocyanin-chrome alum (p. 115). Also see Dutt (1974).

See Mundkur and Brauer (1966) for selective staining of nucleoli.

CONTROL SLIDE TECHNIQUES

In some techniques the reaction is so specific that controls are unnecessary. But in others controls are essential to separate a genuine reaction from one similar in appearance but not specific in nature. An essential ingredient may be omitted from a solution used in the control. This is the common procedure in histochemistry, in which the specific substrate is omitted from the incubating mixture used for the control. In other methods, inactivators, inhibitors, or extractives are used.

Extraction Techniques

In the demonstration of the specificity of a reaction of nucleic acids, the extract of one or both of the nucleic acids is frequently essential. The original preparation can then be compared with another preparation that has been subjected to extraction techniques.

Removal of Both Nucleic Acids

Hot 5% trichloroacetic acid, 90–98°C: 30 minutes.

Removal of RNA

1. 5% trichloroacetic acid, 60°C: 30 minutes.
2. 10% perchloric acid, 4°C: 12 hours. The cold temperature is critical (Aldridge and Watson 1963; Kasten 1965).
3. 1 N hydrochloric acid, 60°C: 6 to 12 minutes (Feulgen reaction, p. 187).
4. Ribonuclease (RNAse).

   RNAse (crystallized, salt-free) .... 1.0 mg
   distilled water .... 1.0 ml

Incubate sections, 40°C: 4 hours.

RNAse usually removes nuclear and cytoplasmic RNA, but chromosomal RNA may be resistant.

Removal of DNA (DNAse Method, Daoust 1964)

DNAse (crystallized) .... 0.05 mg
0.1 M tris maleate buffer, pH 6.5, containing $MgSO_4 \cdot 7H_2O$
at a concentration of 0.2 M .... 1.0 ml

Incubate, 37°C: 24 hours.

Amano (1962) claims the results are poor if the slides are incubated in a horizontal position with a few drops of the enzyme. He recommends putting the slides in a Coplin jar with 40 ml of DNAse solution.

Tandler (1974) Method

1. Transfer slides from water into a 10% solution of formalin in saturated picric acid (10 ml concentrated formaldehyde: 90 ml saturated aqueous picric acid): 12 to 24 hours. (Bouin fixative can be used.)
2. Wash in 2 changes distilled water: 2 minutes total.
3. Treat with 25% acetic acid, aqueous: 2 minutes.
4. Treat with 10% aniline in 25% acetic acid: 1 hour.
5. Wash in 2 changes 25% acetic acid: 2 minutes total.
6. Wash well in distilled water and proceed to stain.

Blocking Methods

A blocking method may become an essential part of a technique, not because it stains a component, but because it prevents its staining and thereby verifies that a certain chemical group is responsible for the staining reaction.

Blocking Protein End Groups, Reactions for Amines

Deamination

In this reaction nitrous acid blocks amino groups, most rapidly the α-amine groups; it converts amino groups into hydroxyl groups. Van Slyke reagent (Lillie 1954b) is best for this.

concentrated sodium nitrite (6 g/10 ml water)	10.0 ml
glacial acetic acid	5.0 ml
distilled water	25.0 ml

Use at room temperature: 1 to 12 hours (10 minutes for blood films).

Acetylation (Lillie 1954b, 1958, 1964a,b)

Acetylation produces acyl derivatives of primary and secondary amines and amine compounds which prevent 1,2 glycols from reacting with PAS. Acetylation and deacetylation are most commonly used for controls with PAS for mucopolysaccharides, even if the validity of the control is not absolutely infallible.

1. Take sections to absolute alcohol.
2. Dip in pyridine.
3. Transfer to following solution, 25°C: 1 to 24 hours, or 58°C: 30 minutes to 6 hours.

acetic anhydride	16.0 ml
anhydrous pyridine	24.0 ml

(The acetylation of glycols is progressive but slow, and it may require 24 to 48 hours for a complete blockage of PAS. It can be somewhat accelerated by adding 0.4–0.5% by volume of concentrated sulfuric acid.)

4. Wash in 2 changes each of absolute, 95% and 80% alcohols. Proceed to method for end groups under investigation.

Rapid acetylation and complete blockage of PAS can be accomplished by use of the following for 4 minutes:

acetic anhydride	25.0 ml
glacial acetic acid	75.0 ml
sulfuric acid	0.25 ml

Alcohol acetylation of phenolic groups (Lillie 1964b) can be done in a mixture of equal volumes of acetic anhydride and absolute ethyl alcohol. Use at 23–25°C: 30 minutes.

Benzoylation (Lillie 1954b; Barnard 1961)
Benzoylation is similar to acetylation, and steps 1 and 2 are the same.

3. Transfer to following, 25°C: 1 to 24 hours; or 58°C: 30 minutes to 6 hours.

    benzoyl chloride .... 2.0 ml
    anhydrous pyridine .... 38.0 ml

4. Wash in alcohol as in step 4 above and proceed to method for group under investigation.

Barnard's method requires anhydrous conditions throughout benzoylation, room temperature, in a desiccator: 3 hours.

    dry acetonitrile .... 50.0 ml
    benzoyl chloride .... 4.2 ml
    anhydrous pyridine .... 2.2 ml

Deacetylation (Saponification)
Acetylation blockage is almost completely reversed by saponification, which is effected by placing sections in the following for 20 to 30 minutes:

    potassium hydroxide .... 1.0 g
    70% alcohol .... 100.0 ml

and then treating for 16 to 24 hours in

    absolute alcohol .... 70.0 ml
    distilled water .... 10.0 ml
    ammonia, 28% concentrated .... 20.0 ml

Keep solution in closed container during action. Wash in 80% alcohol and water and proceed to method under investigation.

Blocking Aldehydes

Aniline-acetic acid (Lillie and Glenner 1957), room temperature: 20 to 30 minutes.

    glacial acetic acid .... 45.0 ml
    aniline .... 5.0 ml

If this fails to block a Schiff reaction, aldehydes are excluded.

If aldehydes are made to combine with sodium sulfite or phenylhydrazine, the reaction with Schiff after hydrolysis (Feulgen) or oxidation (PAS) can be blocked (Pearse 1968).

Phenylhydrazine

phenylhydrazine	5.0 ml
glacial acetic acid	10.0 ml
distilled water to make	50.0 ml

Treat at 60°C: 2 to 3 hours.

Sodium bisulfite

absolute alcohol	10.0 ml
50% aqueous sodium bisulfite	40.0 ml

Treat at room temperature: 2 to 8 hours.

Blocking Carboxyl Groups or Other Acid Groups

This reverses metachromasia of most components and blocks basophilia of nucleic acids; it abolishes reactivity of phosphoryl, carboxyl, sulfate groups of polysaccharide and carboxyl groups of proteins.

Methylation (Lillie 1954b, 1958)

Use 0.1 N HCl (8 ml/1000 ml) made up in methyl alcohol, 25°C: 2 to 3 days. Use screw-cap jars for the procedure. A temperature of 58°C is required to block the PAS of glycogen, epithelial gland mucins, thyroid colloid, collagen, and reticulin.

One percent HCl in methyl alcohol is required to block nuclear and cytoplasmic staining with azure A, and it destroys the reactivity of argentaffin cells in the methenamine silver procedure.

Geyer (1962) says that methanol-thionyl chloride is more effective than methanol-HCl under some conditions. With 100 ml of absolute methanol in an Erlenmeyer flask, add 4 ml thionyl chloride very slowly down a wall of the flask, held at a slant and away from the face. Use transparent thionyl chloride. Mix immediately before use. Use at 56°C: 30 minutes to 2 hours.

Demethylation (Saponification)

Saponification is used to deblock tissue; that is, to return it to the original active form that was blocked by one of the blocking methods. However, it is not always successful in reversing the effect of methylation.

Cover sections with thin nitrocellulose and place in 1% potassium hydroxide in 70% ethyl alcohol, 25°C: 30 minutes. This restores reactivity of methylated sulfonic acid and of carboxyl and phosphoryl groups, but not that of sulfate groups.

Blocking Tryptophane and Cystine

Performic acid oxidation (Pearse 1968) is specific for the indole ring of tryptophane.

98% formic acid	40.0 ml
30% H_2O_2 (must be fresh)	4.0 ml
sulfuric acid	0.5 ml

Use within 8 hours and stir vigorously before using, room temperature: 15 to 60 minutes.

Peracetic acid reagent, see p. 225

Blocking Tyrosine

Treat with 1% 2,4-dinitrofluorobenzene (DNFB) in 90% ethyl alcohol containing 0.01 M NaOH, room temperature: 16 to 20 hours.

Lillie (1965) and Pearse (1968) include detailed discussions of blocking methods.

19
Staining Lipids and Carbohydrates

LIPIDS

Lipids (lipoids, fatty substances) include a large number of substances that have been grouped together because of their solubility properties. They are insoluble in water and soluble in fat solvents such as alcohol, ether, chloroform pyridine, benzene, and acetone. Classification of the lipids is not undertaken here, but a few familiar groups can be mentioned: carotenoids (vitamin A), fatty acids, triglycerides (acetone soluble) (neutral fats), phospholipids, cerebrosides, sterols, and lipid pigment (essentially insoluble in acetone) (lipofuscins—p. 246).

For the fixation of fats, formalin is best, particularly if 1% calcium chloride is added to make the phospholipids insoluble (Gomori 1952). Because of the use of fat solvents, the tissue cannot be embedded in paraffin or nitrocellulose, but it can be embedded in carbowax (Rinehart and Abul-Haj 1951b), polyvinyl alcohol (Feder 1962; Masek and Birns 1961), polyethylene glycol (Hack 1952), or ester waxes (p. 342). Frozen sections are simpler to make and are the most frequently used. During any processing, alcohol higher than 70% must be avoided. (For the fixation of lipids, see Elftman 1958).

The dying of lipids is one of the simplest forms of staining. The coloring agent merely dissolves the lipid contained within the tissues. In addition, it should be emphasized that a dye solvent that does not dissolve the lipid itself should be used. The dye, therefore, must meet certain requirements:

1. The dye must be strongly colored.
2. It must be soluble in the substance that it is intended to show, but it must not be soluble in water, the major constituent of cells.
3. It must not attach itself to any tissue constituents except by solution.
4. It must be applied to tissues in a solvent that will not dissolve the substance to be dyed, and it must be less soluble in the solvent than in the substance.

Baker (1958) suggests the name *lysochromes* for these dyes that dissolve in the tissue elements to be colored. The name is derived from the Greek *lúsis*, meaning solution. The Sudans were the first synthetic dyes of this sort, followed by the Nile blues and reds.

These dyes are used in saturated solutions and often create the problem of dye precipitate on the tissue. Vlachos (1959) makes the sensible proposal that the precipitate is probably formed by the solution becoming oversaturated and that perhaps a saturated solution is unnecessary. He suggests two alternatives:

1. Make up to concentration below the saturation point; for example, 0.25 g Sudan IV per 100 ml 60% alcohol.
2. Desaturate the solution either by dilution of a saturated solution with equal volumes of 60% alcohol or by refrigeration. When the solution has been refrigerated long enough to have acquired the refrigerator temperature, filter. Neither method alters the staining quality of the solution, and the amounts of precipitate produced are negligible.

Lipid Staining

Oil Red O for Lipids

FIXATION
10% neutral buffered formalin or other aqueous general fixatives.

SOLUTION
oil red O .... 0.7 g
absolute isopropanol .... 200.0 ml

Shake; leave overnight. Filter. Dilute 180 ml of oil red O solution with 120 ml of distilled water. Leave overnight at 4°C. Filter. Let stand 30 minutes and filter. Ready for use. Stable for 6 to 8 months.

CONTROL
frozen sections of adipose tissue

PROCEDURE
1. Mount frozen sections (10 to 15 μm) on subbed slides. Allow to dry: 5 to 10 minutes. Loose, unmounted sections also may be stained this way.
2. Rinse in 60% isopropanol: 30 seconds.
3. Stain in oil red O: 10 minutes.
4. Rinse in 60% alcohol: few seconds.
5. Wash in running water: 2 to 3 minutes.
6. Stain in Mayer hematoxylin (or the like): 2 to 3 minutes.
7. Wash in tap water: 3 minutes.
8. Blue in Scott solution (p. 464): 3 minutes.
9. Wash in tap water: 5 minutes. Mount in glycerin jelly. For permanency, ring cover glass.

RESULTS
lipids—orange red or brilliant red
nuclei—blue

Comments

There are aqueous mountants commercially available for oil red O sections that do not require sealing.

This method is the best of the oil red O methods, for it never leaves a stain precipitate on the sections.

For a blue coloration, oil blue N can be used, but the color is not as intense as that of oil red O. Use a red nuclear stain.

Sudan IV or Sudan Black B (Chiffelle and Putt 1951)

Fixation

10% neutral buffered formalin, no alcohol.

Solution

Dissolve 0.7 g Sudan IV or Sudan black B in 100 ml propylene or ethylene glycol. Add small amounts at a time, and heat to 100–110°C, stirring, for a few minutes. Do not exceed 110°C or a gelatinous suspension is formed. Filter hot through Whatman no. 2 paper and cool. Filter again through fritted glass filter with the aid of suction or through glass wool by vacuum.

Control

adipose tissue

Procedure

1. Frozen sections, 15 μm, into water.
2. Place in pure propylene or ethylene glycol: 3 to 5 minutes, 2 changes. Agitate.
3. Stain, 2 changes: 5 to 7 minutes each, agitate occasionally.
4. Differentiate in glycol and water: 3 to 5 minutes. Agitate.
5. Wash in distilled water: 3 to 5 minutes.
6. Counterstain in hematoxylin if desired.
7. Mount in glycerin jelly.

Result

lipids—Sudan IV, orange to red; Sudan black B, blue black to black

Comments

Sudan black B is considered the most sensitive of the lipid dyes.

Chiffelle and Putt recommend glycol as a perfect solvent for a fat stain because it does not extract lipids.

Zugibe et al. (1958, 1959) suggest Carbowax 400 as a solvent for oil red O and Sudan IV.

Gomori (1952) questions the use of glycols because they are solvents for so many water-insoluble substances, and he prefers triethylphosphate. It has a low volatility and is harmless to lipids. His method follows.

1. Frozen sections are rinsed in water and transferred into 50% alcohol for a few minutes.
2. Stain in a saturated, filtered solution of any of the fat dyes in 60% triethylphosphate: 5 to 20 minutes.
3. Differentiate in 50% alcohol: 1 minute.
4. Counterstain in hematoxylin or any preferred stain.
5. Mount in glycerin jelly.

Osmium tetroxide (Mallory 1944)

Oleic acid or triolein react with the osmic acid by oxidation of their double bonds.

Fixation
10% neutral buffered formalin, no alcohol.

Solution
osmium tetroxide	1.0 g ampule
distilled water	100.0 ml

With a triangular file, score a circle around ampule and drop it into bottle with the distilled water. Several sharp shakes will break the ampule and allow the water to dissolve the crystals. This method eliminates the possibility of breathing the fumes. However, new packaging methods have virtually eliminated this problem. Nevertheless, work in a fume hood.

Procedure
1. Frozen sections, 10 to 15 μm, into water.
2. Transfer to osmium solution: 24 hours.
3. Wash in several changes distilled water: 6 to 12 hours total.
4. Treat in absolute alcohol: 4 to 5 hours, for secondary staining of fat.
5. Wash well in distilled water: 5 minutes or more.
6. Mount in glycerin jelly.

Results
lipids—black
background—yellowish brown

Naphthalene Yellow (Sills and Marsh 1959)

Naphthalene yellow in 60% acetic acid was used to restore the yellow color to the fat of formalin-fixed gross specimens. This cannot be used for sections; the color is too pale.

Fluorescent Method (Metcalf and Patton 1944; Peltier 1954)

FIXATION

10% neutral buffered formalin (salts of heavy metals—$HgCl_2$—have a quenching effect on fluorescence), or use fresh tissue.

SOLUTION

phosphine 3R	0.1 to 1.0 g
distilled water	100.0 ml

CONTROL

colon with attached adipose tissue

PROCEDURE
1. Cut frozen sections, 10 μm.
2. Wash in distilled water.
3. Stain in phosphine solution: 5 minutes.
4. Rinse quickly in distilled water.
5. Mount in glycerin or examine as a water mount.

RESULT

Lipids (except fatty acids, cholesterol, and soaps) will fluoresce brilliant white.

Sudan Black B Method for Phospholipids (Elftman 1957)

FIXATION

mercuric chloride, 5% aqueous	100.0 ml
potassium dichromate	2.5 g

Adjust pH to 2.5 with HCl. Mix fresh. Fix for 3 days; oxidizes phospholipids and decreases their solubility in fat solvents such as are used in paraffin embedding. Nonetheless, leave tissues for short periods in all solvents.

SOLUTIONS

Sudan black B, see p. 248

Ethylene glycol solution

ethylene glycol	85.0 ml
distilled water	100.0 ml

PROCEDURE
1. Deparaffinize and transfer slides into absolute alcohol: 2 minutes.
2. Transfer to absolute ethylene glycol.
3. Stain in Sudan black solution: 30 minutes.
4. Differentiate in 95% ethylene glycol: 2 to 3 minutes.

5. Wash in distilled water.
6. Counterstain, if desired.
7. Wash well, mount in glycerin jelly.

Result
phospholipids—black

Comments
Phospholipids are increasingly important in the study of mitochondria and Golgi apparatus and are an important constituent of myelin.

Roozemond (1969) recommends fixing phospholipids in 4% formalin with 1% calcium chloride added and then cutting frozen sections: fixation at 0°C reduces the breakdown of lipids. He suggests that a strong interaction between calcium and phospholipids may aid in the retention of the phospholipids when they are surrounded by a calcium-containing medium.

Acid Hematein for Phospholipids (Hori 1963)

Fixation
Formalin-calcium (10% formalin/1% calcium chloride), formalin-calcium-cadmium (1% cadmium chloride added to above), or glutaraldehyde-formalin-calcium chloride (2.5:25:1) and 0.1 M sodium cacodylate, pH 7.2 (p. 329): 18 hours.

Solutions
Acid hematein
   hematoxylin .... 50.0 mg
   0.01% sodium iodate .... 50.0 ml

Heat until it begins to boil, cool, and add:

   glacial acetic acid .... 1.0 ml

Prepare fresh.

Borax-ferricyanide
   borax ($Na_2B_4O_7 \cdot 10H_2O$) .... 250.0 mg
   potassium ferricyanide ($K_3Fe(CN)_6$) .... 250.0 mg
   distilled water .... 100.0 ml

Keeps well in refrigerator.

Procedure
1. Cut frozen sections 10 to 15 μm.
2. Chromate sections in 5% aqueous potassium dichromate, 60°C: 4 hours. Transfer sections with glass rod.

3. Wash in several changes of distilled water.
4. Stain in acid hematein, 37°C: 30 minutes.
5. Wash briefly in distilled water.
6. Differentiate in borax-ferricyanide, 20–25°c: 18 hours.
7. Dehydrate, clear, and mount.

RESULT

phospholipids and some proteins—blue, blue gray, or blue black.

COMMENTS

This is a simplification of Baker's (1946) original acid hematein test of phospholipids. If it is desired to distinguish between phospholipids and certain proteins that also give a positive reaction in this test, apply Baker's pyridine extraction method:

1. Fix small pieces of tissue in weak Bouin solution: 20 hours.

picric acid, saturated aqueous	50.0 ml
formaldehyde (37–40%), concentrated	10.0 ml
glacial acetic acid	5.0 ml
distilled water	35.0 ml

2. Transfer to 70% alcohol: 1 hour.
3. Transfer to 50% alcohol: 30 minutes.
4. Wash in running water: 30 minutes.
5. Transfer to pyridine, room temperature, 2 changes: 1 hour each.
6. Transfer to pyridine, 60°C: 24 hours.
7. Wash in running water: 2 hours.
8. Transfer to step 2 of acid hematein method.

RESULTS

phospholipids—unstained
nuclei—stain after extraction, not before extraction
erythrocytes—stain before and after extraction

COMMENTS

Luxol fast blue can be used for phospholipids (see p. 172).

For unsaturated lipids, see Norton et al. (1962); for lipoproteins, "masked" lipids, see Carmichael (1963), Holczinger and Bálint (1961), and Serra (1958).

Treatment with potassium dichromate makes phospholipids nonextractable; this may be due to the formation of complexes between the phospholipids and chromium ions. Formol-calcium is preferred as the primary fixative when postchromation is used. The calcium aids in preventing leaching of the phospholipids.

Ellender and Lojda (1973) recommend a modification of Weigert hematoxylin for phospholipids.

1. Postfix fresh frozen sections in formol-calcium.
2. Stain in 3 parts A and 1 part B, freshly mixed: 6 to 8 minutes.
   Solution A

   distilled water .... 298.0 ml
   HCl concentrated .... 2.0 ml
   ferric chloride ($FeCl_3 \cdot 6H_2O$) .... 2.5 g
   ferric sulfate ($FeSO_4 \cdot 7H_2O$) .... 4.5 g

   Solution B

   distilled water .... 100.0 ml
   hematoxylin .... 1.0 g

   Dissolve with low heat.
3. Wash, running water: 5 minutes.
4. Differentiate in 0.2% HCl.
5. Wash, running water: 10 minutes.
6. Dehydrate, clear, and mount.

CARBOHYDRATES (SACCHARIDES)

Polysaccharides are a complex group of substances that are found throughout animal tissues. Complete agreement in nomenclature is still lacking, but in general they are classified as follows:

Polysaccharides (glycans) include glycogen, starch, and cellulose. Glycogen is the only naturally occurring member of this group that remains in animal tissue after aqueous fixation and embedding.

Polysaccharide-protein complexes include, (1) acid mucopolysaccharides—hyaluronic acid (Wharton's jelly), chondroitins (cartilage), and heparin; (2) neutral mucopolysaccharides—chitin; (3) mucoproteins—lens capsule and vitreous humor, part of epithelial, glandular and ductal mucins.

Glycoproteins and glycopolypeptides—ovomucoid and salivary gland mucoid.

Glycolipids—cerebrosides and gangliosides.

Glycolipid-protein complexes—ox brain mucolipid.

Certain specific stains and aldehyde reactions are commonly used to demonstrate carbohydrates. The aldehyde groups must first be liberated by some chemical agent, either oxidized (periodic or chromic acid) or hydrolyzed (dilute hydrochloric acid). Then the specific reagent for aldehydes can be applied. (See the Feulgen and periodic acid-Schiff reactions, pp. 183, 187.)

Alcoholic mixtures are usually recommended as fixatives for saccharides. Most conventional fixatives cause glycogen to migrate to one side of the cell. Gomori (1952) theorized that when glycogen is enclosed in a complex mixture of proteins and lipids, a good protein precipitant may coat the glycogen with a protein membrane. This will be impermeable to the large glycogen molecules and keep them in situ. The fixative should act and harden rapidly. Alcohol or acid fixatives form glycogen in coarse droplets, and fixatives containing formalin distribute the glycogen in fine granules. In some tissues, such as liver, unless the tissue and fixative are chilled, enzymes can cause a loss of glycogen. Although in most tissues glycogen is more stable, Gomori recommended placing the tissue in fixative in the refrigerator. Leske and von Mayersbach (1969) have demonstrated that block fixation does not safely preserve glycogen. Sections should be cut fresh frozen, and fixed in Carnoy fixative or 2.5–10.0% aqueous trichloroacetic acid: 5 to 15 minutes. Murgatroyd (1971) reports that cold formal alcohol or cold Rossman fluid preserve the most glycogen.

After proper fixation paraffin embedding is satisfactory, but after deparaffinizing the mounted slides, rinse them in absolute alcohol and protect the sections with a coat of 1% nitrocellulose to prevent diffusion of the glycogen during staining. Freeze-drying or freeze substitution methods are satisfactory. Lillie (1962) recommends the use of 5% formic acid if a tissue must be decalcified. Check the tissue each day by the oxalate method (p. 40). Stop decalcification as soon as a negative oxalate test is obtained. Avoid any use of heat. Lillie also suggests that 1 to 3 days of infiltration of the tissue with 1% nitrocellulose before the decalcification can improve the resistance of glycogen to acid extraction. Trott (1961a) uses EDTA, pH 7.5, for decalcification; RDO is satisfactory.

For mucoproteins and mucopolysaccharides, any protein precipitant fixative—Zenker, Helly, or Regaud—is satisfactory (Gomori 1952; Hale 1957; Thompson 1966).

Glycogen Staining

Best Carmine

FIXATION

Avoid aqueous media (McManus and Mowry 1958).

```
absolute alcohol ............................................... 9 parts
formaldehyde (37–40%) ........................................ 1 part
```

Use ice cold.

This method starts dehydration for embedding with 95% alcohol, but the loss of glycogen is about 20–30% (Trott 1961a,b; Trott et al. 1962). Double embedding with nitrocellulose and paraffin can be used to reduce the loss of glycogen.

Mounting Slides

Humason reported better localization of glycogen by mounting paraffin sections with 95% alcohol in place of water. As soon as sections are spread, drain off excess fluid and continue to dry.

Solutions

Best carmine stock solution

carmine	2.0 g
potassium carbonate	1.0 g
potassium chloride 5.0 g	
distilled water	60.0 ml

Boil gently: 5 minutes. Cool. Filter. Add:

ammonium hydroxide	20.0 ml

Lasts 3 months at 0–4°C.

Working solution

carmine stock solution	30.0 ml
ammonium hydroxide	25.0 ml
methyl alcohol	25.0 ml

Lasts 2 to 3 weeks.

Differentiating fluid

absolute alcohol	16.0 ml
methyl alcohol	8.0 ml
distilled water	20.0 ml

Procedure

1. Deparaffinize, coat with nitrocellulose, and transfer into absolute alcohol: 3 minutes.
2. Coat with 1% celloidin; dry slightly in air.
3. Dip 2 or 3 times in 70% alcohol and then into water.
4. Stain in Mayer hematoxylin: 5 minutes.
5. Wash, blue in Scott solution, and wash.
6. Stain with Best carmine working solution: 15 to 30 minutes.
7. Treat with differentiating fluid: 5 to 15 minutes.

8. Rinse quickly in 80% alcohol.
9. Dehydrate, clear, and mount.

Results

glycogen—red
mast cell granules, mucin, and fibrin—light red
nuclei—blue

Comments

Pearse (1968) uses Lison "Gendre-fluid" method of fixation, claiming that it shows a localization of glycogen comparing favorably with that in tissue preserved by freeze-drying.

picric acid, saturated in 96% alcohol	85 parts
formalin	10 parts
glacial acetic acid	5 parts

Cool before use to $-73°C$ in an acetone-CO_2 snow mixture. Use small pieces; fix for 18 hours.

Trott et al. (1962) fix in 1% periodic acid in 10% formalin, which they claim shows greater amounts of glycogen than tissues fixed in acetic acid-formalin-alcohol. Schiff reagent methods are replacing the classical staining methods for glycogen.

Since it has been suggested that carmine staining is due to the hydrogen bonding between glycogen and remaining acid (Goldstein 1962), other dyes that carry groups capable of forming hydrogen bonds also should stain glycogen (Murgatroyd and Horobin 1969). Acid alizarine blue SWR and alizarine brilliant blue BS stained glycogen red. Alizarine red S stained it red; gallein, purple; hematein and hematoxylin, red changing to brown. The staining solution was made as follows: 1 g dye, 1 g potassium carbonate, and 5 g potassium chloride are added to 60 ml of boiling distilled water. Let stand 1 hour. To 20 ml of dye solution add 15 ml of ammonia and 15 ml of absolute methanol in that order. The solution is stable only 24 hours. Stain: 2 to 5 minutes. Dehydrate in 2 to 3 changes of absolute methanol, clear, and mount.

Staining of Acid Mucopolysaccharides

For metachromatic methods see p. 271.

Alcian Blue Method, pH 2.8 (Putt 1971)

At this pH the Alcian blue stains both sulfated and nonsulfated mucosubstances.

Fixation

Any general fixative but formalin-99% alcohol (1:9) or Rossman fluid are recommended.

Solutions

Alcian blue, pH 2.5

calcium chloride ($CaCl_2 \cdot 2H_2O$)	0.5 g
Alcian blue 8GX	1.0 g
distilled water	100.0 ml

Dissolve $CaCl_2$ first. Filter; add thymol crystal to prevent mold.
Kernechtrot, see p. 215.

Control

small intestine

Procedure

1. Deparaffinize and hydrate slides to water.
2. Rinse in 3% acetic acid: 3 minutes.
3. Stain in Alcian blue: 30 minutes.
4. Rinse in water and treat with 0.3% sodium bicarbonate, aqueous: 30 minutes.
5. Wash in running water: 10 minutes.
6. Counterstain with Kernechtrot: 5 minutes.
7. Rinse in distilled water.
8. Dehydrate, clear, and mount.

Results

acid mucopolysaccharides—blue green
nuclei—red to bluish red

Comments

Alcian blue solutions must be acid to prevent the staining of other tissue elements. Alcian green may be substituted for the blue and preceded by the nuclei stained with the celestin blue-hematoxylin combination (p. 106) or Slidders hematoxylin (p. 104). These stains will not be destained by the acidic Alcian solution.

Can be followed by PAS (p. 123) for demonstration of both acidic groups and 1,2 glycols, or the Feulgen reaction (p. 187).

Lison (1954) counterstained with chlorantine fast red for efficient differentiation of mucin from collagen (mucin—bluish green; collagen—cherry red). Follow steps 1, 2, and 3 above, and then substitute the following steps.

4. Treat with phosphomolybdic acid, 1% aqueous: 10 minutes.
5. Rinse in distilled water.
6. Stain in chlorantine fast red: 10 to 15 minutes.

>   chlorantine fast red .... 5B 0.5 g
>   distilled water .... 100.0 ml

7. Rinse in distilled water.
8. Dehydrate, clear, and mount.

Williams and Jackson (1956) note possible diffusion of acid mucopolysaccharides under aqueous conditions. The fixative should form complexes insoluble in water and alcohol. They suggest two possible solutions containing organic chemicals that form such insoluble complexes with acid mucopolysaccharides:

1. 0.5% cetylpyridium chloride (CPC) in 4% aqueous formalin. (See also Conklin 1963 and Wolfe 1964).
2. 0.4% 5-aminoacridine hypochloride in 50% ethyl alcohol.

Cêjková et al. (1973) reduced acid mucopolysaccharide leakage to a minimum with unfixed frozen sections followed by alcohol fixation.

Spicer and Meyer (1960) precede staining with Alcian blue by immersing the tissue for 5 minutes in aldehyde-fuchsin (p. xxx), transferring it to 70% alcohol, and washing it in running water.

Mowry (1960) and Scott and Mowry (1970) report that lots of Alcian blue have differed in staining efficiency and fastness. Do comparison staining.

Alcian blue staining is attributed to the presence of caboxyl groups and in some reactions to sulfates. All acidic mucopolysaccharides contain either sulfates alone or sulfate and carboxyl groups, and the stainability should be consistent. Yamada (1963), however, thinks that the basis for staining remains to be determined precisely and that it is better to examine acidic groups by using metachromatic dyes at different pH levels. See below for staining at pH 1.0 for sulfated carbohydrates.

The Alcian green method (Putt and Huskill 1962) substitutes Alcian green for Alcian blue. Since Alcian green stains more rapidly than Alcian blue, the time required for staining (step 3) can be reduced from 30 minutes to 10 minutes. The procedure and results are identical, except that acid mucopolysaccharides stain deep green rather than blue green.

Carlo (1964) identifies sulfated groups with Alcian blue and carboxyl groups with Alcian yellow.

Wolfe (1964) stains extremely water-soluble mucopolysaccharides in a series of toluidine blue dissolved in decreasing concentrations of alcohol. Horn and Spicer (1964a,b) use azure A at a low pH of 4.0 to 5.0.

For fluorescent staining of mucopolysaccharides, see Saunders (1962).

Alcian Blue Method, pH 1.0 (Luna 1968)

FIXATION
10% neutral buffered formalin

SOLUTION
Alcian blue solution
Alcian blue, 8GX .... 1.0 g
0.1 N hydrochloric acid .... 100.0 ml

CONTROL
small intestine

PROCEDURE
1. Deparaffinize and hydrate sections to distilled water.
2. Stain in Alcian blue: 30 minutes.
3. Do not rinse the sections in water; blot them dry with filter paper.
4. Dehydrate in 95% alcohol, absolute alcohol, and clear and mount.

RESULTS
sulfated mucosubstances—greenish blue
nonsulfated mucosubstances—unstained

Iron Diamine for Acid Mucosubstances

N,N-Dimethyl-meta-phenyldiamine dihydrochloride and its paramonohydrochloride isomer form a colored cationic oxidation product which becomes bound to the anionic site of acidic mucosubstances. Addition of ferric ions seems to promote oxidation and improves staining. In low concentration the iron permits staining of sulfated mucosubstances and sialomucins; with a content of high iron only the sulfomucins are stained. Both methods can be followed by Alcian blue. Consult Spicer et al. (1967) for more details.

High Iron Diamine Method

FIXATION
Any general fixative.

SOLUTIONS
N,N-dimethyl-m-phenylenediamine-dihydrochloride .... 0.12 g
N,N-dimethyl-p-phenylenediamine-monohydrochloride .... 0.02 g

Add *together* to

> distilled water .... 50.0 ml

When dissolved pour immediately (within 5 minutes) into a Coplin jar containing NF ferric chloride (Fisher Scientific Co., ferric chloride solution, 40% W/V purified; use undiluted), 1.4 ml.

PROCEDURE
1. Deparaffinize and hydrate slides to water.
2. Rinse in distilled water and place slides in Coplin jar with diamine solution overnight. Do not crowd slides, one slide only to a slot.
3. Rinse off excess solution in distilled water, but do this briefly and one slide at a time.
4. Dehydrate in 95% and absolute alcohol (ethyl or isopropyl), clear and mount.

RESULTS
most sulfated mucosubstances—purple black
acid mucopolysaccharides lacking sulfate esters—unstained

COMMENTS
Following step 3 the slides can be stained with Alcian blue 2.8 (p. 256) and the sialomucins and hyaluronic acid will be stained. It is recommended that when Alcian blue is not used, to counterstain with orange G after step 3. The orange stain gives better contrast to the background than the dull gray of unstained tissue.

The low iron diamine method uses 0.03 g of *m*-diamine and 0.005 g of *p*-diamine with 0.5 ml of ferric chloride. The rest of the procedure is exactly like that of the high iron diamine. The low iron stains sulfated and carboxylic acid mucopolysaccharides gray to purple to purple black.

Warning: The diamine chemicals seem to have a limited life even when stored in the refrigerator. Six months of refrigerated storage is pushing the limit. Diamine chemicals are also highly toxic.

Colloidal Iron Method (Mowry 1958)

SOLUTIONS
Colloidal iron stock solution
> ferric chloride 29% .... 4.4 ml
> distilled water .... 250.0 ml

Bring water to boil. While the water is boiling, pour in the ferric chloride solution. Stir. When solution is dark red, remove from heat and allow to cool. The water must be dark red and clear. It is stable for months.

Working solution

glacial acetic acid	5.0 ml
distilled water	15.0 ml
stock colloidal iron solution	20.0 ml

pH should be 1.1 to 1.3. Nonspecific staining takes place at pH 1.4 or higher. Effective for 1 day.

Hydrochloric acid-ferrocyanide

2% HCl (2 ml/98 ml water)	50.0 ml
2% potassium ferrocyanide (2 g/100 ml water)	50.0 ml

Mix immediately before use.

CONTROL

small intestine

PROCEDURE

1. Deparaffinize and hydrate slides to water.
2. Rinse in 12% acetic acid (12 ml/88 ml water): 30 seconds (to prevent dilution of reagent).
3. Transfer to freshly prepared colloidal iron (working solution): 60 minutes.
4. Rinse in 12% acetic acid, 4 changes: 3 minutes each.
5. Treat with hydrochloric acid-ferrocyanide, room temperature: 20 minutes. (Include a control slide, not treated with colloidal iron.)
6. Wash in running water: 5 minutes.
7. Counterstain if desired: van Gieson Solution or hematoxylin.
8. Optional: dip in aqueous picric acid to color cytoplasm and erythrocytes; rinse for a few seconds in tap water. (Acetic-orange G can also be used.)
9. Dehydrate, clear, and mount.

RESULTS

acid mucopolysaccharides—bright blue; uncolored in control slide
Mucins of connective tissue, epithelium, mast cell granules, capsules of some microbial agents, pneumococci—bright blue

COMMENTS

Hale (1957) was the first to suggest the use of colloidal iron for the demonstration of acid polysaccharides, claiming that acidic groups can react with metals and cations. In order that such acidic groups may be demonstrated, the acid moiety should be a firmly bound part of a fixed and insoluble tissue constituent and be free to react. Also the link between the metal and acid should be strong enough that it will not be disrupted when water is used to wash away excess unbound metal. Using a positively charged colloidal solution is the easiest way to prevent this disruption. The metal is held tightly, probably by the multiple valencies of both reacting particles. Since the range of reactivity of the colloidal iron reagent depends on the pH of the solution, it is possible to stain strongly acid groups selectively by making the solution strongly acid. If strongly acidic groups are blocked by bound iron, it should be possible to stain the weakly acid groups with another colloidal solution. Wolman (1956, 1961) developed his Bi-Col staining technique by using two colloidal metal solutions—colloidal iron and colloidal gold (Wolman 1956).

Mucin

Fluorescent Method (Hicks and Matthaei 1958)

FIXATION
    10% neutral buffered formalin preferred.

SOLUTIONS
    Iron alum
        ferric ammonium sulfate .... 4.0 g
        distilled water .... 100.0 ml

    Acridine orange
        acridine orange .... 0.1 g
        distilled water .... 100.0 ml

CONTROL
    small intestine

PROCEDURE
    1. Deparaffinize and hydrate paraffin sections to water or cut frozen sections and place in distilled water.
    2. Treat with iron alum solution: 5 to 10 minutes.
    3. Wash briefly in distilled water.
    4. Stain in acridine orange: 1.5 minutes.
    5. Rinse briefly in distilled water, blot, and mount in fluorescent mountant.

RESULT
    mucin—brilliant reddish orange fluorescence

COMMENTS
    This is not a permanent slide. The iron inhibits the production of fluorescence with acridine orange in nearly all tissue components except mucin.

Hicks and Matthaei (1955) also suggest that the mucins of acid polysaccharides (rather than mucoproteins) fluoresce.

For a fluorescent antibody technique, see Kent (1961).

Lev and Stoward (1969) use eosin as a fluorescent dye to demonstrate mucus cells. The concentration of eosin can be 0.1% to 10%; the pH should be 9.5 to 5.7 but no lower. Staining time is 5 to 10 minutes. The fluorescence is stronger after Carnoy than after formalin fixation. Isopropyl extracts less stain than does ethyl alcohol, thereby producing a stronger fluorescence. Hematoxylin used with the eosin improves the contrast because of the orange-red color it gives to nuclei.

20
Staining Cellular Elements

THE ARGENTAFFIN REACTION

The argentaffin reaction should not be confused with silver impregnation (p. 143). In argentaffin reaction some substances (ascorbic acid, aldehydes, uric acid, polyphenols, and others) reduce silver solutions under specific conditions. This reaction with the tissue itself can therefore be used histochemically to identify these substances.

The only source of error in the method is found in calcification areas, but only if these are in large masses. Most silver phosphates and carbonates will be dissolved out during the process. If not, treat the slides with 0.2–0.5% nitric acid or hydrochloric acid in absolute alcohol: 2 to 3 minutes before step 3. Wash off the acid with 95% alcohol and proceed with step 3.

ENTEROCHROMAFFIN CELL STAINING

Fontana Method (Culling 1957) **and Methenamine Silver** (Gomori 1952, 1954b)

FIXATION

10% neutral buffered formalin. Do not use alcohol—it dissolves argentaffin granules. During embedding, do not expose to paraffin longer than 30 minutes.

SOLUTIONS

Silver solution A (Fontana method)
silver nitrate, 10% (2.5 g/25 ml water) .... 25.0 ml
ammonia, 28% .... as needed
distilled water .... 25.0 ml

To silver nitrate add ammonia drop by drop until the precipitate that forms is almost dissolved. Add distilled water. Store for 24 hours in brown bottle. Filter before use. Good for 2 weeks.

Methenamine silver solution B (Gomori method)
Solution I

methenamine, 3% (3 g/100 ml water) .... 100.0 ml
silver nitrate, 5% (5 g/100 ml water) .... 5.0 ml

Shake until white precipitate disappears. Keeps for several months.
Solution II

> $M/5$ boric acid (12.368 g/1000 ml water) .... 80.0 ml
> $M/5$ borax (19.071 g/1000 ml water) .... 20.0 ml

The pH should be 7.8 to 8.0.

Working solution
> Solution I .... 30.0 ml
> Solution II .... 8.0 ml

Gold chloride
> gold chloride stock solution, 1% .... 10.0 ml
> distilled water .... 100.0 ml

Safranin (or Kernechtrot, p. 215)
> safranin O .... 1.0 g
> distilled water .... 100.0 ml

Add a few drops of glacial acetic acid.

CONTROL
small intestine and integument

PROCEDURE
1. Deparaffinize and hydrate slides to water.
2. Treat with Lugol solution (p. 463): 30 minutes to 1 hour.
3. Wash in running water: 3 minutes.
4. Bleach in 5% aqueous sodium thiosulfate: 3 minutes.
5. Wash in running water: 4 minutes. (Steps 2 to 4 suppress background staining.)
6. Treat with silver solution A in the dark, room temperature: 18 to 48 hours. *Alternate method:* Use methenamine silver solution B at 60°C for 3.5 hours or at 37°C for 12 to 24 hours—until enterochromaffin (EC) cells stand out black.
7. Rinse in several changes of distilled water.
8. Tone in gold chloride: 10 minutes
9. Rinse in distilled water.
10. Fix in 5% aqueous sodium thiosulfate: 3 minutes.
11. Wash in running water: 5 minutes.
12. Counterstain in safranin (or Kernechtrot).
13. Rinse in 70% alcohol: few seconds.
14. Dehydrate, clear, and mount.

Results

argentaffin granules—black
melanin—black
other tissue elements—reds and pinks
If background is grayish to blackish, then slide is overstained.

Comments

Chromaffin material is found only in the adrenal medulla and in the argentaffin cells of the gastrointestinal tract. It received its name because of its reaction with chromium salts and certain other metals to produce a yellowish to brown color, a brownish chromium oxide. This material can be detected in cells without staining. To preserve this material, either use a dichromate fixative or postchromate with potassium dichromate (p. 40). Melanin, also brownish, is not chromaffin substance.

Methenamine silver methods are also used for fungi and bacteria (see p. 314).

Enterochromaffin cells and granules are argyrophilic ("silver-loving") and are impregnated by silver solutions when a reducer is present. This method, however, shows other cells as well, as it is not specific.

Diazo-Safranin Method (Lillie et al. 1953)

Fixation

10% formalin buffered with 2% calcium acetate: 1 to 3 days.

Solution

Diazotized safranin

The three parts of the diazotized safranin solution can be kept as stock solutions. To prepare the working solution, they must be mixed *immediately before use.*

Acid safranin

safranin O	3.6 g
distilled water	60.0 ml
N HCl	30.0 ml

This solution will remain stable for several weeks.
N sodium nitrite

sodium nitrite ($NaNO_2$)	6.9 g
distilled water	100.0 ml

Keep in refrigerator, stable for 3 months.
Disodium phosphate, $M/10$

disodium phosphate (Na$_2$PHPO$_4$), anhydrous .... 14.2 g
distilled water .... 1000.0 ml

To prepare the working solution, add 0.5 ml of sodium nitrite solution to 4.5 ml of ice-cold safranin solution. The resulting mixture turns deep blue and foams. Keep at 0–5°C for 15 minutes for diazotization. Dilute 1 ml of the solution with 40 ml of ice-cold disodium phosphate solution and use immediately (pH should be about 7.7).

CONTROL
small intestine

PROCEDURE
1. Deparaffinize and hydrate slides to distilled water.
2. Place slides in a previously chilled Coplin jar and pour over them the freshly prepared diazo-safranin solution: 5 minutes.
3. Decant stain and wash slides in 3 changes of $N/10$ aqueous hydrochloric acid or acid-alcohol (1 ml concentrated HCl/99 ml 70% alcohol): 10 to 13 seconds total, to remove most of adherent stain. Longer extraction lightens the colors but does not improve contrast.
4. Wash off acid with water (if $N/10$ HCl is used) or 95% alcohol (if acid-alcohol is used).
5. Dehydrate and mount.

RESULTS
enterochromaffin granules—black
gastric gland, chief cell granules—dark red
Paneth cell granules—red
mucin—unstained
nuclei and cytoplasm—pink to red

Ferric-Ferricyanide Method (Modified Schmorl Technique) (Lasky and Greco 1948; Lillie and Burtner 1953a)

FIXATION
10% formalin buffered with 2% of calcium acetate: 1 to 3 days.

SOLUTION
Ferric-ferricyanide
potassium ferricyanide, 1% aqueous .... 10.0 ml
ferric chloride, 1% aqueous .... 75.0 ml
distilled water .... 15.0 ml

CONTROL
small intestine

Procedure

1. Deparaffinize and hydrate slides to water.
2. Stain in freshly prepared ferric-ferricyanide: 5 minutes.
3. Rinse in 3 changes of distilled water.
4. Counterstain in safranin or the like.
5. Dehydrate, clear, and mount.

Results

enterochromaffin granules—dark blue
nuclei—red

Comments

Lillie and coworkers devoted several years to study of the enterochromaffin granules, melanins, other pigments, and related substances (Glenner and Lillie 1957b; Lillie 1955, 1956a,b,c, 1957a,b,c, 1960, 1961; Lillie et al. 1953, 1957, 1960).

Azo-Coupling Method (Gurr 1958)

Fixation

Formalin or Bouin

Solutions

Garnet reagent

garnet GBC salt	0.5 g
distilled water	100.0 ml
borax, saturated aqueous	2.5 ml

If it is necessary to purify the garnet GBC, see comments on p. 350.

Control

small intestine

Procedure

1. Deparaffinize and hydrate to water.
2. Transfer to garnet solution: 30 to 60 seconds.
3. Wash in running water: 30 seconds.
4. Stain in Mayer (or other) hematoxylin: 3 minutes.
5. Wash in running water: 5 minutes.
6. Dehydrate, clear, and mount.

Results

argentaffin granules—red
nuclei—blue
background—light yellow

AMYLOID STAINING

Congo Red (Puchtler et al. 1962)

FIXATION

10% neutral buffered formalin or alcoholic fixative. Carnoy fixative increases red binding of the Congo red.

SOLUTIONS

Solution A

    1% aqueous sodium hydroxide .... 0.5 ml
    80% ethanol saturated with NaCl .... 50.0 ml

Use within 15 minutes.

Solution B

    80% ethanol saturated with Congo red and NaCl .... 50.0 ml
    1% aqueous sodium hydroxide .... 0.5 ml

Filter. Use within 15 minutes.

CONTROL

known positive tissue

PROCEDURE

1. Deparaffinize and hydrate slides to water.
2. Stain in Mayer hematoxylin: 10 minutes.
3. Rinse in 3 changes distilled water: few seconds each.
4. Treat with solution A: 20 minutes.
5. Stain in solution B: 20 minutes.
6. Dehydrate rapidly in 3 changes absolute alcohol, clear, and mount.

RESULTS

amyloid—red
nuclei—blue

COMMENTS

Considerable research is being done on amyloid, a predominately extracellular deposition of protein-mucopolysaccharide complexes and other substances. The disease occurs principally in aging tissues of the animal species and has been noted in considerable quantity in Alzheimer patients. Staining is not always reliable and may not stain small deposits. Black and Jones (1971) consider frozen sections the best aid in amyloid detection. Freeze immediately without fixation.

The working solutions must be freshly made, but their major components,

80% ethanol saturated with NaCl and 80% alcohol saturated with Congo red and NaCl, will keep for several months. The latter should stand 24 hours before use.

This method is used in preference to Bennhold's or Highman's because in it there is no need for differentiation. Bennhold's method lacks uniformity, and Highman's easily overdifferentiates.

Staining is more intense in alcohol-fixed tissues than in formalin-fixed tissues, but even in the latter small deposits of amyloid are visible.

Puchtler and coworkers suggest that the selectivity of Congo red for amyloid may be due to the polysaccharide moiety of amyloid. They also found that certain direct cotton dyes stain amyloid. See Puchtler and Sweat (1965) and Puchtler and Sweat (1966) for a review of the concepts of amyloid substance. See also metachromatic methods (p. 271). Amyloid may also be detected using polarization microscopy (see Chap. 8).

Navagiri and Dubey (1976) use a Leishman stain; it stains normal parenchymal cells faint to deep blue and the amyloid, purple to violet.

Fluorescent Method

Thioflavine T is recommended for amyloid by Vassar and Culling (1959, 1962), but it colors amyloid only faintly. Puchtler and Sweat (1965) use their Congo red method (see above) with excellent fluorescence and pink to red coloring of amyloid and with other tissues greenish gray. Mount in fluorescent mountant for a black background. Synthetic mountants have only moderate fluorescence, and slides mounted with them can be used for fluorescent observation; the faintness of the fluorescence does not interfere with slide evaluation.

Fluorescent Method (Jagatic and Weiskopf 1966)

FIXATION
   Any general fixative.

SOLUTIONS
   Weigert hematoxylin, see p. 103.

   Acridine orange
      acridine orange .... 0.1 g
      distilled water .... 100.0 ml

CONTROL
   mesentery

PROCEDURE
   1. Deparaffinize and hydrate sections to water.
   2. Stain in hematoxylin: 5 minutes.

3. Wash in running water: 3 to 5 minutes.
4. Stain in acridine orange: 5 to 6 minutes.
5. Rinse in distilled water: 1 minute.
6. Dehydrate, clear, and mount in fluorescent mountant.

RESULT
mast cell granules—bright red orange

COMMENTS
Old H&E slides have been restained by this method; also sections from paraffin blocks 10 to 15 years old.

METACHROMASIA

A few tissue elements are stainable by a particular group of cationic dyes, changing in the tissues from blue (the usual orthochromatic form) to purplish red or reddish purple. Such a dye, called *metachromatic,* is of considerable value in the study of specific elements of connective tissue. Among the metachromatic dyes most commonly used are toluidine blue O, thionine, methylene blue, azures, crystal violet, cresyl violet, methyl violet, safranin O, celestine blue, gallocyanin, and pinacyanole. Some of the tissues identified by this means are mast cells, amyloid, cartilage, and mucus materials (Schubert and Hamerman 1956).

The methods are tricky, and a technician must learn to distinguish a true metachromasia from a false one. The difficulty in preserving metachromasia lies in the dehydration of the tissue after staining. Increasing strengths of ethyl alcohol revert the dye back to the orthochromatic form. Sections can be examined in an aqueous condition, but this produces at best a semipermanent preparation. Some workers use acetone, isopropyl, or tertiary butyl alcohol for dehydration and then follow with one of the clearing agents. Always stain at room temperature; heat can destroy metachromasia.

Amyloid

Crystal Violet (Lieb 1947)

FIXATION
10% formalin or alcohol fixative.

SOLUTION
Crystal violet stock solution
  crystal violet .... 14.0 to 15.0 g
  saturated in 95% ethyl alcohol .... 100.0 ml

Working solution

```
stock solution ................................................. 10.0 ml
distilled water ................................................ 300.0 ml
hydrochloric acid, concentrated ................................ 1.0 ml
```

CONTROL
known positive tissue

PROCEDURE
1. Run frozen sections or deparaffinized sections down to water.
2. Stain in working solution: 5 minutes to 24 hours.
3. Rinse in water.
4. Mount in glycerin jelly.

RESULTS
amyloid—purple
other tissue elements, including hyalin—blue

COMMENTS
Acid in the staining solution makes it self-differentiating, and staining time is flexible. Lieb suggests that, if there is only a small amount of amyloid in the tissue, thicker sections should be cut to make the color reaction more clearly visible.

Mucin

Toluidine Blue (Lillie 1929)

FIXATION
Any general fixative, but an alcoholic one is preferred. Do not use a chromate fixative.

CONTROL
small intestine

PROCEDURE
1. Deparaffinize and hydrate slides to water; remove $HgCl_2$.
2. Stain in toluidine blue: 1 minute.
3. Wash in water: 2 to 3 minutes.
4. Dehydrate in acetone, 2 changes: 3 to 5 minutes each.
5. Clear in xylene and mount.

RESULTS
mucin—reddish violet
nuclei and bacteria—deep blue

cytoplasm, fibrous tissue—bluish green
bone—bluish green
cartilage matrix—bluish violet
muscle—light blue
cell granules—blue violet
hyalin and amyloid—bluish green

COMMENTS

Gomori (1952) uses 0.02–0.05% toluidine blue O in citrate buffer, pH 3.5 to 4.5: 10 to 15 minutes, or until the nuclei are blue and the mucin, intense pink.

For nonmetachromatic staining of mucin, see p. 262.

Thionine (Mallory 1944)

FIXATION

Any general fixative, but an alcoholic one is preferred.

SOLUTION

thionine	1.0 g
25% alcohol	100.0 ml

CONTROL

small intestine

PROCEDURE
1. Deparaffinize and hydrate slides to water; remove $HgCl_2$.
2. Stain in thionine: 15 minutes to 1 hour.
3. Differentiate in 95% alcohol.
4. Dehydrate in absolute alcohol, clear, and mount.

RESULT

mucin—light to dark red or purple

Acid Mucopolysaccharides

Thionine (Gurr 1958)

FIXATION

10% neutral buffered formalin or other general fixative.

SOLUTIONS

Thionine

thionine, saturated aqueous	0.5 ml
distilled water	100.0 ml

Molybdate-ferricyanide solution
    ammonium molybdate, 5% aqueous .... 50.0 ml
    potassium ferricyanide, 1% aqueous .... 50.0 ml

Make up solutions fresh each time.

CONTROL
    small intestine

PROCEDURE
1. Deparaffinize and hydrate slides to water.
2. Stain in thionine: 5 to 15 minutes.
3. Rinse in distilled water.
4. Treat with molybdate-ferricyanide: 2 minutes.
5. Wash in distilled water: 2 to 3 minutes.
6. Dehydrate, clear, and mount.

RESULTS
    acid mucopolysaccharides—purple
    other cell elements—bluish

COMMENTS
    The molybdate-ferricyanide solution prevents loss of metachromasia.
    See Kuyper (1957) for suggestions concerning fluorescent methods for mucopolysaccharides. See also acid mucopolysaccharides, p. 273.

MAST CELL STAINING

Mast cells are of common occurrence in connective tissue. Because of their cytoplasmic granules, however, staining methods for these cells have been included here. The specific staining of these granules is the primary means of identification of mast cells.

Thionine Method (Lillie 1965)

FIXATION
    Any general fixative.

SOLUTION
    thionine .... 0.5 g
    0.01 M acetate buffer (see results) .... 100.0 ml

CONTROL
    small intestine

PROCEDURE
1. Deparaffinize and hydrate slides to water; remove $HgCl_2$.
2. Stain in thionine: 30 minutes for light stain.
3. Rinse in water.
4. Dehydrate, clear, and mount.

RESULTS
mast cell granules—red purple
nuclei—faint blue violet

COMMENTS

At pH 2, metachromasia stains only mast cells and cartilage; at pH 5, muscle and connective tissue stain light green; at pH 3 to 2, cytoplasm stains poorly.

Quick Toluidine Blue Method

FIXATION
Any general fixative.

SOLUTION
toluidine blue O	0.2 g
60% alcohol	100.0 ml

CONTROL
save a previous positive tissue

PROCEDURE
1. Deparaffinize and run slides down to 60% alcohol. Remove $HgCl_2$ if present.
2. Stain in toluidine blue: 1 to 2 minutes.
3. Rinse quickly in tap water.
4. Dehydrate in acetone, 2 changes: 2 to 3 minutes each.
5. Clear in xylene and mount.

RESULTS
mast cells—deep reddish purple
background—faint blue

Toluidine Blue (Conroy and Toledo 1976)

FIXATION

Formalin-alcohol preferred; 10% neutral buffered formalin satisfactory.

Donaldson et al. (1973) observed that, after mercuric chloride fixation and iodine treatment, mast cell staining was suppressed. Shorten the fixing time to 4 to 6 hours, wash in 80% alcohol: 40 hours to 6 days, and omit iodine treatment.

formalin	10.0 ml
95% alcohol	90.0 ml

SOLUTION

toluidine blue O	0.1 g
distilled water	100.0 ml

pH should be 6.8 to 7.2. If this is not possible with distilled water, make up the stain in a buffer, such as McIlvaine (p. 474).

CONTROL
   known positive tissue section

PROCEDURE
1. Deparaffinize and hydrate sections to water.
2. Stain lightly with hematoxylin (1 minute Harris, 3 minutes Mayer, for example), wash, and blue.
3. Stain with eosin and rinse off excess eosin with 95% and absolute ethyl alcohol: 3 dips each.
4. Rinse in tap water: 5 dips.
5. Stain in toluidine blue O: 7 to 10 seconds, agitate gently.
6. Rinse in distilled water: 1 dip.
7. Dip once in 95% alcohol.
8. Dehydrate in absolute alcohol, 2 changes: 4 dips and 6 dips. Isopropanol may be used and the dehydrate time lengthened.
9. Dehydrate in absolute alcohol, 2 changes, 10 dips each.
10. Xylene, 2 changes: 2 minutes each, and mount.

RESULTS
   mast cells—blue to purple
   collagen—pink
   cytoplasm—pink
   nuclei—blue
   erythrocytes—red

ENDOCRINE GLAND STAINING

Pituitary cells

The adenohypophysis (the epithelial component of the pituitary gland) of all vertebrates is composed of cells that range in orientation from masses that are strictly segregated by cell type (bony fishes) to more randomized mixtures of hormone-producing cells (birds and mammals). (See Schreibman 1986 for a detailed discussion of the comparative structure of the vertebrate pituitary gland.) It is essential in analyzing pituitary structure and function to be able to identify the various cell types according to the function that they serve. In the classical studies this was accomplished by applying stains, either singly or in combination, and then recognizing them on the basis of their tinctorial affinities. Staining methods have progressed from the standard hematoxylin and eosin to the trichrome and tetrachrome methods. Eventually, stains were used to localize specific chemical components of the hormones they produced through the application of such procedures as periodic acid Schiff, Alcian blue, aldehyde fuchsin, and lead hematoxylin. Electron microscopy confirmed the heterogeneity of cells as determined by the above methods and introduced the concept of identifying cells on the basis of granule size and ultrastructural characteristics. On the basis of these parameters, it was possible to recognize five or six different cell types that correlated well with the number of hormones suspected to originate in the adenohypophysis (Schreibman 1986). When these methods of analysis and others, such as autoradiography, were coupled to physiological experiments, it was no longer necessary to use the cumbersome and confusing system of nomenclature by Greek letter, and we could now identify the cell according to the functional role of the cell (i.e., the hormone elaborated as, e.g., prolactin cell, somatotrope).

In recent years the application of immunohistochemical techniques has served well in confirming identification or reevaluating the functional role of the endocrine cell (see p. 276). The more recent development of in situ hybridization has taken us a step further in functional cytology by allowing the investigator to distinguish cells that actively produce a substance from those that merely store or incorporate it.

Some of the more useful stains used routinely in the laboratory, each valuable for some particular cell or group of cells, are the trichrome (e.g., Mallory) or tetrachrome (e.g., Herlant) types, PAS, aldehyde fuchsin, and lead hematoxylin (alone or in combination with PAS).

In the anterior pituitary, the glandular cells are frequently classified as either *chromophils* (accept stains) or *chromophobes* (remain unstained). In most animals the majority of the chromophils are normally acidophilic (*acidophils*), and the remainder are basophilic (*basophils*). While these terms are traditionally applied, we hasten to add that these terms are not accurate, since cells classified as basophils stain with one or the other of the trichrome stains, all of which are acid stains.

There are several ways to categorize pituitary cell types; what we present here covers several approaches and is generally accepted. Acidophils are type 1 (stain with azocarmine, acid fuchsin, and erythrosin and produce prolactin) or type 2 (stain with orange G and produce growth hormone). Basophils are type 1 (stain with PAS, aldehyde fuchsin, Alcian blue, and aniline blue and produce thyrotropin) or type 2 (similar staining characteristics as the type 1 cells except that in some vertebrate classes they also stain with aldehyde fuchsin and contain orange G granules; they produce the gonadotropins). Chromophobes or the type 3 cells have little affinity for stains except for lead hematoxylin, which colors them black; they produce adrenocorticotropin (ACTH) or melanocyte-stimulating hormone (MSH). Separating the chromophobes from the chromophils is not always as simple as it may sound; transitional stages, or the state of secretion of the cells, can make the two types of cells difficult to identify.

The cells of the pars tuberalis contain no cytoplasmic granules. Those of the pars intermedia are pale, with a few of the cells containing basophilic granular cytoplasm. The pars nervosa (posterior lobe or neurohypophysis) does not show the well-organized cell structure of the adenohypophysis but reflects, rather, the preponderance of neuronal tracts conveying neurosecretory products (the neurosecretory substance, NSS) and networks of vascular elements. Pituicytes are a class of neuroglial elements whose processes make contact with, and envelope, axon endings and perivascular space. Gomori chrome hematoxylin is perhaps the most commonly used method for its demonstration, but Bargmann's modification is preferred. The substance is easily identified, even though nuclei, Nissl substance, lipofuscin, and the basophils of the anterior pituitary are also stained. Since the NSS is so rich in cystine, the performic acid-Alcian blue method of Adams and Sloper (1955, 1956) and Sloper and Adams (1956) also demonstrates the substance.

Controls for Pituitary Gland Staining

One of the special characteristics of the teleost (bony fishes) pituitary gland is that all of the adenohypophysial cells are separated into clearly defined areas according to function. Therefore, if a stain for a particular cell type or hormone needs to be evaluated, midsagittal sections of fish pituitaries, run alongside the test material, are good controls, for they present the complement of cells to assess staining characteristics or to evaluate the results of an experimental procedure.

Heidenhain Azan (see p. 124 for details of procedure)

This method takes longer to execute than most of the other stains and requires personal attention, especially during the differentiation steps, but the brilliant staining and the ability to identify clearly many of the cells in the adenohypophysis makes it highly desirable. Masson trichrome (p. 126) also permits easy identification of cells with less difficulty than the Heidenhain azan protocol. We have adapted this method for automated staining on the Autotechnicon (although any automatic processor will probably work) because of the consistency in the required times for the individual steps and the uniformity of the results; Schreibman finds this approach extremely useful when large numbers of pituitary sections need to be stained.

Herlant (1960) **Pituitary Stain I**

FIXATION
Any good general fixative, but Zenker formol (Helly fixative) preferred.

SOLUTIONS
Erythrosin

erythrosin B	1.0 g
distilled water	100.0 ml

Mallory II

aniline blue WS	0.5 g
orange G	2.0 g
distilled water	100.0 ml
glacial acetic acid	1.0 ml

Acid alizarine blue

acid alizarine blue	0.5 g
aluminum sulfate	10.0 g
distilled water	100.0 ml

Bring to boil, approximately 5 minutes. Cool. Adjust to 100.0 ml with distilled water and filter. Stable, but add thymol.

Phosphomolybdic acid

phosphomolybdic acid	5.0 g
distilled water	100.0 ml

CONTROL
pituitary gland (preferably from a small fish)

PROCEDURE
1. Deparaffinize and hydrate to water; remove $HgCl_2$, if present.
2. Stain in erythrosin: 5 minutes.
3. Rinse briefly in distilled water.
4. Stain in Mallory II: 5 to 10 minutes.
5. Rinse briefly in distilled water.
6. Stain in acid alizarine blue: 5 to 10 minutes.
7. Rinse briefly in distilled water.
8. Treat with phosphomolybdic acid: 5 to 10 minutes.
9. Rinse briefly in distilled water.
10. Differentiate in 70% alcohol.

11. Clarify stain in 70% alcohol with a few drops of acetic acid added.
12. Dehydrate, clear, and mount.

Results

somatotropes—yellow
gonadotropes—violet or pale blue
thyrotropes—dark blue
prolactin cells—red to rose
nucleus—dark blue to violet

Herlant (1960) Pituitary Stain II

Fixation

Bouin or Hollande Bouin preferred.

Solutions

Permanganate solution

2.5% aqueous potassium permanganate	10.0 ml
5% aqueous sulfuric acid	10.0 ml
distilled water	60.0 ml

Mix just before use.

Alcian blue, pH 3.0

Alcian blue 8GX	1.0 g
distilled water	100.0 ml
glacial acetic acid	1.0 ml

Alcian blue, pH 0.2

Alcian blue 8GX	1.0 g
10% sulfuric acid	100.0 ml

The Alcian blue dissolves less easily in the acid solution than in water. Warm the solution until stain dissolves, cool, and filter. Good for 1 month; filter occasionally.

Sodium metabisulfite

sodium metabisulfite	5.0 g
distilled water	100.0 ml

Periodic acid

periodic acid	1.0 g
distilled water	100.0 ml

Schiff reagent, see p. 182.

CONTROL
small intestine

PROCEDURE
1. Deparaffinize and hydrate slides to water.
2. Oxidize in permanganate solution: 1 to 2 minutes.
3. Rinse in distilled water: few seconds.
4. Bleach in metabisulfite solution: 1 minute.
5. Wash in running water: 5 minutes.
6. Rinse in distilled water.
7. Stain in Alcian blue: 15 to 30 minutes.
8. Wash in running water: 5 to 10 minutes.
9. Treat with periodic acid: 10 minutes.
10. Rinse in distilled water: 1 to 2 minutes.
11. Treat with Schiff reagent: 10 to 20 minutes.
12. Wash in 3 changes of metabisulfite solution: total of 5 minutes.
13. Wash in running water: 15 minutes.
14. Dehydrate, clear, and mount.

RESULTS AND COMMENTS
Step 13 can be followed by staining with hematoxylin if desired.

This method is sometimes used to separate the two kinds of gonadotropes by tinctorial methods; results should be confirmed by immunohistochemical procedures. The potassium permanganate oxidation confers on the somatotropes (growth hormone–producing cells) only a faint affinity for the PAS reaction. They will stain a faint rose, much less marked than the color of the gonadotropes (gonadotropin–producing cells). If step 13 is followed by staining with 1% orange G, the rose color of the somatotropes is masked, and they become yellow; gonadotropes, violet or brick red; thyrotropes (thyroid-stimulating hormone [TSH]), dull blue.

With Alcian blue, increased acidity improves staining specificity. At pH 3, the thyrotropes and gonadotropes show a similar affinity for the stain and react only feebly with PAS. At pH 0.2 the two cells are more easily distinguished; the gonadotropes are paler than the thyrotropes, and PAS does not change the blue of the latter, but colors the gonadotropes violet. Apparently the gonadotropes are so weakly stained by the Alcian blue that they do not mask the PAS.

Ewen (1962) Modification of Cameron and Steele (1959) Aldehyde-Fuchsin Method

This modification increases the method's specificity.

Fixation

Fix in Bouin with 0.5–1.0% trichloroacetic acid instead of acetic acid. Helly is also good.

Solution

Aldehyde-fuchsin

Add 1 g basic fuchsin to 200 ml boiling water: boil 1 minute. Cool and filter. Add 2 ml concentrated hydrochloric acid and 2 ml paraldehyde. Leave tightly capped at room temperature. When mixture has lost reddish fuchsin color and is deep purple (3 to 4 days), filter it and discard filtrate. Dry precipitate on filter paper in an oven. Remove and store in bottle. Makes about 1.9 g. To make stock solution, dissolve 0.75 g in 100.0 ml of 70% ethyl alcohol. Keeps 6 months.

Paraldehyde decomposes readily. Freshly opened paraldehyde will give excellent results, but a bottle that has been open for a long time may give negative results (Gairdner 1969).

Working solution
- stock solution . . . . 0.75 g
- 70% alcohol . . . . 75.0 ml
- glacial acetic acid . . . . 1.0 ml

Potassium permanganate, 0.3%
- potassium permanganate . . . . 0.3 g
- distilled water . . . . 100.0 ml
- sulfuric acid, concentrated . . . . 0.3 ml

Sodium bisulfite, 2.5%
- sodium bisulfite . . . . 2.5 g
- distilled water . . . . 100.0 ml

Halmi mixture (1950)
- distilled water . . . . 100.0 ml
- light green SF, yellowish . . . . 0.4 g
- orange G . . . . 1.0 g
- chromotrope 2R . . . . 0.5 g
- glacial acetic acid . . . . 1.0 ml

Keeps indefinitely.

Control

aorta, integument

Procedure
1. Deparaffinize and hydrate slides to water; remove $HgCl_2$.
2. Oxidize in potassium permanganate: 1 minute.

3. Rinse in distilled water.
4. Bleach in sodium bisulfite until permanganate color is removed.
5. Wash in running water: 5 minutes.
6. Transfer to 70% alcohol: 2 minutes.
7. Stain in aldehyde-fuchsin: 2 to 10 minutes.
8. Wipe off back of slide and rinse in 95% alcohol.
9. Differentiate in acid alcohol (0.5 ml HCl/100.0 ml absolute ethyl alcohol): 10 to 30 seconds.
10. Transfer through 70% alcohol and water.
11. Mordant in the following: 10 minutes.

   phosphotungstic acid .... 4.0 g
   phosphomolybdic acid .... 1.0 g
   distilled water .... 100.0 ml

12. Rinse in water and stain for 1 hour in a modified Halmi's mixture.
13. Wipe off back of slide, differentiate in 95% alcohol plus 0.2% acetic acid: 2 to 3 minutes.
14. Rinse in fresh 95% alcohol.
15. Dehydrate in absolute alcohol, clear, and mount.

RESULTS
granulation of gonadotropes—dark purple
thyrotropes—green
acidophilic granules—orange
nucleoli—bright red

Lead Hematoxylin Method (Solcia et al. 1969)

FIXATION
An aldehyde fixative is best.

SOLUTION
Stock stabilized lead solution
Mix equal volumes of 5% aqueous lead nitrate and saturated aqueous ammonium acetate. Filter. Add 2 ml of 40% formalin for every 100 ml of filtrate. Store at room temperature. Keeps for several weeks.

Working solution
Add 0.2 g hematoxylin dissolved in 1.5 ml 95% ethanol to 10 ml of lead stock solution. Dilute with 10 ml of distilled water. Stir repeatedly. After 30 minutes, filter. Makes up to 75 ml with distilled water. Use immediately.

CONTROL
pancreas

PROCEDURE
1. Deparaffinize and hydrate to water.
2. Stain in lead hematoxylin, 37°C: 2 to 3 hours.
3. Wash in running water: 10 minutes.
4. Dehydrate, clear, and mount.

RESULTS
Pancreatic islet A and D cells, thyroid C cells, pituitary gland MSH and ACTH cells—dark blue black
nucleoli and nuclear chromatin—dark blue
calcium deposits, nerve fibers, A and Z bands of striated muscle—blue

Chrome-Hematoxylin, Bargmann Modification for Neurosecretory Substance (NSS) (Pearse 1968)

The collective term *neurosecretory substance* refers to the various peptides produced in the brain that ultimately make their way to the pituitary gland. They include the hormones vasopressin and oxytocin, which are produced in the hypothalamus but are released from axonal endings that abut blood vessel in the posterior lobe. They also include the factors (hormones) that control the synthesis and release of pituitary hormones.

FIXATION
Bouin preferred. Susa and Stieve satisfactory. If tissue is alcohol-fixed, float sections on Bouin solution instead of on water when mounting sections on slides.

SOLUTIONS
Bouin chrome alum
  Bouin solution (p. 27) .... 100.0 ml
  chrome alum .... 3.0 to 4.0 g

Potassium permanganate-sulfuric acid
  2.5% aqueous potassium permanganate .... 1 part
  5% aqueous sulfuric acid .... 1 part
  distilled water .... 6 to 8 parts

Oxalic acid
  oxalic acid .... 1.0 g
  distilled water .... 100.0 ml

Chromium hematoxylin, see p. 286.

Acid alcohol
  hydrochloric acid, concentrated .... 1.0 ml
  70% alcohol .... 100.0 ml

Phloxine
```
    phloxine B ..................................................... 0.5 g
    distilled water ............................................... 100.0 ml
```

Phosphotungstic acid
```
    phosphotungstic acid ........................................... 5.0 g
    distilled water ............................................... 100.0 ml
```

CONTROL
    pituitary gland

PROCEDURE
1. Deparaffinize and hydrate sections to water.
2. Mordant in Bouin chrome alum solution, 37°C: 12 to 24 hours.
3. Wash in running water until sections are colorless.
4. Oxidize in permanganate solution: 2 to 3 minutes.
5. Wash in distilled water: 1 minute.
6. Bleach in oxalic acid: 1 minute.
7. Wash in running water: 5 minutes.
8. Stain in chrome hematoxylin: 10 minutes.
9. Differentiate in acid alcohol: 30 seconds.
10. Wash in running water: 2 to 3 minutes.
11. Stain in phloxine: 2 to 3 minutes.
12. Treat with phosphotungstic acid: 2 minutes.
13. Wash in running water: 5 minutes.
14. Dehydrate, clear, and mount.

RESULTS
    neurosecretory substance—deep purple
    nuclei—lighter purple
    backgrounds—pinkish red

COMMENTS
McGuire and Opel (1969) oxidize with freshly prepared permanganate-sulfuric acid and oxalic acid (see Bargmann method above), and stain 15 minutes in the following:

```
    resorcin fuchsin ............................................... 1.0 g
    70% ethanol .................................................. 98.0 ml
    HCl concentrated ............................................... 2.0 ml
```

Can be used immediately and is stable 20 days if kept tightly closed. Counterstain with light green.

Tan (1973) oxidized with peracetic acid: 10 to 15 minutes.

acetic acid glacial .... 72.0 ml
hydrogen peroxide .... 226.0 ml
sulfuric acid .... 2.0 ml

Let stand 1 to 3 days before use. At 4–8°C is stable 5 to 6 months.

Stain with aldehyde- or resorcin-fuchsin and counterstain with Halmi mixture (see p. 282).

Shyamasundari and Rao (1975) demonstrate NSS and mucosubstances simultaneously.

Pancreatic Islet Cells

It is now generally accepted that the pancreatic islets (islets of Langerhans) of vertebrates above the cyclostomes are characterized by four major cell types: A cells produce glucagon; B cells, insulin; D cells, somatostatin; and F cells, pancreatic polypeptide. In addition, there are at least 12 peptides other than the four major hormones that can be identified in these cells. (See Epple and Brinn [1986] for a detailed discussion on the comparative structure and function of the islets.)

Various techniques have been used to identify the cells by differential staining of the granules. Mallory's will stain A cells red and B cells blue, but the Azan method is better, allowing the tinctorial separation of the major cell types. Glenner and Lillie (1957a) specifically stain A cells by a postcoupled benzylidene reaction for indoles. The A cells are associated with a strong nonspecific esterase reaction, and the B cells with a strong acid phosphatase reaction. Gomori chrome-hematoxylin is a favorite method for staining the islets.

See Epple (1967) for methods of demonstrating D cells and the use of THF as an aid in overcoming difficulties encountered in embedding pancreas blocks; also Solcia et al. (1968, 1969). See Bussolati and Bassa (1974) for staining of B cells.

Chromium-Hematoxylin-Phloxine (Gomori 1941b)

FIXATION

Bouin preferred. Stieve satisfactory. Zenker, Carnoy, and formalin unsatisfactory.

SOLUTIONS

Bouin solution, see p. 27.

Potassium dichromate-sulfuric acid

This can be made up as separate 0.3% solutions and mixed, or it can be made as follows:

potassium dichromate	0.15 g
distilled water	100.0 ml
sulfuric acid, concentrated	0.15 ml

Hematoxylin solution

hematoxylin	0.5 g
distilled water	50.0 ml

When dissolved, add:

potassium chromium sulfate (chrome alum), 3% aqueous	50.0 ml

Mix well and add:

potassium dichromate, 5% aqueous	2.0 ml
N/2 sulfuric acid (about 2.5 ml/100 ml water)	2.0 ml

Allow to ripen for 48 hours. Can be used until a film with a metallic luster does not form on its surface after 1 day's staining. Store at 0–4°C. Filter before use.

Phloxine

phloxine B	0.5 g
distilled water	100.0 ml

CONTROL
  pancreas

PROCEDURE
1. Deparaffinize and hydrate slides to water; remove $HgCl_2$.
2. Refix in Bouin solution: 12 to 24 hours.
3. Wash in running water: 5 minutes.
4. Treat with potassium dichromate-sulfuric acid: 5 minutes.
5. Decolorize in 5% aqueous sodium bisulfite: 3 to 5 minutes.
6. Wash in running water: 5 minutes.
7. Stain in hematoxylin solution until beta cells are deep blue (check under microscope): 10 to 15 minutes.
8. Differentiate in hydrochloric acid, 1% (1 ml/99 ml water): about 1 minute.
9. Wash in running water until clear blue: 5 minutes.
10. Stain in phloxine: 5 minutes.
11. Rinse briefly in distilled water.
12. Treat with 5% aqueous phosphotungstic acid: 1 minute.
13. Wash in running water: 5 minutes. Sections turn red again.

14. Differentiate in 95% alcohol. If the sections are too red and the alpha cells do not stand out clearly, rinse 15 to 20 seconds in 80% alcohol.
15. Dehydrate in absolute alcohol, clear, and mount.

Results
B cells—blue
A cells—red
D cells (not present in all animals)—pink to red, but indistinguishable from A cells

Comments
See also Heidenhain azan method, p. 124.

If the zymogen granules (acidophilic) are to be preserved in the acinar cells of the pancreatic lobules, avoid fixatives containing acetic acid.

Herlant stain I (p. 279) is also excellent for pancreas, as are most of the tri- and tetrachrome procedures.

Monroe and Spector (1963) use tannic acid, hematoxylin, Alcian blue, and basic fuchsin.

Kallman (1971) stains for A and B cells: 1 hour in aldehyde fuchsin; wash in 2 changes 70% ethanol; a rinse in distilled water; and stain 4 minutes in 0.05% toluidine blue O in 0.2 M McIlvaine phosphate buffer, pH 5. The B cells have violet-red granules and the A cells, light blue. The preferred fixative is Bouin; formalin is good, but Zenker, Helly, and Susa are poor.

Klessen (1972) uses a permanganate-HID technique for zymogen granules.

Thyroid Cells

Outstanding features of the thyroid are the secretory epithelial cells and the extracellular, amorphous colloid within the follicle lumen—the stored secretion. The colloid consists of proteins associated with carbohydrates. Protein reactions can be used; some of the proteins are the basic type and contain arginine, which can be demonstrated by Sakaguchi modifications. The ferric ferricyanide reduction technique (Lillie 1965) reacts with the colloid. Since the colloid is strongly PAS positive, mucopolysaccharides are probably present, and a positive dialyzed iron reaction would indicate acid mucopolysaccharides. The colloid is basophilic. The thyroid uses iodine to make up its hormone and, since radioactive iodine was one of the first isotopes prepared for biological purposes, the isotope tracer technique has been used for some time in thyroid studies. Bélanger and Bois (1964) recommend AFA fixation—1 part acetic acid, 5 parts formaldehyde (37–40%), and 15 parts absolute alcohol: 24 hours. This gives excellent fixation of the colloid, with less vacuolization than after formalin (NBF) alone, and well-stained colloid and cell detail with any of the staining techniques. However, Humason found that preservation of cellular detail and colloid is just as good after fixation in cold

(4–10°C) calcium formalin (p. 328). The F cells (assumed to produce polypeptide hormones) of the thyroid can be demonstrated by MacConaill lead hematoxylin as modified by Solcia et al. (1969) (p. 283).

Ljungberg (1970) uses 0.05% cresyl fast violet aqueous: 1 to 5 minutes, rinse in water, dehydrate, and clear.

Hot HCl eliminates diffuse tissue basophilia and increases the basophilia of certain endocrine cells, particularly thyroid C cells. The basophilia of secretory granules is increased in part to stored proteins. Treat for 6 hours in 1 N HCl, 60°C, and wash well. Stain with basic dyes, such as ethylene blue, toluidine blue O, or azure A (Petkó 1974; Solcia et al. 1968).

Suggested Readings

See Epple and Brinn (1986) and Schreibman (1986).

21
Staining Golgi Apparatus, Mitochondria, and Living Cells

GOLGI APPARATUS STAINING

The Golgi apparatus (Golgi bodies, Golgi substance, Golgi complex) is usually lost in routine fixation and requires special treatment. Since the methods are not always predictable under all conditions, it may be necessary to modify the fixing and/or staining time in order to attain precise and predictable results. In osmium and silver methods, the Golgi appears as either a dark net, a granular mass, a cord, or even a more diffuse condition. These reactions seem to indicate the presence of lipids, principally phospholipids. The Golgi apparatus (or adjacent lysosomes) can have a high level of acid phosphatase activity and lesser levels of alkaline phosphatase and other enzymes. Nucleoside diphosphatase is restricted to the Golgi apparatus.

There is a great deal of controversy about the responses of the Golgi apparatus to different conditions, physiological changes, and techniques. Golgi composition and appearance differs from cell to cell, and this can be related to the specific activity and physiological state of the cell. Silver or osmium tetroxide leaves the Golgi in the form of globules that have been described as spheres with an osmiophilic cap surrounding an osmiophobic center. But a thin lamella also is found adjacent to the spherical elements, suggesting a duplex structure. Most of the fixatives recommended for Golgi fixation contain heavy metal salts, yet it has been claimed that these are not necessary and the light metal or organic salts or acids do just as well (Pollister and Pollister 1957). (See Histochemistry, acid phosphatase, p. 345.) Much of the more informative research on the Golgi apparatus has been derived from electron microscopic studies.

Osmium Tetroxide, Ludford Method (Cowdry 1952; Lillie 1954b, 1965)

FIXATION
   Mann osmic sublimate: 18 hours.

   osmium tetroxide, 1% aqueous .... 50.0 ml
   mercuric chloride, saturated aqueous, plus
      0.37 g sodium chloride .... 50.0 ml

Procedure

1. Wash blocks of tissue in distilled water: 30 minutes.
2. Impregnate:

    2% osmium tetroxide: .... 3 days, 30°C
    2% osmium tetroxide: .... 1 day, 35°C
    1% osmium tetroxide: .... 1 day, 35°C
    0.5% osmium tetroxide: .... 1 day, 35°C

3. Wash in distilled water: 1 day
4. Dehydrate, clear, and embed.
5. Section 6 to 7 μm, mount, and dry.
6. Deparaffinize, clear, and mount.

Results
Golgi apparatus—black
yolk and fat—black (these may be bleached out with turpentine)

Comments
If it is advantageous to have mitochondria stained on the same slide, follow deparaffinization (step 6) by hydrating to water (include cautious treatment with 0.125% potassium permanganate) and stain by the Altmann Method (p. 295). Mitochondria will be crimson.

When using silver and osmium tetroxide techniques, considerable experimentation may be necessary. During the impregnation keep solutions in a dark place and follow instructions for temperature carefully. When the solutions turn dark, renew them.

Nassonov-Kolatchew Method (Nassonov 1923, 1924)

This method does not use the graded osmium tetroxide series.

Fixation
24 hours in:

    3% aqueous potassium dichromate .... 10.0 ml
    1% aqueous chromic acid .... 10.0 ml
    2% aqueous osmic acid .... 5.0 ml

Procedure
1. Wash in running water: 24 hours.
2. Place in 2% aqueous osmium tetroxide, 40°C: 8 hours, or 35°C: 3 to 5 days.
3. Wash in running water: overnight.

4. Dehydrate rapidly, clear, embed, and section, 2 to 4 μm.
5. If sections are too dark, turpentine (preferably old and oxidized) will remove most of the excess color.
6. Proceed to deparaffinize, etc., according to desired process. Mounted sections may be counterstained with Altmann's stain, p. 295.

Results
Golgi apparatus—black
background—yellow
mitochondria after Altmann's—red

Comment
In interpreting the results, bear in mind that the mitochondria may become impregnated if the period in osmium tetroxide is prolonged.

Saxena Method (1957)

Fixation
3 to 6 hours in:

barium chloride .... 1.0 g
distilled water .... 85.0 ml

Add just before use:

formaldehyde (37–40%) .... 15.0 ml

Solutions
Ramón y Cajal reducing solution
hydroquinone .... 1.5 g
formaldehyde (37–40%) .... 15.0 ml
distilled water .... 100.0 ml
sodium sulfite, anhydrous .... 0.5 g

Gold chloride
gold chloride stock solution (p. 463) .... 1.0 ml
distilled water .... 80.0 to 90.0 ml

Procedure
1. Rinse blocks of tissue in distilled water.
2. Impregnate in 1.5% aqueous silver nitrate, room temperature: 1 to 2 days. (Use 1% for very small pieces or embryonic tissues, 2% for fatty tissues and spinal cord.)
3. Rinse, 2 changes, in distilled water.
4. Cut blocks into slices thinner than 2 mm. Reduce in developer: 5 hours.

5. Wash thoroughly in distilled water.
6. Dehydrate, infiltrate, and embed.
7. Section at 6 to 7 μm, mount, and dry.
8. Deparaffinize and hydrate to water.
9. Tone in gold chloride: 2 hours.
10. Rinse in distilled water and fix in 5% aqueous sodium thiosulfate: 3 minutes.
11. Wash in running water: 5 minutes.
12. Counterstain, if desired, in hematoxylin, thionine, carmalum, etc.
13. Dehydrate, clear, and mount.

Results
Golgi—black
cytoplasm—gray
mitochondria—medium to dark gray or black

Comments
The silver preparations depend on fixation with salts of a heavy metal; barium, in the Saxena method. Aoyama (1930) varies with the method, using:

cadmium chloride	1.0 g
formaldehyde (37–40%)	15.0 ml
distilled water	85.0 ml

The rest of the procedure is the same as that of Saxena.

Cold-blooded animal tissues require longer fixation and impregnation than do warm-blooded animal tissues (Aoyama 1930).

McDonald (1964) uses nitrocellulose embedding for silver-impregnated Golgi tissues, because the material is very friable and often difficult to handle when embedded with paraffin.

Direct Silver Method (Elftman 1952)

This is a one-step procedure, silver and fixative in one action.

Procedure
1. Immerse small blocks of fresh tissue in silver nitrate in formalin (2 g/100 ml 15% formalin): 2 hours.
2. Rinse briefly in distilled water.
3. Develop for 2 hours in:

hydroquinone	2.0 g
formalin, 15% (15 ml/85 ml water)	100.0 ml

4. Return to 10% formalin to complete fixation: at least overnight.
5. Wash, dehydrate, clear and embed.
6. Section at 6 to 7 μm, mount, and dry.
7. Deparaffinize, clear, and mount.

RESULT

Golgi—black

COMMENTS

Do not use buffered formalin, which may limit the solubility of the silver salts.

If the silver is too dense, Elftman suggests bleaching it with 0.7% iron alum. Check under the microscope and stop the reaction by washing thoroughly in running water.

Because the silver is readily oxidized, gold toning is usually preferable for a more permanent slide. This can follow deparaffinization; see other Golgi methods. Also counterstaining may be included before dehydrating and clearing slides.

Elftman warns that all the black is not necessarily Golgi.

Sudan Black Method (Baker 1944)

FIXATION

Formalin-calcium (p. 328): 3 days.

EMBEDDING AND SECTIONING

Embed in gelatin (p. 389). Harden the block in formalin-calcium cadmium:

formaldehyde (37–40%)	10.0 ml
10% aqueous calcium chloride	10.0 ml
10% aqueous cadmium chloride	10.0 ml
distilled water	70.0 ml

Wash in running water: 3 to 4 hours. Cut frozen section at 15 μm.

PROCEDURE

1. Sections can be affixed to slides as for frozen sections, or they can be stained first and then mounted on slides.
2. Transfer through 50% and 70% alcohol: 1 to 2 minutes each.
3. Place in saturated solution of Sudan black in 70% alcohol: 7 minutes.
4. Remove excess stain in 3 changes of 50% alcohol: 1 to 2 minutes each.
5. Rinse in distilled water.

6. Counterstain with a light hematoxylin or a red nuclear stain.
7. Mount in glycerin jelly.

RESULT
Golgi—black. The vesicles, not the network, are stained. (See also Malhotra 1961).

MITOCHONDRIA STAINING

Mitochondria (from Greek *mittos*, filament; and *kondria*, granule) are tiny complex cell organelles, which are bounded by a double membrane and include folds of the inner membrane (cristae), a ground substance (matrix), and occasionally dense granulations. Chemically the mitochondria matrix consists mainly of proteins and lipids, and the membranes are high in insoluble protein, phospholipid, and insoluble enzyme content. They also contain DNA for replication. The osmiophilic state of the membranes is due to the lipids component. Mitochondria are able to concentrate a large number of substances, such as proteins, lipids, metals, viruslike particles, and several chemical substances. They are known to contain numerous enzymes—oxidative enzymes, cytochrome oxidase, cytochrome C, succinic dehydrogenase, glutamic hydrogenase, and adenosine triphosphatase to name only a few. Mitochondria participate in forming cell organelles; probably of mitochondrial origin are the granules of granulocytes, platelets, neutrophilic myelocytes, zymogen bodies, yolk platelets, secreting granules in neurohypophysis, etc. The electron microscope has contributed greatly to our knowledge of the ultrastructure of mitochondria and metabolic studies and biochemistry have clarified their physiological roles. (See also histochemistry, succinic dehydrogenase, pp. 353, 355.)

Altmann Method

FIXATION
Regaud fixative (p. 35): change every day for 4 days; store in refrigerator. Mordant in 3% potassium dichromate: 8 days, change every second day. Wash in running water overnight; dehydrate and embed. Cut sections 2 to 4 μm.

SOLUTIONS
Altmann aniline fuchsin
Make a saturated solution of aniline in distilled water by shaking the two together. Filter. Add 10 g acid fuchsin to 100 ml of filtrate. Let stand for 24 hours. Good for only 1 month.

CONTROL
mammalian kidney tubules

Procedure
1. Deparaffinize and hydrate slides to water.
2. Treat with 1% aqueous potassium permanganate: 30 seconds (see comments below).
3. Rinse briefly in distilled water and bleach in 5% aqueous oxalic acid: 30 seconds.
4. Rinse in several changes distilled water: 1 to 2 minutes total.
5. Place in steaming Altmann's solution: 5 to 10 minutes. Remove heat when the slides have been placed in the solution.
6. Differentiate in dilute (0.1%) sodium carbonate until the cytoplasm is pale pink or almost colorless.
7. Stop differentiation and heighten the color by a brief dip in 1% HCl.
8. Wash in distilled water: several dips.
9. Stain in 0.5% methylene blue: 5 to 6 seconds.
10. Rinse in distilled water and a quick dip in 1% HCl.
11. Wash in distilled water: several seconds.
12. Dehydrate, clear, and mount.

Results

mitochondria—bright red
nuclei—blue to bluish green

Comments

Iron hematoxylin can also be used for mitochondria. Best fixative is Regaud; mordant in 3% potassium dichromate: 7 days. Use the long method of staining: overnight in iron alum and overnight in hematoxylin. The method is not as specific as Altmann; other granules also stain.

Osmic Method (Newcomer 1940)

Fixation

Zirkle solution: 48 hours.

potassium dichromate	1.25 g
ammonium dichromate	1.25 g
copper sulfate	1.0 g
distilled water	100.0 ml

Procedure
1. Wash tissue blocks 8 hours to overnight.
2. Impregnate in 2% aqueous osmium tetroxide: 4 to 6 days. Change solutions on alternate days.
3. Wash: 8 hours or overnight.

4. Dehydrate, clear, and embed. Use benzene, not xylene, for clearing.
5. Cut 5 μm sections, mount on slides.
6. Deparaffinize and hydrate slides to water.
7. Bleach in 1% aqueous potassium permanganate: 5 minutes.
8. Rinse in distilled water.
9. Treat with 3% aqueous oxalic acid: 2 to 3 minutes.
10. Wash in running water: 15 minutes.
11. Dehydrate, clear, and mount.

RESULT
mitochondria—black

COMMENTS
Newcomer used his method on plant cells. A counterstain such as acid fuchsin can be added.

Short Acid-Fuchsin Method (Novelli 1962)

FIXATION
10% neutral buffered formalin or 1% osmium tetroxide, room temperature: 24 hours. Formalin-fixed tissues are easier to process. Wash in running water: 4 hours.

EMBEDDING AND SECTIONING
paraffin method, 1 to 4 μm

SOLUTION

acid fuchsin	0.2 g
methyl blue	0.1 g
N hydrochloric acid	100.0 ml

PROCEDURE
1. Deparaffinize and hydrate slides to water.
2. Stain: 5 minutes.
3. Rinse gently in distilled water.
4. Dehydrate quickly through 95% alcohol and 2 changes of absolute alcohol.
5. Clear and mount.

RESULTS
mitochondria—purple red with peripheral blue wall
chromatin and collagen—blue
nucleoli and erythrocytes—brilliant red

Acid-Hematein Method (Hori and Chang 1963)

Solutions

Acetone solution

acetone	50.0 ml
uranyl nitrate	0.01 to 0.02 g
chloral hydrate	0.01 to 0.02 g

Baker acid hematein, see Hori method, p. 251.

Borax-ferricyanide solution

borax ($Na_2B_4O_7 \cdot 10H_2O$)	250.0 mg
potassium ferricyanide ($K_3Fe(CN)_6$)	250.0 mg
distilled water	100.0 ml

Store in refrigerator.

Procedure

1. Prepare and section tissues by freeze-substitution method of Chang and Hori (1961a, b).
2. Transfer the sections into a screw-cap bottle of acetone mixture. Pre-chill the solution in crushed ice in a Dewar flask and leave buried in dry ice overnight. The presence of metal in the acetone intensifies the lipid reaction.
3. Rinse briefly in acetone at room temperature and mount on cover glasses.
4. Chromate in 5% aqueous potassium dichromate, 60°C: 5 to 6 hours.
5. Wash in distilled water: 5 minutes.
6. Stain in Baker acid hematein: 30 minutes.
7. Wash in distilled water.
8. Differentiate in borax-ferricyanide, room temperature: 18 hours.
9. Dehydrate, clear, and mount.

Results

mitochondria—sharply stained blue black
lipids in other cellular elements—may also stain

Comments

Caulfield (1957) reports that tonicity has a direct effect on the appearance of the mitochondria; low tonicity produced swollen mitochondria, and high tonicity shrank them. He recommends fixing 1 hour at 0–4°C in the following.

Caulfield solution

stock buffer (see below)	5.0 ml

0.1 N hydrochloric acid .... 5.0 ml
distilled water .... 2.5 ml
2% osmium tetroxide .... 12.5 ml

Adjust pH to 7.4, if necessary, by adding a few drops of 0.1 N HCl or stock buffer. Then add 0.045 g of sucrose per ml of solution.

Stock buffer
sodium veronal .... 14.714 g
sodium acetate .... 9.714 g

Dilute to 500.0 ml.

See Chang (1956) for method with frozen dried tissues, Benés (1960) for method using amido black 10B, Avers (1963) for fixing and staining for the electron microscope, and p. 353 of this book for demonstrating enzymes in mitochondria.

Takaya (1967) uses the Luxol fast blue method (p. 172) but counterstains with 0.5% aqueous phloxine instead of PAS.

SUPRAVITAL STAINING

Certain dyes will penetrate living cells and stain specific organelles within the cells. Neutral red reacts on certain granulations, products of the cytoplasm, secretory granules, vacuoles of digestion, and others. These vacuoles or granules appear to be heterogeneous and contain a phospholipoprotein complex, acid phosphatase, lipase, and some alkaline phosphatase, and they apparently correspond to lysosomes (Koenig 1963; Ogawa et al. 1961). Neutral red is commonly used in combination with Janus green B for the vital staining of Golgi and mitochondria in a single preparation. Neutral red, however, does not stain a reticular Golgi apparatus, but does stain vesicles in the region of the apparatus.

Janus green B vitally stains the mitochondria and is dependent upon the enzymatic activities of the cell. The staining will take place under partial anaerobic conditions, but it will decolorize when all oxygen is removed, and it is reversibly inhibited by cyanide. Janus green B can be reduced in both mitochondrial and nonmitochondrial portions of the cell, but the localization of a cytochrome oxidase system in the mitochondria slows down the rate of reduction in that organelle, while reduction proceeds more rapidly in the other portions of the cell. It is erroneous to consider all isolated structures taking up Janus green B as mitochondria; secretion granules of the islets of Langerhans and of enterochromaffin and others will stain as well.

Neutral Red-Janus Green Method

The dyes should be certified for vital staining so they are not toxic to living cells. The solutions used for staining are so weak that it is convenient to make them up as a stock solution of greater strength. They will keep indefinitely if stored in glass-stoppered bottles. If using a pipette with a rubber bulb, do not allow the dye to run up into the bulb. Make the diluted solutions in small quantities, and if they are mixed together, use immediately. The mixture is not stable. Slides must be cleaned in dichromate solution, washed well in running water and then in distilled water, and stored in 95% alcohol. Rub dry just before spreading with stain.

Solutions
Neutral red stock solution
neutral red .... 0.5 g
absolute alcohol .... 100.0 ml

Janus green B stock solution
Janus green B .... 0.5 g
neutral absolute ethyl alcohol .... 100.0 ml

Procedure
1. Dilute neutral red stock solution, 1 to 10 ml of neutral absolute ethyl alcohol.
2. Mix 0.4 ml of Janus green B stock with 3 ml of dilute neutral red solution.
3. Flood clean slides with mixed dyes and touch edge of slide to absorbent material to draw off excess solution. Allow to dry horizontally in warm air, but keep them protected from dust. Be sure to mark the stained surface. The stain should be distributed thinly and evenly; this can be accomplished by blowing gently on the surface of the stain. When dry, the slides can be stored in a dust-free box. Sometimes successful smears can be made by placing a drop of solution at the end of the slide and smearing it like a blood film.

   Another method is to make enough solution to fill a Coplin jar or bottle of comparable size. Gently flame a clean slide and plunge it while hot into the stain. Withdraw the slide and drain it upright on absorbent paper. Wave it rapidly in the air to quick-dry it and to preserve even distribution on the slide. If the humidity is high, dry the smears in hot air by using a hair dryer or other drying equipment. The smear may not be as uniform as one following slower drying, but it is necessary to speed up the drying because humidity allows the alcohol to take up moisture and ruin the dye film before the evaporation is complete.
4. Place a drop of live cells or organism on a clean cover glass and carefully lower onto a stained slide. Seal with petroleum jelly.

   Blood, bone marrow, and other tissue fluids or cells may be diluted with an equal amount of homologous heparinized plasma and examined on a stained slide. After a few trials and errors, the proper size of drop will be realized. With blood, the red cells should spread in a single layer. If the preparations are too thick, the leukocytes will round up instead of flatten. If air bubbles are trapped under the cover glass, the cells do not spread properly. If the cover glass is dropped on the stain or if there is too little of the suspension, the cell membranes may rupture.

5. If a warm stage is not available, place slides of tissue cells in a 37°C incubator for 20 minutes. Then examine immediately. Cells begin to round up as the slide cools.

6. If it is desirable to make the slides semipermanent when the mitochondria or neutral red bodies are sufficiently stained, clean off the Vaseline and slip the cover glass toward the edge of the slide, so that the blood or other fluid is spread evenly on both the slide and the cover glass. Slip the cover glass off and rapidly dry both the slides and the cover glass. (Both carry cell preparations.) If dried too slowly, the cells will be distorted. The drying may be completed in a vacuum desiccator (2 to 4 hours) or by shaking in several changes of anhydrous ether. The anhydrous ether is essential to remove fat from bone marrow. Clear in xylene and mount.

RESULTS
mitochondria—green
neutral red bodies—red

Vital-Nonvital Stain (DeRenzis and Schechtman 1973)

Trypan blue alone does not accurately distinguish vital from nonvital cells. Dead cells are stained blue but live cells do not take up color and are difficult to identify or can be overlooked. This two-dye method corrects the problem.

Add 0.5 ml of cell suspension to 0.5 ml of 0.04% neutral red in balanced salt solution. Incubate at 37°C: 10 minutes. Add 0.5 ml of 0.5% trypan blue in balanced salt solution for 2 to 3 minutes. Mix well and place in a hemacytometer. The cells are now ready to be counted and have good color contrast. Red cells are viable; blue ones are not. If a cell contains both colors, it was damaged in processing. The method can be used on cells growing on glass. Remove culture medium and add stain.

22
Staining Microorganisms

In this chapter, to simplify the specificity of staining methods, the parasitic microorganisms will be placed into the following groups: bacteria, spirochetes, fungi, rickettsia, and viruses (Burrows 1954).

STAINING BACTERIA

Bacteria are customarily studied by direct microscopic observation and differentiated by shape, grouping of cells, presence or absence of certain structures, and the reaction of their cells to differential stains. Bacteria may be stained with aniline dyes—in a single dye, in mixed dyes, in polychromated dyes, or by differential methods. One of the most universally used stains was developed by the histologist Gram while he was trying to differentiate the bacteria in tissue. His method separates bacteria into two groups: (1) those that retain crystal violet and are said to be Gram-positive; (2) those that decolorize to be stained by a counterstain and are said to be Gram-negative.

Some bacteria of high lipid content cannot be stained by the usual methods but require heat or long exposure to the stain. They are also difficult to decolorize. Because they resist acid alcohol, they have been given the name acid fast. The spiral forms will stain only faintly if at all and must be colored by silver methods.

Many bacteria form a capsule from the outer layer of the cell membrane, and the capsule appears like a halo around the organism, or over a chain of cells. This capsule will not stain in the customary stains; Hiss stain (p. 308), however, is simple and usually effective in this situation.

Some bacteria are able to form spores that can be extremely resistant to injurious conditions (heat, chemicals). Boiling will destroy some of these spores, but many are more resistant. Some bacteria have flagella—filamentous appendages for locomotion.

Bacteria can be classified according to shape: coccus, spherical; bacillus, rod shaped; spiral, a curved rod.

Gram Staining

When a Gram-staining procedure has been applied, a Gram-positive cell or organism retains a particular primary dye. This process includes mordanting with iodine to form (with the dye) a precipitate that is insoluble in water and is neither too soluble nor insoluble in alcohol, the differentiator. According to Bartholomew and Mittwer (1950, 1951), the mechanism of stainability can be explained by differences in the permeability of the cell membrane. The bacteria stain by linkage between acid groups of the bacteria and alkaline groups of dye. Iodine forms a complex with the dye, and this complex is dissociated by alcohol. If alcohol passes easily through the membrane, decolorization is rapid and the reaction is Gram negative. If the membrane is hardly or not at all penetrated by alcohol, the reaction is Gram positive. The condition of the membrane, therefore, is important, it must be intact.

Pearse (1968) suggests that the initial acidic protein-crystal violet complex is broken by the iodine, which then combines with the basic groups at the ends of the triphenylmethane molecule to form a relatively insoluble crystal-violet-iodine precipitate. This precipitate is not easily removed if the following conditions exist: (1) if more of the crystal violet-iodine complex has formed where originally there had been more crystal violet in combination with protein; (2) if there are physical barriers, consisting of lipid or lipoprotein membranes, that resist extraction of the crystal violet-iodine.

For more details about Gram staining, see Bartholomew et al. (1959).

Gram-Weigert Method (Krajian and Gradwohl 1952)

FIXATION
10% neutral buffered formalin or Helly

SOLUTIONS
Eosin

eosin Y	1.0 g
distilled water	100.0 ml

Sterling gentian violet

crystal violet (gentian violet)	5.0 g
95% alcohol	10.0 ml
aniline oil	2.0 ml
distilled water	88.0 ml

Mix aniline oil with water and filter. Add the crystal violet dissolved in alcohol. Keeps for several weeks to months.

Gram iodine solution, see p. 464.

CONTROL
appendix or known positive

PROCEDURE
1. Deparaffinize and hydrate slides to water. Remove $HgCl_2$.
2. Stain in eosin: 5 minutes.

3. Rinse in water.
4. Stain in Sterling gentian violet solution; 3 minutes for frozen sections; 10 minutes for paraffin sections.
5. Wash off with Gram iodine and then flood with more of same solution: 3 minutes.
6. Blot with filter paper.
7. Flood with equal parts of aniline oil and xylene; repeat flooding until color ceases to rinse out of section.
8. Clear in xylene and mount.

RESULTS

Gram-positive bacteria, and fungi—violet
Gram-negative organisms—not usually stained
fibrin—blue black

Leaver et al. (1977) Substitute for Brown and Brenn

FIXATION

10% neutral buffered formalin preferred.

SOLUTIONS

Crystal violet

crystal violet	1.0 g
distilled water	100.0 ml

Gram iodine, see p. 464.

Sandiford stain

malachite green	0.05 g
pyronine Y	0.15 g
distilled water	100.0 ml

CONTROL

known positive

PROCEDURE

1. Deparaffinize and hydrate sections to water.
2. Stain in crystal violet: 3 minutes.
3. Rinse in tap water.
4. Treat with Gram iodine: 3 minutes.
5. Rinse in tap water and blot almost dry.
6. Differentiate in equal parts of acetone and absolute alcohol until no more blue color comes off.
7. Wash in running tap water: 2 to 3 minutes.
8. Counterstain with Sandiford stain: 2 minutes.

9. Rinse in tap water and blot almost dry.
10. Differentiate for a few seconds in 95% alcohol and dehydrate in absolute alcohol, clear, and mount.

RESULTS

Gram-positive organisms—purple black
Gram-negative organisms—red
background—blue green

COMMENTS

Humason determined this method to show the best Gram-stained organisms of any of the Brown and Brenn type methods.

Acid-Fast Staining

In acid-fast staining, the presence of phenol in the stain solution appears to be essential; it must, in some way, influence the dye or substance that imparts the acid-fast character. Lartique and Fite (1962) propose that phenol decreases the solubility of fuchsin in water and increases its solubility in the lipids of the bacillus. The acid-fastness can be enhanced by artificially coating the bacilli with oil. Actually, no specific chemical role can be assigned to phenol; it does not combine with the fuchsin, nor does it cause capsular disruption or protein denaturation. Harada's (1976) work indicated that adding phenol made water-soluble dyes, such as basic fuchsin, more lipid soluble. He also concluded it was possible that the presence of hydroxyl and free carboxyl groups are necessary in a mycolic acid type of long chain fatty acids to assure acid-fastness.

Staining of lepra bacilli is improved if, when removing paraffin from sections, 15% of mineral oil is added to the xylene (5 minutes). Wash 30 seconds in detergent solution (Haemosol), wash in water, and proceed to stain.

Harada Method (1973), a Ziehl-Neelsen Type

FIXATION

Any general fixative, but 10% neutral buffered formalin is best.

SOLUTIONS

Carbol-fuchsin

basic fuchsin, saturated alcoholic	10.0 ml
phenol, 5% aqueous	90.0 ml

Loeffer alkaline methylene blue

methylene blue	3.0 g
absolute alcohol	30.0 ml
potassium hydroxide, 0.01% aqueous	100.0 ml

CONTROL
known positive

PROCEDURE
1. Deparaffinize and hydrate sections to water.
2. Oxidize in 1% potassium permanganate, aqueous: 1 hour.
3. Rinse in tap water and bleach in 1% oxalic acid, aqueous: 3 minutes.
4. Stain in carbol-fuchsin, warmed to steaming: 5 minutes (or stain in a Coplin jar at 60°C in a water bath or oven for 30 minutes).
5. Rinse in distilled water.
6. Decolorize in 1% HCl in 70% alcohol: approximately 20 seconds.
7. Wash in running water: 2 to 3 minutes.
8. Counterstain in Loeffler methylene blue diluted 1:9 with distilled water: 15 seconds. Do not overstain. Dip slide individually. Sections should be pale blue.
9. Rinse in tap water.
10. Dry smears and mount. Dehydrate sections, clear, and mount.

RESULTS
acid-fast bacteria—red
red blood corpuscles—pink
mast cell granules—deep blue
other bacteria—blue
nuclei—blue

COMMENTS
Pottz et al. (1964) use a wetting agent, dimethylsulfoxide (DMSO). Their solution is:

basic fuchsin	4.0 g
95% alcohol	25.0 ml
phenol, liquefied	12.0 g
glycerol	25.0 ml
DMSO	25.0 ml

Bring solution to 160 ml with distilled water.

See Harada (1976) for a methenamine-silver method. Carson (1990) states that in order to counterstain with methylene blue it is necessary to wash the acid from the tissue.

Fluorescent Method (Bogen 1941; Richards 1941; Richards et al. 1941)

FIXATION
10% neutral buffered formalin for sections; smears by heat.

SOLUTION
Auramine stain

auramine O	0.3 g
distilled water	97.0 ml
melted phenol (carbolic acid)	3.0 ml

Shake to dissolve dye or use gentle heat. Solution becomes cloudy on cooling but is satisfactory. Shake before using.

Decolorizing solution

hydrochloric acid, concentrated	0.5 ml
70% alcohol	100.0 ml
sodium chloride	0.5 g

CONTROL
known positive

PROCEDURE
1. Deparaffinize and hydrate sections to water.
2. Stain in auramine, room temperature: 2 to 3 minutes.
3. Wash in running water: 2 to 3 minutes.
4. Decolorize: 1 minute.
5. Transfer to fresh decolorizer: 2 to 5 minutes.
6. Wash in running water: 2 to 3 minutes.
7. Dry smear and examine. Mount sections in fluorescent mountant (p. 99).

RESULT
bacilli—golden yellow

COMMENTS
Bogen counterstains with Loeffler alkaline methylene blue solution before mounting (p. 305).

See also Braunstein and Adriano (1961), Moody et al. (1958), and Yamaguchi and Braunstein (1965).

Carter and Leise (1958) use a single fluorescent antiglobulin for various bacteria. Sheehan and Hrapchak (1980) describe an auramine-rhodamine fluorescence technique.

Capsule Staining

Hiss Method (Burrows 1954)

FIXATION
Any good general fixative; 10% neutral buffered formalin is satisfactory

SOLUTIONS
Stain solution
Either of the following may be used:

> A.
> basic fuchsin .... 0.15 to 0.3 g
> distilled water .... 100.0 ml
>
> B.
> crystal violet .... 0.05 to 0.1 g
> distilled water .... 100.0 ml

Copper sulfate solution
> copper sulfate crystals .... 20.0 g
> distilled water .... 100.0 ml

CONTROL
known positive

PROCEDURE
1. Deparaffinize and hydrate slides to water.
2. Flood with either staining solution and heat gently until the stain steams, or stain in Coplin jars in a 60°C oven or water bath for 30 minutes.
3. Wash off the stain with copper sulfate.
4. Blot, but do not dry.
5. Dehydrate through alcohols.
6. Clear and mount.

RESULTS
capsules—light pink (basic fuchsin) or blue (crystal violet)
bacterial cells—dark purple surrounded by the capsule color

STAINING SPIROCHETES

These are spiral-shaped organisms, which multiply by transverse fission. They cause relapsing fever, syphilis, and yaws.

Dieterle Method (Beamer and Firminger Modification 1955)

FIXATION
10% neutral buffered formalin.

SOLUTIONS
Dilute gum mastic
  gum mastic, saturated in absolute alcohol .... 30 drops
  95% alcohol .... 40.0 ml

Developing solution
  distilled water .... 20.0 ml
  hydroquinone .... 0.5 g
  sodium sulfite .... 0.06 g

While stirring add:

  formaldehyde (37–40%) .... 4.0 ml
  glycerin .... 5.0 ml

When thoroughly mixed add, drop by drop, with constant stirring:

  gum mastic saturated in absolute alcohol .... 4.0 ml
  absolute alcohol .... 4.0 ml
  pyridine .... 2.0 ml

CONTROL
known positive

PROCEDURE
1. Deparaffinize and hydrate slides to water.
2. Remove any formalin pigment in 2% ammonium hydroxide (2 ml/ 98 ml water): few minutes.
3. Transfer to 80% alcohol: 2 to 3 minutes.
4. Wash in distilled water: 5 minutes.
5. Treat with uranium nitrate 2 to 3% (2 to 3 g/100 ml water), previously warmed to 60°C: 10 minutes.
6. Wash in distilled water: 1 to 2 minutes.
7. Transfer to 95% alcohol: 1 to 2 minutes.
8. Treat with dilute gum mastic: 5 minutes.
9. Wash in distilled water, 3 to 4 changes, until rinse is clear.
10. Impregnate with silver nitrate, 2% (2 g/100 ml water), previously warmed to 60°C: 30 to 40 minutes.
11. Warm developing solution to 60°C. Dip slide up and down in solution until sections turn light tan or pale brown.

12. Rinse in 95% alcohol, then in distilled water.
13. Treat with 2% silver nitrate: 1 to 2 minutes.
14. Wash in distilled water: 1 to 2 minutes.
15. Dehydrate, clear, and mount.

RESULTS
spirochetes—black
background—yellow

COMMENTS
Use thin sections, 4 to 5 μm

The uranium nitrate prevents the impregnation of nerve fibers and reticulum.

Warthin-Starry Silver Method
(Bridges and Luna 1957; Faulkner and Lillie 1945a; Kerr 1938)

Caution: All glassware must be cleaned with potassium dichromate-sulfuric acid. Avoid contamination. Coat forceps with paraffin. Solutions must be fresh (no more than 1 week old) and made from distilled water. Carry a known positive control slide with test slide through the process, or, preferably, a control section on the same slide.

FIXATION
10% neutral buffered formalin

SOLUTIONS
Acidified water
triple distilled water .... 1000.0 ml
citric acid .... 10.0 g

The pH can range from 3.8 to 4.4.

2% silver nitrate
silver nitrate .... 2.0 g
acidified water .... 100.0 ml

1% silver nitrate
Dilute a portion of the 2% silver nitrate solution with equal volume of acidified water.

0.15% hydroquinone
hydroquinone .... 0.15 g
acidified water .... 100.0 ml

5% gelatin
```
gelatin ................................................. 10.0 g
acidified water ......................................... 200.0 ml
```

Developer
```
2% silver nitrate ....................................... 1.5 ml
5% gelatin .............................................. 3.75 ml
0.15% hydroquinone ...................................... 2.0 ml
```

Warm solutions to 55–60°C and mix in order given, with stirring. A hot plate stirrer combination would be useful. Use immediately.

CONTROL
  known positive

PROCEDURE
1. Deparaffinize and hydrate slides to acidified water.
2. Impregnate in 1% silver nitrate, 55–60°C: 30 minutes.
3. Place slides on glass rods, pour on warm developer (55–60°C). When sections become golden brown or yellow, and the developer brownish black (3 to 5 minute), pour off. The known positive can be checked under microscope, for black organisms.
4. Rinse with warm (55–60°C) tap water, then distilled water.
5. Dehydrate, clear, and mount.

RESULTS
  spirochete—black
  background—yellow to light brown
  melanin and hematogenous pigments—may darken

Underdevelopment will result in pale background, very slender and pale spirochetes.
Overdevelopment will result in dense background, heavily impregnated spirochetes, obstructed detail, sometimes precipitate.

COMMENTS
  Faulkner and Lillie (1945a) use water buffered to pH 3.6 to 3.8 with Walpole $M/5$ sodium acetate-$M/5$ acetic acid buffer (p. 471). Swisher (1987) uses a modified method for spirochetes and nonfilamentous bacteria.

Levaditi Method for Block Staining (Mallory 1944)

FIXATION
  10% neutral buffered formalin.

SOLUTIONS
Silver nitrate
    silver nitrate .... 1.5 to 3.0 g
    distilled water .... 100.0 ml

Reducing solution
    pyrogallic acid .... 4.0 g
    formaldehyde (37–40%) .... 5.0 ml
    distilled water .... 100.0 ml

PROCEDURE
1. Rinse blocks of tissue in tap water.
2. Transfer to 95% alcohol: 24 hours.
3. Place in distilled water until tissue sinks.
4. Impregnate with silver nitrate, 37°C, in dark: 3 to 5 days.
5. Wash in distilled water.
6. Reduce at room temperature in dark: 24 to 72 hours.
7. Wash in distilled water.
8. Dehydrate, clear, and infiltrate with paraffin.
9. Embed, section at 5 μm, mount on slides, and dry.
10. Remove paraffin with xylene, 2 changes, and mount.

RESULTS
spirochetes—black
background—brownish yellow
For a fluorescent method, see Kellogg and Deacon (1964).

STAINING FUNGI

Open or draining lesions are difficult to examine for fungi because of heavy bacterial contamination, but dermatophytes are easily demonstrated. Scrapings from horny layers or nail plate or hair can be mounted in 10–20% hot sodium hydroxide. This dissolves or makes transparent the tissue elements and then the preparation can be examined as a wet mount. Fungi in tissue sections are readily stained.

Hotchkiss-McManus Method (McManus 1948)

FIXATION
10% neutral buffered formalin or any general fixative.

SOLUTIONS
Periodic acid
    periodic acid .... 1.0 g
    distilled water .... 100.0 ml

Schiff reagent, see p. 184.

Differentiator
```
potassium metabisulfite, 10% aqueous .......................... 5.0 ml
N HCl (p. xxx) ............................................... 5.0 ml
distilled water .............................................. 100.0 ml
```

Light green
```
light green SF, yellowish .................................... 0.2 g
distilled water .............................................. 100.0 ml
glacial acetic acid .......................................... 0.2 ml
```

CONTROL
known positive

PROCEDURE
1. Deparaffinize and hydrate slides to water; remove $HgCl_2$ if present.
2. Oxidize in periodic acid: 5 minutes.
3. Wash in running water: 15 minutes.
4. Treat with Schiff reagent: 10 to 15 minutes.
5. Differentiate, 2 changes: total 5 minutes.
6. Wash in running water: 10 minutes.
7. Stain in light green: 3 to 5 minutes. If too dark, rinse in running water.
8. Dehydrate, clear and mount.

RESULTS
fungi—red. Not specific, however; glycogen, mucin, amyloid, colloid, and others may show rose to purplish red.
background—light green

COMMENTS
To remove glycogen, starch, mucin, or RNA, see p. 185.

DePalma and Young (1963) use 0.1% basic fuchsin for 2 minutes with agitation instead of Schiff reagent. Differentiate only 1 to 2 minutes with agitation.

Gridley Method (1953)

FIXATION
Any good general fixative.

SOLUTIONS
Chromic acid
```
chromic acid ................................................. 4.0 g
distilled water .............................................. 100.0 ml
```

Stable for 2 months.
Schiff reagent, see p. 184.

Sulfurous rinse
    sodium metabisulfite, 10% (10 g/100 ml water) .... 6.0 ml
    N hydrochloric acid .... 5.0 ml
    distilled water .... 100.0 ml

Aldehyde-fuchsin, see p. 28.

Metanil yellow
    metanil yellow .... 0.25 g
    distilled water .... 100.0 ml
    glacial acetic acid .... 2 drops

CONTROL
  known positive

PROCEDURE
1. Deparaffinize and hydrate slides to water; remove $HgCl_2$ (see p. 27).
2. Oxidize in chromic acid: 1 hour.
3. Wash in running water: 5 minutes.
4. Place in Schiff reagent: 15 minutes.
5. Rinse in sulfurous acid, 3 changes: 1.5 minutes each.
6. Wash in running water: 15 minutes.
7. Stain in aldehyde-fuchsin: 15 to 30 minutes.
8. Rinse off excess stain in 95% alcohol.
9. Rinse in water.
10. Counterstain lightly in metanil yellow: 1 minute.
11. Rinse in water.
12. Dehydrate, clear, and mount.

RESULTS
  mycelia—deep blue
  conidia—deep rose to purple
  background—yellow
  elastic tissue, mucin—deep blue

Gomori Methenamine-Silver Nitrate Method (Grocott Adaptation, 1955; Mowry Modification, 1959), **also called GMS**

FIXATIVE
  10% neutral buffered formalin or any good general fixative.

Solutions
Methenamine-silver nitrate stock solution
    silver nitrate, 5% aqueous .... 5.0 ml
    methenamine, 3% aqueous .... 100.0 ml

Silver nitrate is considered to be stable for 2 weeks in refrigerator, methenamine for 1 month, but it is recommended that freshly prepared solutions be used for superior results.

Working solution
    borax, 5% aqueous .... 2.0 ml
    distilled water .... 25.0 ml
    methenamine-silver nitrate stock solution .... 25.0 ml

Use within 24 hours.

Light green stock solution
    light green SF, yellowish .... 0.2 to 0.5 g
    distilled water .... 100.0 ml
    glacial acetic acid .... 0.2 ml

Working solution
    light green stock solution .... 10.0 ml
    distilled water .... 100.0 ml

Control
known positive

Procedure
1. Deparaffinize and hydrate slides to water; remove $HgCl_2$ (see p. 27).
2. Oxidize in periodic acid, 0.5% (0.5 g/100 ml water): 10 minutes.
3. Wash in running water: 3 minutes.
4. Oxidize in chromic acid, 5% (5 g/100 ml water): 45 minutes.
5. Wash in running water: 2 minutes.
6. Treat with sodium bisulfite, 2% (2 g/100 ml water): 1 minute to remove chromic acid.
7. Wash in running water: 5 minutes.
8. Rinse in distilled water, 2 to 3 changes: 5 minutes total.
9. Place in methenamine-silver nitrate, 58°C: 30 minutes. Do not use metal forceps. Sections appear yellowish brown.
10. Wash thoroughly, several changes distilled water.
11. Tone in gold chloride (10 ml stock solution/90 ml water) until sections turn purplish gray, fungi are black.

12. Rinse in distilled water.
13. Fix in sodium thiosulfate, 5% (5 g/100 ml water): 3 minutes.
14. Wash in running water: 5 minutes.
15. Counterstain in light green: 30 seconds.
16. Dehydrate, clear, and mount.

RESULTS
fungi—black
background—light green

COMMENTS
The Mowry modification uses oxidation with both periodic and chromic acid (former methods use only the latter) with the result that the final staining is stronger and more consistent than that of the original Gomori method.

Gold toning may be omitted and the methenamine stain followed by 8 minutes in each of 3 changes of 10% thiosulfate to give a brown instead of a black stain. An aldehyde-fuchsin stain may also be used for fungi.

Schneider (1963) simplifies a Gram stain for fungi.

Fluorescent Method (Pickett et al. 1960)

FIXATION
10% neutral buffered formalin, Zenker, alcohol, and other general fixatives.

SOLUTIONS
Acridine orange
  acridine orange .... 0.1 g
  distilled water .... 100.0 ml

Weigert hematoxylin, see p. 103.

CONTROL
known positive

PROCEDURE
1. Deparaffinize and hydrate sections to water.
2. Stain in Weigert hematoxylin: 5 minutes.
3. Wash in running water: 5 minutes.
4. Stain in acridine orange: 2 minutes.
5. Rinse in tap or distilled water: 30 seconds.
6. Dehydrate in 95% alcohol: 1 minute.
7. Dehydrate in absolute alcohol, 2 change: 2 to 3 minutes.
8. Clear in xylene, 2 changes: 2 to 3 minutes.
9. Mount in a nonfluorescing medium.

RESULTS

All fungi fluoresce except *Nocardia* and *Rhizopus*; colors of the fluorescing genera appears as follows:

Coccidiodes, Rhinosporidum—red
Aspergillus—green
Actinomyces, Histoplasma—red to yellow
Candida, Blastomyces, dermatides, Monosporium—yellow green
Blastomyces brasilienis—yellow

COMMENTS

Old hematoxylin and eosin slides may be decolorized and restained as above.

Weigert hematoxylin staining is necessary as a quenching agent because some fungi are difficult to see against a background which also fluoresces. With the hematoxylin, the fungi fluoresce brightly against a dark setting.

For a fluorescent method, see Clark and Hench (1962) and Mote et al. (1975).

For fluorescent antibody techniques, see Batty and Walker (1963) and Procknow et al. (1962).

STAINING OF RICKETTSIAE AND INCLUSION BODIES

Rickettsiae are very small, Gram-negative coccobaccillary-type microorganisms associated with typhus and spotted fever and related diseases. They may appear as cocci or short bacilli, and may occur singly, in pairs, or in dense masses. Most of them are intracellular. Some species are found only in the cytoplasm; others prefer the nucleus. Rickettsiae stain best in a Giemsa-type stain or by the Ordway-Machiavello method.

Viruses are microorganisms too small to be visible under the microscope, and they are capable of passing through filters. Viruses are responsible for many diseases, including yellow fever, poxes, poliomyelitis, influenza, measles, mumps, shingles, rabies, colds, infectious hepatitis, infectious mononucleosis, trachoma, psittacosis, foot-and-mouth disease, and, of course, AIDS (acquired immune deficiency syndrome). In tissue sections and smears, viruses are characterized by elementary bodies and inclusion bodies. Elementary bodies are infectious particles, and inclusion bodies are composed of numerous elementary bodies. Both types of bodies vary in size and appearance; some, such as rabies, psittacosis, and trachoma, are located in the cytoplasm of infected cells; some, such as poliomyelitis, are intranuclear. Special staining methods can demonstrate them effectively.

Modified Pappenheim Stain (Castañeda 1939)

Fixation
Any general fixative; Regaud recommended.

Solutions
Stock Jenner solution

Jenner stain	1.0 g
methyl alcohol, absolute	400.0 ml

Stock Giemsa stain

Giemsa stain	1.0 g
glycerin	66.0 ml

Mix and place in oven 2 hours, 60°C. Add:

methyl alcohol, absolute	66.0 ml

Working solution A

distilled water	100.0 ml
glacial acetic acid	1 drop
Jenner stock solution	20.0 ml

Working solution B

distilled water	100.0 ml
glacial acetic acid	1 drop
Giemsa stock solution	5.0 ml

Control
known positive

Procedure
1. Deparaffinize and hydrate slides to water.
2. Stain in solution A, 37°C: 15 minutes.
3. Transfer directly to solution B, 37°C: 30 to 60 minutes.
4. Dehydrate quickly, 2 changes absolute alcohol.
5. Clear and mount.

Result
rickettsiae—blue to purplish blue

Castañeda Method (Gradwohl 1963)

Fixation
Any general fixative; Regaud recommended.

Solutions

Buffer solution
Solution A

    dibasic sodium phosphate ($Na_2HPO_4 \cdot 12H_4O$) .... 23.86 g
    distilled water .... 1000.0 ml

Solution B

    monobasic sodium phosphate anhydrous (NaH_2PO_4) .... 11.34 g
    distilled water .... 1000.0 ml

Working solution

    solution A .... 88.0 ml
    solution B .... 12.0 ml
    formalin .... 0.2 ml

Methylene blue solution
    Dissolve methylene blue .... 2.1 g
    in 95% alcohol .... 30.0 ml
    Dissolve potassium hydroxide .... 0.01 g
    in distilled water .... 100.0 ml

Mix the two solutions and let stand 24 hours.

Staining solution
Mix buffer working solution with 1.0 ml methylene blue solution.

Control
known positive

Procedure
1. Deparaffinize and hydrate slides as far as 50% alcohol.
2. Stain in methylene blue: 2 to 3 minutes.
3. Wash in running water: 30 seconds.
4. Counterstain in 1% aqueous safranin: 1 to 2 minutes.
5. Dip briefly in 95% alcohol.
6. Dehydrate in 2 changes absolute alcohol, clear, and mount.

Result
rickettsiae and inclusion bodies—light blue

Comments

Burrows (1954) recommends this as one of the best methods for rickettsiae. The Giménez (1964) is a good one, too.

For a fluorescent method, see Anderson and Grieff (1964).

Ordway-Machiavello Method (Gradwohl 1963)

Fixation
Regaud recommended.

Solution

Poirier blue 1% aqueous	20.0 ml
eosin bluish 0.45% aqueous	15.0 ml

Mix just before use. Add slowly with constant shaking:

distilled water	25.0 ml

Use within 24 hours.

Control
known positive

Procedure
1. Deparaffinize and hydrate slides to water.
2. Stain: 6 to 8 minutes.
3. Decolorize in 95% alcohol until slides appear pale bluish pink.
4. Dehydrate in absolute alcohol: 1 minute.
5. Clear and mount.

Results
rickettsiae and inclusion bodies—bright red
nuclei and cytoplasm—sky blue

Giemsa Method (American Public Health Association 1956)

Fixation
10% neutral buffered formalin; also any general fixative.

Solutions
Giemsa stock solution, see p. 203.

Buffer solutions
Solution A

dibasic sodium phosphate, anhydrous (Na_2HPO_4)	9.5 g
distilled water	1000.0 ml

Solution B

> monobasic sodium phosphate ($NaH_2PO \cdot H_2O$) .... 9.2 g
> distilled water .... 1000.0 ml

Working solution, pH 7.2

> solution A .... 72.0 ml
> solution B .... 28.0 ml
> distilled water .... 900.0 ml

Giemsa working solution
Dilute 1 drop Giemsa stock solution with 5 ml of buffer working solution.

CONTROL
known positive

PROCEDURE
1. Deparaffinize and hydrate slides to water. (Smears can be carried directly to water: wash well in running water: 5 to 10 minutes.)
2. Leave slides in Giemsa working solution overnight, 37°C.
3. Rinse thoroughly in distilled water. Dry between sheets of bibulous paper.
4. Dip rapidly in absolute alcohol. If overstained, use 95% alcohol to decolorize; dip in absolute alcohol.
5. Clear and mount. (Smears, after treatment with absolute alcohol, can be washed in distilled water: 1 to 2 seconds, and blotted dry. Examine with oil or mount with a cover glass.)

RESULT
rickettsiae and inclusion bodies (psittacosis)—blue to purplish blue

Modified Gomori Method for Trachoma

SOLUTION
Dilute 1 drop stock Giemsa with 2 ml of buffer working solution above.

CONTROL
known positive

PROCEDURE
1. Hydrate slides to water (these will be smears).
2. Stain in working solution, 37°C: 1 hour.
3. Rinse rapidly, 2 changes, 95% ethyl alcohol.
4. Dehydrate in absolute alcohol, clear, and mount.

RESULT
inclusion bodies—blue to purplish blue

Schleifstein Method for Negri Bodies (1937)

FIXATION
Zenker preferred.

SOLUTION

basic fuchsin	1.8 g
methylene blue	1.0 g
glycerin	100.0 ml
methyl alcohol	100.0 ml

For use add about 10 drops to 15 to 20 ml of dilute potassium hydroxide (1 g/40 liters water). Alkaline tap water may be used. Keeps indefinitely.

CONTROL
known positive

PROCEDURE
1. Deparaffinize and hydrate slides to water; remove $HgCl_2$.
2. Rinse in distilled water and place slides on warm electric hot plate.
3. Flood amply with stain and steam for 5 minutes. Do not allow stain to boil.
4. Cool and rinse in tap water.
5. Decolorize and differentiate each slide by agitating in 90% ethyl alcohol until the sections assume a pale violet color. This is important.
6. Dehydrate, clear, and mount.

RESULTS
Negri bodies—deep magenta red
granular inclusions—dark blue
nucleoli—bluish black
cytoplasm—blue violet
erythrocytes—copper

COMMENTS
Schleifstein outlines a rapid method of fixing and embedding so the entire process can be handled in 8 hours, including fixing, embedding, and staining.

Massignani and Malferrari Method for Negri Bodies (1961)

FIXATION
10% neutral buffered formalin or saturated aqueous mercuric chloride-absolute alcohol (1:2).

Embedding and Sectioning
    Paraffin method, sections 4 µm.

Solutions
    Harris hematoxylin, see p. 102.

    Dilute hydrochloric acid
            hydrochloric acid, concentrated .... 1.0 ml
            distilled water .... 200.0 ml

    Dilute lithium carbonate
            lithium carbonate, saturated aqueous .... 1.0 ml
            distilled water .... 200.0 ml

    Phosphotungstic acid–eosin stain
    Grind together 1 g eosin Y and 0.7 g phosphotungstic acid. Mix thoroughly into 10.0 ml of distilled water and then bring volume up to 200.0 ml with distilled water. Centrifuge at 1500 rpm: 40 minutes. Pour off supernatant solution but do not throw it away. Dissolve the precipitate in 50 ml of absolute alcohol. When dissolved, add to the supernatant solution. The solution is ready to use. If the eosin and phosphotungstic acid are not ground together and then dissolved, it requires 24 hours for the dye-mordant combination to form.

Control
    known positive

Procedure
    1. Deparaffinize and hydrate sections to water; remove $HgCl_2$.
    2. Stain in hematoxylin: 2 minutes.
    3. Wash in running water: 5 minutes.
    4. Dip 8 times in dilute hydrochloric acid.
    5. Wash in running water: 5 minutes.
    6. Blue in dilute lithium carbonate: 1 minute.
    7. Wash in running water: 5 minutes.
    8. Dehydrate to absolute alcohol.
    9. Stain in phosphotungstic acid–eosin: 8 minutes.
    10. Rinse in distilled water.
    11. Dehydrate by quick dips in 50%, 70%, 80% and 90% alcohol; then follow with 95% alcohol: 1 second.
    12. Complete dehydration, clear and mount.

Result
    Negri bodies—deep red

Comments

Massignani and Refinetti (1958) adapted the Papanicolaou stain (p. 406) for Negri bodies. Because further study led to the discovery that eosin combined with phosphotungstic acid is responsible for Negri body staining, the Massignani and Malferrari stain above was developed specifically for these bodies.

PART IV

Histochemistry and Miscellaneous Special Procedures

23
Histochemistry

By definition the field of histochemistry is concerned with the localization and identification of a chemical substance in a tissue. In a broad sense this might include staining, combining chemical and physical reactions, and using chemical methods to demonstrate basic and acidic properties of tissue. But strictly speaking, histochemistry applies only to chemical methods that immobilize a chemical or enzyme at the site it occupies in living tissue. These methods can apply to inorganic substances, such as calcium, iron, barium, copper, zinc, lead, mercury, and others, as well as to organic substances, such as saccharides, lipids, proteins, amino acids, nucleic acids, enterochromaffin substance, and some pigments. Some substances are soluble and react directly; others are insoluble and must be converted into soluble substances before a reaction takes place. Occult or masked materials are part of a complex organic molecule, which has to be destroyed by an unmasking agent before the chemical can react. Some chemicals may be fixed in place; others that are soluble or diffusible have to be frozen quickly and prepared by the freeze-drying method, without a liquid phase. It is always essential to make control slides (i.e., by excluding substrate or enzyme from the reaction, by using enzyme poisons, or by competitive reactions), thereby preventing confusion between a genuine reaction and a nonspecific one that gives a similar effect.

It is difficult and somewhat impractical to make a sharp distinction between staining and histochemical methods. Some will disagree with the arrangement of methods in this book, but a sequence according to similar tissue or cell types seems most adaptable to general laboratory application—the primary intent of the book. This section on histochemistry will, therefore, cover mainly procedures used to identify enzymatic activity.

The field is a tremendous and exciting one of increasingly extensive activity. It is impossible and impractical to include all histochemical methods and their variations in this type of manual. Because many excellent books and journals on histochemistry describe the newest developments, only methods most familiar (at least to the authors) and practical are incorporated in this text. The procedures can be adapted to a minimum of glassware, equipment, chemicals, and time and yet demonstrate some of the more cooperative of the enzymes. Students who enter research in any aspect of histochemistry or enzymology will have to consult the more specialized and the most recent literature as their interest continues to progress. Many companies offer kits that facilitate histochemical and enzymatic procedures.

For greater details start with the following standard references: Burstone (1962), Casselman (1959), Danielli (1953), Davenport (1960), Glick (1949), Gomori (1952), Gurr (1958), Lillie (1965), Pearse (1968, 1972), and Thompson (1966).

FIXATION

The use of fixed tissue for enzyme demonstration is debatable, but cutting frozen unfixed tissues may result in a greater loss of enzyme into the incubating medium and more cell damage than if fixed tissue is used. Many enzymes will tolerate some exposure to fixatives and should be so treated, if possible, for ease of processing and reduction of diffusion. Even succinic dehydrogenase can tolerate calcium-formalin fixation at 5–10°C for 8 to 16 minutes, if the pieces are kept small. The addition of sucrose (10–15%) to fixatives will often improve the preservation of some enzymes and should be tried.

Aldehydes

Calcium formalin (Baker 1944) is used cold, 4°C, and is recommended for the preservation of many enzymes (acid phosphatase, esterase, and lipase, but not for alkaline phosphatase or aminopeptidase).

calcium chloride, anhydrous	1.0 g
distilled water	60.0 ml
formaldehyde (37–40%)	10.0 ml

Adjust pH to 7.0 to 7.2 with 1 N NaOH. Add distilled water to make a total of 100.0 ml. Store in refrigerator. Check pH frequently, or just before use, and adjust if necessary.

Glutaraldehyde

Sabatini et al. (1964) discovered that 2–6% glutaraldehyde in a phosphate buffer can be used for some enzymes (alkaline phosphatase and partial preservation of acid phosphatase and esterase). Sea water can be used as diluent for marine tissues, and tyrode solution can be used for tissue culture cells.

50% solution of glutaraldehyde (biological grade)	4.0 to 12.0 ml
0.1 to 0.2 M phosphate buffer	100.0 ml

The pH should be 7.4. If shrinkage occurs in the tissue, use the dilution at 2% (4 ml of glutaraldehyde per 100 ml of buffer, not a higher concentration). One percent sucrose can be added to the glutaraldehyde solutions for better preservation of tissue constituents. Store in the refrigerator.

Cacodylate buffer, pH 7.2

1 M cacodylic acid (137.99 g/1000 ml)	54.6 ml
1 N NaOH (40 g/1000 ml)	50.0 ml
distilled water	895.4 ml

Glutaraldehyde-fixed tissue frequently is postfixed in 1% osmium tetroxide buffered with 0.1 M phosphate buffer and glucose or veronal-acetate-sucrose after cacodylate-buffered glutaraldehyde.

Physiological buffer
Solution A. 2.26% $NaH_2PO_4 \cdot H_2O$
Solution B. 2.52% NaOH
Solution C. 5.4% glucose
Solution D. Isotonic disodium phosphate

solution A	41.5 ml
solution B	8.5 ml

Adjust pH to 7.4 to 7.6 with solution B.

Phosphate-buffered fixative

solution D	45.0 ml
osmium tetroxide	0.5 g
solution C	5.0 ml

Stable for several weeks at 4°C.

Wash tissues in buffer several hours or overnight. The maximal stability of glutaraldehyde is at a concentration of 2–10%. Low temperatures and storage in the dark also are advisable. See Anderson (1967) about the purification of glutaraldehyde.

Paraformaldehyde

0.2 M s-Collidine buffer
Solution A

s-collidine (2,4,6-trimethylpyridine)	2.64 ml
distilled water	47.36 ml

Solution B

1.0 N HCl	9.0 ml
distilled water	41.0 ml

Working solution

Equal parts of A and B, pH 7.4 to 7.45

Paraformaldehyde solution

paraformaldehyde	4.0 mg
distilled water to make	66.0 ml

Add distilled water to the paraformaldehyde in a graduated cylinder until the combined total is 66.0 ml. Warm in flask to 60°C (use water bath). Depolymerize by adding 0.1 N NaOH, drop by drop; shake until solution clears.

Working solution

paraformaldehyde solution	66.0 ml
0.2 M s-collidine buffer	33.0 ml
0.5 M $CaCl_2$ (5.55 g/100 ml)	1.0 ml

Adjust pH to 7.4 if necessary.

Acetone (Absolute)

Acetone, 4°C or colder (store in refrigerator or freezer), is used for some enzymes, such as aminopeptidase and alkaline phosphatase.

Buffered acetone (Kaplow and Burstone 1963) is recommended for several enzymes, such as alkaline phosphatase, esterase, peroxidase, and dopa oxidase.

acetone	300.0 ml

Add, with stirring, to a mixture of

0.03 M sodium citrate (dehydrate)	32.0 ml
0.03 M citric acid (monohydrate)	168.0 ml

The pH should be 4.2. Store in freezer.

Preparation for Fixation and Storage of Tissues

For the best results always fix very small pieces of tissue, no larger than 1 to 2 mm thick. They can be left in calcium formalin or glutaraldehyde in the refrigerator overnight. Then proceed to sectioning or store for several days or weeks in a gum sucrose solution or glycerin.

Gum Sucrose (Holt et al. 1960)

Store tissue at 0° to −2°C.

sucrose	30.2 g
distilled water	100.0 ml
gum acacia	1.0 g

Prepare only enough for immediate use. If prepared in excess, stock mixtures should be kept frozen to prevent growth of mold. This can be done in small vials; then only single vials are melted as needed.

The gum sucrose preserves the enzymes, facilitates cutting, and produces good morphological detail.

Glycerin (Turchini and Malet 1965)

Store tissue at $-20°C$.

glycerin .... 50.0 ml
distilled water .... 50.0 ml

Place fresh tissue immediately into the glycerin solution. It can be stored in this condition for at least 9 months. For processing, remove the block of tissue and fix for routine histological staining in calcium formalin. Nuclear and cytoplasmic alterations may occur, but enzymatic activities will be well preserved.

See specific enzymes for additional fixation and preservation recommendations.

DEHYDRATING AND EMBEDDING
Freeze-Drying Method

If the apparatus is available and controllable, freeze-drying is considered the proper tool for the preparation of tissues for enzyme study. There are many types of freeze-drying apparatus on the market; many have been expensive, but cheaper ones are now available. Temperatures sometimes are difficult to control, and the drying time may be long. The principle of freeze-drying is that frozen tissue is kept under high vacuum until all water molecules are removed and condensed onto a cold surface or collected by a desiccant. Small pieces (approximately 1 mm) are frozen solid instantly with isopentane or SUVA-MP66 cooled by liquid nitrogen to a temperature of $-150°C$. [Note: Freon 12, which has been used for many years, is no longer available. It belongs to a class of chlorofluorocarbons (CFCs) developed more than 60 years ago with a variety of uses, including use as refrigerants. However, the stability of these compounds, coupled with their chlorine content, has linked them to the depletion of the earth's protective ozone layer. As a result production of CFCs is being phased out and environmentally acceptable hydrofluorocarbon (HFC) alternatives are being introduced.] This tissue must be frozen rapidly to prevent large crystal formation, which would disrupt the cells. The initial freezing is commonly called *quenching;* it stops all chemical reactions in the tissues. Immediately after freezing and while still frozen, the tissue is dehydrated in a drying apparatus in vacuo at a temperature of −30° to −40°C, and the ice is sublimed into water vapor and removed. There is no liquid phase and therefore no diffusion of enzymes.

When the material is dry, allow it to rise to room temperature, infiltrate with paraffin, and embed.

Because of sensitivity to water, the sections cannot be floated on water but must be applied directly to warm albumenized slides. Embedding blocks must be stored in a desiccator.

Many types of freeze-dry apparatus are offered by the manufacturers. Thieme (1965) gives directions for the construction of an inexpensive and convenient model that is easy to operate and dries tissue overnight. Wijffels (1971) describes a model that is easy to construct and operate.

Freeze-Substitution Method

A simpler and cheaper method that produces excellent results is the freeze-substitution procedure. The tissue is rapidly frozen, and then the ice formed within the tissue is slowly dissolved in a fluid solvent such as ethyl or methyl alcohol. When the tissue is free of ice and completely permeated by the cold solvent, it is brought to room temperature and embedded.

Advantages of this method are these: there are no disruptive streaming movements in the tissue—the ice simply dissolves—and the substitution solution can sometimes be chosen to contain a specific precipitant for a specific substance; and the method can be used for autoradiography, for fluorescent preparations, for studying many enzymes, and for other cytohistochemical preparations. There can, however, be cases in which the freeze-drying procedure is preferred—for instance, when an enzyme might be damaged by the denaturing action of the alcohol as it warms.

Freeze-Substitution Technique (Feder and Sidman 1958; Hancox 1957; Patten and Brown 1958)

PROCEDURE

1. Cool a beaker of liquid propane-isopentane (3:1) or SUVA-MP66 to −170° to −175°C with liquid nitrogen.
2. Place tissue specimen (small piece, 1.5 to 3 mm) on thin strip of aluminum foil and plunge into the cold propane-isopentane. Stir vigorously: 30 seconds.
3. Transfer frozen specimen to ice solvent, previously chilled to −60° to −70°C (see comments below). After a few minutes the tissue can be shaken loose from the foil. Store in dry-ice chest for 1 or 2 weeks; tissues have been stored for 52 days without harm (Patten and Brown 1958).
4. Remove fluid and tissue to refrigerator, and wash 12 hours or longer in 3 changes of absolute alcohol in refrigerator.

5. Transfer to chloroform: 6 to 12 hours in refrigerator, to remove any traces of water. (This step may be omitted in most cases.)
6. Transfer to cold xylene (or similar agent), $-20°C$, and remove to room temperature for 10 minutes.
7. Transfer to fresh xylene, room temperature: 10 minutes.
8. Transfer to xylene-paraffin (50:50), 56°C, vacuum: 15 minutes.
9. Infiltrate, paraffin no. 1, vacuum: 15 minutes.
10. Transfer to paraffin no. 2, vacuum: 15 minutes.
11. Transfer to paraffin no. 3, vacuum: 15 minutes.
12. Embed.

COMMENTS

Ice solvents that can be used are absolute methyl, absolute ethyl alcohol, and *n*-butyl alcohol. Feder and Sidman include complete directions for obtaining, storing, and disposing of liquid nitrogen, isopentane, and propane; also details about equipment. *Warning:* There are potential explosion hazards.

See Bullivant (1965) concerning the use of glycerin preservation prior to freeze-substitution.

Blank et al. (1951) use glycols if fixation is to be avoided, alcohol or acetone if fixation is permissible. This method is as follows:

1. Place slides (2 to 3 mm) of tissue in cold propylene glycol in deep freeze, below $-20°C$: 1 to 2 hours.
2. Transfer to mixture of Carbowaxes, 55°C: 2 to 3 hours.

   Carbowax 4000 .... 9 parts
   Carbowax 1500 .... 1 part

3. Prepare blocks.
4. Cut sections and float on slides according to Blank and McCarthy method (Blank et al. [1951]).
5. Stain or use for autoradiographs. Water-soluble substances will not have leached out. For autoradiographs, before each cut coat the block face with thin molten wax. Press wax-enforced sections against emulsion for exposure. See also Feder and Sidman (1958), Freed (1955), Masek and Birns (1961), Meryman (1959, 1960), and Woods and Pollister (1955).

Freeze-Substitution of Sections (Chang and Hori 1961; Patten and Brown 1958)

PROCEDURE

1. Quench 1- to 2-mm slices in liquid nitrogen or in isopentane chilled in a bath of liquid nitrogen.

2. Section at −15°C in cryostat.
   3. Sweep sections into screw-cap vial of acetone prechilled in dry ice.
   4. Return vial to dry ice and complete dehydration.

Chang et al. Method (1970)

PROCEDURE
   1. Cut small pieces of fresh tissue and section in cryostat. Transfer sections to cold cover glasses and immediately thaw by placing a finger on the glass under the section.
   2. Immediately immerse in acetone at dry-ice temperature for freeze substitution: 2 hours to overnight.

This method permits the cutting of many tissue samples to be preserved at dry-ice temperature until they are processed or incubated. A number of enzymes can be preserved in this way.

CRYOSTAT SECTIONING

Paraffin sections can be cut at room temperature, but many enzymes and other tissue constituents must be sectioned at low temperatures of −25° to −20°C or they are lost. Since embedding is impractical at these temperatures, tissue is frozen in isopentane or SUVA-MP66 (−160°C) or in dry ice-ethyl alcohol mixtures (−70°C). The tissue must be maintained at least at dry-ice temperature (−20°C) and sectioned and mounted at this temperature. This is performed inside a cold-chamber cryostat (−25° to −20°C). The object holder is precooled (object holders can be stored in the cold chamber), and the tissue is frozen on it with a layer of embedding compound (e.g., OCT, Fisher Scientific). This medium also supports and protects the tissue and facilitates sectioning. The embedding medium melts, leaves a thin transparent film against the slide, and does not interfere with enzymatic reactions. *Caution:* If the slides are to be used for autoradiography, remove the embedding compound by washing slides in distilled water before dipping in emulsion.

The knife used in sectioning must be kept sharp and dry. Carefully wiping with a finger tip before each section keeps it dry and clean and reduces compression in the section. If the embedding compound is used and the sections are to be mounted on slides, cut through the compound and through the tissue just until the knife edge reaches the block along the opposite edge of the tissue. This means that the cut is complete through the tissue but not through the embedding compound (Fig. 23-1). Stop sectioning. (As the cut is being made, lightly hold the section against the knife with a small cold brush. [Brushes are kept in the cold chamber below the frost line.] This prevents curling. Most microtomes are now equipped with at least one device that prevents the rolling of sections as they are cut.) Flatten the section with the brush and finish cutting through the block. With the cold (!) brush (sometimes two brushes), lift the section to a cold slide. If serial sections are required, it is possible to collect several sections on the blade and then brush them onto a cold slide in correct sequence. Thaw sections by placing the undersurface of the glass against a finger or the cushiony part of the palm. Cold slides can be used to pick up tissue sections; place a finger on the back of the slide to warm slightly and the tissue sections will adhere to the slide. Slides kept at room temperature can be used to adhere sections to slides at the moment the two are brought together.

When thawed, the sections are ready for processing. If the sections are to go directly into a substrate or fixative, lift them, curled or uncurled, with the cold brush and drop them into the solution. Usually a slight tap of the brush against the side of the container will shake them loose. Frozen tissues can be mounted on cover glasses in the same way as on slides. In some cases the section can be picked up by touching it with the cover glass. If the tissues have been suspended in gum sucrose, 6- to 10-μm sections are easily cut and then dropped into the proper substrate for incubation or mounted on cover glasses or slides. Gelatin embedding (Burkholder et al. 1961; Taylor 1965) can be tried for better tissue support; it does permit easier sectioning and handling.

Fitz-Williams et al. (1960) recommend that, immediately before each section is cut, the tissue block be coated with 20% polystyrene dissolved in methylene chloride (volatile at cryostat temperature). A highly reflecting surface will form; wait until it disappears and then cut the section. Curling of sections is minimal and they can be picked up with fine forceps. The polysty-

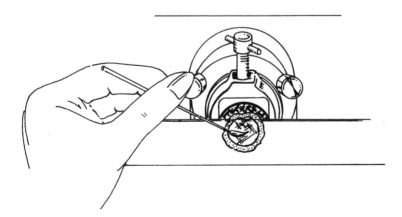

Figure 23-1. Flattening and correcting folds in section: the cut through the embedding compound (*stippled area*) is not yet complete.

rene solution should be stored in the cryostat. These authors also include complete directions on the care of the microtome used in the cryostat.

An innovative set of methods for making cryostat sections with cytological quality close to that achieved with paraffin-embedded material was developed by Leonard Ornstein in 1985. This approach not only deals with the cutting of sections, but presents a full complement of methods of tissue preparation for routine histology and histochemistry. The essence of the cryostat frozen section aid (CFSA) process is (1) the capture of an undistorted, thin, frozen section on a tape made with special cold temperature adhesive; (2) the lamination of the section on a cold (13°C) slide coated with an ultraviolet light-curable, pressure-sensitive adhesive; (3) the curing of the adhesive on the slide with an 8-ms ultraviolet flash and the subsequent removal of the adhesive tape, leaving a still frozen section firmly adhered to the microscope slide; and (4) finally, the ice in the section undergoing freeze-substitution at the cryostat temperature with an appropriate solvent such as acetone. All of these steps are carried out in a routine laboratory cryostat that has had the necessary components (i.e., CFSA Kit; obtainable from Instrumedics, 61 South State Street, Hackensack, NJ 07601) fitted into the cryostat chamber. The company provides instructions for a variety of quick freezing, fixing, and staining methods that are particularly applicable.

Consult the following references for more about cryostat sectioning: Blank et al. (1951), Feder and Sidman (1958), Freed (1955), Glick and Malstrom (1952), Hancox (1957), Hanzon and Hermodsson (1960), Ibanez et al. (1960), Jennings (1951), Mellors (1959), Meryman (1959, 1960), Patten and Brown (1958), Woods and Pollister (1955), Zlotnik (1960), Sheehan and Hrapchak (1980), Proffitt et al. (1992), and Carson (1990). Additionally, directions are usually provided by cryostat manufacturers.

SECTIONING WITHOUT A CRYOSTAT

If a cryostat is not part of the laboratory equipment, frozen sections of many of the enzymes can be made on the clinical freezing microtome or by the method developed by Adamstone-Taylor.

Adamstone-Taylor (1948) Cold Knife Technique

PROCEDURE
1. Quickly freeze fresh tissue (2-mm blocks) on freezing block of microtome or in liquid nitrogen, isopentane, Freon 12 chilled with nitrogen, or petroleum ether containing solid carbon dioxide.
2. Store in dry ice ($-75°C$) or in deep freeze ($-25°C$).
3. Chill knife by fastening solid blocks of dry ice on each end. Thin sheets of metal foil generally work well.

4. When the knife is chilled, freeze the frozen block of tissue in place on freezing head.
5. Cut sections and hold them flat on the knife with a camel's hair brush. Do not allow them to thaw.
6. Remove section to a small metal spatula, which is kept cold with a few bits of dry ice. Transfer the section to a slide. If the slide is cold, speed of transfer is not as essential as it is if the slide is at room temperature. Sections can be transferred directly from the knife to the slide.
7. Press the section in place and, as it begins to soften but before it completely thaws, immerse slide in fixative.

COMMENTS

Unthawed sections can be removed rapidly to a beaker of cold acetone surrounded by dry ice. With needles chilled in the dry ice, mount the sections in the acetone onto chilled cover glasses. Remove, allow to dry, then fix.

This method is difficult in warm and humid atmospheres; the sectioning preferably should be done in a cold room.

ACETONE FIXATION AND EMBEDDING

Some enzymes may be carried through rapid paraffin embedding with good results, but great care must be taken to keep the duration of time in solutions to a minimum and temperatures as low as possible. Even with extreme care, results can occasionally be disappointing.

Gomori Method (1952)

PROCEDURE
1. If possible, chill tissue for a short time before immersing in acetone. This may prevent some shrinkage artifacts. Fix within 10 minutes of death for best results. Cut very thin slices, 1 to 2 mm, no larger.
2. Fix in cold acetone, 2 to 3 changes: 12 hours each.
3. Transfer to ether-absolute alcohol (1:1): 2 hours in refrigerator.
4. Infiltrate with dilute celloidin or nitrocellulose (approximately 10%): overnight in refrigerator. *Note:* Nitrocellulose is difficult to obtain; what is presented here is for historical and heuristic value.
5. Drain on paper towels. Place in benzene or chloroform: 30 minutes to 1 hour. Stirring with a magnetic stirrer aids penetration of fluid.
6. Infiltrate with paraffin 52–56°C (52°C is best): 15 to 20 minutes in oven without vacuum, 30 minutes with vacuum, and 10 to 15 minutes without vacuum.
7. Embed. Store in refrigerator.
8. Section. Float on lukewarm water, briefly (10 minutes maximum).

338 Histochemistry and Special Procedures

9. Drain off water and dry.
10. If to be kept for some time, put in oven 5 to 10 minutes to melt paraffin, forming a protective coating against atmospheric influence. Store in cold room or refrigerator. The safest way to store is in uncut paraffin block in refrigerator.

COMMENTS

Step 4 can be reduced to 2 to 3 hours in nitrocellulose, and the entire procedure following fixation can thus be completed in 1 day with good results.

Alternate Short Gomori Method

1. Fix in cold acetone overnight.
2. Transfer through 2 changes of cold petroleum ether: 1 hour each.
3. Remove petroleum ether and tissues to room temperature: 15 minutes.
4. Infiltrate in 45°C paraffin vacuum: 30 minutes.

Novikoff et al. (1960) find acetone fixation is good for some cryostat preparations.

Frozen Sections Fixed in Acetone (Burstone 1962)

PROCEDURE

1. Cut unfixed frozen sections. Mount on slides.
2. Immerse in acetone: 1 to 5 minutes.
3. Transfer to 90–95% acetone water: 1 to 3 minutes.
4. Transfer to 80–85% acetone-water: 30 seconds to 1 minute.
5. Transfer to 70% acetone-water: 30 seconds.
6. Rinse in water: 10 to 20 seconds.
7. Incubate.

COMMENTS

The quality of fixation is not the best, but the method is simple and useful when cytological detail is not of great importance.

GENERAL SUGGESTIONS

Stock Solutions

The following solutions can be kept in storage and ready for immediate use; in the refrigerator—calcium-formalin and glutaraldehyde; in the freezer—

Histochemistry 339

buffered acetone, absolute acetone, absolute methyl alcohol, and gum sucrose. (See p. 342 concerning methyl alcohol.)

Read labels on all substrates and diazonium salts. Some must be kept in the refrigerator or freezer. If in doubt, store such materials in the refrigerator for longer life.

Control Slides

Most enzyme reactions should be accompanied by a control slide to help detect false positives (see above). An enzyme can be inactivated by any one of the following methods:

—Place slide in distilled water and bring to boil: 10 minutes.
—Immerse in 1 N HCl: 15 minutes; wash in tap water: 0.5 hour.
—Incubate slide in substrate medium containing an inhibitor (NaCN or KCN).
—Incubate untreated slide in a medium lacking substrate.

For additional information, see each specific enzyme method.

Preservation of Incubation Media

Klionsky and Marcoux (1960) prepared 12 different media, froze them in a mixture of dry ice and acetone, and stored them in dry ice chests. The solutions remained as good as freshly prepared solutions for at least 6 months. The types of media tested were:

—Azo dye and metal precipitate methods for acid and alkaline phosphatase.
—Alpha-naphthyl and indoxyl acetate for esterase.
—5-Nucleotidase.
—Nitro-BT for succinic dehydrogenase.
—Postcoupling technique for beta-glucuronidase.
—Triphosphopyridine nucleotide and diphosphopyridine nucleotide diaphorase.

Storage of Slides and Tissues

Storage for long periods is not generally recommended for enzymes; even at $-40°$C they can be kept only about 6 hours. If, however, storage is unavoidable, quenching at $-160°$C and storage at $-85°$C will preserve many enzymes in the tissue. Some enzymes on slides can be kept for 3 months at $-25°$C, but $-85°$C is even better and good for a year.

ALKALINE PHOSPHATASE

Sections containing active phosphatase are incubated in a mixture of calcium salt and phosphate ester, usually sodium glycerophosphate, so that calcium phosphate is precipitated at the sites. The enzyme liberates the phosphate ions, which are held on the spot by salts of metals whose phosphates are insoluble. (Magnesium ions are often required as an activator and are added in the form of magnesium sulfate or chloride. The mechanism of activation is unknown.) The insoluble phosphatase is made visible by cobalt nitrate and ammonium sulfide, which produces cobalt phosphate and finally black cobalt sulfide.

Phosphatase is dissolved or destroyed by ordinary fixatives, but alcohol or acetone will preserve it. Paraffin embedding can be used safely. Freeze-substitution preserves it well. See Burstone (1960) for fluorescent demonstration of alkaline phosphatase.

Gomori Method (Frankel and Peters Modification 1964)

FIXATION

Thin slices (1 to 2 mm) are fixed at once in chilled acetone, 3 changes: 24 to 48 hours. Other fixatives may be 80%, 95%, or absolute alcohol or neutral buffered formalin, followed by freezing cryostat sectioning.

INFILTRATION AND EMBEDDING

For paraffin embedding use chloroform or petroleum ether and paraffin at no higher than 56°C melting point, preferably 52°C. The short methods are recommended. Cut sections 6 to 10 μm. Do not dry slides at higher than 37°C. Store all blocks in the refrigerator until needed.

INCUBATING SOLUTION

0.8% paranitrophenyl phosphate	2.0 ml
2% sodium barbital	2.0 ml
distilled water	1.0 ml
2% calcium chloride	4.0 ml
5% magnesium sulfate	0.2 ml

PROCEDURE
1. Deparaffinize in light petroleum or chloroform: 1 minute.
2. Hydrate with 3 or 4 dips in each of absolute acetone, 80% acetone-water, 50% acetone-water.
3. Rinse with several dips in water or until surface of slide appears homogeneous.
4. Incubate in incubating solution, 37°C: 30 to 45 minutes.
5. Wash well in distilled water.

6. Treat with 2% cobalt nitrate: 3 to 5 minutes.
7. Rinse well in distilled water.
8. Treat with 2% yellow ammonium sulfide (in a laboratory hood because of the obnoxious odor): 1 to 2 minutes.
9. Counterstain with hematoxylin, safranin, Kernechtrot, or the like.
10. Dehydrate, clear, and mount.

RESULTS

sites—black, sharp and clear, nondiffuse
nucleus and nucleolus—show a minimum of gray deposit of cobalt sulfide

COMMENTS

Paranitrophenyl phosphate is split by alkaline phosphatase at a faster rate than other phosphates (phenyl phosphate, glycerol phosphate, and phenolphthalein phosphate) that have been used for this reaction. Since the paranitrophenyl phosphate is so rapidly hydrolyzed, the time of the reaction is reduced to 30 minutes and diffusion artifact formation is minimized. The sites correspond to those demonstrated by the azo-coupling method below. The color in the coupling method tends to fade within a year, but both methods should be tried.

Freeze-substituted and cryostat-cut sections can be used.

The pH optimum is 8.5 to 10.0. The specific activator of alkaline phosphatase is Mg^{++}; other divalent cations, such as Mn^{++}, Zn^{++}, and Co^{++}, may activate in very low concentrations. Inhibitors include EDTA, zinc chloride, formaldehyde, oxidizing agents such as potassium permanganate and iodine, cysteine, and sodium arsenate.

False positives can be due to hemosiderin or melanin, and the sites should be checked for the presence of these pigments.

For additional reading about the Gomori method, see Ackerman (1958), Gomori (1941a, 1946), Kabat and Furth (1941), Moffat (1958), and Pearse (1968).

Azo-Coupling Method

One of the very important naphthols is 3-hydroxy-2-naphthoic acid. It is made when sodium naphtholate is treated with carbon dioxide and used to produce the so-called naphthol AS (naphthol anilid-säure) compounds. Naphthol AS phosphates are phosphate esters of these complex compounds and release highly insoluble naphthols upon enzymatic hydrolysis. These naphthols couple immediately with diazonium salts to form insoluble azo dyes, thereby demonstrating microscopic localization of enzymes in properly fixed tissues. Several of these naphthol phosphate compounds are suitable for demonstrating alkaline phosphatase, but probably the most popular is AS-MX (others are AS-KB, AS-TR, AS-BS, AS-AN, AS-E, and AS-BI).

The azo-coupling method can be used on tissues prepared in the same manner as those for the modified Gomori technique. After deparaffinizing and hydrating, transfer the sections into the incubation solution below.

For the beginner in enzyme methodology, smears and touch preparations are simple to prepare and can be processed in conjunction with sections of the tissue to verify the presence or absence of an enzyme. For example, if tissue blocks are made of liver, also make and fix touch preparations from the same organ. Stain sections and touch preparations in the same incubating medium. If the sections are negative, but the touch preparations are positive, the enzyme is being lost somewhere in the processing for sectioning.

The occurrence of alkaline phosphatase in the neutrophils of the peripheral blood of most animals (not mice or chickens) makes these cells excellent and easily prepared subjects for the demonstration of this enzyme. Spin whole blood in a small tube (hematocrit or bacteriological culture tube, 11 mm in diameter) at 2500 to 3000 rpm for 10 minutes. With a Pasteur pipette collect the buffy coat of white cells lying on top of the packed red cells and place a *small* drop on a slide. Immediately smear like a blood smear (p. 193), or place another slide on top of the drop. It should be placed so the drop is close to one end of the slide, as shown in Figure 23-2. Wait until drop stops spreading; then quickly pull off the top slide. When preparing smears, do not allow the cells to be overcrowded; the cells must be well flattened to demonstrate precisely the alkaline phosphatase granules. Allow the smears to dry thoroughly; slight heat, such as that from a slide warmer, for a few seconds, causes no damage to the enzyme and helps flatten the cells.

When the smears are dry, place them in cold fixative ($0°$ to $-10°C$) consisting of 5.0 ml of formaldehyde (37–40%) and 45.0 ml of absolute methyl alcohol, and leave them overnight. The alcohol must be fresh and in a tightly closed container. Alcohol from the final third of a gallon bottle may not preserve the enzyme adequately; therefore, purchase only 1-lb bottles, and store them in a refrigerator or freezer. Mixed fixative can be kept in a freezer, but not longer than 1 week.

Temperature is very important; if it is too high or too low, the fixation of the enzyme will be imperfect and either diffuse or reduced activity will result. Incubation of smears and touch preparations can be undertaken on the same day after 30 minutes of fixation. Such preparations can be kept in the cold fixative for 2 nights and then be processed with sections prepared from tissue blocks that have been fixed overnight, rapidly embedded, and sectioned. Do not leave smears and touch preparations in the fixative more than 2 days; to do so reduces the activity. Storing in gum-glucose does not preserve alkaline phosphatase activity.

When drawing blood for buffy coats, use heparin, not EDTA, as an anticoagulant. The EDTA is an inhibitor of alkaline phosphatase.

Although a phosphatase score will drop very slightly within the first couple

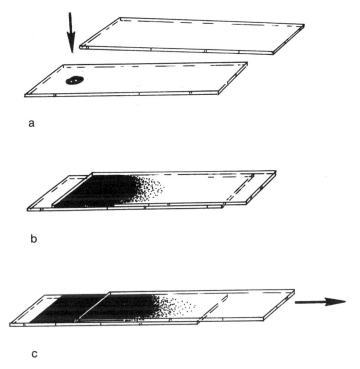

Figure 23-2. Type of smear useful for enzyme preparations: (a) Place a slide on top of drop of white cells; (b) allow cells to spread between slides; (c) when spreading stops, pull top slide length of and off bottom slide.

of hours after blood withdrawal, good scores can be obtained within 6 hours after withdrawal. Keep the blood at room temperature or 37°C. Do not refrigerate it; low temperatures inhibit the activity. If a centrifuge is not handy, allow a buffy coat to form by sedimentation for 1 to 2 hours. Then collect the white cells.

Kaplow Method, Modified (1963)

SOLUTIONS

Propanediol stock solution

2-amino-2-methyl-1,1,3-propanediol	10.5 g
distilled water	500.0 ml

Buffer working solution, pH 9.7

propanediol solution	25.0 ml
0.1 N HCl	5.0 ml
distilled water to make	100.0 ml

Incubating solution

    naphthol AS-MX phosphate .... 5.0 mg

Dissolve in:

    N,N-dimethyl formamide (DMF) .... 0.2 to 0.3 ml

Add:

    buffer working solution .... 60.0 ml
    fast red violet salt LB .... 30.0 to 40.0 mg

Shake for 30 seconds and filter onto slides.

Procedure

1. Deparaffinize sections in chloroform: 1 minute. Hydrate through acetone: few dips; through 70% acetone-water: few dips; through 50% acetone-water: few dips; then wash in distilled water: 30 to 60 seconds. Wash smears and touch preparation in distilled water: 20 to 30 seconds, and blot.
2. Incubate, room temperature: 20 to 25 minutes.
3. Wash in running water: 1 minute.
4. Stain in Mayer hematoxylin: 3 minutes.
5. Wash in running water: 1 to 2 minutes.
6. Blue in Scott solution: several dips.
7. Wash in running water: 1 minute.
8. Stain in 0.5% aqueous wool green S: 1/2 minute.
9. Rinse in water (tap or distilled): 3 to 4 seconds.
10. Dehydrate in 2 changes of 95% ethyl alcohol: few seconds in each.
11. Dehydrate, clear, and mount.

Results

sites—red
nuclei—blue
cytoplasmic staining—pale green
eosinophilic granules—brilliant green

Comments

Fast green FCF or light green can be substituted for wool green, but the stain is paler and less interesting. Kaplow does not use the green, but its addition provides background cytoplasmic staining and distinctive green color to the eosinophilic granules.

Kaplow blots and dries the slides after step 5, and mounts in glycerol jelly or Permount. Bluing in Scott solution and counterstaining does not alter the score of the phosphatase in the neutrophils.

For scoring the amount of alkaline phosphatase in the neutrophils, see Kaplow (1963).

Coupling can be done with other salts, such as fast blue RR, which gives blue sites.

For references about alkaline phosphatase, see Burstone (1962), Elias (1990), Gomori (1951), Manheimer and Seligman (1949), and Pearse (1968).

ACID PHOSPHATASE

This enzyme exhibits optimal activity at pH below 7.0, usually between 3.8 and 6.0. The best tissues for its demonstration are spleen, liver, kidney, and blood leukocytes. In many cells, acid phosphatase seems to be associated almost exclusively with the lysosomes, so much so that acid phosphatase is considered to be a "marker" for them. Lysosomes are rich in hydrolytic enzymes—more than 10 of them, all showing an acid pH optimum. Esterase also can be demonstrated in lysosomes. The small granules of concentrated acid phosphatase found in the Golgi region are tentatively considered to be lysosomes.

Care must be taken in the preparation of tissues for acid phosphatase. The fact that the enzyme is readily made soluble and released from the tissue is a probable cause of failures in past methods. The simplest and most dependable fixation is to place small pieces (2 mm) in cold calcium-formalin (p. 328), 4–10°C, overnight. Large organs from large animals often require 1 to 2 minutes of perfusion with fixative before a small block of tissue is removed. The tissue should then be stored in gum sucrose, 4°C, for 24 hours.

Section on freezing microtome and drop sections into incubation medium. Mount sections on slides and allow to air dry: 2 hours. Proceed to substrate or store overnight in refrigerator. If sectioning cannot be done immediately after overnight fixation, transfer the blocks of tissue into gum sucrose (p. 330) and refrigerate. Krus et al. (1961) says it can be preserved for 40 days. (Freeze-substitution preserved 75–80% of the enzyme.)

Tissues can be acetone-fixed and paraffin-embedded (short method), but deparaffinizing the sections and mounting them on slides may inactivate the enzyme.

Cold 2% glutaraldehyde can be used for good preservation of vacuoles and other structures, but it should be accompanied by calcium-formalin-fixed blocks. There are disadvantages to using glutaraldehyde. It penetrates slowly; even after overnight fixation the center of a block may remain unfixed, and diffusion of the enzyme into the outer fixed area may have taken place. (Never cut blocks thicker than 1 to 2 mm for enzyme studies.) Glutaraldehyde leaves a yellow background color, which may be considered undesirable. See Janigan (1965) for studies on aldehyde fixation for acid phosphatase.

DMSO (dimethylsulfoxide) is being used in enzyme research because it protects against freezing damage and increases cell permeability. But it must be used with caution; a more than desirable diffuse reaction can take place. Brunk and Ericsson (1972) use it and sucrose in their solutions and thereby reduce the incubation time. Sucrose is added to a total of 0.1 M to the fixative, and the wash contains 0.25 M sucrose and 10% DMSO. The incubation medium is Gomori's made 0.22 M with sucrose plus 1% of DMSO.

As recommended for alkaline phosphatase, touch preparations and smears are useful adjuncts to sections. Two methods for fixing such preparations are suggested. Prepare duplicate slides; leave one dry and unfixed, and fix the other for 20 to 30 seconds (no longer; do not store in fixative) in cold buffered acetone (p. 330) or cold calcium-formalin. Drain against filter paper or cleansing tissue, and allow to dry. Incubate both slides with tissue sections. Postfix the unfixed slide before counterstaining (see step 4, azo-coupling method).

Protozoans make fine subjects for lysosome studies and for other enzymes. Prepare some on subbed slides for better adherence, but always make control slides that have no adhesive on them, since it is possible that subbing material or albumen may inhibit an enzyme.

Gomori Method (1952)

SOLUTIONS

Sodium acetate buffer, pH 5.0
0.6% acetic acid	300.0 ml
sodium acetate (27 g/1000 ml water)	700.0 ml

Substrate
acetate buffer	25.0 ml
distilled water	100.0 ml
lead nitrate	0.12 g
3% sodium glycerophosphate (freshly prepared)	10.0 ml

Keep at 37°C for 24 hours. Filter. Add 2 to 5 ml of distilled water to prevent precipitation on evaporation. Keep in refrigerator. If the solution becomes turbid, discard. It is best when used within 3 to 4 days.

PROCEDURE
1. Deparaffinize sections as for alkaline phosphatase and hydrate to water. Smears and frozen sections can be placed directly into substrate.
2. Incubate in substrate, 37°C: 1 to 24 hours; average is 4 hours.
3. Rinse in distilled water: few seconds.
4. Wash in 1–2% aqueous acetic acid: 1 minute (see comments below).

5. Wash in distilled water: 1 minute.
6. Transfer to 1% aqueous ammonium sulfide (in laboratory hood): 2 minutes.
7. Wash in distilled water: several minutes.
8. Counterstain in hematoxylin, Kernechtrot, or the like.
9. Wash well in tap water to remove excess stain.
10. Mount in glycerin jelly; or dehydrate, clear, and mount.

Result
sites—black

Comments
Gomori followed step 3 with a wash in dilute acetic acid to remove nonspecific bound lead, but Goldfischer et al. (1964) felt that the rinse can remove all of the reaction product from an enzyme site of low activity. One slide should therefore be processed as a test slide with the acetic acid rinse omitted.

Gomori method is analogous to the alkaline phosphatase technique in that phosphate is released by enzymatic activity and is precipitated as lead phosphate, which is finally converted to a black lead sulfide.

Controls: Since sodium fluoride inhibits acid phosphatase, add the fluoride to the incubating solution to a final concentration of 0.005 M. The sodium glycerophosphate can then be left out of the incubating solution as a control.

Thin blocks of tissue can be incubated, embedded in paraffin or some other medium, 2- to 3-μm sections cut, mounted on slides, cleared, and coverslipped.

For unfixed frozen sections, see Bitensky (1963) for modification of Gomori method.

Azo-Coupling Method (Kaplow and Burstone 1964)

This also is analogous to the alkaline phosphatase method but uses a buffer at pH 5.0.

Solution
Substrate
    naphthol AS-MS phosphate .... 5.0 mg

Dissolve in:

    N,N-dimethyl formamide (DMF) .... 0.2 to 0.3 ml

Add:

    distilled water .... 25.0 ml
    0.2 M acetate buffer (see Gomori method above) .... 25.0 ml
    fast red violet salt LB .... 30.0 to 40.0 mg

Shake for 30 seconds and filter onto slides. One may substitute 60 ml of a 0.1 M citrate buffer, pH 5.2, for the water and acetate buffer.

PROCEDURE
1. Deparaffinize sections in petroleum ether or chloroform, and hydrate quickly through acetone solutions to water. Frozen sections, smears, and touch preparations go directly into substrate.
2. Incubate, 37°C: 2 hours.
3. Wash in running water: 1 minute.
4. Fix unfixed smears in 10% formalin: 30 seconds, and wash in running water: 2 to 3 minutes. This step may be omitted, but better nuclear staining is obtained after brief fixation.
5. Stain in Mayer hematoxylin: 3 minutes.
6. Wash in running water: 5 minutes.
7. Mount in glycerin jelly.

RESULTS
sites—red
nuclei—blue

COMMENTS
Since the lysosomal membrane is less permeable to glycerophosphate than to naphthol AS phosphates, the reaction is slower in the glycerophosphate technique (Maggi and Riddle 1965).

Schajowicz and Cabrini (1959) say that, if decalcification is necessary, 2% formic acid and 20% sodium citrate (1:1), pH 5, is satisfactory up to 15 days. There is loss of enzyme if Versene is used.

For additional reading about acid phosphatase, see Burstone (1959, 1960, 1962), de Duve (1959, 1963a,b), Gomori (1950a, 1956), Novikoff (1960, 1961b), Pearse (1968), Weissmann (1964), and Zugibe (1970).

AMINOPEPTIDASE (PROTEOLYTIC ENZYME)

The proteolytic enzymes hydrolyze proteins and amino acid compounds. The aminopeptidase (and Exopetidase) reaction depends on the hydrolysis of a peptide bond adjacent to a terminal alpha amino group, resulting in the liberation of a chromogenic moiety. The enzyme hydrolyzes L-leucinamide at a highly preferential rate, which makes leucyl naphthylamide a favored substrate. It can be activated by bound metal ions such as Mg^{++}, Mn^{++}, and Co^{++}. It can be destroyed by a 1-hour immersion in water at 70°C; it can be inhibited by heavy metal ions Zn^{++}, Cu^{++}, Hg^{++}, Cd^{++}, Pb^{++}, and Fe^{++} or by fixing in absolute methanol and 40% formaldehyde.

High activity will be found in the kidney, parathyroid, small intestine, and uterus; some activity in salivary glands, thyroid, liver, and thymus. Buffy coats make excellent preparations: fix in cold acetone: 1 hour. Fix tissue blocks in cold acetone ($-10°$ to $0°C$) overnight and embed by the short method (p. 338). Xylene, 1 to 2 hours, can be substituted for petroleum ether. Freeze-dried cryostat sections or fresh tissue cut frozen can be used.

A large number of diazotates are stabilized as double zinc salts and, since aminopeptidase is sensitive to metallic ions like zinc, it is important to use the proper diazonium salt in the coupling reaction. Garnet GBC is available as a stabilized diazonium sulfate.

Burstone and Folk Method (1956)

SOLUTIONS

Substrate stock solution

L-leucyl-*b*-naphthylamide	1.0 g
distilled water	100.0 ml

Tris buffer, pH 7.19

0.2 M tris (hydroxymethyl) aminomethane	25.0 ml
0.1 N HCl	45.0 ml

Dilute with distilled water to 100.0 ml.

Working solution

stock substrate	1.0 ml
distilled water	40.0 ml
tris buffer	10.0 ml
garnet GBC (diazotized *o*-aminoazotoluene)	30.0 mg

Shake and filter.

For substrate, the alanyl compound may be substituted: 10 mg in 40 ml hot distilled water ($90°C$). Cool and add 10 ml of tris buffer and 30 mg garnet GBC.

PROCEDURE

1. Chloroform or petroleum ether, 2 changes: 4 minutes each.
2. Absolute acetone: 1 minute.
3. 95% acetone: 1 minute.
4. 85% acetone, several dips.
5. Distilled water, few dips.
6. Substrate: 15 minutes to 1 hour, room temperature, or $37°C$.
7. Tap water: 5 minutes.
8. Hematoxylin: 3 minutes.
9. Tap water: 10 minutes.
10. Mount in glycerin jelly.

RESULT
sites—red

COMMENTS
If necessary to purify GBC salt:

1. Dissolve 10 g salt in 50 ml methyl alcohol. Let stand 30 minutes. Filter.
2. Repeat with residue.
3. Dissolve second residue in 500 ml ether. Let stand 2 to 3 hours.
4. Filter. Wash residue 2 to 3 times with ether. Store dry powder in refrigerator.

Aminopeptidase preparations are only semipermanent. Bubbles develop in glycerin jelly mounts within a few weeks, and the reaction fades in PVP.

For further information about aminopeptidase, see Burstone (1962), Pearson et al. (1963), and Zugibe (1970).

See Lojda and Havránková (1975) for a method using bromindolyl leucinamide.

ESTERASES (NONSPECIFIC) AND LIPASES

These terms are used in reference to enzymes that hydrolyze esters of carboxylic acids. There is an overlapping between enzyme types. Some seem to occupy intermediate positions between lipases and esterases and react like both forms (Burstone 1962; Chessick 1953; Gomori 1952). In general, esters of short-chained fatty acids are acted on by esterases; long-chained esters, by lipases. A large number of presumably specific enzymes have been reported and demonstrated in lysosomes.

Esterases or lipases are not destroyed by acetone fixation, are resistant to heat, and can be embedded in paraffin if not used excessively (3 hours maximum, 58°C). Sections fixed overnight in calcium-formalin or 10% formalin and frozen sections can be used for esterase and lipase. Infiltration with gum sucrose, 12 to 24 hours or longer, facilitates sectioning. Freeze-drying gives good preservation.

Fix buffy coats in formalin vapor (1 to 2 ml of concentrated formaldehyde in a Coplin jar), add dry smears, and cover tightly: 30 minutes. Store dry until use. Unfixed smears can accompany fixed smears through incubation; then, before counterstaining, postfix the unfixed smears in 10% formalin: 60 seconds; or methyl alcohol: 30 seconds.

Lipases are incubated with the long-chained fatty acid esters or sorbitan and mannitan, water-soluble commercial products by the name of Tween.

Histochemistry 351

On hydrolysis, the liberated fatty acids form insoluble calcium soaps deposited at the sites. These are converted into lead soaps and then into brown sulfide of lead.

Esters of naphthols also are used, and naphthol is liberated and demonstrated by azo-coupling.

Different esterases react to different inhibitors. For nonspecific esterases, use E600 (diethyl-p-nitrophenyl phosphate), 10^6 to 10^{-3} M, as an inhibitor, or leave the substrate out of the solution.

Esterase (Moloney et al. Modification 1960)

SOLUTIONS
Buffer, pH 7.4

    sodium barbiturate (1.03 g/50.0 ml water) .... 50.0 ml
    0.1 N HCl .... 31.0 ml
    distilled water to make .... 100.0 ml

Substrate

    naphthol AS-D chloroacetate .... 10.0 mg

Dissolve in:

    acetone .... 0.5 ml

Add:

    distilled water .... 25.0 ml
    buffer .... 25.0 ml
    fast blue RR .... 30.0 mg

Shake 30 seconds and filter onto slides.

PROCEDURE
1. Deparaffinize and quickly hydrate paraffin sections. Wash fixative out of fixed smears or frozen sections.
2. Incubate in substrate, room temperature: 30 seconds.
3. Rinse in running water: 1 minute (2 to 3 changes of water for loose sections).
4. Counterstain in Mayer hematoxylin: 3 minutes.
5. Wash in running water: 3 minutes (2 to 3 changes of water for loose sections). Do not blue; the hematoxylin will remain purplish and present fair contrast to the dark blue of fast blue RR.
6. Mount in glycerin jelly.

Comments

Postfixing in 10% neutral buffered formalin improves cellular detail and nuclear staining of both fixed and unfixed smears.

Kernechtrot can be substituted for hematoxylin, but it fades in glycerol jelly. The fast blue RR salt is preferred by some since the garnet GBC produces an undesirable yellowish background. Glycerin jelly mounts keep well; PVP and Gelva mounts show leaching and fading after standing for a few weeks.

Burstone (1962) uses a tris buffer at pH 7.1.

0.4 M tris (24.2 g/500 ml water)	500.0 ml
N HCl	189.0 ml
distilled water to make	1000.0 ml

See also Holt (1956) concerning the indoxyl methods.

Lipase (George and Ambadkar 1963; George and Ipye 1960)

Solution

5% Tween 85 (see comment below)	2.0 ml
barbiturate buffer, pH 7.4 (p. 351)	5.0 ml
10% calcium chloride	2.0 ml
distilled water	40.0 ml

Incubate mixture, 40°C, for 8 to 10 hours before use to precipitate free fatty acids. Filter. Add crystal of thymol.

Procedure

1. Quickly freeze fresh tissue and cut 40-μm sections. Drop into cold (4°C) 6% neutral formalin.
2. Wash in cold distilled water: 1 hour.
3. Mount on albumenized slides and dry.
4. Coat slides with a thin layer of 5–10% aqueous gelatin. Preserve gelatin with thymol; phenol inactivates the enzyme. (See comments below.)
5. Place in freezer to solidify gelatin.
6. Fix in 6% cold formalin: 1 hour.
7. Wash in running water: 2 hours. Rinse in distilled water.
8. Incubate, 37°C: 8 to 12 hours.
9. Wash well in warm (40°C) distilled water.
10. Treat with 1% lead nitrate, 20 minutes.
11. Wash in warm distilled water.
12. Wash in running water: 10 minutes. Rinse in distilled water.
13. Treat with 1% aqueous yellow ammonium sulfide in fume hood: 1 to 2 minutes.

14. Wash with 1% aqueous acetic acid to remove yellow color few minutes.
15. Wash in distilled water and mount in glycerin jelly (p. 98).

Result
   sites—brownish black

Comments
Eapen (1960) proved that lipase is activated by gelatin; he coated sections to give better results.

If Tween 80 is used, it is acted on by nonspecific esterases as well as lipase. Counterstaining with hematoxylin can be done.

Bokdawala and George (1964) refine the method by using alizarine red S stain for the calcium soap that is formed.

Additional pieces of tissue can be fixed in calcium-formalin and embedded in gelatin, and the frozen sections can be cut and stained for lipids.

To inactivate the lipase, use heat or Lugol iodine solution.

Sources of error: Brownish pigment or calcareous deposits, pigments visible in unincubated section. These may be hemosiderin (perform Prussian blue reaction) or calcium deposits (remove before incubation in citrate buffer of pH ± 4.5: 10 to 20 minutes).

SUCCINIC DEHYDROGENASE

Oxidation-reduction reactions, which take place in biological systems, are characterized by hydrogen atoms or electrons being transferred from molecule to molecule. Oxidation refers to the loss of an electron. It follows that the substance donating hydrogen or electrons is oxidized and the one accepting is reduced. These reactions are catalyzed by enzymes, and the living cell contains a series of electron donors and acceptors, which act in a chain. Molecular oxygen is the final electron acceptor in most reactions, and hydrogen combines with the oxygen to form water. Some enzymes are unable to use oxygen directly, and the hydrogen atoms must be transferred by intermediate compounds. Dehydrogenases are the first compounds of the chain of enzymes, which transfer the hydrogen atoms from the substrate to the molecular oxygen. Succinic dehydrogenase acts as an intermediate carrier.

Among the techniques for the demonstration of dehydrogenases, the tetrazolium technique is the preferred one. The tetrazolium salts, characterized by a heterocyclic ring structure containing one carbon and four nitrogen atoms, are colorless or pale and are readily reduced to form intensely colored water-insoluble formazans. The tetrazolium salts, however, do not accept the electrons from the dehydrogenases, but from subsequent links in the chain.

The tetrazolium salts are:

TPT-2,3,5-triphenyl tetrazolium chloride: simplest, but with limited sensitivity.
BT-2,2',5,5'-tetraphenyl-3,3' (3,3'-dimethoxy-4,4'-biphenylene) ditetrazolium chloride: better pigment qualities.
NitroBT-2,2'-di-p-nitrophenyl-5,5'-diphenyl-3,3'(3,3'-dimethyoxy-4,4'-biphenylene) ditetrazolium chloride.

Most methods require anaerobic (absence of oxygen) conditions to prevent atmospheric oxygen from competing with the tetrazolium for the electrons of the dye and thus to decrease reduction. Sodium succinate is the substrate.

Mitochondria are complete biochemical units involved in biological oxidations and have the succinic oxidase (also cytochrome oxidase) system built within their structure. If mitochondria are well preserved and the substrate is not oxidized by nonmitochondrial components, the staining could be interpreted as mitochondrial. But not all sites are necessarily mitochondria; formazan can form crystals on the surface of lipid droplets, lipofuscin granules, fat vacuoles, etc.

For further details concerning dehydrogenases and tetrazolium salts, see Burstone (1962).

Most sections are fresh frozen and cut at 30 μm (not less than 20 μm) in a cryostat. They may be floated on a beaker of the incubating fluid inside the cryostat mounted on slides and air dried in the cryostat. For anaerobic conditions, remove air from the solution with vacuum or boil the substrate and then cool to 37°C. In either method keep the solution in full, tightly closed bottles (small weighing bottles are excellent). Drop the sections in and close the bottle immediately.

Warning: The freezing and thawing of sections may disrupt the morphological structure of the tissue and may result in small formazan crystal artifacts, which can be confused with mitochondria. Cardiac muscle is the best tissue to use for an unfixed frozen preparation.

Novikoff (1961a) relates that localization is not possible unless the cell structure is better preserved than is that of cryostat-cut unfixed sections. Try fixing small (1 mm) slices of tissue in cold calcium-formalin, 0–5°C: 15 minutes. Quench in isopentane, −70°C: 1 to 2 minutes, and cut in cryostat at −25°C. Transfer to cover glasses and incubate or transfer directly into incubation solution.

Touch preparations and smears make beautiful demonstrations of succinic dehydrogenase. Some cells may be broken and produce some diffusion, but those with unbroken membranes can always be recognized. Both fixed and unfixed slides can be incubated together. Fix the touch preparations and smears in cold calcium formalin: 15 to 20 seconds.

Nitro BT Method

Solutions

Tris buffer, pH 7.4

N HCl	17.0 ml
0.4 M tris (24.2 g/500 ml water)	83.0 ml

Substrate

Nitro BT	5.0 mg

Dissolve in:

N,N-dimethyl formamide (DMF)	0.25 ml

Add:

sodium succinate	500.0 mg
distilled water	10.0 ml

Add:

tris buffer	20.0 ml

Shake the solution after each addition. Filter onto slides.

Procedure

1. Incubate fixed or unfixed smears, frozen or cryostat-cut sections, 37°C: 30 to 45 minutes.
2. Wash in running water: 1 minute. Following wash, unfixed preparations may be fixed in 10% neutral buffered formalin or calcium formalin, room temperature: 10 to 30 seconds.
3. Counterstain in Mayer hematoxylin: 2 to 3 minutes.
4. Wash in running water: 3 minutes. Do not blue the hematoxylin.
5. Mount in glycerin jelly, or dehydrate, clear, and mount.

Results
sites—blue
nuclei—purplish blue

Comments

Control slides: Replace succinate with distilled water and incubate as above.

Aronson and Pharmakis (1962) observed irregular staining in hearts and fatty degeneration. By adding 25.0 ml of 0.05% potassium cyanide to 75.0 ml of substrate, they improved the staining.

A 0.1 M to 0.06 M phosphate buffer, pH 7.5, may be substituted for the tris buffer.

For more information about succinic dehydrogenase, see Burstone (1962), Farber and Bueling (1956), Farber and Louviere (1956), Morrison and Kronheim (1962), Nachlas et al. (1957), Novikoff (1961a), Novikoff et al. (1960), Pearson (1958), Rutenburg et al. (1950, 1953), and Zugibe (1970).

THE OXIDASES

The oxidases are enzymes that catalyze the transfer of electrons from a donor substrate to oxygen. Most oxidases contain iron or copper. Tyrosinase contains copper and catalyzes the aerobic oxidation of tyrosine and of some phenols. When tyrosine is acted upon by tyrosinase, dihydroxyphenylaline (dopa) is formed. Dopa is converted to O-quinone and is then condensed to form hallochrome, a red pigment, which is spontaneously transformed into a black pigment, melanin. The enzyme tyrosinase, therefore, is responsible for the conversion of tyrosine into melanin, but the so-called tyrosinase system is very complicated, and care should be taken in interpreting its reaction (Burstone 1962; Pearse 1968).

Laidlaw Dopa Oxidase Method (Glick 1949)

FIXATION

Only fresh tissue with no fixation (or 5% formalin for no longer than 3 hours, but results are inferior to those from fresh tissue).

SOLUTIONS

Dopa stock solution
    DL-b-(3,4-dioxphenylalanine) phenyl tyrosine .... 0.3 g
    distilled water .... 300.0 ml

Store in refrigerator. Discard when it turns red.

Buffer solution A
    disodium hydrogen phosphate .... 11.0 g
    distilled water .... 1000.0 ml

Buffer solution B
    potassium dihydrogen phosphate .... 9.0 g
    distilled water .... 1000.0 ml

Buffer dopa solution
    stock dopa solution .... 25.0 ml
    buffer solution A .... 6.0 ml
    buffer solution B .... 2.0 ml

Filter thorough fine filter paper. Should be pH 7.4.

Buffered control solution
distilled water	2.0 ml
buffer solution A	6.0 ml
buffer solution B	2.0 ml

Cresyl violet
cresyl violet acetate	0.5 g
distilled water	100.0 ml

PROCEDURE
1. Cut frozen sections, 10 µm.
2. Wash in distilled water: few seconds.
3. Treat with buffered dopa solution, 30–37°C. The solution becomes red in about 2 hours and gradually turns sepia brown in 3 to 4 hours. Do not allow sections to remain in a sepia-colored solution; they may overstain. After first 30 minutes, change to a new solution.
4. Wash in distilled water.
5. Counterstain in cresyl violet.
6. Dehydrate, clear, and mount.

RESULTS
dopa oxidase—black
leukocytes, melanoblasts—gray to black
melanin—yellow brown

COMMENTS
Freeze-dried and embedded tissues give a good dopa reaction.

Tyrosinase is inhibited by compounds that form a stable complex with copper; potassium cyanide and hydrogen peroxide are good examples.

PEROXIDASE METHODS

Peroxidases are hemoproteins that catalyze the transfer to two electrons from substrates to hydrogen peroxide to form water and oxidized dyes. Benzidine was introduced in 1904 as a reagent for peroxidase and is used as an acceptor in the transfer of oxygen from hydrogen peroxide in the reaction catalyzed by this enzyme. This produces a blue product that in a short time fades into brown and diffuses. Treatment with nitroprusside stabilizes this blue reaction product by forming a more stable salt than is formed by the benzidine and peroxidase alone. Peroxidase activity occurs in the mitochondria of striated muscle and heart and the granules of myeloid and mast cells.

358   Histochemistry and Special Procedures

It takes place also in the acinar cells of thyroid and salivary glands, the medulla of the kidney, the Kupffer cells of the liver, and the hair follicles.

Wachstein and Meisel (1964) use fresh frozen sections (cut 10 to 20 μm), which they spread on lukewarm water and then put directly into the incubation mixture.

Smears, air-dried, can be fixed in formalin-95% alcohol (1:9) or acetone: 30 seconds.

Straus Method (1964)

FIXATION
Fix thin slices of tissue in 10% neutral buffered formalin plus 30% of sucrose, 0–4°C: 18 hours, not less than 15 hours.

SOLUTIONS

Benzidine solution
50–70% ethanol	50.0 ml
benzidine	0.1 to 0.2 g
hydrogen peroxide	0.015 ml

Warning: benzidine is considered a carcinogen.

Nitroprusside solution A
sodium nitroprusside	4.5 g
70% ethanol	50.0 ml

Nitroprusside solution B
30% aqueous sodium nitroprusside	15.0 ml
absolute methanol	35.0 ml
0.2 M acetate buffer, pH 5.0 (p. 470)	2.5 ml

PROCEDURE
1. Cut frozen sections, 4 to 6 μm, attach to slides, and dry 1 hour.
2. Treat in ice-cold benzidine solution until peroxidase is distinctly blue. Stop reaction before blue pigment crystals form.
3. Transfer into nitroprusside solution A, ice-cold: few seconds.
4. Transfer into nitroprusside solution B, ice-cold in refrigerator: 1 to 2 hours.
5. Dehydrate through 3 changes 70% alcohol, clear, and mount.

RESULTS
sites—gray green to blue black

COMMENTS
For blood pictures the smears or sections can be counterstained in Giemsa: 10 minutes.

    Giemsa stock solution (p. 203) .... 4.0 ml
    acetone .... 3.0 ml
    0.1 M phosphate buffer, pH 6.5 .... 2.0 ml
    distilled water .... 31.0 ml

Then dehydrate through isopropanol. Do not use a buffered fixative or a calcium fixative; the use of salts appears to increase the formation of crystals in the blue reaction product. Washing the fixed sections for 24 hours in 30% sucrose also reduces this tendency. Storage in glucose at 0–4°C for several weeks preserves the enzyme.

If catalase is added to the incubation mixture, no reaction occurs because the catalase destroys the peroxide. Boiling the tissue also destroys activity. If the slides are kept in the refrigerator, they are usable for several months. If kept at room temperature, they fade in a few days.

Wachstein and Meisel (1964) have found that frozen sections are satisfactory, but the red blood cells hemolyze in the aqueous medium and do not react. Calcium-formalin fixation, 4°C: 30 minutes, preserves a positive stain in the hemoglobin and improves activity in muscle. Longer fixation, however, suppresses the activity.

A peroxidase method can be used to demonstrate microbodies (cytoplasmic bodies; type II, peroxisome) in hepatocytes of livers and renal proximal tubules of kidneys. See Essner (1970), Roels et al. (1970), and Straus (1967).

Cytochrome Oxidase

Cytochrome oxidase is an iron-porphyrin enzyme that catalyzes the reduction of oxygen to water and accounts for the reduction of most of the oxygen used by tissues. It is closely related to cytochrome C, a hemoprotein capable of undergoing a reversible oxidation reduction. Cytochrome oxidase is demonstrated by the indophenol blue or "nadi" reaction. It was named *nadi* from the first two letters of *na*phthol and *di*amine (alpha-naphthol and N,N-dimethyl-*p*-phenylenediamine), the two chemicals that are used to produce the reaction (Burstone 1962).

For information about oxidases, see Burstone (1962), Lillie and Burtner (1953b), and Wachstein and Meisel (1964).

SUBSTRATE FILM METHODS

In Daoust's method (1968), a film of gelatin containing a substrate is placed in contact with tissue sections. During the exposure, the tissue areas that contain the appropriate enzyme attack the substrate in the film. After exposure, either the unaltered substrate or the products of the reaction are visualized by staining. An "autograph" is made in the film and can be compared with the corresponding tissue sections to reveal the sites of enzyme activity. For example, if checking for DNAse activity, add DNA to the gelatin. After exposure the DNA remaining in the film is stained with toluidine blue. Unstained regions correspond to areas of tissue in which DNAse activity occurs (Daoust 1957, 1961; Daoust and Amano 1960; Mayner and Ackerman 1962, 1963).

In another method, a blackened photographic plate is used, and the gelatin film is digested by protease enzymes, thereby leaving an autograph (Adams and Tuqan 1961).

OSMIUM BLACK METHODS

Seligman and Hanker and their coworkers have developed substrates that use groups capable of reacting with osmium tetroxide and can be used for demonstrating enzymes, such as esterase, phosphatases, dehydrogenase, and other substances. These osmiophilic reagents can be used for stable preparations that are nonvolatile and electron-opaque for the electron microscope and have good pigment qualities for the light microscope. The end product is called osmium black. A new principle of osmium bridging uses thiocarbohydrizide (TCH) as a ligand to bond osmium black to other metals. The application of this principle has been described in numerous reports; start with Hanker et al. (1972), Seligman et al. (1968), and their references.

24
Immunohistochemistry

Immunohistochemical (IHC) staining (occasionally also referred to as immunocytochemistry [ICC]) has emerged as a powerful investigative tool, for it extends the basic techniques of histology and histochemistry to permit the identification and localization of biologically active specific amino acids and proteins within tissue sections.

Immunohistochemistry exploits one of the most specific binding reactions in all of biochemistry, that of an antibody to its antigen. It is founded in the fundamental biological principal that an animal will use its immune system to produce specific antibodies against foreign proteins (antigens) that have been introduced into its body, naturally or by experimental means. The function of the antibody is to locate the antigen and physically and chemically bind to it, thus rendering it innocuous and ready for destruction or removal from the infected animal's body.

Two classes of antibodies can be produced experimentally, polyclonal and monoclonal antibodies. Polyclonals are manufactured by different immunologically competent lymphoid cells. As a result the antibodies are immunochemically dissimilar and will bind at several different structural (chemical) parts (called *epitopes*) of the antigen. The animals most frequently used to raise these antibodies are rabbits, goats, and sheep. Monoclonal antibodies, on the other hand, are produced by clones of lymphoid cells (plasma cells); the antibodies are therefore immunologically similar and react with a specific epitope on the antigen against which they are raised. After an immune response has been achieved (usually in mice), B-type lymphocytes are taken from the spleen or lymph nodes and fused with mouse myeloma cells to produce hybrid (hybridoma) cells. These cells can now be propagated either in culture medium or by transplanting them into the peritoneal cavity of genetically similar mice, where unlimited quantities of identical antibodies may now be produced. Among the advantages of monoclonal antibodies are their homogeneity from batch to batch and the absence of nonspecific antibodies. These advantages are based on the supposition that the targeted epitope maintains its integrity after processing (e.g., fixation) of tissues and that it is unique to a given antigen.

Thus, immunohistochemistry adds an important dimension in relating structure to function, for it permits inferring dynamic cellular activities from the evaluation of "static" tissue sections. To keep within the scope and limitations of this book, we are providing essentials to understand the mechanisms of the procedure and details of the protocol. The reader is referred to the literature citations at the end of the chapter for additional details and concise discussions.

As in basic histological and cytological techniques, preservation of cells and tissues is essential for successful immunohistochemistry. The details provided in other sections of this book for securing of tissues, fixation, and subsequent processing into an embedding medium or preparation for cryostat sectioning are basically similar, and the same concerns and precautions must be stringently adhered to when employing immunohistochemical techniques. This is especially so for the choice of fixative and the duration of fixation, which could markedly affect the achievement and interpretation of results. Bouin solution is frequently used, as is phosphate-buffered neutral formalin and B5 (mercuric chloride-based) fixative. Fixation by perfusion (administration of the fixative through the circulatory system of an anesthetized animal) is preferred by many. In this case cold 4% paraformaldehyde, 1–3% glutaraldehyde, or 5% carbodiimide (followed by 5% glutaraldehyde) is the fixative of choice, especially when studying the central nervous system.

For the most part, paraffin, frozen, and Vibratome sections, as well as cell smears and crushes, may be successfully used—the choice of method is suggested by expediency and the detail required by the questions being posed.

IMMUNOSTAINING METHODS

Immunological methods "stain" an antigen by localizing it with a compatible antibody; the site of the antigen-antibody binding is then visualized with an appropriate chromogen marker on one of the components of the antigen-antibody complex. The following discussion introduces the several immunoenzymatic staining approaches available for the localization of antigens. Selection of an appropriate method depends upon the level of sensitivity desired, processing time available, specimen being studied, and, of course, cost.

Direct Method

In this method, which is the least popular, an enzyme-labeled primary antibody reacts with the tissue antigen (Fig. 24-1). The advantages of this method, speed and little nonspecific staining, do not compensate for a low level of signal amplification, which reduces the ability to localize the antigen-antibody complex.

Indirect Methods

1. In the two-step approach, the sequence of events is the application of an unlabeled (unconjugated) primary antibody, followed by an enzyme-labeled secondary antibody directed against the primary antibody (which now serves as its antigen), and finally by the substrate-chromogen solution (Fig. 24-2). If the primary antibody is generated in a rabbit, for example, the

Immunohistochemistry

■ KEY

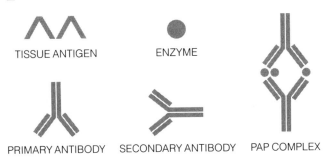

Figure 24-1. See text and key (*above*).

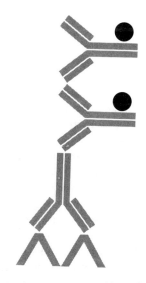

Figure 24-2. See text and key (*above*).

364 Histochemistry and Special Procedures

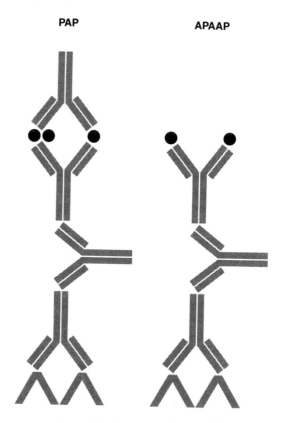

Figure 24-3. See text and key (p. 363).

secondary antibody must be made in a second animal species directed against the rabbit's immunoglobulins. This method has greater versatility, sensitivity, and amplification of signal than does the direct method.

2. In the three-step approach, a second enzyme-conjugated antibody (the tertiary antibody) is used (Fig. 24-3). Application of the unlabeled primary antibody is followed by an enzyme-conjugated secondary antibody, which is, in turn, followed by an enzyme-conjugated tertiary antibody specific to the secondary antibody. Both conjugated antibodies must be labeled by the same enzyme. This method is desirable because an increased number of enzyme molecules at the reaction site allows for the intensification of the final color response. This method provides an acceptable alternative to the soluble immune complex and avidin-biotin procedures described below.

Soluble Enzyme Immune Complex (Unlabeled Antibody) Method

This method is one of the most sensitive of the immunocytochemical techniques and gives excellent results on fixed, paraffin sections studied with light microscopy. It utilizes a soluble enzyme and an antibody directed against the enzyme, which together form an enzyme-antienzyme complex incorporating several molecules of enzyme. The primary antibody and the enzyme complex must be made in the same animal species. The secondary antibody is directed against antigens of the animal in which both the primary antibody and the enzyme-antienzyme complex originate, so that this secondary antibody serves as a "bridge" or "link" between the two. The enhanced sensitivity of the soluble enzyme immune complex method may be attributed to the greater number of enzyme molecules that can be localized at each antigenic site by virtue of the enzyme complex.

The soluble enzyme-antienzyme immune complex techniques are named for the particular enzyme complex used; for example, the PAP method uses a peroxidase-antiperoxidase complex, APAAP uses an alkaline phosphatase-antialkaline phosphatase complex, and GAG uses glucose oxidase-antiglucose oxidase complex (Fig. 24-4). The details for immunostaining are now presented for the PAP method; they are essentially similar to the other soluble enzyme-antienzyme complex techniques.

Description and Staining Steps for the Peroxidase-Antiperoxidase (PAP) Immunostaining Method

The PAP technique was developed essentially by Sternberger and co-workers (Sternberger et al. 1970; Sternberger 1979). A chromogen, such as diaminobenzidine (DAB) or 3-amino-9-ethylcarbazole (AEC), is oxidized in the presence of peroxidase (the enzyme) and hydrogen peroxide (the substrate) at the site of the antigen to form a colored precipitate, which may be now viewed with a light microscope.

IMMUNOSTAINING STEPS
1. Bring the frozen tissue to room temperature or hydrate paraffin sections to water. Apply a drop of phosphate-buffered saline (PBS) or tris-buffered saline (TBS) (see p. 477) to cover each tissue section and allow the drop to remain in place for 5 to 10 minutes at room temperature.
2. Remove any excess buffer around the tissue section with an absorbent tissue.
3. Apply one or two drops (50 to 100 μl) of preblocking (preimmune) agent to the tissue section and allow it to remain in place for 10 to 20 minutes at room temperature.

■ KEY

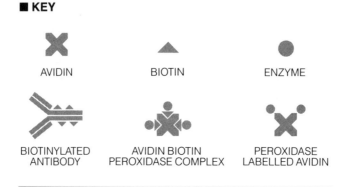

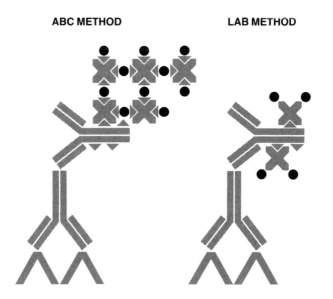

Figure 24-4. See text and key (*above*).

4. Remove preimmune serum by tapping slide on paper towels and wiping away any excess from around the tissue. Do not rinse.
5. Cover each section with one or two drops (50 to 100 μl) of appropriately diluted primary antibody or solution selected as a negative control (see end of this chapter). Incubate the slides at 4°C for 12 to 24 hours in a closed moist incubation chamber (see General Immunohistochemical Directions and Hints).
6. Remove the slides from the incubation chamber and rinse the excess antiserum from the tissues by dipping into a staining dish containing cold PBS or tris/saline buffer. Continue washing in buffer for 2 changes of 5 to 10 minutes each.

7. Blot the excess buffer from the slide with an absorbent tissue (e.g., Kimwipe). Cover each section with 1 to 2 drops (50 to 100 µl) of the appropriately diluted secondary antibody. Incubate the slides in a moist incubation chamber at room temperature for 20 to 60 minutes or 37°C for 15 to 30 minutes. Use the recommended incubation times suggested by the manufacturer of the secondary antibody until you have determined appropriate times for your materials and conditions.
8. Repeat step 6.
9. Blot excess buffer from the slides with absorbent tissue. Cover each section with 1 or 2 drops (50 to 100 µl) of horseradish peroxidase-antiperoxidase (PAP) complex, appropriately diluted. Incubate slides in moist chamber for 60 minutes at room temperature or for 15 to 30 minutes at 37°C.
10. Remove the slides from the incubation chamber and rinse the excess antiserum from the tissue with a gentle stream of room temperature PBS or tris/saline buffer. Place slides in glass slide carriers and continue washing sections in glass staining dishes containing this buffer for 2 changes of 5 to 10 minutes each.
11. Immerse the slides in a freshly prepared chromogen/substrate reagent solution, such as DAB (diaminobenzidine)/H_2O_2 or AEC (3-amino-9-ethylcarbazole)/H_2O_2 for 2 to 30 minutes. Staining development can be monitored by visual inspection of the increasing coloration (DAB, brown; AEC, red). After appropriate time (see General Immunohistochemical Directions and Hints, p. 375). Wash in tap water for three 2-minute changes.
12. Counterstain (optional).
13. If Permount or other organic-based mounting media are to be used, dehydrate sections with increasing concentrations of alcohol followed by xylene. However, AEC is organically soluble; therefore, sections should not be dehydrated or counterstained in alcoholic stains.
14. Coverslip.

Always run specificity controls (see p. 376).

Avidin/Biotin Peroxidase Immunostaining

These methods utilize avidin or strepavidin (the "avidins"). The avidins have very high affinity for biotin, and this, in part, accounts for the increased sensitivity of this method compared to the direct and indirect methods described above. The strepavidin is more widely used. Two avidin-biotin methods may be used; the avidin-biotin peroxidase complex (ABC) method and the peroxidase-labeled avidin-biotin (LAB) method. It has been suggested by the authors of these procedures that the ABC method is more sensitive than the PAP method (Hsu et al. 1981) and that the LAB technique is considerably more sensitive than ABC (Giorno 1984). In both techniques a biotinylated secondary (link) antibody is required to link the primary antibody to an avidin complex. The open sites on the avidin in both the ABC and LAB approaches bind to the biotin on the secondary antibody.

Antigenic sites are made visible by the addition of DAB or AEC. The chromogen reacts with peroxidase in the presence of the hydrogen peroxidase to form a visible, colored precipitate (see above). Alkaline phosphatase is frequently used in place of peroxidase.

IMMUNOSTAINING STEPS

1. Bring the frozen tissue to room temperature or hydrate paraffin sections (see p. 108). Apply a drop of PBS or tris-buffered saline (TBS) to cover the tissue section, and allow the drop to remain in place for 5 to 10 minutes at room temperature.
2. Remove any excess buffer around the tissue section with an absorbent wipe.
3. Apply 1 or 2 drops (50 to 100 μl) of preblocking (preimmune) serum to the tissue section, and allow it to remain in place for 10 to 20 minutes at room temperature.
4. Tap off preimmune serum and wipe away excess around the tissue.
5. Cover each section with one or two drops (50 to 100 μl) of diluted primary antibody or negative control solution. Incubate the slides at 4°C for 12 to 24 hours in a closed, moist incubation chamber.
6. Remove the slides from the incubation chamber and wash in cold PBS or tris/saline buffer for 2 changes of 5 to 10 minutes each.
7. Blot the excess buffer from the slide with absorbent tissue. Cover each section with 1 to 2 drops (50 to 100 μl) of diluted biotinylated secondary antibody. Incubate the slides in a moist incubation chamber at room temperature for 20 to 60 minutes at 37°C for 15 to 30 minutes. Use incubation times recommended by manufacturer of the second antibody.
8. Remove the slides from the incubation chamber and wash in PBS or TBS at room temperature for 2 changes of 5 to 10 minutes each.
9. Blot the excess buffer from the slide with absorbent tissue. Cover each section with 1 to 2 drops of diluted avidin peroxidase complex. Incubate the slides in a moist incubation chamber at room temperature for 20 to 60 minutes or 37°C for 15 to 30 minutes.
10. Remove the slides from the incubation chamber. Wash slides in PBS or tris/saline buffer at room temperature for 2 changes of 5 to 10 minutes each.
11. Immerse the tissue sections in a freshly made chromogen/substrate reagent solution for 2 to 30 minutes. See p. 376 for staining times.

12. Counterstain (optional).
13. Dehydrate, clear, and mount.

Don't forget to run the appropriate controls.

Using Free-Floating Sections and Antiperoxidase or Avidin-Biotin Peroxidase Immunostaining

This technique enhances the immunostaining reaction by creating a greater tissue surface area for immunochemicals to penetrate. Tissue sections are cut thicker (30 to 300 μm) and placed into small volumes of reagents.

IMMUNOSTAINING STEPS

1. Cut sections with a cryostat, freezing microtome, or Vibratome.
2. Collect sections with a soft artist's brush (camel's hair) and place in incubation wells containing PBS. A sufficient volume of fluid should be used so that tissues are free floating and do not adhere to each other—a minimum of 300 μl of diluted antibody for 3 to 4 large sections. Greater volumes should be used for wash buffers and preincubation reagents. Multiple-well plastic and polystyrene dishes in various sizes are commercially available. It is recommended to use a shaker or similar device to agitate sections mildly during incubations and washes.
3. Remove buffer from the wells by siphoning with a Pasteur pipette, and replace with preimmune serum containing 0.3% Triton X-100 (a detergent that emulsifies membrane phospholipids). Incubate tissue sections for 30 minutes at room temperature.
4. Remove preimmune serum and replace with PBS. Wash tissue twice for 15 minutes.
5. Replace buffer with diluted primary antibody. (Antibody diluent should contain 0.3% Triton X-100.) Incubate for 12 to 48 hours at 4°C.
6. Remove primary antibody and replace with PBS. Wash tissue with two 15-minute PBS buffer baths.
7. Replace buffer with diluted secondary antibody. (Antibody diluent should contain 0.3% Triton X-100.) With the PAP method use a working dilution 3 times greater than that used for tissue on slides. Incubate for 1 hour at room temperature.
8. Remove secondary antibody and replace with PBS. Wash tissue with two 15-minute PBS buffer baths.
9. Replace buffer with diluted peroxidase reagent (peroxidase-antiperoxidase or avidin peroxidase). We suggest for the PAP method using a dilution 5 times greater than that used for tissues on slides. Incubate for 1 hour at room temperature.

10. Remove peroxidase reagent and replace with PBS. Wash tissue with two 15-minute PBS buffer baths.
11. Replace buffer with freshly prepared chromogen (DAB)/substrate (H_2O_2) reagent. (Prepared by adding 20 μl of 30% H_2O_2 to 200 ml of cold buffer. Add 150 mg of DAB. Stir and filter through 2 papers into a glass staining dish. When working with DAB, use a fume hood; see precautions below.) Incubate for 5 to 20 minutes until tissue turns a light caramel color (see step 7 under in general directions, p. 276).
12. Terminate the reaction by replacing the chromogen solution with PBS. Wash tissue with three 5-minute PBS buffer baths.
13. Mount sections on gelatinized slides, and dry for 3 to 12 hours on a slide warmer.
14. Dehydrate, clear, and mount.

Indirect Immunofluorescence

This technique involves the use of a primary antibody directed against a specific antigen in the tissue, followed by a secondary antibody tagged with a fluorescent substance, such as fluorescein- or rhodamine-isothiocyanate (Coons 1958; Hartman 1973; Hokfelt ct al. 1975), which binds to the primary antibody. This method works best with frozen material. Final preparations are observed with fluorescence or dark-field microscopy. One is cautioned not to use glutaraldehyde with fluorescein-isothiocyanate label. The relatively rapid loss of fluorescence that accompanies microscope viewing can be slowed by using Vectashield according to the manufacturer's instructions (Vector Laboratories, Burlingame, CA).

IMMUNOSTAINING STEPS
1. Bring frozen sections to room temperature.
2. Apply a drop of PBS to cover the tissue for 10 to 30 minutes.
3. Remove the PBS drop with a Pasteur pipette or with an absorbent tissue.
4. Cover each section with 1 to 2 drops of the diluted primary antisera or negative control. Incubate the slides at 4°C for 12 to 24 hours in a closed, moist incubation chamber.
5. Remove the slides from the incubation chamber and wash by immersing slides in PBS for 15 minutes at room temperature.
6. Remove the excess PBS from the slide with an absorbent tissue. Apply 1 to 2 drops of diluted fluorescein- or rhodamine-labeled secondary antibody. Incubate the slides in an incubation chamber at room temperature for 1 hour or 37°C for 30 minutes.
7. Remove the slides from the incubation chamber and wash by immersing slides in PBS for 15 minutes at room temperature.
8. Drain and blot the excess PBS from all areas of the slide, except that

area containing the tissue. Place a drop of glycerin/PBS over the tissue and apply a coverslip.
9. Observe the tissue staining using a dark-field microscope with filters appropriate for the fluorescein (excitation 440–490 nm; emission 520 to 560 nm) or rhodamine (excitation 546 nm; emission 580 to 600 nm) labeling. If slides will not be observed immediately, freeze them horizontally in a dark environment because fluorescence fades rapidly.

Do not forget the appropriate specificity controls.

Quick Staining
At times, rapid processing of slides for immunohistochemistry is needed and occasionally essential, as in histopathological evaluations during surgical procedures. Prime prerequisites for the rapid method are good antibody quality and a highly sensitive method for detection. Once reagents and techniques have been determined, consistency of protocol with properly fixed tissues is of utmost importance. Generally, all reagents are used in higher concentrations and for significantly shorter incubation periods. One example of a rapid staining procedure that uses DAKO Corporation's labeled avidin-biotin "Quick Staining Kits" (K685, K686, K687) follows:

1. Hydrogen peroxide quench (optional): 3 minutes
2. Primary antibody: 1.5 minutes
3. Biotinylated secondary antibody: 3 minutes
4. Peroxidase labeled avidin: 1.5 minutes
5. Substrate: 3 minutes
6. Counterstain (optional): 2 minutes

Rinse gently and briefly with buffer between steps. All reagents should be ready for use before beginning the staining procedure.

PAP Pre-embedding Immunostaining for Use with the Electron Microscope

In this method for electron microscopy, tissues are immunostained before plastic embedding and ultrathin sectioning (Priestly and Cuello 1983). This greatly facilitates selecting small areas from larger ones for subsequent processing for electron microscopy.

PROTOCOL
Specimen Preparation
1. Perfuse anesthetized animals with 4% paraformaldehyde and 0.1% glutaraldehyde in 1.1 M phosphate buffer, pH 7.4.
2. Continue fixation by immersion in the same fixative for an additional 3 to 12 hours. Collect 50- to 100-μm-thick Vibratome sections in incubation wells containing cold PBS.

Immunostaining Steps

Mildly agitate sections during incubations and washes with a suitable mechanical device. Remove solutions from wells with a Pasteur pipette.

1. Incubate tissue with preimmune serum. In place of adding detergent (such as Triton X-100) to the diluent for the preimmune serum and primary antibody, tissues can be subjected to 2-minute incubations in a graded series of alcohols (10%, 25%, 50%, 25% and 10%) followed by a 15-minute rinse in 0.1 M phosphate buffer, pH 7.4, before preimmune serum is applied.
2. Remove preimmune serum from the wells and replace with PBS. Wash tissue with one or two 15-minute PBS buffer baths.
3. Replace buffer with diluted primary antibody. Incubate sections at 4°C for 12 to 48 hours.
4. Remove the primary antibody and replace with PBS. Wash tissue with three 15-minute PBS buffer baths.
5. Remove buffer and replace with secondary antibody diluted, to begin with, 3 times greater than that used for tissue sections on slides. Incubate for 1 hour at room temperature.
6. Remove secondary antibody and replace with PBS. Wash tissue with three 15-minute PBS buffer baths.
7. Remove buffer and replace with PAP complex diluted in PBS at a dilution recommended by the manufacturer. Incubate for 1 hour at room temperature.
8. Remove the PAP complex and replace with PBS. Wash tissue with three 15-minute PBS buffer baths.
9. Remove buffer and replace with 0.05% DAB in PBS and incubate for 15 minutes.
10. Incubate in wells containing DAB/H_2O_2 until the tissue turns a light caramel brown color (usually within several minutes). Do not allow the incubation to exceed 30 minutes.
11. Terminate the reaction by replacing the DAB solution with PBS. Continue washing the tissue by rinsing several times in buffer.
12. Trim tissues to no more than 3 mm square. Rinse in PBS.
13. Place tissue in 1% osmium tetroxide in 0.1 M phosphate buffer for 2 hours. *Caution:* do this in a functional fume hood.
14. Remove osmium tetroxide and replace with phosphate buffer. Wash tissue in phosphate buffer for two 15-minute baths.
15. Remove buffer and dehydrate tissues in a graded series of alcohols: 70%, 95%, and 100% ethanol. For each alcohol concentration use two baths, 10 minutes each.
16. Remove ethanol and replace with propylene oxide. Use two baths, 10 minutes each.

17. Replace propylene oxide with propylene oxide/resin (1:1) for 30 minutes.
18. Replace propylene oxide/resin with plastic (Spurr's resin) for several hours. [Spurr's resin is made as follows: add together 10.0 g vinylcyclohexane dioxide, 6.0 diglycidyl ether of propylene glycol, 26.0 g of nonenyl succinic anhydride, and 0.4 g of dimethylaminoethanol. Mix for 15 minutes. Cure mixture in 60°C oven for 12 to 16 hours.] Replace with fresh plastic and allow an additional 12 to 16 hours to elapse. (Other resins are available for use.)
19. Embed sections. We suggest using slides that have been dipped in dimethyldichlorosane (DDS) and allowed to air dry for at least 10 minutes. Place a tissue on the slide with a small drop of resin. Cover with another slide dipped in DDS.
20. Polymerize.
21. Pop the slides apart with a razor blade.
22. Glue the sections to blank resin blocks using a cyanoacrylate adhesive.
23. Section as generally done for electron microscopy.

Silver-intensified Immunogold Staining

Immunogold staining is a highly sensitive immunohistochemical technique that utilizes an immunoglobulin absorbed to colloidal gold rather than an enzyme marker such as horseradish peroxidase. Having been used extensively in electron microscopy, immunogold staining has recently gained importance at the light microscope level but with the use of a silver-intensification step (Holgate et al. 1983; Meyer 1983; Van den Pol 1986; see, too, the special issue of the *Journal of Histotechnology* [vol. 16, no. 3, 1993] on immunogold-silver staining for light and electron microscopy).

TROUBLESHOOTING (OR DEALING WITH THE UNEXPECTED)
High Background
Common Remedies

- Check the dilution of your antiserum; a higher dilution may eliminate some (or all) of the background problem while maintaining specific staining.
- Incubate tissues with a nonimmune serum from the species providing the secondary antibody serum to bind nonspecific sites. A recommended concentration varies from 1% to 10% normal serum in rinses and diluted antisera.

- Preincubate tissue with bovine serum albumin (BSA) or gelatin. Each of these proteins has reactive amino groups that can "prebind" nonspecific sites.
- Preabsorb the primary antisera with its carrier molecule if you are using antibodies generated against haptens. For example, some antigens may be conjugated to BSA to generate antibodies; therefore, the antibody would be preabsorbed with BSA at 4°C for at least 12 to 24 hours before actually staining with the antiserum.
- If autofluorescence is a problem using the fluorescein-isothiocyanate method, switch to a rhodamine or peroxidase label. (Autofluorescence and fluorescein are in the same emission range.)
- If using a peroxidase label, you may need to remove the endogenous peroxidase present. Treat sections with either of the following:
  —3% H_2O_2 in distilled/deionized water at room temperature for 5 to 20 minutes.
  —0.01% phenylhydrazine at room temperature for 15 to 30 minutes.
- Up to 0.5 M sodium chloride may be added to the primary antiserum diluent to reduce background.
- Treatments that will increase penetration of the sample, improve signal, and simultaneously reduce background:
  —Use smaller or thinner pieces of tissue.
  —Increase Triton concentration.
- Preabsorb the primary antiserum with tissue from the same animal that you are studying but that does not contain the antigen of interest.
- Change fixatives.
- Change buffers. In some systems, cacodylate buffer causes high background.
- Check the storage conditions of the antibodies. Antiserum that has been frozen and thawed excessively may deteriorate and result in increased background. Avoid frost-free freezers.
- Increase the number and length of washes between antisera incubations to reduce background.

Absence of Staining

If you have reason to believe the antigen is present in your tissue:

- Try a different fixation; you could be overfixing and thereby losing antigenicity or underfixing and thereby losing the antigen itself during rinses and incubations.
- Try a lower antibody dilution (more concentrated antibody). Caution should be used here, as antiserum that is too concentrated can also give negative results.

- Try proteolysis to "unmask" antigenic activity. However, excess proteolytic digestion may destroy antigenicity. Check antibody specification sheets or other accompanying instructions.
- Eliminate any treatment that can sometimes damage antigenicity:
  —Dehydration/extraction before immunocytochemical processing
  —Proteolysis (see above)
- Try different stocks of ancillary antibodies.
- If using the DAB reaction, check the potency of DAB and (especially) hydrogen peroxide.
- Examine procedure for proper antibody sequence.
- Do not allow tissue to dry during the experiment.
- Eliminate preservatives, such as sodium azide, in the wash buffer baths. Sodium azide and mercury-containing preservatives are harmful to the peroxidase enzyme. Always use a buffer that is fresh and free of preservative.
- Do not allow tissues to be exposed to temperatures greater than 60°C as in paraffin infiltration. Excessive heat destroys proteins.
- Check the solubility of the chromogen. If using the chromogen AEC, do not use organic-based solutions after the chromogen has been applied. AEC is soluble in these solutions. Therefore, tissue cannot be dehydrated through alcohol and xylene, stained with a counterstain based in an organic solvent, or mounted with an organic-based mounting medium.
- Recently, microwave oven heating has been used to block antibody cross reactivity and to restore immunoreactivity by retrieving antigens (Lan et al. 1995).

GENERAL IMMUNOHISTOCHEMICAL DIRECTIONS AND HINTS

1. *Locating sections on slides.* Using fingernail polish, a diamond pencil, or a china-marking pencil, draw a circle around the section on the microscope slide to facilitate the location of the section and to contain solutions added to the tissues. (Note: For IHC, few sections are placed on a slide rather than the longer ribbons of sections used in routine histology. Four sections per slide, the procedure used in our laboratory, permits testing for more than one antibody at a time and for applying the appropriate controls.)
2. *Never* allow sections to dry throughout the entire procedure.
3. *Never* touch sections.
4. Setting up and using *moist chamber and slide-holding tray.* Slides are best placed on flat, plastic, slide-holding trays placed into shallow plasticware that can be sealed with a tight fitting cover (e.g., Tupperware).

Table 24-1. A Comparison of the Use of Monoclonal and Polyclonal Antibodies

	Polyyclonals	Monoclonals
primary antibody generated	in rabbit (R)	in mouse (M)
secondary antibody generated	in sheep against R	in goat against M
PAP generated	in R	in M
blocking (preimmune)	sheep serum	goat serum

Moistened paper toweling placed beneath the slide-holding tray serves to maintain the essential moist environment.

5. Exert care in using *compatible immunologicals;* primary antibody and PAP complex should be generated in the same animal species. Secondary antibody is generated in another animal but directed against the species from which the first antibody came. Blocking serum derives from the same animal species as the origin of the secondary antibody. Examples are shown in Table 24-1.

6. *Making dilutions* and determining the correct one to use (i.e., serial dilutions). To select the appropriate working dilutions for primary antisera and for the tissues you are studying, start with the concentration recommended by the supplier. Bridge this concentration with several (serial) dilutions on either side of the recommendation. For example, if 1:400 is suggested, use 1:100, 1:200, 1:600, 1:800, and 1:1000. Choose the final dilution that gives you the most intense immunoreactive response with the lowest background staining.

7. Determining *staining time with chromogens.* Once the time for development of stain has been determined for the concentration of DAB used, it is usually repeatable for subsequent runs. For example, at a concentration of 150 mg per 200 ml, the DAB response is adequate after 1 minute but should be allowed to continue for a total of 2 minutes. *Most chromogens (e.g., DAB) are carcinogenic. Handle carefully wearing latex gloves while working in a functioning laboratory fume hood; dispose of in an appropriate manner in compliance with local and federal regulations.*

8. If *many slides* are to be studied, they should be processed simultaneously using standard histological glass slide racks and staining dishes of reduced volume.

9. *Controls* are essential and not to be avoided! In addition to traditional ones, plan to process control and experimental slides at the same time

---

Figures and tables are being used with permission from the Dako Corporation.

so that an effective internal control to deal with variations in antisera and other conditions is obviated. Some deposits of reaction products may be due to nonspecific staining ("false positive"). Therefore, it is *absolutely essential* in all immunocytochemical procedures to run control slides with the experimental ones to determine the specificity of the reaction. Antibody specificity may be tested by using a "negative" control serum (ideally serum taken from the same animals before being used to generate the specific antibodies) on adjacent sections and preferably on the same slide. Any staining that occurs using the negative control is due to nonspecific background and helps to differentiate specific positive staining from nonspecific background staining on the specimens under investigation. Another test of specificity that should be conducted is to incubate for 12 hours after mixing 1 ml of the diluted primary antibody with 100 μg of the same antigen used to generate the antibody. Structures that are stained using untreated antiserum but unstained with pretreated antiserum are considered "specific" localizations by present standards. (See Petrusz et al. [1976] and Swaab et al. [1977] for further discussion.)
10. To determine appropriate incubation times, it is always prudent to begin by using the time recommended by the manufacturer of the substance in question. There is an inverse relationship between incubation time and antibody titer—the higher the antibody titer, the shorter the incubation time. Incubation times may vary greatly; therefore, it is suggested that preliminary tests be performed to determine the appropriate schedule for staining. In addition, it is important to recognize that reaction times will be shorter, all things being equal, at 37°C than at room temperature.
11. *Double labeling:* It frequently becomes important to determine whether more than one antigen occurs in the same location (co-localization). Several enzymatic methods have been developed to accomplish this. Essentially, there are two approaches; use the same enzyme label with different substrates for the detection of each antigen or use different enzymes for each antigen. In both cases, the intent is to produce two different colors to identify the two antigens they label. As with other immunocytochemical procedures, staining kits are commercially available. There are many companies that sell kits for double labeling, which always come with user-friendly instructions.
12. Make sure paraffin is completely removed from the paraffin-embedded tissue sections.
13. When studying organs, the use of *consecutive (serial) sections* is strongly suggested, for it facilitates analysis by direct comparison. Placing four sections on a 1- × 4-inch microscope slide with ample space between (the width of one tissue section) will permit the testing

of different antisera at the same or different dilutions or control tests on adjacent sections.
14. *Additional treatment of tissues* after processing and before IHC staining [I-4,5]:
    a. Mercury removal (with I-2,70).
    b. Trypsin digestion (can be used to "unmask" antigenic sites that have been "concealed" by excessive cross linkage after formalin fixation).
    c. Deactivation of endogenous peroxidase by incubating the sections in 3% hydrogen peroxide for 8 to 10 minutes. Sometimes 3% hydrogen peroxide (1 part) is combined with absolute methanol (4 parts) for 20 minutes. The methanol and hydrogen peroxide treatment is problematic in loosening frozen sections from slides and preventing the localization of cell surface markers.
15. Unless otherwise indicated, all incubations are to be carried out at room temperature.
16. Making dilutions: Generally, dilutions greater than 1:100 are best made in two steps. The first step should yield 100 μl of an intermediate solution; in the second step, the intermediate solution is further diluted to yield the final required dilution.

A STANDARD TABLE

Specific Examples

EXAMPLE I: CONVERTING A 1:1 SOLUTION TO 1:5

Each microliter of a 1:1 dilution will yield 5 μl of a 1:5 solution because each microliter of a 1:1 dilution contains 1 μg of solute. Therefore, in a proportion:

$1 \mu g / x \mu l = 1 \mu g / 5 \mu l$; $x = 5 \mu l$ as the final volume

Therefore, one would take 1 μl of 1:1 dilution and add 4 μl of diluent for 5 μl of 1:5 dilution.

EXAMPLE II: IF STARTING WITH A 1:10 STOCK SOLUTION, EACH MICROLITER CONTAINS 0.1 μG OF SOLUTE (I.E., 1 ÷ 10).

To prepare a 1:200 dilution, set the proportion:

$0.1 \mu g / x \mu l = 0.1 \mu g / 200 \mu l$

$x = 20 \mu l$ final volume

Therefore, for a 1:200 dilution you take 19 μl of diluent for every 1 μl of the 1:10 stock solution.

EXAMPLE III. IF STARTING WITH A 1:200 STOCK SOLUTION, EACH MICROLITER CONTAINS 0.005 µG OF SOLUTE (I.E., 1 ÷ 200).

To prepare a 1:900 dilution set the proportion:

$0.005 \, \mu g \, / \, x \, \mu l = 1 \, \mu g \, / \, 900 \, \mu l$

$x = 4.5 \, \mu l$ final volume

Therefore, for a 1:900 dilution take 3.5 µl of diluent for each microliter of the 1:200 stock solution.

References

See Childs (1986), Elias (1990), Incstar Corp. (1989), and Naish (1989).

25
Microwave Histology

The use of the microwave oven was introduced in histology and pathology laboratories in the early 1980s. The use of microwaves to achieve rapid fixation, dehydration, and staining of biological material has progressed rapidly in recent years. While there is still a need for the refinement of microwave technology and general acceptance by the histotechnologist, it is clear that its place in the laboratory has been secured. It has been demonstrated that microwave processing accelerates diffusion, can stabilize proteins, and can speed up chemical linkage. Part of the problem of universal acceptance and establishing routine procedures has been the inadequacy of kitchen microwave ovens for laboratory use. The laboratory microwave oven offers features that the kitchen oven cannot and allows reproducibility, which is very important in quality control.

As a general rule, be prepared to modify procedures to accommodate the specifications of your oven. For example, if a procedure calls for a microwave with 450 watts and the oven in your laboratory is 750 watts, a time adjustment will have to be made. An excellent reference is Kok and Boon's (1992) *Microwave Cookbook for Microscopists*. The book is very thorough and covers all aspects of microwave use in the histology laboratory.

Basic Rules for Microwave Oven Usage

1. The laboratory oven should never be used to heat food.
2. Read the manual and be familiar with the operations of the oven.
3. Lids on containers should always be loose or vented.
4. Metallic objects should not be used (except the temperature probe).
5. Avoid heating flammables, such as alcohols and most clearing agents.
6. Monitor frequently for leakage.
7. Use rectangular containers whenever possible. (Curved ones act as a lens and can affect temperature.)
8. Use plastic and nylon containers whenever possible. (Glass can crack or break.)
9. Keep clean; wipe all spills immediately. By putting the staining containers inside plastic bags, spills can be prevented.
10. Be consistent, use the same amount of reagent (e.g., 50 ml) each time.
11. Use with caution in vicinity of people fitted with pacemakers.

Recommended Microwave Components

1. Temperature probe.
2. Extraction system for venting fumes into a fume hood.
3. Nitrogen or air bubble system for agitation and rotation of load.
4. Temperature control within $\pm 1°C$.
5. 500 or more watts.
6. Avoid microwave ovens with the grill option.

Drying Slides

To dry the tissues after they are sectioned and mounted, place the slides in a wooden slide box with lid on. The box actually becomes a convection oven. A recommended drying time is 2.5 to 3 minutes (depending on the wattage). A temperature setting of 60°C should be used.

Fixation

Fixation time can be decreased significantly with the aid of the microwave oven. A routine batch of cut and ready-to-process specimens can be microwaved for 5 to 15 minutes. One can then proceed with the hydration process and eliminate 2 to 8 hours of fixation time.

Large specimens such as brain can be prepared in a very short time using the microwave oven. This method is described by Kok and Boon (1992).

The oven temperature should be monitored at 50°C and 450 watts (80% of power).

1. Sprinkle the entire, unsectioned brain with normal saline and place in a plastic bag for 30 minutes.
2. Remove brain, unwrap plastic, and section the brain in 2- to 3-cm slices.
3. Place slices on a plate, sprinkle with saline, and cover with plastic. Microwave for 15 minutes.
4. Take section for paraffin sections from the microwaved slices, 2 cm × 2 cm maximum.
5. These smaller sections are then placed in 10% neutral buffered formalin for 3.5 hours or microwaved for 5 minutes.
6. Continue with regular processing.

This method can be used also with a previously perfused brain. A jar containing the saline can be used as well. Use 280 watts of power (60%) for 15 minutes.

After washing brain in 10% sucrose in 0.1 M citrate buffer (pH 7.1) 6 to 12 hours, frozen sections can be cut.

Routine Tissue Processing

The formalin should be removed from the tissue before microwave processing. After rinsing thoroughly in running water, place the tissues in 50% alcohol for 1 hour. Tissues have to be processed in plastic capsules or cassettes.

Vacuum paraffin impregnation is recommended for the final infiltration, so the use of a vacuum infiltrator or oven is necessary.

1. Place tissue in absolute alcohol for 30 minutes at 67°C. With a small batch of samples, the time should be 5 to 15 minutes and with a larger batch, 60 minutes.
2. Place tissues in isopropanol for 30 minutes at 74°C. With a small batch, the time should be 3 to 15 minutes and with a large batch, 45 minutes.
3. Place in melted paraffin and microwave for 30 minutes at 67°C. With a small batch 2 to 10 minutes and with a large batch 30 minutes is still recommended. (Paraffin wax itself is microwave transparent and will absorb little energy.) Paraffin with added synthetic plastic polymers (e.g., Paramat; Gurr BHD Chemicals, Poole, England) or Merck waxes (Prod. Nr. 7165 and 7158) may need to be used and at higher temperatures.
4. After a 30- to 60-minute vacuum infiltration, embed as usual.

Methods for Microwave Stains

Grizzle and Staples (1996) recommend using the microwave for heating the solutions only. When the optimum temperature is reached, remove the solution and place the slide into it for the desired time. Using wooden applicator sticks to stir the solution can eliminate hot spots and uneven staining. Practicing the stain with extra slides can help with the necessary modifications to the procedure that may be necessary for your individual microwave oven. Constant checking of the staining progress with a microscope is essential.

Hematoxylin and Eosin Staining

The routine H&E stain can be accelerated with use of the microwave. Bear in mind, however, that standard staining procedures are usually not very time consuming and, with other problems of reproducibility of response, the 2- to 6-minute short cut may not be worth the time saved.

1. Hydrate slides to water.
2. Place slides in microwave in 50 ml hematoxylin (Gills or Harris): 20 to 30 seconds.
3. Proceed as usual with your regular staining method.
4. Rinse in water.
5. Place slides in microwave with 50 ml eosin: 20 to 30 seconds.
6. Dehydrate, clear, and mount.

RESULTS
nuclei—dark blue
cytoplasm—shades of pink, red

Grocott's Methenamine Silver Nitrate (Brinn modification 1983)

This method is used to demonstrate fungi and *Pneumocystis carinni*.

FIXATION
10% neutral buffered formalin

TECHNIQUE
paraffin (5 μm) sections, touch preps, or frozen sections

CONTROL
known positive

PROCEDURE
1. Remove paraffin and hydrate slides.
2. Place slides in a plastic Coplin jar with 50 ml of 10% chromic acid with a loosely applied lid for 50 seconds.
3. Remove and wash in running water.
4. Clear slides in 1% sodium metabisulfite: 30 to 60 seconds.
5. Rinse in distilled water.
6. Place slides in plastic staining jar with 50 ml of working methenamine solution: 60 to 70 seconds.
7. Remove slides, check microscopically for desired intensity, and return to solution for 5 to 10 seconds, if necessary.
8. Rinse in distilled water.
9. Tone in gold chloride: 10 seconds.
10. Rinse in distilled water.
11. Place in sodium thiosulfate: 1 minute.
12. Wash in water.
13. Counterstain in 0.2% nuclear fast red (or 1% light green): 2 minutes.
14. Wash in water.
15. Dehydrate, clear, and mount.

RESULTS
fungi and *Pneumocystis*—black
background—pink and red (or green)

Perl's Iron Stain (1867) **Kok and Boon Modification** (1992)

FIXATION
neutral buffered formalin

SOLUTIONS
   Working solution
      potassium ferrocyanide .... 0.25 g
      distilled water .... 50 ml

   Counterstain
   1% nuclear fast red

CONTROL
   save previous positive

PROCEDURE
   1. Hydrate slides to water.
   2. Place slides in the working potassium ferrocyanide solution (using loosely capped plastic jars): 45 seconds.
   3. Remove from microwave and wash in distilled water.
   4. Counterstain in nuclear fast red: 2 minutes.
   5. Wash in water.
   6. Dehydrate, clear, and mount.

RESULTS
   iron—dark blue

Luxol Fast Blue (Kennedy Modification)

PROCEDURE
   1. Hydrate slides to water.
   2. Stain in Luxol fast blue solution: 1 minute (power level 2; low or defrost).
   3. Rinse in water.
   4. Stain in 1% neutral red: 1 minute (no microwave).
   5. Rinse in water.
   6. Dehydrate, clear, and mount.

RESULTS
   myelin—blue
   nuclei—red

Rapid Papanicolaou Method

The conventional method does not require much more time than the microwave method and, as with the H&E stain, the microwave oven may not be a real time saver.

1. Cover the thin smear with fixative and microwave: 20 seconds at 80% power.
2. Rinse in distilled water.
3. Cover slides with hematoxylin (Gill's recommended) and microwave: 20 seconds.
4. Rinse in tap water.
5. Rinse in 50% alcohol.
6. Cover slides with eosin azure and microwave for 20 seconds at 80% power level.
7. Dehydrate, clear, and mount.

RESULTS
nuclei—blue
acidophilic cells—red to orange
blood vessels—orange, red

Other stains that can be accelerated are numerous, and it would be impossible to list them all here. Attempts to use the microwave for immunohistochemical reactions should be made with care. Overheating the reagents can be detrimental to the process. Leong and Milios (1986) were the first to use microwave exposure to shorten the incubation times. Kok and Boon (1992) illustrate many of these procedures in their book. More recently, the microwave oven has been utilized before staining to unmask the antigens in formalin-fixed tissue sections, thus making them available for immunohistochemical staining. This method, referred to as *antigen retrieval*, is still being refined.

Suggested Readings

Additional references that may be of use include Brinn (1983) for routine staining of histological tissues, Hafiz et al. (1985) for acid and alcohol fast staining, Leong and Milos (1986) for immunoperoxidase staining of lymphocyte antigens, Login et al. (1987) for immunoperoxidase, Lan et al. (1995) for antigen retrieval and Login and Dvorak (1994) for microwave equipment and safety concerns.

26
Special Procedures I

FREEZING TECHNIQUES

The freezing technique for preparing specimens is unsurpassed in certain situations. It is rapid, making it the best choice for diagnostic purposes where speed is essential. It can be used for the preparation of sections containing such elements as fats, enzymes, and some radioisotopes that would be lost in alcohol or paraffin solvents or by heat. One disadvantage is that some distortion may be caused by the freezing and cutting.

Freezing methods formerly used ether, ethyl chloride, or other volatile liquids. Present methods employ a cryostat, which is a stainless steel rotary microtome in a temperature-controlled cabinet. Most of these instruments have a cooler area for specimen preparation and use a heat extractor to assist in freezing the tissue. There are several models available on the market, including one with motorized specimen advance and with motorized hand wheel control. There are models for all budgets and applications.

With frozen sections, ribbons are difficult to obtain, although not impossible with available mounting media (e.g., OCT; Fisher Scientific). Tissues usually curl as they pass over the knife edge; however, most cryostats have at least one antiroll system mounted on the knife holder to assist with flattening the sections. With practice, a fine camel's hair brush can be used to pull the sections gently away from the knife edge. Disposable blades also work well with frozen sections.

Temperature is another important variable with frozen sections. Different tissue types cut easier at certain temperatures, so usually a medium range temperature is set, for example, $-15°C$ to $-20°C$. Also, the temperature should always be measured at the level of the specimen, near the knife edge. Some recommended cryostat chamber temperatures are:

—Brain, lymph nodes, kidney, liver, bone marrow, testes, blood clots: $0°C$ to $-15°C$.
—Muscle, pancreas, skin without fat, uterus, adrenal, intestine, tongue: $-15°C$ to $-30°C$
—Skin with fat, adipose tissues: $-30°C$ to $-60°C$

FIXATION, BLOCKING, AND SECTIONING

Fresh tissues may be used; however, it is preferable to fix the material. Formalin is frequently used, for it gives the tissue an ideal consistency for the freezing technique. Formalin is an aqueous solution and requires no washing. Hartz (1945), however, recommends Bouin as better than formalin but says that the tissue must be washed for a short time before freezing. If the fixative contains mercuric chloride, the crystals may be removed immediately after sectioning with an iodine solution, such as Lugol's, which contains no alcohol. The samples of tissue for processing should be no more than 2 to 4 mm thick for rapid and uniform freezing. One surface should be flat to provide sufficient contact with the tissue carrier. If, during freezing, there is a tendency for the tissue to break loose from the carrier, cut a piece of filter paper to fit the top of the carrier and place it, wet, under the tissue to hold the tissue in place while freezing and sectioning. Young (1962) suggests, for extremely small tissues, the use of a small piece of sponge as a base to raise the tissues high enough to prevent the knife from striking the metal of the object holder. Pieces of cork can also be useful if the tissue must be stored and sectioned again later. The cork facilitates removal of the tissue from the object holder and, when wrapped in aluminum foil, it can be sectioned later if necessary. The most frequently used medium for mounting tissues on the microtome chuck is OCT compound (VWR Scientific Products). OCT facilitates mounting and sectioning and aids in retaining sections so that they may be mounted in sequence.

Consider the shape and type of tissue when it is oriented on the carrier or in the mounting medium. If one side is narrower than the other, place that side, or any straight side, parallel to the knife. If there is a touch membrane on one side of the tissue, place that side toward the knife to prevent the membrane from breaking away from the rest of the block during sectioning. Usually the amount of water carried over with the tissue is sufficient for firm attachment to the carrier. (Use normal saline for fresh tissues.) Avoid too much water, and do not allow it to settle around the tissue in a wall of ice, which can deflect the knife or tear through the tissue during sectioning. Uneven or torn sections can result. If, however, an embedding compound is being used, it is necessary to build the compound around the tissue while freezing it. In this case the tissue must be completely encased by the medium.

With forceps, gently press the tissue onto the tissue specimen holder. Allow the specimen to freeze. With the environmental concern, use of aerosols should be limited. Good sections depend on proper tissue temperature. Material frozen too hard forms white brittle fragments on the knife; if too soft it forms a mushy mass. In both cases the sections break up when placed in water. These difficulties can be avoided by slightly overfreezing the block and then, as it warms and reaches the correct temperature, a number of sections can be cut in close succession. When these sections pass the knife edge, allow several to accumulate and pick up the group with one sweep of the slide. Or, as soon as the section is cut, remove it from the knife with a cold, fine brush and place it in a dish of water or collect on a microscope slide that has been stored in the freezer chamber of the cryostat. Do not allow water to collect on the knife.

Keep both knife and brush dry. Some tissues, such as adipose, tend to stick to the brush and will not shake loose in water. Transfer them to 70% or 80% alcohol or place directly on slide.

MOUNTING

The preferred technique for transferring sections to slides is quite simple. With a brush, arrange the tissue sections on the cold knife, lower the room temperature slide, and gently touch the sections. The tissue will quickly adhere to the slide. Another method is to store the slides in their original container in the cryostat cabinet; with your thumb, warm the back of the slide and then approach the tissue sections. We have also found that the use of cold slides (above) facilitates mounting and orientation if the cold sections are brushed onto the cold slides in the cryostat chamber and then brought to room temperature for defrosting.

If the sections are collected in a glass dish with water, place the dish on a black surface, such as a table top or a piece of black paper. (Thicker sections, 10 to 20 μm, are often collected in this manner.) The sections are more easily manipulated over the black background, and the best ones can be selected. Dip one end of an albumenized or subbed slide into the water. With a needle or preferably a small, round-tipped glass rod, gently bring a section against the slide. Hold the section at one corner; by maneuvering under water, creases can be unfolded and the section unrolled as the slide is drawn out of the solution.

If the wrinkles refuse to straighten, drop the section in 70% alcohol and then remove it to a slide. Drain off excess water or pull it off with filter paper. Press the section in place with filter paper moistened with 50% alcohol. With a pipette add absolute alcohol directly on top of the section; let stand for approximately 30 seconds. Replace with more absolute alcohol. Immerse in 70% or 80% alcohol, 5 to 10 minutes or longer; the section can be left here for several hours or days. (*Caution:* Use an old solution of alcohol, either one from the staining series or one that has been mixed for at least a day. A freshly prepared solution forms bubbles between section and slide.)

Fresh tissue sections may be removed from normal saline to the slide and thoroughly drained. Add 95% alcohol to remove the water; let stand 30 to 60 seconds. Drain off and allow the section almost to dry. Place in 95% alcohol for 1 to 2 minutes and proceed to stain.

A gelatin fixative can be used to affix the sections. Spread fixative on the slides and dry them in an incubator. (Slides can be stored for some time in a dust-free box.) Float the sections on water on the gelatin-coated slides. Drain off excess water and place the slides over formalin in a covered dish. This converts the gelatin into an irreversible gel, holding the sections in place. After the slides have been over the formalin for 30 minutes, wash them in running water for 10 minutes and proceed to stain. Subbed slides are preferred by many over other methods. Mount the sections out of the water onto the subbed slides; tilt the slides to drain off excess water, and allow to dry in a vertical position.

To facilitate freezing many laboratories quick freeze specimens in a small Dewar flask of isopentane. This quenches the tissue and causes little artifact. There is also a system by Micrometrics that makes it easier to section difficult specimens. In this procedure tape is placed over the face of a trimmed specimen block and then a section is cut. The section, adhered to the tape, is placed on a slide and processed with an ultraviolet light source; after the tape is removed, the slide can be stained as usual. (See Orenstein apparatus and technique on p. 336.) This system also works well for bones.

Gelatin Embedding

This is recommended for tissues as difficult to handle as lung and bone marrow.

Pearse Method (1968)

1. Fix in cold formalin (15%) at 4°C if enzymes are to be preserved: 10 to 16 hours.
2. Wash in running water: 30 minutes.
3. Infiltrate with gelatin, 37°C: 1 hour.

gelatin	15.0 g
glycerin	15.0 ml
distilled water	70.0 ml
small crystal of thymol	

4. Cool and harden in formalin (40%), 17–22°C: 1 hour.
5. Wash in water.
6. Store at 4°C or below until sectioned.

RAPID STAINING METHODS

Hematoxylin and Eosin

Staining and mounting time: 3 to 4 minutes.

PROCEDURE
1. After cutting 4- to 6-μm sections, mount on a slide. Place in alcoholic formalin for 15 seconds.
2. Dehydrate in 70% alcohol (10 dips).
3. Rinse well in distilled water.

4. Stain with Harris (Gills may be substituted) for 45 seconds to 1 minute.
5. Rinse in tap water.
6. Dip 3 times in saturated lithium carbonate or until the section turns blue.
7. Rinse in water.
8. Counterstain with eosin/phloxine solution for 30 seconds.
9. Dehydrate, clear, and mount.

Results

nuclei—deep blue
cytoplasmic structures—pink, rose

Humphrey Method (1936)

This is perhaps the fastest method.

1. Place section on a slide (and wipe off excess water).
2. Add 1 drop of 0.5% brilliant cresyl blue in saline.
3. Cover with cover glass and examine.

This is never a permanent mount.

Thionine or Toluidine Blue Method

1. Remove section from water to a solution of either 0.5% thionine or toluidine blue O in 20% alcohol plus a few drops of acetic acid: 30 seconds.
2. Rinse in water and float on slide.
3. Drain off excess water, blot around edges of section, add drop of glycerin, and cover glass.

This is not a permanent section.
The above two methods stain the tissue elements in shades of blue and purple.

Pinacyanol Method (Humason and Lushbaugh 1961; Proescher 1933)

Staining and mounting time: 3 to 4 minutes.

Procedure

1. After cutting 4- to 6-μm sections, mount on a slide. Place in formalin for 15 seconds.
2. Dehydrate in 70% alcohol (10 dips).
3. Rinse well in distilled water.
4. Stain with Pinacyanol (0.5% in 70% alcohol) for 5 seconds.

5. Rinse in tap water.
6. Dehydrate, clear, and mount.

RESULTS
chromatin—well-differentiated blue to reddish blue
connective tissue—pink
elastic tissue—dark violet
muscle—violet to purple
plasma cells—red cytoplasm
hemosiderin—orange
hemoglobin—neutrophil and eosinophil granules, unstained
neutral fat—colorless to faint blue-violet
lipoids—blue-violet to purple
amyloid—carmine red

NITROCELLULOSE METHOD

This form of embedding is often referred to as the *celloidin technique*. Celloidin has become somewhat of a generic term including the various cellulose compounds such as nitrocellulose and soluble guncotton or collodion. These are solutions of pyroxylin consisting chiefly of cellulose tetranitrate. Obviously, a purified nonexplosive form of pyroxylin was necessary, and several were available: Parloidin, celloidin, and Photoxylin. Parloidin is the most common type of nitrocellulose used today. Neurologists prefer this method over paraffin because it avoids the damaging effects of heat in the process.

The nitrocellulose method requires 7 to 10 days.

FIXATION
Any fixative.

SOLUTIONS
Stock celloidin (12%)
Dissolve 24 g Parloidin (strips) in 100 ml of absolute alcohol with frequent stirring. Add 100 ml of ether. Store in air-tight containers and avoid contamination with water.

Follow safety procedures with ether. Use a fume hood.

2% celloidin: 1 part 12% celloidin to 5 parts absolute alcohol/ether (equal parts)
4% celloidin: 2 parts 12% celloidin to 4 parts absolute alcohol/ether (equal parts)
6% celloidin: 3 parts 12% celloidin to 3 parts absolute alcohol/ether (equal parts)
8% celloidin: 4 parts 12% celloidin to 2 parts absolute alcohol/ether (equal parts)

10% celloidin: 5 parts 12% celloidin to 1 part absolute alcohol/ether (equal parts)

Specimen size should be no more than 5 mm in thickness.

PROCEDURE
1. Fix specimen and dehydrate.
2. Ether/absolute alcohol (equal parts): 24 hours.
3. Celloidin (4%): 2 to 3 days. Vacuum can be used to facilitate infiltration.
4. Celloidin (8%): 2 to 3 days.
5. Celloidin (12%): 2 to 3 days.
6. Embed in 12% celloidin.
7. Place the molds in a desiccator (or bell jar) until all of the bubbles have disappeared. Hardening can be accelerated by adding a chloroform-soaked cotton ball (do not touch the specimen). The blocks must be firm and are considered hard when you can no longer leave a finger print on the surface.
8. Remove the blocks and place them in 80% alcohol (upside down) until ready to cut.
9. To mount the block for sectioning, use 2% celloidin to coat the wooden block (or specimen holder) and press the embedded specimen block on the holder. A small vise can be used. Harden the block in 80% alcohol before sectioning. These blocks are usually sectioned on a sliding microtome. Alcohol is used to coat the block surface and knife edge while sectioning.

ALTERNATE METHOD, WALL (1936)
1. The dry celloidin method involves steps 1 to 6 of the above method.
2. Place tissue in equal parts cedarwood oil and chloroform for 24 hours.
3. Place specimen in 3 parts cedarwood oil and 1 part chloroform for 24 hours. Blot block.
4. The blocks can now be sectioned "dry," without the 80% alcohol necessary for regular wet celloidin sectioning.

ESTER WAX EMBEDDING

The ester waxes were introduced by Steedman (1947) as embedding media that combine the advantages of both paraffin and celloidin, but reduce their disadvantages. Tissue structure is well preserved, and structural lipids are retained perhaps better than when either paraffin or celloidin is used. The ester wax method is based on diethylene glycol distearate, whose hardness compares favorably with that of paraffin. Ester wax melts at lower temperature and does not require the use of hardening hydrocarbons. The wax is soluble in the usual paraffin solvents and is also miscible with n-butyl alcohol, isopropyl alcohol, Cellosolve, glycol monobutyl ether, acetone, and aniline. It supports soft tissues adjacent to hard ones and adheres to smooth surfaces, and it can be sectioned on a rotary microtome.

Steedman Method (1960)

SOLUTION

diethylene glycol stearate	60.0 g
glyceryl monostearate	30.0 g
300 polyethylene glycol distearate	10.0 g

Melt the diethylene glycol distearate and heat until clear. Add glyceryl monostearate. When this is dissolved, add the 300 polyethylene glycol distearate. Filter.

PROCEDURE
1. Transfer fixed tissues to 70% alcohol: 1 to 2 hours (time depends on size of tissues).
2. Transfer to 70% alcohol-Cellosolve (ethylene glycol-monoethyl ether; 50:50): 1 to 2 hours.
3. Transfer to pure Cellosolve: 1 to 3 hours.
4. Transfer to Cellosolve-ester wax (50:50): 1 to 3 hours.
5. Infiltrate in pure polyester wax, 3 baths: 24 hours each.
6. Block.

ALTERNATE PROCEDURE
1. Transfer fixed tissues to 95% alcohol: 1 to 3 hours.
2. Transfer to 95% alcohol-ester wax (50:50) in oven (35–40°C): 1 to 3 hours.
3. Transfer to pure wax: 24 hours.
4. Block.

Have the melted wax about 10°C above the melting point (48°C) for blocking. Use a cold mold. Cool block on ice but do not submerge it in water. The more quickly it cools, the better the final texture of the wax.

SECTIONING

Polyester waxes must be cut very slowly with a sharp knife. The wax is so hard that the knife must be held firmly in the clamps, and all parts of the microtome must be in good repair and tight or the sections will be of uneven thickness. Thick sections may be cut the same day that the tissue is blocked, but thin sections will cut more easily the second or third day after the blocking.

Mount as for paraffin sections on albumenized slides. Drain and dry at room temperature.

Modification of Sidman et al. (1961)

After receiving some poor batches of wax, Sidman and his coworkers tested a number of products and found that polyethylene glycol 400 distearate gave the most consistent results. It melts at 35°C and begins to solidify at 30°C. Melt it overnight (no longer) at 56°C and then transfer it to a 37°C oven for use.

1. Fix and dehydrate the tissue as desired.
2. Infiltrate in 3 changes of wax, 37°C: 24 hours at least.
3. Embed. Pour wax in mold and set it on a cube of ice until a solid base of wax forms. Remove from ice and add tissue. Orient it and place mold and tissue in refrigerator: 1 hour.
4. Allow to warm to room temperature: 1/2 hour. Trim and mount on object disc as for paraffin blocks. Section under cool conditions. If the room is warmer than 25°C, solid CO_2 held in a kitchen strainer on a ring stand about 6 inches above the knife will provide adequate cooling. Cut slowly.
5. Float sections on 0.1% gelatin in water (25°C) and onto clean slides, or use subbed slides. If the sections are thicker than 10 μm, use 30–32°C water. Drain. Dry at room temperature for several hours or overnight. Sections should not float too long on water because the wax is soluble and loses its cohesiveness. In addition, the sections will hydrate and thus destroy the tissues. Remove water within a minute with a pipette and absorbent tissue (e.g., Kimwipe). A wet brush is easier to use than needles because of the stickiness of the wax. Blocks should be stored in a cool place, and unused sections should be stored in the refrigerator.
6. For staining, remove the wax with absolute alcohol, hydrate, and stain.

Distearate 300 or 400 is not satisfactory for warmer climates, so Sage (1972) uses the longer chain polyethylene glycol, distearate 600, with the addition of 1-hexadecanol wax (cetyl alcohol). Melt only the amount to be used at one time; storing in a melted condition lowers the melting point. Sections can be cut at 2 μm at 21° with stearate 600 and 1% hexadecanol wax added, and at 24° with 10% hexadecanol added. The sections are mounted on albumenized slides with 10% formalin. Excess formalin is blotted off, and the sections are allowed to dry for 2 days.

Capko (1984) finds that the modification of Steedman's polyester by Norenburg and Barrett (1987) is excellent for infiltrating yolk platelets of amphibian eggs in light microscope studies.

Polyester Wax Method of Norenburg and Barrett (1987)

FIXATION

Any routine fixation followed by appropriate washing.

SOLUTIONS
    polyethylene glycol-400 distearate (mp 35–37°C;
        Aldrich catalog #30,541-3) .... 90 parts
    1-hexadecanol (cetyl alcohol) (mp 49°C;
        Aldrich catalog #25,874-1) .... 10 parts

Preparation of polyester wax

Heat the above combined two components only one time to mix and then aliquot into disposable 50-ml tubes for each use, and allow to harden. The polyester wax can only be melted one more time at 40°C without changing its sectioning characteristics. Initial heating should be at 57°C using a water bath, since the wax mixture is flammable. Stir until cetyl alcohol is dissolved and mixed with the polyethylene glycol (PEG).

PROCEDURE
1. Dehydrate tissues in graded series of alcohols to absolute (95% alcohol will also work).
2. Heat aliquot tube to 40°C (never more than 45°C) to melt and to use for infiltration.
3. Dehydrate tissues to 100% ethanol. Infiltrate with a minimum of two steps; first with 50% polyester wax/50% ethanol and then change to 100% polyester wax. Change polyester wax a minimum of 2 times. Use impregnation times similar for those used with paraffin.
4. Embed as discussed above. Capko (personal communication) recommends embedding in a modified Beem capsule to facilitate handling of the block and for its removal from the embedding mold (see diagrams).
5. Section with steel blade set at 10°. Set at same angle when using glass knife in an ultramicrotome. Cool block, especially when sections are desired with thicknesses less than 5 μm, as described above. Do not use water-filled boat for ultramicrotome because of hydration problems. Sections from conventional microtome are floated on drop of water on a subbed slide. Remove water rapidly, as described above. Dry slides for at least 24 hours; do not use slide warmer because of low melting point of the wax.
6. For staining, remove wax with 100% ethanol.

Diethylene glycol distearate (DGD) is a useful alternative to PEG as a removable embedding mixture (see detailed discussion by Capko [1984]). It too is waxlike and can be used for preparation of sections that can be observed with both light and electron microscopes. DGD has a broad spectrum of applications in cytological studies. DGD is now available in consistently excellent quality from Polysciences Inc. For light microscope studies, DGD prepared sections can be used for the same applications for which one uses paraffin. For ultrastructural studies, DGD is used as easily as plastic-embedded material; however, the ability to remove DGD permits the analysis of sections much thicker than those prepared with plastic. DGD will not penetrate hard substance as does PEG (see above for discussion of yolky eggs). Recently, it has been suggested that DGD be referred to as "Pentament" in recognition of Sheldon Penmen's rediscovery in the early 1980s and for his use of this material in cytological studies (cf. Capko 1984). In more recent times DGD has also been successfully applied in immunohistochemistry and in situ hybridization studies.

PROCEDURES WITH DGD (CAPKO 1984)
1. Fix and wash as desired.
2. Dehydrate through graded series of ethanol; time depends on thickness of specimen: a minimum of 2 hours in several changes of absolute alcohol.
3. After a stay in a 1:1 mixture of absolute ethanol and 1-butanol, place specimen in several changes of 100% 1-butanol. During the last butanol, place sample into a 70°C oven for 15 minutes to warm tube and the butanol before infiltration in the embedding medium. DGD should be melted in a 70°C oven before using. DMSO 0.3% added to DGD before use will reduce the shrinking that occurs when the block hardens. Mix thoroughly after adding the DMSO; in this state the mixture may be resolidified and remelted several times.
4. Add a premixed solution of 1:1 1-butanol and molten DGD to the samples and allow to incubate for at least 45 minutes. Replace with several changes of DGD (with 0.3% DMSO) and incubate for a minimum of 2 hours. Vacuum oven will aid infiltration of the DGD.
5. Embed in flat molds that have a depth of 4 mm. Lamps around the embedding dish will keep embedment soft to facilitate orientation of the tissues.
6. Permit block to cool at room temperature (not on ice or in refrigerator).
7. Block may be trimmed with a razor blade. Allow a small layer of DGD around tissue to facilitate ribbon formation. Keeping the block cool will keep the block hard. It is recommended that the room temperature be less than 22°C.
8. Sections floated onto water-filled troughs provide interference colors to indicate section thickness. One of the advantages of DGD is that sections can be floated onto water whereas PEG sections have to be prepared with a dry knife. The sequence of colors runs from silver, to gold, to green, to blue, then to dark gold, to dark green, and to dark blue. A similar sequence will repeat itself for an additional cycle. The series of sections from gold to dark gold are suitable for transmission electron microscopy. Sections thicker than 1 μm can be prepared for light microscopy.

Sections are collected on grids that are Formavar coated, carbon stabilized, and treated with 0.1% polyvisine. Specimens may also be collected on subbed slides.
9. After specimens are dried, the DGD is removed with 1-butanol; up to 1 hour with thick sections. Transfer to a mixture of 1-butanol and absolute ethanol for 15 minutes. Then transfer to absolute ethanol for a minimum of 1 hour.
10. Sections can now be processed for ultrastructure study or histochemical staining.

GLYCOLMETHACRYLATE PROCESSING FOR THIN SECTIONS

Methacrylate processing for thin sections for light microscopy was pioneered by Zambernard in 1969. The procedure now used has been modified considerably. This method allows thin 0.5- to 3-μm sections as well as sectioning of undecalcified bone.

Fixation

Any routine fixation followed by appropriate washing. Formalin must be thoroughly removed.

Processing

There are several kits available that make the process easy.

The JB-4 water soluble kit, available from Polyscience (Warrington, PA), is one of the more widely used products. It is a reliable product and gives consistent results. The company also sells the appropriate molds and accessories.

The tissues are trimmed to a proper size, usually 5 mm × 10 mm × 2 mm. If a microtome with a tungsten carbide-tipped steel knife is used, the specimen size can be somewhat larger. However, the thinner sections (0.25 to 0.5 μm) must be obtained on a glass knife. The specimens are then dehydrated through 50%, 80%, 95%, and 100% alcohol (2 changes of each for 15 minutes). This can be done on a routine tissue processor or manually.

The infiltration process requires 3 to 12 hours. Follow the manufacturer's instruction for mixing the components of the plastics, adding the catalyst to the final infiltration. Polymerization usually takes 1 to 2 hours or, preferably, overnight.

Sectioning

Sectioning is best done on a heavy duty retracting microtome with glass (Ralph or triangle) knives. The sections can be floated on a water bath (at room temperature) and picked up on slides. The slides are then placed on a warming plate (80°C for 20 minutes).

Staining

Hematoxylin and Eosin

Harris hematoxylin, p. 102

    acid alcohol-glacial acid: .... 1.0 ml/95% alcohol 100 ml

Eosin-phloxine
- eosin, 1% .... 100 ml
- phloxine, 1% .... 20 ml
- alcohol, 95% .... 780 ml
- glacial acetic acid .... 4.0 ml

Scott's water, p. 464

Procedure
1. Harris hematoxylin: 15 minutes.
2. Rinse in running water.
3. Differentiate in acid alcohol: 1 second (one quick dip).
4. Rinse in running water.
5. Blue in Scott's water for 1 minute.
6. Rinse in running water.
7. Counterstain in eosin-phloxine for 8 minutes.
8. Dehydrate, clear, and mount.

Results
nuclei—blue
background—pink-red

Modified Maynard's Trichrome (1986)

Solutions
1% glacial acidic acid
hematoxylin—Gill's III (purchased commercially)
Scott's water
- water .... 100 ml
- magnesium sulfate (anhydrous) .... 1 g
- sodium bicarbonate .... 0.2 g

Use gentle heat to mix (60°C).

Biebrich scarlet/acid fuchsin
  1.0% Biebrich scarlet .... 90 ml
  1.0% acid fuchsin .... 10 ml
  glacial acetic acid .... 1 ml

2.5% aniline blue solution
  aniline blue .... 2.5 g
  distilled water .... 100 ml

Heat gently to dissolve. Add 2.5 ml glacial acetic acid. Filter.

Working aniline blue solution
  2.5% stock aniline blue .... 1.0 ml
  1% glacial acetic acid solution .... 20 ml

Saturated picric acid/alcohol
  saturated picric acid .... 50 ml
  absolute alcohol .... 50 ml

Phosphomolybdic/phosphotungstic acid solution
  phosphomolybdic acid .... 5.0 g
  phosphotungstic acid .... 5.0 g
  distilled water .... 200 ml

CONTROL
  small intestine

PROCEDURE
  1. Place in saturated picric acid solution for 10 minutes, 5 minutes at 60°C and 5 minutes at room temperature.
  2. Running water: 2 minutes.
  3. Gill's hematoxylin III: 2 minutes.
  4. Running tap water: 2 minutes.
  5. Scott's water substitute: 2 minutes.
  6. Dry slides at 60°C for 2 minutes.
  7. Biebrich scarlet/acid fuchsin: 2 minutes.
  8. Distilled water: 8 to 10 quick dips.
  9. Phosphomolybdic/phosphotungstic acid: 2 minutes.
  10. Aniline blue: 2 minutes.
  11. 1% acetic acid: 20 to 30 dips (check slide microscopically for desired intensity).
  12. Distilled water: 2 to 3 quick dips.
  13. Dehydrate in absolute alcohol, clear, and mount.

RESULTS

cytoplasm, keratin, muscle, and intracellular fibers—red
nuclei—purple-blue
collagen—blue

EPOXY RESIN PROCESSING

The epoxy resins, popular for electron microscopy, can be used as well for light microscopy. They produce uniformly hard blocks with very little shrinkage of the tissue. The resin is cured (polymerized) with anhydrides; DDSA (dodecenyl succinic anhydride) produces a soft block, while NMA (nadic methyl anhydride) forms a very hard one. Varying proportions of the anhydrides, therefore, can be used to adjust the resin block to the desired hardness. Dibutyl phthalate, a plasticizer, sometimes is added to impart flexibility to the casting and to reduce viscosity. At final embedding, an accelerator such as DMP [2,4,6-tri(dimethylaminomethyl)-phenol] or BDMA [benzyl dimethyl amine] is added. Propylene oxide is commonly used as the clearing agent after dehydration through alcohol. Poly/Bed 812 is similar to Epon 812.

Humason preferred Mollenhauer's mixture, which follows.

Epoxy Resin Processing for Light Microscopy (Mollenhauer 1964)

FIXATION

Any routine fixative, but preferably one of the aldehydes.

SOLUTIONS

Resin mixture

Poly-Bed 812	30.0 ml
DDSA	55.0 ml
Arladite 502	15.0 ml
DMP-30	2.0 ml

Mix thoroughly; a Teflon stirrer is recommended. To eliminate the difficulties of cleaning resins from glassware, use disposable beakers for mixing and disposable syringes for measuring the reagents. Store for short periods in the refrigerator, but the resin can be kept for several months in a freezer.

Catalyzed resin

resin mixture	5.0 ml
DMP-30	3 drops

Mix just before use.

PROCEDURE

1. After washing out fixative, dehydrate through 70% and 95% alcohol and 4 changes of absolute alcohol: 30 minutes (or more) each. (Tissue blocks should be no larger than 4 to 5 mm in thickness.)

2. Transfer to propylene oxide, 2 changes: 30 minutes (or more) each.
3. Transfer to propylene oxide and catalyzed resin (2:1): 1 hour (or longer).
4. Transfer to propylene oxide and catalyzed resin (1:2): 1 hour (or longer).
5. Transfer to catalyzed resin in "Peel-A-Way" molds and polymerize overnight at 60°C, or leave overnight at 35°C, the next day at 45°C, and then once again overnight at 60°C. The latter forms a more easily sectioned block.
6. Remove mold and section at 1 to 3 μm with a very sharp steel knife. The sections can be removed from the knife with fine forceps and placed in a blackened box, as in paraffin sectioning.
7. With fine forceps and a dissecting needle, pull out any folds in the sections, straighten them as much as possible, and place them on albumen-covered (or subbed) slides. The albumen is not essential, but it does make the mounting solution puddle more uniformly.
8. Place on 45°C warm plate and immediately add 10% aqueous acetone under the sections. Allow the solution to evaporate and continue to straighten the sections with needles, if necessary. As soon as the slide is dry, it can be stained.

COMMENTS

As with the Zambernard method, Humason preferred to mount the tissue blocks on wooden blocks so that the microtome clamps on the wood rather than on the plastic. Sliding or rotary microtomes can be used, but the sliding microtome must be a heavy-duty model to prevent the same problem as described for the Zambernard blocks. A hard-grade steel or tungsten-carbide steel knife with a 55° cutting bevel edge is recommended. Plastic dulls the average microtome knife; only three or four sections can be cut successfully in a single area of the knife.

If, after curing, the block feels "sticky" soft and will not cut smoothly, return it to the 60°C oven overnight. Quickly peel off the mold and trim off the sloping sides of the block. Do this immediately after removal from the oven and while the block is still warm. It trims more easily in this condition than after it has cooled. Cut the first sections with an old knife, as recommended for paraffin sectioning, until the best part of the tissue is reached. Change to the good knife and collect the desired sections. Keep the edge of the knife clean; wipe it after every section.

Plastic sections usually adhere to the slides through all solutions. Should loose sections develop, however, pass the slides through a Bunsen burner flame 3 or 4 times. Allow them to cool before staining.

A student considering plastic embedding and sectioning should consult the article by Bennett et al. (1976) as well as the basic introductory manual by King (1983).

Sanderson and Bloebaum (1993) use the Exakt grinding system to facilitate bone sections even when the bone contains steel implants. Sterchi and Eurell (1989, 1995) are capable of sectioning large undecalcified bones and in addition can carry out milling and polishing to obtain thin sections.

Staining Epoxy Sections

Tissues embedded in plastic will not respond to all staining methods; most stains cannot penetrate plastic. Generally speaking, these sections will stain satisfactorily in alcoholic solutions and in basic dyes in alkaline solutions in the pH 11.0 range. An alcoholic and a buffered aqueous method are described below. Other aids to staining are pretreatment with mordants, hydroxide, peroxide, or iodine. The hydroxide method gives excellent staining results; only the PAS stain may give less than satisfactory results (see comments below).

Pinacyanol Stain (Humason and Lushbaugh 1961)

1. Cover sections with several drops of pinacyanol (0.5 g/100 ml 70% ethanol): 10 to 15 minutes. Cover the slides to prevent evaporation of alcohol.
2. Wash off excess stain in running water: 2 to 3 minutes.
3. Blot and allow to dry or dehydrate in isopropanol: 2 minutes in each of 3 changes.
4. Clear in xylene and mount.

Toluidine Blue Stain (Trump et al. 1961)

1. Flood slide with freshly prepared and filtered toluidine blue O (0.1 g/100 ml 2.5% aqueous sodium carbonate, pH 11.1): 30 to 120 minutes, or until cells are dark reddish purple.
2. Tip slide and draw off extra stain with filter paper.
3. Flood slide gently with tap water from a medicine dropper and draw off water with filter paper.
4. Wash briefly with 90% alcohol and then absolute. This differentiates the stain, so check it under microscope.
5. Clear in xylene and mount.

Sodium Hydroxide Treatment (Lane and Europa 1965)

1. Prepare a saturated solution of sodium hydroxide in absolute alcohol. Allow to stand: 2 to 3 days.
2. Immerse slides in solution: 1 hour. Keep container tightly closed to prevent evaporation and deposits of sodium hydroxide (NaOH) on slides.
3. Drain well; do not blot.

4. Transfer through absolute alcohol, 4 changes, 5 minutes each. If the first change becomes milky with accumulated NaOH, discard it.
5. Treat with phosphate buffer, pH 7.0 (p. 473): 5 minutes.
6. Wash in 3 changes of distilled water: several seconds in each.
7. Treat with phosphate buffer, pH 4.0 (p. 474): 5 minutes.
8. Wash in running water: 5 minutes.
9. Stain.

COMMENTS

Berkowitz et al. (1968) use a dilution of saturated NaOH 1:5 with absolute alcohol.

Snodgrass et al. (1972) describe a method involving treatment with acetone, benzene, and picric acid subsequent to the NaOH-alcohol.

Litwin et al. (1975) and Litwin and Kasprzyk (1976) made a precise study of the PAS reaction on Epon-embedded tissues treated with NaOH saturated in absolute alcohol. Treatment beyond 10 minutes produces a diffuse staining, and it must be reduced to 1 to 5 minutes depending on each batch of NaOH-alcohol. Mixtures vary due to water absorbed and ripening time of the solution. They use 10 minutes in periodic acid and 30 minutes in Schiff reagent.

Pool (1973) uses hydrogen peroxide oxidation for PAS. His solution is 15 ml of hydrogen peroxide (H_2O_2) and 30 ml distilled water. Adjust the pH to 3.2 with approximately 0.18 ml of 0.1 N H_2SO_4. Make fresh; pH must be 2.9 to 3.2. Oxidize for 1 to 2 minutes, rinse in distilled water, stain in Schiff 1 to 3 minutes, wash, drain, dry, and mount.

For the peroxide method, see Aparicio and Marsden (1969) or Pool (1969). For the iodine treatment, see Yensen (1968), and for the permanganate mordanting, see Grimley et al. (1965) and Shires et al. (1969).

Alsop (1974) stains epoxy sections by heating a polychrome stain dissolved in PEG 200. Chang (1972) heats his hematoxylin and eosin stains on plastic sections. Sato and Shamoto (1973) use a warmed polychrome stain, an alkaline solution of basic fuchsin, and methylene blue. Jha (1976) modifies their method.

Suggested Readings

For descriptions of other stains, see Buijs and Dogterom (1983), Flax and Caulfield (1962), Grimley (1964), Grimley et al. (1965), Hendrickson et al. (1968), Hoefert (1968), Konno and Takahashi (1985), Martin et al. (1966), Richardson et al. (1960), Schantz and Schechter (1965), Schenk et al. (1984), Spurlock et al. (1966), and Trump et al. (1961).

27
Special Procedures II

EXFOLIATIVE CYTOLOGY

Exfoliative cytology is the study of normal and disease-altered desquamated cells derived from various body sites. The process of exfoliation is a continuous, natural process of cell renewal whose rate is dependent on tissue type, function, and metabolic capacities. Some of the desquamated cells that collect in natural body recesses and cavities and among hairs are easily sampled and prepared for microscopic study by a variety of techniques. The collection and examination of exfoliated cells provides a rapid and effective method for the diagnosis of malignant and infectious processes. By screening slides made from gynecological smears (vaginal, cervical) or from body fluids (peritoneal, pleural, pericardial) and specimens from other body sites (bronchial washings, urine, cerebral spinal fluids, etc.), rapid diagnosis of malignancies is possible. The cells in question degenerate rapidly, and smears should be made and fixed immediately in 95% alcohol for a minimum of 15 minutes. The slides may remain in this fluid for as long as a week before staining.

GYNECOLOGICAL CYTOLOGY

The most widely utilized technique for the diagnosis of cancers of the female genital tract is the "PAP" smear, a colloquial term used to identify the procedure named after the American researcher Dr. George F. Papanicolaou, who, with Dr. Herman Traut, described the procedures and benefits of cytological examination of cells exfoliated from the female genital tract, especially in diagnosing lesions that were not clinically suspected. There is little variation among the methods used to collect and prepare cell samples in humans and animals; however, the preferred method of staining cytological slide preparations from animals are the Romanovsky-type stains (e.g., Wright's, Giemsa) on air-dried slides. Cell samples that are stained with the PAP method must be fixed in 95% alcohol immediately after preparation.

Exfoliated cells from the vagina and cervix can be sampled with relative ease using swabs or scrapers. The sample is smeared on a glass slide and, depending on the preferred method of staining, the slide can be air-dried or immediately fixed in 95% alcohol. The cells can then be examined microscopically for the presence and/or absence of premalignant or malignant processes. Vaginal cell samples are also invaluable for determining the stage of the estrus cycle of laboratory as well as farm and domestic animals.

NONGYNECOLOGICAL CYTOLOGY

Exfoliated cell samples can be collected from nongynecological body sites with minimum difficulty by using a variety of techniques, which include scrapings, imprints, washings, and aspirations. These are easy to obtain and generally provide sufficient quantity of quality cells for examination. Specimens that are processed soon after collection do not require fixation. However, if specimens must sit for any length of time before preparation, they should be prefixed in a 50% alcohol solution and refrigerated to minimize cell deterioration. (Body fluids and spinal fluids are exceptions). *Note:* Alcohol-fixed specimens and air-dried specimens have different morphological criteria in accordance with the type of stain used. For optimal differentiation alcohol-fixed preparations should be stained with the Papanicolaou stain and air-dried preparations should be stained with Romanovsky-type stains.

Body fluids and pleural, peritoneal, pericardial, and spinal fluids must be collected without fixation ("fresh"), prior to preparation, and can be stored refrigerated without significant cell deterioration for several days. Bladder urine samples are easy to collect, and preparation of the specimen can be either by the smear technique or by cytocentrifugation, depending on the amount of sediment. A standard table-top centrifuge can be used if a cytocentrifuge is not available.

Generally, slide preparation involves centrifugation of a specimen for 5 to 10 minutes at 2000 rpm. The supernatant is poured off, and smears are made from the cell sediment. One or two drops of the sediment are dropped on one slide, and the material is spread using a second slide until a uniform distribution of cells on both slides is achieved. The specimen must be fixed immediately in alcohol if the PAP stain is to be used. (Fully frosted slides can be used for specimens that will be stained with the PAP stain. The etched surface of the slides helps to retain the specimen.) After fixation, the slides should not be allowed to dry at any time before or during the staining procedure. Transporting slides requires minimal preparation, but care must be taken to ensure the preservation of the material and prevention of slide breakage. There are a variety of nonaerosol fixatives on the market and other types of fixatives using polyethylene glycol that can be used on fresh preparations to ensure adequate preservation. According to Ehrenrich and Kerpe (1959), the addition of 5% polyethylene glycol to the fixative will aid in cell preservation. After complete fixation, allow the slides to dry 5 to 10 minutes and then prepare them for transporting to the laboratory. Papanicolaou (1957) protects smears by covering the fixed smears with Diaphane solution: 2 parts of Diaphane to 3 parts of 95% alcohol. Allow to dry 20 to 30 minutes, and prepare for transporting to the laboratory. Several types of plastic slide holders and cardboard mailing tubes are available that will protect the slide(s) from damage. As we have seen, the preparation and staining of cellular material differs according to the body site from where it was collected. Preparation techniques, as well

Papanicolaou Method (1942, 1947, 1954, 1957)

Harris hematoxylin (Mallory 1944)
hematoxylin	5.0 g
absolute methanol	50 ml
distilled water	1000 ml
mercuric oxide	2.5 g
aluminum ammonium sulfate (alum)	100 g

Orange G-6 (OG-6)
orange G-6, 0.5% (0.5 g/100 ml 95% alcohol)	50 ml
95% alcohol	950 ml
phosphotungstic acid	0.015 g

Eosin-azure 36 (EA-36 or EA-50)
light green SF yellowish 0.5% (0.5 g/100 ml 95% alcohol)	45 ml
Bismarck brown 0.5% (0.5 g/100 ml 95% alcohol)	10 ml
phosphotungstic acid	0.2 g
lithium carbonate (saturated aqueous)	1 drop
eosin Y 0.5% (0.5% g/100 ml 95% alcohol)	45 ml

EA-65 is the same except for the light-green content. However, differentiation between acidophilic and basophilic cells is better with EA-36 (EA-50) and, therefore, it is preferable for vaginal, endocervical, and endometrial smears.

Control
Check for quality of stains with an unstained slide from a prior successful staining procedure. Any type of smear can be used.

Procedure
1. Remove slides from fixative and hydrate slides to water.
2. Stain in Harris hematoxylin, either of the following methods:
    a. Papanicolaou dilutes Harris with an equal amount of distilled water: 8 minutes.
    b. Harris hematoxylin (without acetic acid): 4 minutes.
3. Wash in gently running tap water (do not wash off or loosen parts of smear): 3 to 5 minutes.
4. Differentiate nuclei in 0.5% hydrochloric acid in 70% alcohol (0.5 ml/100 ml) until nuclei are distinct against a pale blue cytoplasm.

5. Wash in gently running tap water, 5 minutes, or until nuclei are a clear blue.
6. Rinse in distilled water.
7. Transfer through 70%, 80%, and 95% alcohol: a few seconds in each until the surface appears homogenous.
8. Stain in orange G-6: 1 to 2 minutes.
9. Rinse in 95% alcohol, 3 changes: a few seconds each.
10. Stain in EA 50: 2 to 3 minutes.
11. Rinse in 95% alcohol, 3 changes: a few seconds each.
12. Dehydrate, clear, and mount.

RESULTS

nuclei—blue
acidophilic cells—red to orange
basophilic cells—green or blue green
cells or fragments of tissue penetrated by blood—orange or orange green
blood vessels—orange red

COMMENTS

Papanicolaou cautions against agitating slides excessively or crowding them while staining. If cells should float off from a positive slide onto a negative one, a false positive can occur. Nongynecology specimens are more prone to "float" than are gynecology specimens, especially respiratory and fluid specimens. It is advisable when processing PAP-stained specimens to run a quick stain on body fluids before staining with other specimens to determine whether the fluid is positive. These specimens can be stained separately to avoid cross contamination. Filter the stains and all solutions before using again. Gill's hematoxylin (1974) may be used instead of Harris and, with the more concentrated solution, the staining time is reduced.

Establishing a method for measuring stain quality is advisable, particularly when large numbers of slides are processed or when additional material is not available to repeat slides that are not microscopically optimal. It is recommended that a few test slides be stained and examined each day to monitor stain quality. Adjustments in staining time and so forth can easily be made before processing large numbers of slides in order to assure optimal differentiation and clarity of cellular detail. The original Papanicolaou staining procedure uses a regressive method of staining in which the nuclei are overstained with an unacidified hematoxylin and the excess stain is removed with dilute HCl. The HCl is removed with running water. In a second, progressive method of staining, the nuclei are stained in hematoxylin until the optimum color intensity is reached. This is followed by an alkaline bath (e.g., lithium carbonate), which changes the stain from red to blue. The bluing agent must be removed by rinsing the slides in water. The staining time in the counterstains varies accordingly. The modified staining method shortens the overall staining time. Commercially made stains are available for both Papanicolaou and modified procedures and are very reliable. At one time acridine-orange (AO) fluorescence techniques were used as a replacement for the Papanicolaou stain in screening of cervical smears, but the method was not successful. However, the staining methods have been retained, since the method is a valuable research tool (Koss 1992). For further reading see *Cytology and Hematology of the Horse* by Cowell and Tyler (1992).

SEX CHROMATIN

Barr and Bertram's (1949) demonstration of sex differences in cat neurological tissue created interest in the cells of all parts of the body. They found that a tiny dark granule migrated from the nucleolus to the nuclear membrane and appeared only in female cats and not male. Further investigation revealed that this difference was present in other tissues and in other mammals, including humans. This granule is now referred to as Barr's body and represents a condensed X chromosome presenting as a dark-staining mass (Barr body) usually lying against the nuclear membrane.

The buccal smear can be used to determine genetic sex. A smear can be prepared by gently scraping cells from the buccal mucosa with a flat toothpick or swab, spreading them evenly on a slide, and fixing in 95% alcohol. Cells from the vaginal mucosa can also be prepared in the same manner. Several slides should be prepared and stained because the reproducibility of the process is sometimes hard to demonstrate.

Klinger and Hammond Method (1971)

FIXATION

Fix blood smears (see p. 193) in methyl alcohol for 1 to 2 minutes. Fix oral mucosal smears in absolute alcohol-ether (1:1) for 2 to 24 hours. Fix tissue blocks in Davidson's fixative:

95% alcohol	30 ml
formaldehyde (37–40%)	20 ml
glacial acetic acid	10 ml
distilled water	30 ml

Pinacyanol solution

pinacyanol	0.25 g
70% alcohol	100 ml

Wright buffer

monosodium potassium phosphate	6.63 g
dibasic sodium phosphate	3.20 g
distilled water	1000 ml

CONTROL
slide (unstained) from prior batch or known positive or negative

PROCEDURE
1. Hydrate slides to water.
2. Extract slides in 5 N HCl: smears, 2 minutes; sections, 3 to 6 minutes.
3. Wash in running water for 2 minutes.
4. Stain in pinacyanol for 45 seconds.
5. Differentiate in buffer for 45 seconds.
6. Wash in running water for 5 seconds.
7. Dehydrate in 2 changes isopropanol, 1 minute each.
8. Clear and mount.

RESULTS
chromatin, blue

Guard's Method (1959)

FIXATION
Fresh smears are placed in ether-alcohol 1:1 for 15 minutes.

SOLUTIONS

Biebrich scarlet solution

Biebrich scarlet	1 g
phosphotungstic acid	0.3 g
glacial acetic acid	5 ml
50% alcohol	95 ml

Harris hematoxylin, p. 102

Fast green solution

fast green FCF	0.5 g
phosphomolybdic acid	0.3 g
phosphotungstic acid	0.3 g
glacial acetic acid	5.0 ml
50% alcohol	95 ml

CONTROL
duplicate slide from previous batch (unstained) known positive

PROCEDURE
1. Remove slides from fixative and place in 70% alcohol for 2 minutes.
2. Stain in Harris hematoxylin for 15 seconds.
3. Stain in Biebrich scarlet solution for 2 minutes.
4. Place in 50% alcohol for 5 minutes.

5. Stain in fast green solution for 24 hours.
6. Dehydrate, clear, and mount.

RESULTS

sex chromatin (Barr bodies)—red
nuclear chromatin—blue
cytoplasm—red

Six cells out of 100 must have the Barr body to be considered female; if fewer than 4 are found, the specimen is considered male.

COMMENTS

The Feulgen technique (p. 187) may be used for Barr body identification.

Skin biopsies (paraffin sections) can be reliable but more difficult to obtain and stain than are oral scrapings. Other methods are those of Moore (1962), Moore and Barr (1955), Moore et al. (1953), Klinger (1958), and Barr et al. (1950). The Barr body is difficult to confirm in species that have dense granular nuclei (Rowson 1974), and the sex chromatin study is not considered a routine procedure for sex differentiation in clinical situations.

CHROMOSOMES

The allied area of chromosome study has grown intensively over the last 10 years, as has the study of molecular biology. The amassed information could easily fill many volumes. It is apparent, therefore, that this book could only present limited and basic methods. For more comprehensive discussion, we recommend *Cytogenetics of Animals* by Halnan (1989) and texts by Alberts et al. (1994) and Darnell et al. (1994).

Chromosome Squashes

Giant chromosomes are found in some of the somatic tissues of Diptera (two-winged flies) and reach the largest size in the salivary glands of the larvae. In well-made squashes, the cross striations of the chromosomes (often referred to as *banding*) stain in varying degrees of intensity. *Drosophila* larvae are frequently used to demonstrate these striations. Barley (1964) found the larvae of black flies (*Simulium vittatum*) easy to process, and Martin (1966) used the larvae of bloodworms (*Chironomus*) found in decaying vegetation along pond or steam edges.

Evans et al. (1964) have presented a now preferred method for the "squash technique." Their method, first described using mouse testes, is equally applicable with insect larvae.

PROCEDURE
1. Remove testes and place in a sodium citrate solution (2.2% wt/vol). Pierce or remove the tunica and swirl the testes to remove fat.
2. Place in fresh sodium citrate solution. Tease the contents from the tubules with curved forceps and transfer the supernatant fluid to a centrifuge tube.
3. Centrifuge the cell suspension at 500 rpm for 5 minutes. Discard the supernatant and resuspend in 1% sodium citrate.
4. Leave the cells for 10 minutes and then centrifuge at 500 rpm for 5 minutes.
5. Remove as much of the supernatant as possible and resuspend with 0.25 ml of fixative (3:1 methanol/glacial acetic acid). Mix thoroughly. Add more fixative (fill to one-third full). Wait 5 minutes and then centrifuge for 5 minutes. Repeat this step.
6. Pour off supernatant and place one drop of the suspension on a slide. Allow to dry and examine with phase microscopy.

A Giemsa stain may also be used at this point.

1. Stain for 5 minutes.
2. Rinse in distilled water.
3. Air dry and coverslip.

RESULTS
chromosomes—shades of blue
Other methods of staining are discussed by Chandley (1988).
Humason's preferred methods are listed below.

Acetocarmine or Aceto-Orcein Staining

SOLUTIONS
Acetocarmine stock solution
Boil an excess (approximately 0.5 g/100 ml) of carmine in 45% acetic acid, aqueous: 2 to 4 minutes. Cool and filter.

Working solution
stock solution .... 25 ml
45% acetic acid, aqueous .... 50 ml

An iron-mordanted stain is often favored because of the darker, bluish-tinged red that is obtained. Belling (1926) adds a few drops of ferric hydrate in 50% acetic acid, but only a few drops; too much iron produces a precipitate in a short time.

Moree (1944) determined quantitatively the amount of ferric chloride to add and includes tables of various normalities and volumes. One percent (by volume) of lactic acid can be added to the solution to intensify the staining of pachytene chromosomes (Yerganian 1963).

Aceto-orcein solution
              orcein .... 1 to 2 g
              glacial acetic acid .... 45 ml

Warm to 56°C (use a water bath).
When cool, add 55 ml of distilled water.
LaCour (1941) used 2% orcein in 70% acetic acid.

Alternate solution
Mix 85% lactic acid and acetic acid in equal amounts. Heat to boiling before adding stain (1 to 2 g/100 ml solution). Cool and filter. This stain does not precipitate (Yerganian 1963).

PROCEDURE
1. Choose large, sluggish *Drosophila* larvae, which have crawled up the side of the jar. Use physiological saline (see p. 464) to keep specimen moist. Place larva in several drops of saline on a glass slide. Hold the posterior end of the larvae with forceps, and with a dissecting needle pull away the mouth parts. The salivary glands will be attached to the mouth parts.
2. Trim off the fat bodies and place the glands in Carnoy fixative or directly into a small amount of stain on an albumenized clean slide.
3. Place a cover glass on top of preparation and cover with bibulous paper. Apply pressure. (See comments below.)
4. Seal edges with paraffin or dental wax, and allow to stand 1 day.

RESULTS
       acetocarmine—red
       iron acetocarmine—deep bluish red
       aceto-orcein—dark purple

COMMENTS
Sometimes it is advantageous to allow material to remain in stains 5 to 10 minutes. Then remove to a fresh batch of stain before squashing.
Fragments of animal tissue (ovaries, testes, biopsies) can be put into a tube with a large excess of stain for 2 to 7 days. Move to fresh solution on slide and squash.
Barley (1964) stores in Carnoy. Before making squashes, he treats the glands with acid alcohol (hydrochloric acid-absolute alcohol, 1:1) for 2 to

3 minutes. He returns glands to Carnoy's for 5 minutes and proceeds to stain. This method facilitates rupture of the cells.

The proper application of pressure on the cover glass is very important for good squashes. Press the thumb in the middle of the cover glass (covered with bibulous paper). Gently roll the thumb toward all edges of the cover glass, so a small amount of stain oozes out and is absorbed by the paper. Continue to roll the thumb toward the edges of the cover glass until no more fluid squeezes out. Always roll from the center toward the edges; the fluid must not roll back toward the center and disrupt the cells. Berrios (1994) described a device to facilitate and give more standardized squashes.

PERMANENT MOUNTS

Use subbed slides if permanent mounts are to be made, for there is a chance of losing the specimens.

Smith Method (1947)

1. Remove paraffin or petrolatum seal with razor blade and xylene.
2. Soak in equal parts of glacial acetic acid and 95% alcohol until cover glass comes off.
3. Place in equal parts of 95% alcohol and tertiary butyl alcohol: 2 minutes.
4. Transfer to tertiary butyl alcohol, 1 to 2 changes: 2 minutes each.
5. Briefly drain slide and cover glass against absorbent paper or blotter.
6. Add thin resin mountant to slide and place a clean cover glass to same position as before.

Step 2 can be followed by 2 changes of 95% alcohol; coverslip.

Nolte Method (1948)

1. Place slide in covered dish so edge of cover glass dips into 95% alcohol: 6 to 12 hours.
2. Immerse in 95% alcohol: 1 to 2 hours.
3. With sharp needle, gently pry cover glass free while immersed.
4. Drain off excess alcohol, add drop of mounting medium, and replace with a clean cover glass.

Other Permanent Methods

Delameter (1951) describes a freezing method for permanent slides.

Conger and Fairchild (1953) freeze smears while they are in stain on a block of dry ice: 30 seconds. Pry off cover glass with razor blade. Place before thawing in 95% or absolute alcohol, 2 changes: 5 minutes and coverslip.

Restoration of Deteriorated Slides (Persidsky 1954)

If dye has precipitated as dark crystals because the preparation has dried:

1. Remove sealing material.
2. Place 1 drop of 2 N HCl at one edge of cover glass and apply blotter to opposite edge to help draw the acid under: 3 to 5 minutes.
3. Replace HCl with acetocarmine by same method.
4. Heat gently; do not boil. Crystals are redissolved and specimens restained.
5. Reseal.

Pachytene Chromosomes

Pachytene chromosomes, which occur during the prophase stage of meiosis, are filaments and somewhat irregular in outline. They are, however, clearly double, forming two parallel strands (chromatids) and are united at one point (the centromere). Squash preparations must be processed so that the chromosomes are well spread and all chromosomes can be identified. See Clendenin (1969), Gardner and Punnett (1964), Ohno (1965), Schultz and St. Lawrence (1949), Welshons et al. (1962), Williams et al. (1970), and Yerganian (1957).

Somatic Chromosomes: Culturing and Spreading Techniques

Again we note that progress in cytogenetics and molecular biology has advanced very rapidly and that the presentation of the many refinements in technique and technology will be limited in this discussion. "Squash" techniques have been used for many years, especially with the chromosomes of plants and invertebrates. However, tissue culture techniques have been simplified. Cultures can now be made from bits of tissue, bone marrow, peripheral blood, and amniotic fluids.

Other methods have been developed for small samples: Brown and Fleming (1965), Edwards and Young (1961), Hungerford (1965), Shelley (1963), and Rønne et al. (1987).

At metaphase, individual chromosomes are more condensed and can be most easily recognized and counted. In their natural state they lie closely packed and must be separated for identification. A mucoprotein found in plant extracts, phytohemagglutinin (PHA), was used originally to agglutinate the erythrocytes as a means of separating the leukocytes from whole blood. It was then found to be a specific initiator of mitotic activity, as well. Lymphocytes appeared to be altered to a state in which they were capable of division (Nowell 1960). PHA is now used to stimulate mitosis as well as to precipitate red cells. Several forms of PHA can be purchased. After a period of incubation, a single inhibitor, such as colchicine or its analogue, is added to the culture to interfere with the formation of the mitotic spindle, arresting the dividing cells and allowing them to accumulate at a metaphase. The concentration of colchicine and the length of the treatment should be carefully controlled. The degree of chromatin condensation varies according to duration of treatment: the longer the action, the more contracted the chromosomes; the longer the chromosomes, the more affected they are by condensation action. Highly contracted chromosomes tend to have thin centromeres, which are located more medially than are those in less contracted chromosomes. Thus the degree of condensation can affect their relative lengths when considered in terms of percentage of the whole complement (Sasaki 1961).

The use of hypertonic solutions (or water) swells the cells and aids in the dispersal of the chromosomes.

Identification and analysis for karyotypes are made from photographs of suitably spread metaphases (Figs. 27-1 and 27-2). The chromosomes are cut individually from the photographic print and arranged in decreasing size, measured, and paired. Human chromosomes are identified numerically and alphabetically according to a combination of the Denver System (Denver 1960) and that of Patau (1960, 1961, 1965); this system of identification is called the London System (London Conference 1964).

Culture methods vary greatly, but space does not permit inclusion of all of them. Variations on techniques should be tried. Failures must be expected but

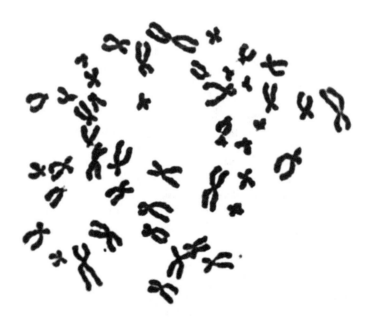

Figure 27-1. Chromosome spread of human male.

416   Histochemistry and Special Procedures

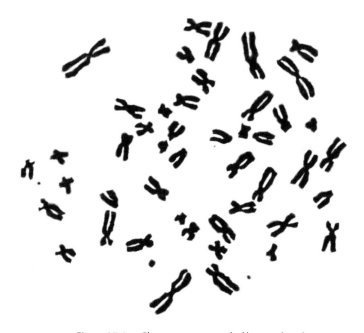

Figure 27-2.   Chromosome spread of human female.

should not discourage the beginner. When preparing cultures from abnormal blood, try two types of cell suspensions, one with and one without PHA. Experiment with cultures: vary the concentration and combination of the constituents of the medium; try different incubation temperatures; or sample the culture for metaphases at the end of 48, 72, and 96 hours, particularly if abnormal cells are present.

Use only glassware or plasticware designed for tissue culture work. (It is best to use disposables.) For washing, use nontoxic cleaners sold especially for such containers. Rinse in distilled water. All containers must be maintained in sterile condition. Keep a constant pH control. Make certain the containers are gas-tight by using screw caps with rubber liners. Peripheral blood cultures are relatively simple and require a short term for culture. Sterile conditions are necessary.

CULTURE METHODS

PROCEDURE

1. The recommended quantity of blood is 0.3 to 5 ml of culture medium. Arterial or venous blood can be used. Use sodium heparin to

prevent the blood from clotting. The blood is allowed to stand for 1 to 2 hours or until the red blood cells can be separated from the leukocytes and plasma.
2. Inject the plasma and suspended leukocytes into sterile culture medium. Most are purchased commercially, and there are many variations; if necessary, the medium recommended by Goh (1965) may be used:

TC 199 .... 7.0 ml
PHA-M .... 0.1 ml
autologous plasma .... 2.0 ml

3. Add 100 units of penicillin/streptomycin combination or gentamicin to prohibit the growth of bacteria. Sodium bicarbonate (2.8%) can be used to correct pH when all components (including specimen) are mixed together.
4. Incubate in closed culture bottles at 37.5°C for 68 to 72 hours. Set up at least 2 cultures per specimen.
5. Add 0.1 ml colchicine (0.0004% in phosphate-buffered saline; see p. 473) per 1 ml of culture and incubate 1 hour (up to 2 hours).
6. Centrifuge 1000 rpm for 5 minutes and discard supernatant fluid.
7. Add 1.0 ml of 37% sodium citrate and shake well (a Vortex mixer can be used).
8. Incubate at 37°C for 4 minutes.
9. Centrifuge for 1000 rpm for 5 minutes.
10. Pour off supernatant fluid.
11. Resuspend with 5 ml chilled 3:1 fixative (sodium citrate); add few drops at a time to begin.
12. Centrifuge at 1000 rpm for 5 minutes.
13. Pour off supernatant fluid.
14. Resuspend with 3 ml of fixative and centrifuge at 1000 rpm for 5 minutes.
15. Pour off supernatant fluid.
16. The last 3 steps can be repeated, if desired.
17. Make slides from the remaining concentrate.

COMMENTS

Humason recommended putting the slides in ice water; remove a chilled slide, then shake off the excess moisture and immediately add a drop of the cell suspension. Tilt slide against blotter to draw off excessive fluid and to help spread the cells. Dry rapidly over a flame or hot plate. Dry thoroughly before staining. The quality of the slide can be verified with phase microscopy before staining.

The chromosome banding phenomenon has revolutionized cytogenetics.

418 Histochemistry and Special Procedures

In salivary gland cells, repeated replication of chromosomes, but without cell division, leads to multiple side-by-side copies, thus accentuating the banding of chromosome segments. The banding is a reflection of base composition and their sequence in DNA. This permits identification of individual chromosomes and parts of chromosomes and allows detection of structural changes within the chromosome.

Other Methods of Culturing for Animals

Dogs: Ford (1965)
Opossums: Shaver (1962)
Macaques: Egozcue and Egozcue (1966), Sanders and Humason (1964)
Spider monkeys: Bender and Eide (1962)
Pigs: Srivastava and Lasley (1968)
Mouse and embryos: Wróblewska (1969)
Birds: Krishan (1962)
Domestic fowl: Newcomber and Donnelly (1963)
Snakes and other cold-blooded vertebrates: Beçak et al. (1962, 1964)
Rana pipiens: Picciano and McKinnell (1977)
Fish: Channel catfish, Wolters et al. (1981); trout, Hartley and Horne (1983); carp, Blaxhall (1983)
Drosophila: Horikawa and Kuroda (1959)
Insects: Crozier (1968)

Giemsa Method for Specific Banding (Drets and Shaw 1971)

SOLUTIONS

Alkaline solution, pH 12.0

sodium hydroxide	2.8 g
sodium chloride	6.2 g
distilled water	1000 ml

Saline-citrate solution

sodium chloride	105 g
trisodium citrate	52.9 g
distilled water	1000 ml

Adjust to pH 7.0 with 1 N HCl if necessary.

Buffered Giemsa

Giemsa stock solution (p. 320) or purchased solution	5.0 ml
absolute methanol	3.0 ml
0.1 M citric acid	3.0 ml
distilled water	89.0 ml

Adjust pH to 6.6 if necessary (use 0.2 M sodium hydroxide).

Procedure
1. Treat fixed and flamed slides in alkaline solution for 30 seconds.
2. Rinse in saline-citrate solution, 3 changes for 5 to 10 minutes.
3. Incubate in saline-citrate, 65°C: 24 hours.
4. Treat with 3 changes of 70% alcohol: 3 minutes each.
5. Treat with 3 changes of 95% alcohol: 3 minutes each.
6. Air dry.
7. Stain in buffered Giemsa for 5 minutes.
8. Rinse briefly in distilled water.
9. Air dry and mount.

Results
chromosome—dark blue

Stains for Squashes and Spreads

Aceto-basic fuchsin: Tanaka (1961)
Aceto-iron hematoxylin: Lowry (1963), Wittman (1962, 1963)
Acridine orange: Schiffer and Vaharu (1962)
Carbol-fuchsin: Carr and Walker (1961)
Iron hematoxylin with acetocarmine: Austin (1959)
Pinacyanol: Klinger and Hammond (1971)
Sudan black B: Cohen (1949)

Other chromosome band staining methods may include C-banding (for heterochromatin), nucleolus organizer staining (with silver staining), Q-banding (with fluorescent dyes), G-banding, and replication banding (see Thorgaard and Disney 1990).

CHROMOSOME ANALYSIS

Human chromosomes are classified in 7 groups on the basis of total length and arm ratio. Certain indices are relied on for analysis: (1) the length of each chromosome relative to total length of a normal X-containing haploid set (the sum of the lengths of 22 autosomes and the X chromosome expressed per 1000); (2) arm ratio of the chromosomes expressed as the length of the longer arm relative to the shorter one; (3) centromeric index (used sometimes), expressed as the ratio of the length of the shorter arm to the length of the whole chromosome; and (4) the presence of additional distinguishing features such as satellites and secondary constrictions.

All is not as simple and reliable as it may sound. Considerable variation does exist even between the homologs in the same nucleus. One may be more contracted than the other, and chromosomes lying near the periphery of a spread tend to be larger than those in the center. Thus, the value of measurements is questionable. Each arm should be measured along its axis, but the choice of exact termination of the arm can be arbitrary. In the best of photographs, the ends of the arms are always fuzzy. However, if one person makes all measurements on a set of chromosomes, some constancy in results can be achieved. Measurements may have to be relied on when other distinguishing features are absent. Labeling with tritiated thymidine will identify some chromosomes. Chromosome analyses should utilize two or more methods to corroborate final results.

The members of each of the 7 groups are easily distinguished from those of the other 6, but it is not always easy to distinguish those within a group from one another. Often it becomes necessary to speak of a chromosome as a C chromosome, for example, rather than to designate it by number. Chromosomes should be arranged according to length, decreasing in size. This form of alphabetical-numerical arrangement is being adopted for other animals.

Exaggerating the constrictions has been tried by Kaback et al. (1964), Patau (1965), Saksela and Moorehead (1962), Sasaki and Makeno (1963), and Upadhya (1963).

For the measurements of normal human chromosomes, see Burdette (1962).

Preparation of Karyotypes

A karyotype is the systematized arrangement of the chromosomes of a single cell prepared by camera lucida drawings or photographs. An idiogram is a diagrammatic representation of a karyotype and may be based on the measurements of the chromosomes of several or many cells.

Kodak Technical Pan 2415 (ASA 100) film can be used for photographing the spreads. Generally, high-contrast photographs are not desirable. Use Kodak HC110 developer for more controlled development and print on Kodachrome II RCF. Enlargement prints can be made, depending on the contrast required for the film. See Christenson (1965) for photographic suggestions.

Two enlargements are developed and printed, one for determining arm measurements and one to cut up for the karyotype. The chromosomes can be cut from the enlarged photograph and arranged and pasted on Bristol board or the like. Photographic mounting tissue can be lightly attached to the back of the photograph with a mounting iron. Then cut out the chromosomes and apply them to the Bristol board with a mounting iron. Cut-out chromosomes can be lightly mounted by one edge with double-sided tape on cardboard. Then, if necessary, they can be pulled loose with fine forceps and rearranged.

With today's image enhancement computer-assisted microscopes, precise measurements, total lengths, arm ratios, centromeric index, percentage total, and complement lengths can be determined. Select metaphases that show a minimum of contractions; slender chromosomes with well-defined centromeres are easier to measure and have more constant arm ratios than do short thick chromosomes. Final matching, however, still requires some visual evaluation, as in the above method.

If chromosomes are spread well, with reasonably straight arms and with a minimum of curls and wiggles, and if precise measurements are not necessary, calipers with fine pointed tips can check the arm lengths or total chromosome length of one chromosome against that of another. This method is still used in classroom situations.

Suggested Readings

See Alberts et al. (1994), Darnell et al. (1994), and Thorgaard and Disney (1990).

28
Special Procedures III

PREPARATION OF INVERTEBRATES FOR WHOLE MOUNTS AND SECTIONS

It is impossible to cover specifically all the members of this huge group of organisms, but certain generalizations can be made that will be applicable to the common forms and perhaps be adaptable to the less common ones. Fixatives can be applied directly to many of the invertebrates, but not to others that, because of their propensity to contract or ball up, pull in their tentacles or other appendages and thereby make whole mounts or sections practically worthless. Careful anesthetizing or narcotizing of these invertebrates must precede killing and fixation. As soon as the narcotization is complete and before death, if possible, fixation can be successful.

Anesthetizing and Narcotizing Agents
(See Laboratory Safety, p. 7, concerning drugs.)

MAGNESIUM CHLORIDE OR MAGNESIUM SULFATE

Either is widely and successfully used on sea anemones, corals, annelids, tunicates, and nudibranchs, to name a few. Crystalline magnesium sulfate can be tied in a bag suspended above and just touching the water surface, or a 33% aqueous solution siphoned slowly in, controlled by a screw clamp. When the organisms are anesthetized (no reaction to the touch of a needle), siphon off the water until the animals are barely covered and carefully add fixative. Disturb the animals as little possible. When partially hardened, transfer to fresh fixative.

COCAINE

Cocaine is an illegal drug and will be difficult to obtain or keep in the laboratory without special permission and security provisions. We present the following for heuristic value.

Cocaine can be used for ciliates, rotifers, bryozoans, hydras and some worms, and nudibranchs. A 1% aqueous solution is added to the water in proportions of about 1.0 ml to 100.0 ml of the water containing the animals. Eucaine hydrochloride can be used in the same manner. Check for contraction and fix.

MENTHOL

Sprinkle on the water surface and leave overnight. Good for difficult-to-narcotize sessile marine animals, coelenterates, some bryozoans, hydroids, and also flukes. It is more efficient when combined with chloral hydrate (a prescription drug, may be unobtainable) in proportions of 45.0 g menthol and 55.0 g chloral hydrate. Grind together in a mortar with a little water. Drop on surface of the water. Fix animals when they no longer contract. Large marine forms probably will require overnight treatment. Chloral hydrate can be used alone, sprinkled on water surface for annelids, mollusks, tunicates, bryozoans, and turbellarians.

CHLORETONE
A 0.33–1% solution can be used, but it is slow acting.

CHLOROFORM
This can be dropped on the water surface for many aquatic forms and is used in special bottles for insects and arachnids (see below). Use in laboratory fume hood.

ETHER AND ALCOHOL
These can be used by dropping on the water, or alcohol can be added gradually to the water by a tube controlled with a screw clamp until the proportion of alcohol to water is approximately 10%. It is particularly good for freshwater forms and earthworms. Ether can be used like chloroform for insects. (See Chap. 2 for use of these substances.)

ASPHYXIATION
Boil water to remove the air and seal it in a jar. Particularly good for gastropods (snails); place them in the boiled water after it has cooled.

Leave overnight.

COLD
Partially freeze organisms in salt and ice mixture, in freezing compartment of refrigerator, or in ice water until they are relaxed. Good for tapeworms. Transfer to lukewarm water and then fix.

HANLEY SOLUTION (GRAY 1954)

water	90.0 ml
ethyl Cellosolve	10.0 ml
eucaine hydrochloride	0.3 g

Add 1 drop per 10 ml of water in which animals are living. Good for rotifers and bryozoans.

PROPYLENE PHENOXETOL (OWEN 1955; OWEN AND STEEDMAN 1956, 1958; ROSEWATER 1963)
Introduce a large globule of the compound into the water containing the animals in an amount equally 1% of water volume, or shake vigorously 5 ml of

the compound with 15 to 20 ml of sea water. Add to water containing animals. Good for mollusks.

Tips on Special Handling

Porifera

Small forms can be dropped directly into osmium tetroxide-mercuric chloride (water, 250.0 ml; osmium tetroxide, 2.5 g; mercuric chloride, 9.0 g). Large forms fix better in alcoholic sublimate (Gilson, Carnoy).

Calcareous sponges can be decalcified in 70–80% alcohol plus 3% of hydrochloric or nitric acid. Siliceous sponges can be desilicified in 80% alcohol plus 5% hydrofluoric acid added gradually. Perform the latter in a glass dish coated inside with paraffin. After a few hours, transfer to 80% alcohol. Do not breathe hydrofluoric acid flames; perform steps in fume hood. Small spicules will section easily without desilification.

Coelenterates

Hydrae

Place in a small amount of water (few drops) in a shallow dish. When animals are extended, rapidly pipette warm mercuric chloride-acetic acid (5.0 ml saturated aqueous mercuric chloride, 5.0 ml acetic acid) at them. Work the fixative from base toward oral region, thereby preventing tentacles from contracting or curling.

Sea anemones

Anesthetize overnight with menthol; or 30% magnesium chloride (50 to 100 ml) can be added gradually over a period of 1 hour; leave in the solution until there is no more contraction of tentacles. Siphon off the solution until anemones are just barely covered. Add fixative (Susa, Bouin, mercuric chloride combinations, 10% formalin-sea water) slowly down the side of the container. Also pipette some directly into the "throats" of the anemones. When the anemones are partially hardened, transfer into fresh undiluted water.

Jellyfish

Place in sea water; while stirring, add fixative (10% neutral buffered formalin) down side of container until proportions are approximately 10 ml or more to 100.0 ml of sea water. Stir for 3 or 4 minutes. After 2 to 3 hours, change to fresh formalin.

Medusae

These must be anesthetized; magnesium solution. Allow them to expand in a small dish of water; treat with magnesium solution. When they no longer contract, add fixative (10% neutral buffered formalin) and stir.

Corals with extended polyps
Narcotize with magnesium and sulfate and fix in hot saturated mercuric chloride plus 5% acetic acid.

Coelenterates
These are sometimes frozen.

PLATYHELMINTHS
Planaria
Starve for a few days before killing. Place in a small amount of water on a glass plate. When the worm is extended, add a drop of 2% aqueous nitric acid directly on it. Follow by pipetting fixative (Gilson or saturated mercuric chloride in saline) on it and then transfer into a dish of fixative. Combined relaxing and fixing agent (Dawar 1973). The animals will contract and roll at first but soon will relax and become flat.

distilled water	200.0 ml
nitric acid, concentrated	2.0 ml
formaldehyde (37–47%)	4.5 ml
magnesium sulfate (MgSO$_4$)	2.5 g

Use at room temperature: 24 hours; temperatures lower than 20–30°C cause mucus formation.

Trematodes
Anesthetize large trematodes with menthol: 30 minutes. Then flatten by the following method: saturate filter paper with Gilson and lay on a plate of glass. Lay worms on paper and quickly cover with another sheet of saturated paper and second plate of glass. Add a weight, but not such a heavy one that the worms are crushed: 8 to 12 hours. Remove glass and paper and transfer worms to fixative. A single specimen can be flattened between slides, tied together, and dropped into fixative. After 2 to 3 hours remove the string, but do not disturb the slides. Leave overnight. Remove slides and transfer animal to fresh fixative. Steaming (not boiling) 5% formalin can be used for trematodes.

Small trematodes can be shaken in a small amount of 0.5–1.0% salt solution: 3 minutes. Add saturated mercuric chloride plus 2% acetic acid. Shake for several minutes. Change to fresh fixative: 6 to 12 hours. If they should be flattened, they can be laid on a slide and covered with another slide or, if they are very delicate, with a cover glass. Pipette fixative carefully along the side of the cover and then lower carefully into fixative. (See Gower [1939]).

Cestodes
Place a large tapeworm in ice water until relaxed (overnight). Then flatten between glass plates for microscope-slide whole mounts (as for trematodes;

see above). The worm can be wound spirally around slides. Then place a slide against both sides, press, and tie together (Demke 1952). For museum mounts, wind the worm around a tall bottle or 100-ml graduated cylinder and pour fixative over it. Their powers of contraction make them difficult to handle. Sometimes (after chilling) they can be relaxed in alcohol, 70%, or the steaming 5% formalin recommended above.

Small cestodes can be stretched from an applicator stick supported over a tall container and fixative poured down the length until they hang straight. Immerse them completely; then they can be flattened between microscope slides.

Nemertines
Drop into saturated mercuric chloride—acetic acid (95:5).

Rotiferae
Narcotize by adding 3 to 5 drops of the following solution for every milliliter of culture:

2% benzamine hydrochloride	3 parts
distilled water	6 parts
Cellosolve	1 part

When the animals are narcotized so they do not contract when touched, but the cilia are still beating, add 10% neutral buffered formalin to the culture.

Nemathelminths

Nematodes (hookworms)
Shake 3 minutes in physiological saline. Pour off and drop worms into hot glycerin alcohol (70–80% alcohol plus 10% of glycerin). A sublimate fixative can be used. (This procedure may be used also for ascarids.) Formalin 3% at 70°C penetrates and fixes within 3 minutes.

Very small nematodes can be relaxed and killed in a depression slide or watch glass by gentle heat. An incubator regulated to 50–52°C is most reliable. If using a flame or hot plate, be careful not to boil the worms. Transfer to fixative.

Formalin-acetic-alcohol is an excellent fixative for small nematodes.

95% alcohol	12.0 to 20.0 ml
glacial acetic acid	1.0 ml
formaldehyde (37–40%)	6.0 ml
distilled water	40.0 ml

If a few drops of saturated aqueous picric acid are added, the worms take up a little color.

Slide whole mounts are easy to prepared. Place the worms in the alcohol-glycerin mixture in an incubator (35–37°C) or on top of an oven to permit evaporation of the alcohol. When the solution is almost pure glycerin, mount in glycerin jelly (p. 98). Helminth ova can also be mounted in this way. For other methods, see whole mounts (p. 433).

Yetwin's (p. 99) is an excellent mountant.

Friend (1963) makes a microstainer for handling small ova. Electron microscope grids are attached to the ends of 3.5 mm (internal diameter) glass tubing with an insoluble resin. This method can also be used for processing small larvae. Helminth ova can be fixed in the steaming 3% formalin mentioned above.

Microfilaria

Blood infected with microfilaria is smeared and dried as for any blood smear. Then fix in any mercuric chloride fixative and stain with Delafield hematoxylin. If it is preferable to have the slides dehemoglobinized, after fixation treat them with 2% formalin plus 1% acetic acid: 5 minutes. Wash and stain. Alternate method: smear and dry the blood; dehemoglobinize in 5% acetic acid and air dry; fix in methyl alcohol and stain with Giemsa (p. 201).

BRYOZOA

Anesthetize with menthol. Fix saltwater forms in chromo-acetic solution (7 to 10 ml 10% chromic acid; 10 ml 10% acetic acid; water to make 100 ml). Fix fresh water forms in 10% formalin.

BRACHIOPODA

The shells of brachiopods are so fragile that the animals must be treated to remain open without forcing the shells. Leave them just covered with sea water, and gradually add 95% alcohol until the alcohol concentration is approximately 5%. Let stand for 30 to 60 minutes. When the shell is easily opened, prop it open with a piece of glass rod and transfer to fixative.

MOLLUSCA

Snails

Place in boiled water or add propylene phenoxetol to the water until snails are limp. Fix.

For decalcifying snail shells, Anderson (1971) uses a mixture of decalcifier-fixative (2:1), and the concentration can be as high as 7:1 for 6 hours. RDO (p. 40) is his choice for best cellular detail and staining properties. This can be used for other mollusks with calcareous exoskeletons.

Mussels

Sections of undecalcified shell can be made by grinding. Decalcification can be undertaken with 3–4% nitric acid. If the soft parts are to be fixed, wedge the valves apart with a small length of glass rod and place entire animal in fixative. Perform dissection after fixation. Gills can be removed and fixed flat in a mercuric chloride fixative.

Nudibranchs
Add 1% neutral buffered formalin, a few drops at a time every 15 minutes, to the sea water containing them.

Freshwater mollusks
Gradually warming the water will make them extend their feet. Fix when they no longer respond to the prick of a needle. Bouin, Zenker, or Gilson is satisfactory.

ANNELIDA

If sections of worms (aquatic or terrestrial) are desired, the intestines must be freed of grit and other tough particles.

Cocke (1938) feeds earthworms on cornmeal and agar (1:1) and some chopped lettuce for 3 days, changing the food every day. Becker and Roudabush (1945) recommend a container with the bottom covered with agar. Wash off the agar twice a day for 3 to 4 days. Moistened blotting paper can be used. When the animals are free from grit, place them in a flat dish with just enough water to cover them. Slowly siphon in 50% alcohol until the strength of the solution is about 10% alcohol. Chloroform can also be used for narcotizing. Fix in Bouin or mercuric chloride saturated in 80% alcohol plus 5% acetic acid. The worms may be dipped up and down in the fixative and then supported by wire through a posterior segment, hanging down in the fixative, or placed in short lengths of glass tubing in fixative to keep them straight. This is necessary if perfect sagittal sections are to be cut. Embedding is probably most successful by the butyl alcohol method. After removal of mercuric chloride in iodized 80% alcohol, transfer to n- or tertiary butyl alcohol: 24 hours (change once). Transfer to butyl alcohol saturated with paraffin (in 50–60°C oven): 24 hours; pure paraffin: 24 hours; and embed.

Sea worms can be kept in a container of clean sea water, changed every day for 2 or 3 days, then anesthetized with chloroform and fixed, using fast-penetrating fixative (Bouin or a mercuric chloride fixative).

ARTHROPODA
Insects and arachnids
Ether, chloroform, or potassium cyanide is used for killing. Simplest method: Place a wad of cotton in the bottom of a wide-mouthed jar and cover the cotton with a piece of wire screen. Dampen the cotton with ether or chloroform or lay a few lumps of potassium cyanide on it before adding the screen wire. Keep tightly closed with a cork or screw cap. A piece of rubber tubing soaked in chloroform until it swells and placed under the screen wire will hold

chloroform for several days. If the appendages should be spread when fixed, as soon as the insect is dead place it on a glass slide with another slide on top of it. Run fixative in between slides.

For whole organisms, rapidly penetrating fixatives should be used: picro-sulfuric, sublimate fixatives, mixtures containing nitric acid (Carnoy), alcoholic and ordinary Bouin, or Sinha (1953) fixative.

For whole mounts, clearing of the exoskeleton is sometimes difficult. Body contents have to be made transparent or have to be removed. Lactophenol (p. 99) mounting will serve the purpose in the first case, but heavily pigmented arthropods probably will have to be bleached in hydrogen peroxide for 12 hours or longer. Fleas, ticks, and the like make better demonstrations if not engorged; in any case they should be treated with 10% potassium hydroxide, 8 to 12 hours, to swell and dissolve the soft tissues, thus clearing out the body contents. Wash well to remove the potassium hydroxide. Acid corrosives, such as that which follows, are preferred by some because they do not soften the integument as much as do alkaline corrosives.

glacial acetic acid .... 1.0 ml
chloral hydrate .... 1.0 g
water .... 1.0 mg

Do not use a fixative containing alcohol or formalin if a corrosive is necessary. The organism will not clear.

Vyas (1972) preserves the soft parts as well as the exoskeleton with the following:

glycerin .... 12.0 ml
formaldehyde (37–40%) .... 2.0 ml
distilled water .... 100.0 ml

Add a few crystals of thymol.

This can be injected into the body cavity of large specimens.

Avoid the use of potassium hydroxide for very small or delicate insects. Transfer them into equal parts of chloral hydrate and phenol for about 2 weeks. If they are not cleared at the end of this period, place insects and solution in a 40°C oven for 2 days. This should complete the clearing. Transfer to absolute alcohol and mount in lactaphenol (p. 99).

Because of chitin, sectioning insects can be difficult. Avoid the high-ethyl alcohols. Soaking the tissue blocks overnight or for several days in water simplifies sectioning.

A modified Carnoy is recommended for fixation by Bilstad (personal communication):

absolute isopropyl alcohol .... 6 parts
chloroform .... 3 parts
formic acid .... 1 part

430 Histochemistry and Special Procedures

Fixation, dehydration, and infiltration with Paraplast are all performed under reduced pressure (De Giusti and Ezman 1955).

For mounting heavily chitinized sections, subbed slides should be used.

See Beckel (1959) for a sectioning method for heavily sclerotized insects. Barrós-Pita (1971) is able to section exoskeleton as though as praying mantis. Roden (1975) uses a nitrocellulose method. For other methods see Nelson (1974) and Kimmel and Jee (1975). Farnsworth (1963) uses Scotch tape to transport insect eggs through processing.

Echinodermata

Narcotize echinoderms by placing them in fresh water or sprinkle menthol on the water surface. Magnesium chloride or magnesium sulfate can be gradually added to the water.

Inject fixative into the tip of the rays of starfish. This will extend the feet. Then drop the animal in fixative; mercuric chloride-acetic acid is good.

See Moore (1962) for collecting and preserving starfish, ova, and larvae.

Staining Invertebrates

Sections can be stained in any manner, depending on the fixatives and the desired results.

For beautiful, transparent, whole mounts, the carmine stains (p. 84) are the usual choice for most of the invertebrates; obelias, hydras and hydroids, daphnids, Bryozoans, medusae, flukes, tapeworms, small annelids, tunicates, and ammocoetes (larvae), to name a few. But Kornhauser hematein (p. 106) is an excellent substitute, particularly for flukes and tapeworms.

If the cuticle and muscle of flukes and tapeworms tend to remain opaque and obscure the details of internal anatomy, either of the two methods can be tried to correct the condition. Dehydrate and clear in an oil, such as cedarwood oil or terpinol. Place in a flat dish under a dissecting microscope and carefully scrape away some of the tissue from both surfaces. With care, this can free the animal of some of the dense tissue material and not harm the internal structures. An alternate method is to finish staining and then wash in water. Transfer to a potassium permanganate solution made from a few drops of 0.5% solution added to water until only a pink color develops. When the worm begins to show a greenish brown sheen, remove it immediately to distilled water: 5 minutes. Transfer the worm to 2–3% aqueous oxalic acid until it is bleached and the sheen is lost. Wash thoroughly in running water at least 1 hour. Dehydrate, clear, and mount.

References for penetrating invertebrates are Becker and Roudabush (1945), Galigher and Kozloff (1964), Gatenby and Beams (1950), Gray (1954), Guyer (1953), Hegner et al. (1927), Mahoney (1968), and Pantin (1946).

PREPARATION OF CHICK EMBRYOS

The following procedure is a standard and simple one for removing, fixing, and staining chick embryos of any size, from primitive streak to age 96 hours.

With the handle of the scissors, break the shell at the air space end, turn scissors, and cut the shell around the long axis, being careful to keep the tip of the scissors pressed against the inside of the shell while cutting. This will prevent penetration of the yolk or the embryo. With egg submerged in physiological saline at 37°C in a finger bowl, remove top half of the shell. (With rapidity developed from experience, warm water serves just as well as saline, with no ultimate harm to the chick if it is to be fixed immediately.) The chick will be found floating on the upper surface of the yolk. It can be a waste of time to remove the lower half of the shell, but many workers prefer to do so. This is a matter of individual preference. With small scissors, cut quickly around the outside of the vascular area. Except for large embryos, do not cut through the vascular area. Keep one edge gripped by forceps, and slip a Syracuse watch glass under the chick. Withdraw the watch glass and embryo with a little of the saline. Bring as little yolk as possible with the chick.

The vitelline membrane most likely will come with the embryo; occasionally it will float free. If not, wave the embryo gently back and forth in the saline until the membrane begins to float loose. Release grip on embryo and remove the membrane. Make certain that the embryo is wrong side up, that is, the side that was against the yolk is now uppermost. Straighten out the chick, taking care that there are no folds. Sometimes washing with saline in a pipette across the top helps to clean and flatten it. Pull off remaining saline and yolk, carefully, not just from one edge but slowly and with gentle pressure on the pipette bulb around the vascular area. Do not get so close that the embryo is drawn up into the pipette.

Have a circle of filter paper cut with the inside hole approximately the size of the vascular area if it is to be retained or a little larger than the embryo if it only is to be used. Being careful not to disturb the embryo, drop the circle of paper around it. Press gently around the paper to make it adhere to the blastoderm. With a pipette apply fixative (Gilson or Bouin), carefully dropping it first directly on the embryo; ease it on (do not squirt it on) and then work outward toward and over the paper circle, finally adding enough fixative to immerse embryo and paper completely. Leave overnight. The embryo should remain adhering to the paper, and it can be transported in this fashion from solution to solution. Follow fixation by proper washing, depending upon choice of fixative. Stain the embryo in a carmine, hematoxylin, or hematein stain (see whole mounts, p. 422), and dehydrate. Remove the chick from the circle, clear, and mount. The cover glass may have to be supported (p. 434) to protect the delicate embryo from destructive pressure. After washing, the embryos can be dehydrated, cleared, and embedded in paraffin.

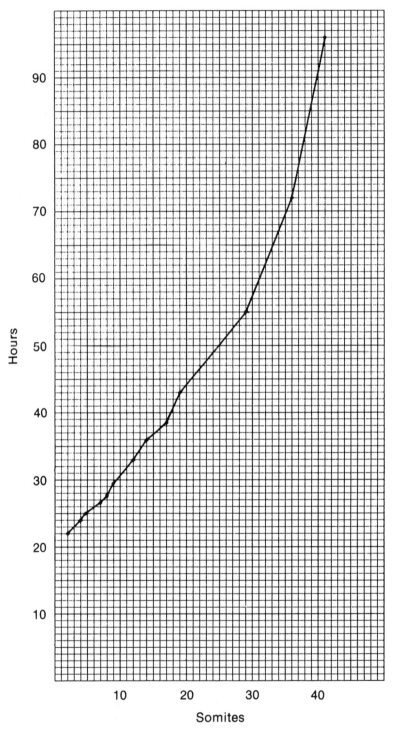

Figure 28-1. Development of a chick embryo. Data from Patten (1952).

Incubation time and determination of age can be puzzling. To determine the precise embryonic age of the chick may be difficult because it is determined not only by the number of hours in and the temperature of the incubator, but the length of time the egg was in the hen and the temperature at which the egg was stored. The temperature of the incubator should be 37.5°C, and about 4 hours are usually required to warm up the egg once it is placed in the incubator. Thus, a 36-hour chick would require about 40 hours in the incubator for proper development. The accompanying chart (Fig. 28-1) indicates how counting the number of pairs of *somites* in the chick will determine its age; thus, 10 pairs indicates an age of 30 hours. This, rather than the number of hours in the incubator, is the best criterion for development.

Any bird or reptilian embryo can be fixed as outlined for the chick. Amphibian larvae can be dropped directly into fixative, after any surrounding yolk or jelly has been dissected off.

Mammalian embryos should be dissected from the uteri and fixed in Bouin. If the uterine site also is to be preserved, cut through the uterus on either side of the embryo and fix. A little fixative can be injected into the uterine cavity to quicken the fixing action.

Ascarid uteri are fixed in alcoholic Bouin; sometimes Carnoy will give desired results.

WHOLE MOUNTS

Whole mounts are slide mounts of entire specimens small enough to be studied in toto and mounted in resins in toluene or water, in gums or glycerin as aqueous mounts, or as dry mounts. Some examples from the innumerable methods will follow, with the hope that a technician can adapt them to almost any whole mount venture.

General Preparation

FIXATION

Ninety-five percent alcohol, followed by a bleaching if necessary. From 70% alcohol into bleaching fluid (3 to 4 drops of Clorox in 10 ml of 70% alcohol, or water plus H_2O_2 (1:1) with a trace of ammonia): 24 hours.

PROCEDURE
1. Wash with 70% alcohol, 3 changes.
2. Dehydrate in 95% and absolute alcohol.
3. Clearing is often difficult. Try mixture of xylene and beechwood creosote (or aniline or phenol) (prevents whole mounts from becoming brittle), 2 changes. If opaque patches persist, return them to absolute alcohol. Vacuum processing is recommended.

4. Impregnating with mountant can be difficult. Specimen may collapse and turn black or opaque when taken directly from alcohol or clearer to mountant. Impregnate gradually by adding each day a drop or two of mountant to the clearer containing the specimen. Carefully stir the mountant into the solution. (Galigher [1934] uses a cone of filter paper containing the mountant and allows it to enter the solution gradually and continually.) When the clearer has obviously thickened, allow concentration to continue by evaporation.
5. Mounting sometimes requires considerable experience (and patience) in judging the correct method and amount of mountant. Large specimens should have supports under the cover glass to prevent tilting of the cover glass; delicate specimens should have supports under the cover glass for protection against smashing. Various materials may be used to support the cover glass: bits of cover glass or slide, thin glass rods, circles, squares, or strip stamped or cut out of heavy aluminum foil, bits of glass wool.

Place a drop of mountant on slide and place specimen in it. Lower cover glass carefully; keep it flat (not tilted); have specimen in center of mount. If several specimens are to be mounted under one cover glass, the chance of their slipping out of place is lessened by allowing the mountant to air dry for a few hours. Place a bit more mountant over them and add cover glass. Warming the cover glass aids in bubble prevention.

Comments

Demke (1952) embeds helminths in celloidin to support them. Dehydrate through absolute alcohol, through alcohol-ether, and into thin celloidin. Pour celloidin and specimens into flat dish (Petri dish); allow solvent to evaporate slowly. Cut out squares of celloidin containing specimens, dehydrate, clear, and mount. No other support of cover glass is required.

Rubin (1951) uses PVA mounting medium. See other mounting media in Chapter 10. Courtright (1966) uses polyester resins.

Glycerin Jelly Mounts

Many materials, including frozen sections, can be mounted directly from water into glycerin jelly. If there is danger of the object collapsing, transfer it from 70% alcohol or water into a mixture of 10–15% glycerin in alcohol or water. Leave the dish uncovered until most of the alcohol or water has evaporated. Mount in glycerin jelly. Sections are mounted without cover glass support. Thicker specimens probably will require a supported cover glass. With a turntable, spin a ring of gold size on the slide. Allow it to dry. If it is not high enough, add more layers to the height desired.

Melt glycerin jelly in water bath or in oven. Add a drop or two of jelly onto specimen inside ring. Use just enough to fill ring. (Warm the slide briefly, and the glycerin jelly will spread readily under the cover glass.) Warm glass and ease horizontally into place. Carefully wipe off any glycerin jelly that works out from cover glass. After a few tries, it becomes relatively easy to estimate the correct amount of jelly to make a clean mount. Seal cover glass, supported or unsupported.

Stained Whole Mounts: Hydras, Embryos, and Flukes

Many fixatives are suitable: Carnoy or Gilson for worms. Bouin or Zenker for embryos, formal acetic or saturated $HgCl_2$ acetic (95/5) for flukes. Follow by proper washing and extraction of any undesirable pigments or crystals. But before fixing consult the previous section on invertebrates or an authoritative source concerning the problem at hand.

Choice of stain and method also depends upon the type of material. A few of wide usage are included here.

Grenacher Borax Carmine (Galigher 1934)

SOLUTION

    carmine .... 3.0 g
    borax .... 4.0 g
    distilled water .... 100.0 ml

Boil approximately 30 minutes until carmine is dissolved. Mixture may be allowed to stand until this occurs. Add:

    70% alcohol .... 100.0 ml

Allow to stand 1 to 2 days. Filter.

PROCEDURE
  1. Transfer from 50% alcohol to borax carmine: 3 to 4 hours or overnight.
  2. Add concentrated HCl, slowly, a drop at a time, stirring, until carmine has precipitated and is brick red. Let stand 6 to 8 hours or overnight.
  3. Add equal volume of 3% HCl in 70% alcohol and thoroughly mix. Let stand 2 to 3 minutes until specimens have settled. Draw off precipitated carmine with pipette. Add more acid alcohol, mix, allow to settle, and draw off fluid. Repeat until most of carmine is removed.
  4. Add fresh acid alcohol and allow tissue to destain, checking at intervals under microscope: may require 2 hours or more. If destaining is exceedingly slow, increase percentage of acid in alcohol.
  5. When destained, replace acid alcohol with 80% alcohol, several changes, to remove acid: over a period of 1 hour.

Table 28-1. Procedure for Creosote Processing

Absolute alcohol	Creosote (aniline or carbol) xylene
80 parts	20 parts
60	40
50	50
40	60
20	80
0	100

6. Dehydrate in 90% alcohol: 30 minutes or more, depending on size of specimen; absolute alcohol: 30 minutes or more.
7. Clear. For delicate objects clearing probably should be done gradually through the following concentration of absolute alcohol and creosote (or aniline or carbol) xylene, and then into pure creosote xylene.
8. Mount. For exceedingly delicate objects and roundworms, heed the warning above concerning slow impregnation with the mounting medium. Judgment concerning its use rests with experience, but slowness is invariably necessary for roundworms.

Mayer Carmalum (Cowdry 1952)

SOLUTIONS
Stock solution
   carmine .... 1.0 g
   ammonium alum .... 10.0 g
   distilled water .... 200.0 ml

When dissolved, filter and add:

   formalin .... 1.0 ml

Working solution
   carmalum stock .... 5.0 ml
   glacial acetic acid .... 0.4 ml
   distilled water .... 100.0 ml

PROCEDURE
Stain for 48 hours, no destaining is necessary with Carmalum. Dehydrate as above.

Hematein (Kornhauser 1930)

SOLUTION

Stock solution

hematein	0.5 g
95% alcohol	10.0 ml
potassium aluminum sulfate, 5% aqueous	500.0 ml

Grind hematein with alcohol and add to the aqueous sulfate.

Working solution
Dilute above 1:10 with distilled water.

PROCEDURE
1. Stain overnight.
2. Transfer to 70% alcohol.
3. Destain in acid alcohol, 5% HCl in 70% alcohol.
4. Blue in alkaline alcohol, ammonia, or sodium bicarbonate in 70% alcohol.
5. Dehydrate and mount as above.

Hematein is good for flatworms. Alum cochineal used to be popular but has been largely replaced by carmine stains. Alum hematoxylin may be used for small organisms if they are not too dense. The celestine blue method (Demke 1952) and trichrome stain (Chubb 1963) give good results.

Cochineal-Hematoxylin

SOLUTION

alum cochineal	3 parts
potassium alum	30.0 g
cochineal	30.0 g
distilled water	100.0 ml
Delafield hematoxylin	1 part
distilled water	25 parts

Can be used immediately but is better after a few hours. Filter before use.

PROCEDURE
1. After fixation wash in 50% alcohol: 1 hour.
2. Wash in distilled water: 10 minutes.
3. Stain: 4 hours or overnight.
4. Transfer to 70% alcohol: 2 minutes.
5. Differentiate in acid alcohol (70% alcohol: HCl, 99:1) until internal structures are visible.

6. Place in 70% alcohol: 1 to 2 minutes.
7. Blue in 70% alcohol plus a few drops of saturated aqueous lithium bicarbonate: 1 to 2 minutes.
8. Dehydrate in 95% alcohol: 1 hour, 3 changes of absolute alcohol: 1 hour, 30 minutes, 1 hour.
9. Clear in absolute alcohol-cedarwood oil (1:1): ½ to 1 hour.
10. Finish clearing in cedarwood oil: 1 hour, and mount in one of the synthetic mountants (Permount or like).

Protozoa

An easy method for handling protozoa is the following: it gives excellent results and rarely fails.

Chen (1944a) Cover Glass Method

FIXATION

Hot Schaudinn, 50°C: 5 to 15 minutes. Other fixatives that can be used are Bouin, Champy, Flemming, or Worcester. See Merton (1932) for fixing stentors, spirostomids, and vorticellas.

Organisms such as paramecia should be concentrated by centrifuging. Quickly pour off some of the solution and add fixative. Amoebae will settle on the bottom of a clean culture dish. Decant off most of the culture medium, leaving the bottom barely covered. Quickly pour hot Schuadinn or Bouin (50–60°C) over the organisms. After a few minutes add an equal amount of 85% alcohol. Carefully loosen any amoebae clinging to the bottom and collection the solution in a centrifuge tube. For both paramecia and amoebae, allow to settle or centrifuge down at low speed. Pour off fixative and wash several times with 70% alcohol, centrifuging after each wash. If a mercuric chloride fixative was used, one of the washes must contain some iodine.

PROCEDURE

1. Follow the 70% washing with 80–85% alcohol: 10 minutes.
2. Smear cover glass with albumen fixative. Cover glasses can be subbed like slides and used here. Place it albumen-side up on slides with a bit of edge projecting beyond the slide. The projecting edge can be easily grasped with forceps when it becomes necessary to remove the cover glass. With a pipette pick up a few drops of alcohol containing organisms. Drop them in the center of the cover glass, where they will spread over its surface. The alcohol will begin to evaporate and bring the specimens in contact with the albumen. Avoid complete drying; the edges may dry, but the center will remain slightly moist. Carefully add a couple of drops of 95% alcohol onto the specimens, and then transfer the cover glass to a Petri dish of 95% alcohol. Slides lying in the bottom of the dish will help to handle the cover glass, as above.

3. Carefully remove the cover glass from the 95% alcohol, dip it gently in absolute alcohol, and flood it with 1% celloidin or nitrocellulose. Drain off excess celloidin against filter paper. Wave it back and forth a few seconds until it begins to turn dull and place it in 70–80% alcohol. It can remain in this solution until ready for staining.
4. Any stain can be applied—hematoxylin, carmine, Feulgen—depending upon study to be undertaken.
5. The cover glasses are finally dehydrated, cleared, and mounted with the specimens albumen-side down on a drop of mounting resin.

COMMENTS

Subbed slides (p. 469) are excellent. Cells adhere well and rarely rupture. Apply a drop of cells, drain off excess fluid, and allow to dry.

Smyth (1944) uses a quicker method and directly on slides. After fixation, he carries the organisms through graded alcohols into absolute alcohol. This is dropped on a film of albumen on a slide. Place the slide in absolute alcohol and continue from there to the stain.

Agrell (1958) places suspensions of minute embryos on albumen-covered slides. Allowed to almost, but not quite, dry, they become flattened and in close contact with albumen. Dip into absolute alcohol, and then into fixative; place them horizontally in 95% alcohol vapor: 1 minute. Then fix. This coagulates the embryos and attaches them.

Paramecia can be fixed to preserve their normal shape without contraction by first adding copper sulfate or acetate to the cell suspension.

Merton Method (Kirby 1947)

1. Add a drop of paramecia to an albuminized or subbed slide: 30 seconds.
2. Add an equal-sized drop of 1% copper sulfate: 7 to 8 minutes, or 3% copper acetate: 45 seconds.
3. Draw off part of fluid and suspend slide over 2% osmic acid: 45 seconds.
4. Add saturated aqueous mercuric chloride to the drop of organisms: 10 minutes.
5. Transfer to 70% alcohol plus small amount of iodine: 10 minutes.
6. Wash with distilled water.
7. Stain.

Prescott and Carrier Method (1964)

Amoebae, well flattened for tritium or other special studies, can be prepared by this method.

1. Place a small drop of some amoebae on a subbed slide (p. 469).
2. Add a small drop of fixative (70% alcohol or acetic alcohol) to a cover glass and place on the amoebas.

3. Immediately freeze in liquid nitrogen: 15 seconds. Flip off cover glass, rinse slide in 95% alcohol, and air dry. If liquid nitrogen is not available, fix in 50% aqueous acetic acid, and freeze in dry ice. (Alcohol fixative will not freeze in dry ice.)

Staining Protozoa

There are many special and routine stains adapted to protozoa. Trichrome and hematoxylin stains are always good. To stain fibrillar elements with iron-hematoxylin, use a fixative containing chromium, warm the stain to 50°C, and destain with 10% commercial H_2O_2 (Kidder 1933). Rothenbacher and Hitchcock (1962) use a Giemsa method for flagella. Loeffler's stain also can be used (Kirby 1947). In ciliates, the surface pattern, cilia, basal granules, and connecting filaments are clearly shown by the Gelei osmium-toluidine blue method (Kirby 1947; Pitelka 1945). For the silver line system, the classic techniques of Chatton and Lwoff (1930, 1935, 1936), Frankel and Heckmann (1968), Gelei (1932, 1935), and Klein (1926) are outstanding. Also see Corliss (1953) for a silver method and Bodian (1937) for an activated protein silver method. Protozoans can be tagged for autoradiographs and are excellent subjects for vital staining and enzyme techniques. Schiff et al. (1967) use a safranin-fast green stain to differentiate the micro- and macronuclei of protozoa. Nigrosin demonstrates various organelles.

Borror (1968) Nigrosin Stain

SOLUTIONS

Solution A

saturated aqueous $HgCl_2$	10.0 ml
glacial acetic acid	2.0 ml
formaldehyde (37–40%)	2.0 ml
tertiary butanol	10.0 ml

Solution B

formaldehyde (37–40%)	20.0 ml
nigrosin, water soluble	4.0 g
distilled water	100.0 ml

Working solution

solution A	12 parts
solution B	1 part

PROCEDURE

1. Place a drop of concentrated organisms on slide.
2. Add, from a height of 2 to 3 cm, a drop of working solution. Leave 3 to 4 seconds.

3. Gently wash culture and stain to one end of slide with additional drops of working solution. Stained fixed organisms will remain attached to slide. Leave 15 seconds.
4. Dehydrate, clear, and mount.

Results
ciliary organelles—black against gray background

Comments

A drop of 10% aqueous nigrosin can be mixed with a culture of living organisms and the staining of the various organs observed under the microscope. After allowing the slide to dry, the dye will adhere to some of the structures and the slide can be mounted with resin and cover glass (Repak and Levine 1967).

The food vacuoles of paramecia are demonstrated by adding a small amount of carmine to the culture. After about 5 minutes the food vacuoles are filled with carmine. Mix a couple of drops of the culture with an equal amount of saturated aqueous solution of nigrosin. Air dry and cover. Do not use too much fluid; it slows down the drying (Wilhelm and Smoot 1966).

Sectioning Protozoa

Stone and Cameron (1964) Agar Method
(Kimball and Perdue Modification 1962)

1. Pour a small amount of melted agar (p. 464) into a short length of glass tubing sealed at one end. Chill until agar is hardened.
2. Pipette a concentration of organisms on top of agar and add fixative.
3. When cells are fixed and settled on the agar, draw off the fixative.
4. Add more agar mixed with a little eosin or other counterstain and chill to solidify agar.
5. With a pipette, force water down side of agar and under it, thereby forcing the block loose and out of the tubing.
6. Dehydrate, clear, and embed block of agar like a piece of tissue.
7. When the tissue is being sectioned, the cells will be found at the junction of colored agar with the uncolored layer.

Alternate method

Form a block of agar and hollow out a small cavity in it. Pipette a drop of concentrated fixed protozoa into the cavity and allow the organisms to settle. Remove as much fluid as possible and seal the cavity with warm agar. Embed in paraffin. Eosin can be added to the organisms to make them more visible in the block. Eosin in the added agar also helps to locate them.

Dry Mounts of Foraminifera and Radiolaria

An opaque type of slide mounting is used for dry objects that are to be examined by reflected light. Glue the shells on a black background, place a cover glass and supporting ring around them, and cement the cover glass and ring in place. Gray (1964b) and Kirby (1947) suggest a simpler and perhaps more practical method. Cut two pieces of cardboard to slide size. Punch a ⅝-inch hole in the center of one piece. Paint a black ⅞-inch square in the center of the other piece or paste a ⅞-inch square of black paper on it. Stick the two pieces of cardboard together with dry mounting tissue (used in photography) and a flatiron. The cavity thus formed can be covered with a cover glass glued on with household cement. Gray recommends gum tragacanth as the best adhesive for sticking the shells to the background. Place little dabs of the gum on the black background, and with a moistened fine brush (red sable) pick up a good specimen and place it in the center of a drop of gum. Breathe on it; moisture is necessary to make it adhere. When all dabs of gum have a shell applied, breathe again several times on the preparation. Then let the slide dry. Turn it over and tap it several times. If some of the shells are not adhering, it may be necessary to add a drop of water to the gum to make the shells stick. But make certain the slide is dry before covering it. Special opaque slides can be purchased for making dry mounts, or regular depression slides can be used if the depression is painted black.

ANIMAL PARASITES

Animal organisms parasitic in or on humans include protozoans, platyhelminths, nemathelminths, and arthropods. Clinical parasitology is extensive and therefore only a few methods are incorporated here. Since the Romanovsky-type stains are used on blood parasites (malaria, trypanosomes, filaria, etc.), the preparation of smears is included in the chapter on hematological elements (p. 193). Tissue sections can be stained in a similar manner. Parasitic roundworms (pinworms, *Brichuris, Ascaris,* hookworms), flatworms (lung, intestinal, bile duct, and blood flukes, tapeworms), and arthropods (ticks, lice, mites) are discussed in the sections on invertebrates and whole mounts in this chapter. Methods given there can be adapted for most parasitic helminths and arthropods. Sections of tissue parasitized by protozoa or helminths are effectively stained by hematoxylin methods, also by periodic acid-Schiff; protozoa and worms are strongly PAS positive because of stored glycogen in both forms and PAS-positive cuticle in the helminths. A methenamine silver method is excellent on protozoans, particularly flagellates. The scolices hooks in hydatid disease (*Echinococcus*) do not show with PAS and are better demonstrated against a hematoxylin background. Kenney et al. (1971) use acid-fast staining to demonstrate the hooklets. Ova and larvae can be handled according to directions given on p. 100 or p. 433. Intestinal protozoa (amoebae, flagellates, ciliates, and coccidia) require the following special methods, both for smears and tissue sections.

Only permanent slide mounts are described. Consult clinical laboratory manuals for temporary and rapid examination methods for immediate diagnosis. Gradwohl (1963) and Lynch et al. (1969) are comprehensive.

Intestinal Protozoa: Smear Techniques

Preparing Concentrate Smears (Arensburger and Markell 1960)

1. Add 1 ml of feces to 10 to 15 times its volume of tap water. Mix well and strain through 2 layers of wet gauze in a funnel. Collect in a small centrifuge tube. Add 1 to 2 ml of ether. Using a cork or thumb for a stopper, cautiously shake the tube. Fill with water to 1 cm from top.
2. Centrifuge 45 seconds, 2500 rpm. Break up any plug at top and decant supernatant fluid.
3. Add 2 to 3 ml of normal saline and shake to resuspend the sediment. Fill tube with normal saline to within 1 cm of top and centrifuge.
4. Decant supernatant fluid. Take a small quantity of the original fecal specimen on the end of an applicator stick and mix well with sediment at bottom of tube.
5. With an applicator stick, transfer as much as possible of the material to a clean slide. Smear as for a conventionally made slide and fix immediately in Schaudinn fluid.
6. Stain as preferred, or with one of the following stains.

Kohn Stain, Combined Fixation and Stain (Faust et al. 1970)

SOLUTION
Basic solution
  90% alcohol .... 170.0 ml
  methanol .... 160.0 ml
  glacial acetic acid .... 20.0 ml
  phenol, melted .... 20.0 ml
  1% phosphotungstic acid, aqueous .... 12.0 ml
  distilled water .... 618.0 ml

Grind 5 g chlorazol black E in a mortar at least 3 minutes, add a small amount of basic solution, and grind until a smooth paste. Add more solution and grind 5 minutes. Allow to settle a few minutes and pour off supernatant fluid into a separate container. Add more solution to mortar and continue grinding and mixing until all dye is in solution. Add remaining basic solution and allow to ripen 4 to 6 weeks. Filter through No. 2 Whatman paper before use. Keep in tightly capped bottle. The prepared solution may be also obtained commercially.

Table 28-2

Stain	Basic solution	Hours
undiluted	—	2–3
1 part	1 part	2–4 to overnight
2 parts	1 part	2–4
1 part	2 parts	2 to overnight
1 part	3 parts	4 to overnight

PROCEDURE
1. Stain according to Table 28-2. Tissue sections require twice as long as smears, and water may be used for dilutions for the sections.
2. Dehydrate in 95% alcohol: 10 to 15 seconds.
3. Dehydrate in absolute alcohol, 2 changes: 5 minutes each.
4. Clear in xylene and mount.

RESULTS
protozoa—gray green, gray, or black
cysts—gray green to dark gray
nuclei, chromatid bodies, karyosomes, cell membranes—dark green to black
ingested red cells—pink to dark red

Kessel (1925) and Chen (1944a) Smears (Modified)

FIXATION
Schaudinn (p. 36), 40°C: 10 to 15 minutes or more.

SOLUTIONS

Iron alum

ferric ammonium sulfate	4.0 g
distilled water	100.0 ml

Hematoxylin stock solution

hematoxylin	1.0 g
absolute alcohol	10.0 ml

Allow to ripen several months or hasten process (p. 102)

Hematoxylin working solution

hematoxylin stock solution	0.5 ml
distilled water	99.5 ml

Add 3 drops of saturated aqueous lithium carbonate. If the hematoxylin working solution looks rusty or muddy brown, it is unsatisfactory and will not stain efficiently.

PROCEDURE
1. Transfer slides into 70% alcohol (from fixative): 2 to 3 minutes.
2. Treat with Lugol solution: 2 to 3 minutes.
3. Wash in running water: 3 minutes.
4. Decolorize with 5% sodium thiosulfate: 2 minutes.
5. Wash in running water: 3 to 5 minutes.
6. Mordant in iron alum, 40°C: 15 minutes.
7. Wash in running water: 5 minutes.
8. Stain in hematoxylin, 40°C: 15 minutes.
9. Wash in running water: 5 minutes.
10. Destain in iron alum 2% (dilute 4% stock with distilled water, 1:1) until nuclei and chromatoidal bodies are sharp against colorless cytoplasm. Check under high dry objective.
11. Wash thoroughly in running water: 15 to 30 minutes.
12. Dehydrate, clear, and mount.

RESULT
nuclei—sharp blue black

COMMENTS
Saturated aqueous picric acid may be used for destaining. Follow it with a rinse in dilute ammonia (2 to 3 drops/100 ml water) and thorough washing in water.

Diamond (1945) added 1 drop of Turgitol 7 to the diluted hematoxylin solution just before use (1 drop per 30 to 40 ml solution). He substituted the Turgitol treatment for heat; it reduces surface tension and increases cell penetration. Staining time was reduced to 5 minutes.

This method may be used for amoebae in tissue as well as in smears.

Lawless Rapid Method (1953)

SOLUTIONS

Schaudinn-PVA fixative (Burrows 1967)

Dissolve 4.5 g $HgCl_2$ in 31 ml of 95% ethanol in stoppered flask. Shake at intervals. Add 5 ml acetic acid and mix. Set aside until needed. Add 5 g PVA powder to 1.5 ml glycerin, and mix thoroughly until all particles of PVA appear to be coated with glycerin. Transfer to 125 ml flask, add 62.5 ml distilled water. Stopper and leave at room temperature: 3 hours to overnight. Occasionally swirl the mixture. Place flask in 70–75°C water bath: 10 minutes, swirling frequently. When PVA is mostly dissolved add fixative mixture. Continue to mix in bath: 2 to 3 minutes, or until PVA is dissolved and solution clears. Cool at room temperature. If solution gels during storage, warm in 56°C water bath. Some protozoans fix better in warm PVA mixture.

Stain

chromotrope 2R	0.6 g
light green SF, yellowish	0.15 g
fast green FCF	0.15 g
phosphotungstic acid	0.7 g
glacial acetic acid	1.0 g
distilled water	100.0 ml

Add acetic acid to dyes and phosphotungstic acid; let stand 15 to 30 minutes. Add water.

PROCEDURE

A small portion of stool is fixed in PVA fixative (1 part stool to 3 parts fixative): 15 minutes to 1 hour or more. In this form it can be shipped in a vial. When ready to make slides, decant off excess PVA solution. Replace cap of vial and shake the emulsion. Remove cap and cover vial opening with gauze. Place 3 or 4 drops of strained material on cleansing tissue or blotter. Allow absorption of PVA for about 5 minutes. Scrape up moist residue and spread with an applicator stick on slide or cover glass. Dry thoroughly and drop into iodine-alcohol (p. 27). If smears wash off, too much PVA was carried over; leave material for longer time on cleansing tissue. (In the dry condition, smears can be stored for several months.)

Smears may be fixed directly on slides. Take 1 drop of fecal material to 3 drops of PVA fixative. Smear over a large area of the slide and dry in oven overnight. Immerse in iodine-alcohol.

1. Leave in iodine-alcohol: 1 minute.
2. Decolorize in 70% alcohol, 2 changes: 1 to 2 minutes each.
3. Stain: 5 to 10 minutes.
4. Differentiate in acidified 90% alcohol (1 drop acetic acid/10 ml alcohol): 10 to 20 seconds or until stain no longer runs from smear.
5. Dehydrate in absolute alcohol: rinse twice, dipping up and down.
6. Dehydrate in second change of absolute alcohol: 1 minute.
7. Clear and mount.

Warning: Press the cover glass in place very gently; the smear is somewhat brittle and is easily loosened. Too much pressure on the cover glass may cause parts of the smear to float around on the slide.

RESULTS

background—predominantly green
cysts—bluish green cytoplasm, purplish-red nuclei; green is more intense than background stain

engulfed red cells—vary; green, red, or black
chromatin—intense green, but may be reddish
karyosomes—ruby red
helminth eggs and larvae—usually red

COMMENTS

If cysts do not stain, or do not stain predominantly red, fixation is incomplete. Warming the fixative may help, although cold fixation yields more critical staining. A newly prepared stain is predominantly red in color and reaction; older stains show more violets and greens. The stain is more transparent than hematoxylin stain, but it fades after a few years.

Hajian (1961) fixes in Bouin for better staining of karyosomes.

Tomlinson and Grocott (1944) describes a phloxine-toluidine blue stain for malaria, leishmania, microfilaria, and intestinal protozoa in tissue.

Verrling and Thompson (1972) use hematoxylin and a polychromatic stain.

Yang and Scholten (1976) use celestine blue B:

Solution A

celestine blue B	0.6 g
glacial acetic acid	20.0 ml
distilled water	80.0 ml

Keeps indefinitely in a brown bottle.

Solution B

ferric alum	4.0 g
distilled water	100.0 ml

Keeps indefinitely in refrigerator.

Mix equal parts. Deteriorates in 2 weeks. From water transfer to stain: 5 minutes, but is not overstained if left an hour. Rinse in tap water, dehydrate, clear, and mount. It can be followed by a trichrome stain.

For immunofluorescent techniques see Hoffman and Miller (1975) and Nayebi (1971).

FISH AND REPTILES

The major problem in handling small fish and reptiles rests with getting appropriate fixing solutions into chambers or spaces in the organs being processed. This is especially problematic when processing entire heads. The use of the vacuum chamber (sometimes with low heat) facilitates penetration of solutions required for working the tissues down and into paraffin. Larger specimens whose organs can be dissected free of the animal, of course, present no unusual problems.

HANDLING SMALL TISSUE SAMPLES

There are several devices commercially available that ensure the safety of tiny samples or tissue fragments while they are being processed. Stainless steel carrier baskets with mesh of varying size are easily handled and can used over and over. Cassettes of different sizes are also available that permit the handling of tissues through all processing steps until final embedding in the very same container without ever handling the tissues. In a pinch, wrapping tissue samples in lens paper could also save the day.

29
Special Procedures IV

AUTORADIOGRAPHY

Autoradiography is a photographic method for determining the spacial distribution of radioisotope-labeled substances within a specimen. The presence and location of radioactivity in the specimen is determined by the formation of metallic silver grains in the photographic material, thereby providing a visible image of the radiation exposure. When photographic materials are placed in direct contact with the labeled specimen, as for example a tissue section adhered to a glass slide, a means for locating and quantifying the radiation emanating from the specimen is provided.

Radioactive isotopes of stable compounds can be incorporated in some tissue compounds; autoradiographs (ARGs) will reveal the location of the compound that was metabolically active during the time of the application of the labeled material. Autoradiographs have become one of the most useful methods for studying biochemical reactions in cells. Many of the most dependable tracer isotopes are those of elements normally found in the body, and the selection of isotope will depend on the study to be undertaken: for example, strontium or calcium for bone, iodine for thyroid, thymidine for DNA.

After proper preparation, the specimen containing the radioactive material is placed in contact with a photographic emulsion. Ionizing radiations are emitted during decay of the isotope, and change the emulsion to produce blackening during its development. The number of developed grains will depend on the emulsion and the kind of ionizing particles emitted from the specimen. Alpha particle tracks are large and dense, are not easily deflected, and move in a straight path only a few micrometers in length. Beta particles are small and produce longer tracks, but they are easily deflected and produce grains far apart and more scattered. Gamma rays are useless with long random paths, but many isotopes that give off gamma rays also produce low energy beta particles, and these will give the picture on film. High-energy beta particles (^{32}P) may travel a millimeter or so before producing a grain, but low-energy beta particles (^{131}I and ^{14}C) have short paths and give very satisfactory resolution on photographic emulsion.

Many varieties of material can be used—tissue sections, chromosome squashes and spreads, microorganisms, whole cells, smears, cultures, sections of whole animals, and so forth. The specimen and film are placed in contact for a certain length of exposure time. The radioactive atoms are decaying, and the emitted radiation strikes the photographic emulsion.

Monochrome film emulsions consist of mixtures of gelatin and crystalline silver halides, primarily bromides. The halide crystals are imperfect and provide electron traps to produce a concentration of electrons. The positively charged silver atoms are attracted by the negatively charged electrons and form a latent image. Development reduces the silver halide to metallic silver, and the unreduced grains of silver are dissolved away in the hypo solution. Also, the emulsion is hardened by the hypo. Ordinary photographic emulsions cannot be used because they contain only about 30% silver halide and have large grains. Special emulsions have been developed to intercept the maximum number of rays. The nuclear emulsions contain up to 90% and 95% silver halide and have small grains. Kodak nuclear emulsions are NTA, NTB, NTB2, and NTB3, arranged according to ascending sensitivity. The last two are the most commonly used for all isotopes; however, NTB2 is especially recommended for use with ^{3}H, ^{14}C, and ^{35}S, and NTB3 for use with ^{32}P, ^{131}I, and ^{125}I. Emulsions of Ilford Ltd., England, are also popular for use because of their sensitivity and grain size.

As mentioned above, the grains are not uniform; some carry a silver speck, and others are relatively insensitive. Some grains develop in an unexposed emulsion, and some become accidentally developed by darkroom light, cosmic rays, or other sources of ionizing radiation. When evaluating autoradiographs, such "background" values have to be considered and checked. Take care to avoid unnecessary background accumulation.

The following are reference texts on autoradiography: Baserga and Malamud (1969), Gude (1968), Rogers (1967), and Roth and Stump (1969).

Kodak offers a variety of data books and technical pamphlets on the subject of autoradiography. Foremost among these is their publication, "Autoradiography of Microscopic Specimens" (N-911; published in 1988). Other publications, many of them complimentary, can be received by writing to Eastman Kodak Co., Department 412-L, Rochester, NY 14650. The latest in technical information may also be obtained more quickly by calling 800-225-5352.

PROCEDURES FOR AUTORADIOGRAPHS

Fixation and Slide Preparation

Knowing the half-life of the isotope that will be used for autoradiographs helps to determine the approximate time of exposure required for ARGs and the length of time that the tissue can be stored before processing. The half-life is the time required for disintegration of one-half of the atoms present. For example, ^{32}P has a half-life of 14.2 days and must be processed immediately. Within 42 days only 12% of the original radioactivity will remain. ^{3}H has a half-life of more than 12 years and loses only 0.5% of its activity each month. ^{14}C has a half-life of 5568 years, so you have your lifetime plus for processing.

^{125}I has 60 days. ^{131}I has only 8.07 days. For the latter, fix in ethanol and process as soon as possible.

The aldehydes are usually satisfactory for fixation, but avoid solutions containing mercuric chloride, which will produce artifacts on the emulsion. If a fixative such as Zenker has been used, treat the sections with iodine and thiosulfate, followed by 0.2 M cysteine for 10 minutes. This will reduce desensitization of the emulsion. Carnoy, methanol, or acetic-alcohol (1:3) can be used for smears and tissue cultures. Allow smears to dry and flatten the cells before fixation. In all cases the fixative must be washed out of the tissue; some fixatives cause air bubbles. This applies as well to smears; wash them in several changes of 70% alcohol and distilled water.

Solubilities of isotopes must be checked carefully; some leach out in water or paraffin solvents and must be prepared by freeze-drying methods (Holt and Warren 1953). In some cases Carbowax can be used, but it will dissolve water soluble isotopes (Holt et al. 1949, 1952; Holt and Warren 1950, 1953). Mounting must be handled with care to prevent leaching (Gallimore et al. 1954).

Witten and Holmstrom (1953) freeze the tissue at the microtome immediately after excision. The knife is kept cold with dry ice, keeping the section frozen after it is cut. Then the section is carried still frozen to the photographic emulsion. As the section thaws it produces a bit of moisture and thereby adheres to the emulsion.

Bone must *not* be decalcified for this process. In some of the best techniques for bone, methacrylate or epoxy embedding is used before sectioning (Arnold and Jee 1954a,b; Norris and Jenkins 1960; Woodruff and Norris 1955).

If the isotope leaches out slowly and quantitative results are not required, paraffin embedding can be undertaken, but keep the periods in fixative and alcohol solutions to a minimum. Butyl alcohol extracts less of some isotopes (^{125}I and ^{131}I) than does isopropyl or ethyl alcohol. Use xylene, not toluene, for clearing before paraffin infiltration. A brownish discoloration of the sections indicates improper infiltration with paraffin. Vacuum infiltration will correct this, but if vacuum is not available when the sections are mounted on slides later, leave them on a hot plate or run them through a Bunsen burner just long enough to melt the paraffin. Check for leaching by counting the reagents after use in an appropriate counter. If a significant count is given off, avoid the use of that reagent. When the tissue is embedded in paraffin, leaching is no longer a problem.

Slides used for mounting must be clean; dust can cause air bubbles and distortion of the emulsion. To clean, soak the slides in cleaning solution. Wash them overnight in running tap water; one hour in distilled water, changing the solution several times; and in 2 changes of 80% alcohol. Dry. They also can be cleaned by boiling in Alconox. Treat with subbing solution (p. 469) and dry. Store the subbed slides in dust-free boxes and use within a month. In the darkroom slides with single-side frosted ends help to determine the side holding the sections. Slides with frosted ends can be used and identification applied with pencil. Fisher Scientific Prob-on slides (cat. no. 15-188-152) are excellent for emulsion attachment and they can be used without cleaning.

Mount the sections, drain off the water, dry for 10 to 15 minutes on a slide warmer, and store in dust-free boxes until application of the emulsion. Subbed slides do not require as long a drying period as do albumenized slides, and the sections and emulsion adhere better to the subbed surface than to albumen. (*Suggestion:* Mount subbed slides from a water bath or place a puddle of water on slide before adding sections. Do not try to place sections on dry subbed slides and then add water.)

Sometimes staining is used before emulsion application, but many stains are affected by photographic processing and some stains may remove the tracer. If staining is done first, a protective coating (celloidin, Formvar, or nylon) usually has to be applied to prevent decolorization during development. This layer will reduce the amount of radiation that reaches the emulsion from weak emitters, for example, tritium and ^{14}C. Feulgen stain is commonly used before emulsion is applied, but most staining is done after development.

Emulsion Application

The two methods of application are the *stripping film* and the *liquid emulsion* or *dipping* methods. Stripping film is a little more difficult to master than dipping, but it produces a more uniform emulsion. The dipping technique is given in detail here, since it is simpler and less likely to go wrong.

Stripping Film Method

Stripping film emulsion must first be removed from its glass plate. To accomplish this, cut the edges of the emulsion with a razor blade, peel off the emulsion from the glass plate, and float it on distilled water with 1% Dupanol (wetting agent) added. Then slip the slide with the section uppermost under the emulsion and lift it out, with the emulsion. To hold the emulsion in place without slipping, fold it under three sides of the slide, dry it in place, and make the exposure (Bogoroch 1951; Pelc 1956; Simmel 1957).

Liquid Emulsion (Dipping) Method

1. Store liquid emulsion at 5°C; 4 months is maximum time. Remove to room temperature, but do not open until in darkroom. The emulsion is a gel at room temperature and has to be melted in a water bath at 42–45°C. This must be done in the darkroom under Wratten series no. 1 safelight, red filter, 15 watt bulb. Work no closer than 3 feet from light.
2. When emulsion is melted pour the amount to be used into the dipping vessel. Never reuse emulsion. For a few slides a Coplin jar is convenient;

if it is filled to the top of the slots, the depth will be sufficient for most of the slides. Leave at 42–45°C for 30 to 45 minutes to attain proper fluidity and to allow bubbles to escape.
3. Test each new bottle of emulsion. Dip a blank slide slowly into the emulsion and withdraw it with a slow uniform motion (4 to 5 seconds total). Check it for lumpiness, uneven coating, and bubbles. Drain against tissue or gauze. Wipe the back dry and air dry in a vertical position in a light-tight box: 1 to 2 hours. Develop and fix (see below) and check for background. If the background is objectionable, see Caro (1964) and Brenner (1962) for ways of reducing it.
4. Squashes, spreads, and so forth can be dipped as is, but paraffin must be removed from sections. Use 2 changes of xylene, remove the xylene with 2 changes of absolute alcohol, and hydrate the sections to water. If Carbowax has been used, remove it in water. The slides with mounted tissue are dipped as described above in step 3. Two slides can be dipped back to back. Coleman (1965) dips 5 at a time by using a molded plastic holder and completes all processing in the holder (Fig. 29-1). To reduce cooling of the emulsion, arrange the slides back to back in an empty Coplin jar in the water bath to warm them before dipping.
5. Drain and separate slides if they have been dipped back to back. Slides in plastic holder can be stood on end, holder uppermost, to drain. Prescott (1964) arranges them in a neoprene-coated test tube rack, where they fit diagonally across the square openings. Place in a light-tight box for drying overnight. Allow them to dry slowly. Do not use hot air; fast drying increases the background (Sawicki and Pawinska 1965) and 40–50% humidity is desirable. An 80% relative humidity incubator for slow drying is recommended by Kopriwa and Leblond (1962).

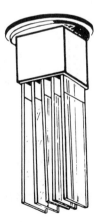

Figure 29-1. Slide holder for autoradiograph processing.

For supporting slides while drying, use lengths (12 to 15 inches) of wood or plastic (1.5 to 2 inches square) lying on large blotters. The slides lean in a vertical position, backside against the wood, and drain on the blotter. For a small number of slides, the side of a closed wooden or plastic slide box is a satisfactory back rest. A cool darkroom with water running in the sink usually provides an adequate drying atmosphere, but must be completely dark.

6. The slides must be dry before they are boxed for exposure, and they must be kept dry, because a wet emulsion is practically insensitive. When dry, transfer them to Bakelite slide boxes (25 slide capacity) no more than 8 to 9 to a box, allowing 2 to 3 slots of spacing between slides. Place 15 to 25 g of Drierite in a tissue or gauze bag in the bottom of the box. Do not allow Drierite to get on the emulsion. Encapsulated desiccants are commercially available and they can be rejuvenated in a 60–70°C oven or discarded after use. Seal the four sides of the box with two lengths of black photographic tape and wrap in aluminum foil. To prevent static electricity, put the lid on the box of dipped slides before unrolling the tape, or do it before you turn off the light. Store in refrigerator, 4°C during exposure. Keep the boxes standing on end with the slides in horizontal position, emulsion side up.

Slides in a molded plastic holder (Fig. 29-1) can be wrapped in several layers of heavy foil for exposure.

7. At intervals of 1, 2, 3, or more weeks, test slides can be removed from the exposure box and be developed. If the slides are cold, allow them to reach darkroom temperature before development. All solutions should be of uniform temperature, 17–18°C. At temperature above 20°C, reticulation occurs in the emulsion.

   a. Develop in D11, D19, or Dektol: 5 minutes.
   b. Rinse in water to stop development (Note: An acid stop bath, such as 1% acetic acid, is not recommended for use with nuclear emulsions. The acetic acid may react with the sodium carbonate in developers to form carbon dioxide gas, thereby forming microscopic bubbles within the emulsion layer): 10 seconds.
   c. Transfer to Kodak Rapid Fixer: 5 minutes. (Some stains react better with the tissue if a hardener is not included in the fixer.)
   d. Wash in running water: 10 to 15 minutes.
   e. Rinse in distilled water.
   f. Dry. Drying is accomplished most effectively by natural evaporation at standard room temperature and humidity conditions in a dust-free environment to lessen positional changes of the silver grains. Rapid drying could lead to cracking and loss of adhesion of the emulsion.
   g. Store in dust-free box, or proceed to stain.

   Autoradiography is also used in in situ hybridization methods. In this technique, a radioactive probe (usually DNA complimentary to a specific messenger RNA) is applied to the tissue section already on the slide. The slides are subsequently dipped in Kodak NTB2 or Ilford K5 emulsion and processed as above.

General references for autoradiography are Boyd (1955), Caro (1964), Controls for Radiation instruction manual, Curtis et al. (1981), Edwards (1988), Fitzgerald et al. (1953), Joftes (1959), Joftes and Warren (1955), Pelc (1958), Perry (1964), Prescott (1964), and Sacks (1965).

For dry mounting of water-soluble materials, see Branton and Jacobson (1961), Fitzgerald (1961), Gallimore et al. (1954), Miller et al. (1964), Sterling and Chichester (1956), and Stumpf and Roth (1964).

For double emulsion techniques, see Baserga (1961), Baserga and Nemeroff (1962), and Pickworth et al. (1963).

See Durie and Salmon (1975) for high-speed scintillation autoradiography.

Staining

For most experimental work, 5 to 10 minutes staining with Mayer hematoxylin or Kernechtrot provides satisfactory backgrounds for ARGs. The red of Kernechtrot is a good contrast color for the black of the developed silver grains. Sams and Davies (1967) add Alcian blue and tartrazine to their staining sequence for additional colors. Gude stain (below) is excellent. Keep all solutions cool; never use warm ones.

After mounting ARGs in synthetic resin, dark opaque areas occasionally develop. The resin, instead of penetrating the emulsion, is drawing the clearing agent out of the tissue and leaving it opaque. If this happens, return the slides to the clearing agent (xylene or toluene) long enough to remove all the mounting resin, and the tissues will look clear again. Then transfer the slides through the following solutions:

1. Xylene/resin (2:1) for 1 to 2 hours.
2. Xylene/resin (1:1) for 1 to 2 hours.
3. Xylene/resin (1:2) for 1 to 2 hours.

Mount.
The slides can be left overnight in any of the solutions.

Giemsa Stain (Modification of Gude et al. 1955)

PROCEDURE
1. Air dry slides of smears, sections, chromosomes, etc.
2. Stain 1 to 2 hours in freshly mixed:

    Giemsa stock solution .... 2.5 ml
    absolute methyl alcohol .... 3.0 ml

            distilled water .... 100.0 ml
            $M/10$ citric acid .... 11.0 ml
            $M/5$ disodium phosphate .... 6.0 ml

   Use only once.
3. Rinse off stain in distilled water.
4. Air dry and cover with synthetic medium and cover glass.

RESULT

The light blue staining of nuclear material furnishes a good contrast for the black-silver grains of the chromatin tag.

COMMENTS

If this stain is used on tissue sections, follow step 3 with 3 changes of isopropanol, 2 to 3 minutes each; xylene-isopropanol, 2 to 3 minutes; xylene, 2 changes, 2 to 3 minutes each (xylene-resin mixes if necessary); and mount. Humason recommended Gurr's stain as the best Giemsa for ARGs.

Excess emulsion can be removed from the back of slides with a razor blade.

Other stains may be used: Darrow red-light green (Wolberg 1965), methylene blue-basic fuchsin (Bélanger 1961), nuclear fast red-indigo carmine (Mortreuil-Langlois 1962), peroxide (Popp et al. 1962), toluidine blue-methyl green-pyronine (Stone and Cameron 1964), and metanil yellow-hematoxylin (Simmel et al. 1951). Celestine blue and gallocyanin have been recommended, but Deuchar (1962) reports a dense overall graining of film with celestine blue, and Stenram (1962) found that gallocyanin causes a loss of silver grains. See Thurston and Joftes (1963) for a review of staining.

The Use of Isotopes in Chromosome Studies

The isotope in popular use for chromosome studies is tritium (^{3}H). It emits only beta radiation, and in water the maximum range of its emitted electrons is 3 μm. If an emulsion lies over a point-like source containing tritium, only an area averaging 3 μm in diameter will be exposed. Tritium, therefore, has become the isotope of choice for tagging chromosomes.

Mitotic chromosomes cannot be labeled, since DNA is not replicated during mitosis. But mitosis is preceded by the synthesis of DNA, and if radioactive DNA precursor is administered during synthesis the nucleus becomes radioactive and retains the tag during mitosis. It also passes the tag on to its daughter cells. Thymidine (deoxyriboside of thymine, one of the bases of DNA) can be effectively incorporated into DNA, and when it is labeled with tritium it enters the synthetic chain to label the DNA. To obtain efficiently labeled metaphases, therefore, the radioactive thymidine must be given to the cells several hours before they divide, that is, during the DNA-synthetic or S-period. This period varies but is species-specific. Different types of cultures may influence the length of periods—the S-period and the lag period that follows. The amount of thymidine influences the rate of uptake, and incubation temperature is critical. It is necessary to explore all these influences when first using thymidine on cultured cells.

Chromosome Cultures

PROCEDURE
1. Add tritiated thymidine (Nuclear Chicago Corp., New England Nuclear Corp., Schwarz BioResearch Inc.) to a final concentration of 1 μc/ml (specific active 1.9 curies/mmole): 5 to 6 hours before termination of the culture. Maintain culture at 37°C.
2. Harvest cells at 3, 4, 5, and 6 hours (also 7 and 8 if desired) after introduction of thymidine, but add colchicine or Colcemid 2 hours before each harvesting.
3. The thymidine is removed with the medium. Treat with hypotonic solution: 10 minutes.
4. Fix, squash, or spread as usual (p. 414). Stain some preparations if desired.
5. Dried spreads or squashes can be stored at 4°C for about 4 weeks. At room temperature background develops.
6. Coating or stripping techniques may be done with NTB2 (Kodak) or Ilford K5 emulsion.
7. Expose 2 days.
8. Develop: 5 minutes. Rinse, fix, and wash.
9. Dry and stain in Gude's Giemsa.

COMMENTS

Caution: Handle the tritium thymidine with care. Since it does not emit gamma radiation, contamination cannot be easily detected with scintillation counters. Work with the labeled thymidine only on protected areas, covered with absorbent paper backed with neoprene or a large sheet of heavy aluminum foil or plastic under a large blotter.

For details about the use of tritiated thymidine, see Frøland (1965), Leblond et al. (1959), Schmid (1965), and Sisken (1964).

PART V
Solution Preparation and General Laboratory Aids

30
Solution Preparation

ABBREVIATIONS AND TERMS

Laboratory solutions, defined in terms of the methods of measurement used in their preparation, can be grouped as molecular solutions, normal solutions, and percentage solutions. In discussing these the following abbreviations are used:

ml = milliliter
cc = cubic centimeter
g = gram

mg = milligram
M = molecular solution
N = normal solution

Molecular Solutions

A molecular solution (M) contains the molecular weight appearing on the manufacturer's label or can be determined from the periodic table of the elements) in grams of the substance made up to 1 liter with distilled water (Cowdry 1952). For example: M oxalic acid $COOH_2 \cdot 2H_2O$ is 126 (molecular weight) grams in 1 liter of water.

Molecular weight expressed in grams is called the gram-molecular weight or mole. A millimole is $1/1000$ of a mole.

Normal Solutions

A normal solution (N) contains 1 gram-molecular weight of dissolved substance divided by the hydrogen equivalent of the substance (i.e., 1 gram equivalent) per liter of solution. N oxalic acid is half the concentration of the M solution above.

Percentage Solutions

In the preparation of percentage solutions, it should be made clear whether the percentage is determined by weight or volume. The percentage should be written out in grams and milliliters, or it should be expressed as follows:

w/v = weight in grams in a 100-ml volume of solution
v/v = volume in milliliters in a 100-ml volume of solution

Although it is erroneous, a long-established practice of technicians is to dilute liquids as though the reagent solution was 100% concentration. In most procedures the correct dilution is indicated in parentheses after the percentage desired. That is, a 1% solution of acetic acid is 1 ml of glacial acetic acid in 99 ml of distilled water. Or consider the formula:

100 ml = final volume
× 0.01 = required percentage
therefore, 1.00 amount of concentrated acid needed

The table below gives dilutions for 1 N solutions that are sufficiently accurate for histological preparations. To make a normal solution, add enough distilled water to the number of milliliters shown in the right-hand column to make a combined total of 1 liter.

Table 30-1. Dilutions for 1 N Solutions*

Reagent	Molecular weight	Percentage assay	Specific gravity	g per liter	ml per liter
acetic acid (CH$_3$COOH)	60.05	99.7–*100*†	1.0498	1050	57.2
		99.0	1.0524	1042	57.6
		98.0	1.0549	1034	58.0
ammonium hydroxide (NH$_4$OH)	17.03	26	0.904	235	72.4
		28†	0.8980	251.4	67.7
		30	0.8920	267.6	63.63
formic acid (HCOOH)	46.03	96	1.217	1180	39.3
		98†	1.2183	1194	38.4
		99	1.2202	1208	38.0
		100	1.2212	1221	37.6
hydrochloric acid (HCl)	36.46	36	1.1789	424.4	85.9
		37–38†	1.188–1.192	451.6	80.4
		40	1.1980	479.2	76.0
nitric acid (HNO$_3$)	63.02	69	1.4091	972.3	64.8
		70†	1.4134	989.4	63.6
		71	1.4176	1006	62.6
		72	1.4218	1024	61.5
sulfuric acid (H$_2$SO$_4$)	98.075	95	1.8337	1742	28.1
		96†	1.8355	1762	27.8
		97	1.8364	1781	27.5

*Calculated from Norbert Adolph Lange, *Handbook of Chemistry* (Sandusky, Ohio: Handbook Publishers, Inc., 1956). *Italicized* figures were used in calculations.
† Indicates most common form of the reagent. In addition, a calculation has been made for the nearest lower and higher percentages.

STOCK SOLUTIONS

Solutions normally found on histology laboratory shelves are referred to as stock solutions. Included here are solutions for general use, physiological so-

General Purpose Solutions

Acid Alcohol

> 70% alcohol .... 100.0 ml
> hydrochloric acid, concentrated .... 1.0 ml

Alkaline Alcohol

> 70% alcohol .... 100.0 ml
> ammonia, concentrated .... 1.0 ml

Or use saturated sodium or lithium bicarbonate.

Carbol-Xylol

> phenol (carbolic acid), melted .... 1 part
> xylene .... 3 parts

During use, keep covered to reduce evaporation of the xylene.
For creosote-xylol, use beechwood creosote in place of phenol. For aniline-xylol use aniline.

Gold Chloride Stock Solution

> gold chloride (15 grains) .... 1.0 g
> distilled water .... 100.0 ml

Use only 3 to 4 times; gold content is reduced at each use.

Lugol Solution

There are various formulas to which this name has been applied; the following are frequently used.

> *Strongest concentration*
> iodine crystals .... 1.0 g
> potassium iodide .... 2.0 g
> distilled water .... 12.0 ml

> *Weigert variation*
> iodine crystals .... 1.0 g
> potassium iodide .... 2.0 g
> distilled water .... 100.0 ml

Earlier in the passage: lutions, environmental solutions, a few adhesive and buffer solutions, and some stain solubilities.

Gram variation

iodine crystals	1.0 g
potassium iodide	2.0 g
distilled water	300.0 ml

For all solutions, first dissolve the potassium iodide, and then the iodine will go into solution readily.

Chatton Agar for Blocking Small Organisms (Gray 1954)

agar	1.3 g

Add agar to:

boiling distilled water	100.0 ml

Stir until dissolved, then add:

formaldehyde (37–40%)	2.5 ml

Store in refrigerator.

Scott Solution

sodium bicarbonate	2.0 g
magnesium sulfate	20.0 g
distilled water	1000.0 ml

Add a pinch of thymol to retard molds.

Sodium Thiosulfate Solution (Hypo), 5%

sodium thiosulfate ($Na_2H_2O_2$)	5.0 g
distilled water	100.0 ml

Physiological Solutions (Balanced Salt Solutions, BSS)

Several physiological solutions should be available to satisfy the needs of different types of research. These solutions are useful for land vertebrates and can be imperative at times. A normal or *isotonic* solution is one that contains the right amount of salts to maintain tissues in a normal condition. For example, red blood cells in isotonic fluid will remain unaltered in form and will not lose their hemoglobin, because the osmotic pressure and salt content of both the solution and the blood fluid are the same. If the solution is *hypotonic*, the blood cells will swell, because the osmotic pressure and salt content are less than that of the blood fluid. If the solution is *hypertonic*, the blood cells will shrink, because the osmotic pressure and salt content are greater than that of the blood fluid. For example, distilled water is hypotonic without the addition of the correct quantity of salt.

Physiological Saline

mammals	0.9% (9 g NaCl/1000 ml water)
birds	0.75% (7.5 g NaCl/1000 ml water)
salamanders	0.8% (8 g NaCl/1000 ml water)
frogs	0.64% (6.4 g NaCl/1000 ml water)
teleosts	0.67% (6.7 g NaCl/1000 ml water)
insects	0.6–0.8% (6 to 8 g NaCl/1000 ml water)
invertebrates	0.75% (7.5 g NaCl/1000 ml water)

Earle Solution

sodium chloride	0.68 g
calcium chloride	0.02 g
magnesium sulfate	0.01 g
potassium chloride	0.04 g
sodium bicarbonate	0.014 g
sodium phosphate, monobasic	0.22 g
glucose	0.1 g
distilled water	100.0 ml

Hank Solution

	(1)	(2)
sodium chloride	0.8 g	0.8 g
calcium chloride	0.02 g	0.014 g
magnesium sulfate	0.02 g	0.02 g
potassium chloride	0.04 g	0.04 g
potassium phosphate, monobasic	0.01 g	0.04 g
sodium bicarbonate	0.127 g	0.035 g
sodium phosphate, dibasic	0.01 g	0.006 g
glucose	0.2 g	0.1 g
distilled water	100.0 ml	100.0 ml

If formula 1 gives too strong a solution, try formula 2.

Locke Solution

	(1)	(2)
sodium chloride	0.9 g	0.95 g
potassium chloride	0.042 g	0.02 g
sodium bicarbonate	0.03 g	0.02 g
calcium chloride	0.024 g	0.02 g
glucose	0.1 g	0.1 g
distilled water	100.0 ml	100.0 ml

Which of the two formulas is used is a matter of personal preference. In both, decrease the sodium chloride content to 0.65 g for cold-blooded organisms, to 0.85 g for birds.

Ringer Solution

sodium chloride	0.9 g
potassium chloride	0.042 g
calcium chloride	0.025 g
distilled water	100.0 ml

Best prepared fresh. For cold-blooded organisms, use 0.65 g of sodium chloride; for birds, 0.85 g.

Environmental Solutions

Just as physiological solutions are useful for land vertebrates, environmental solutions are convenient for some invertebrates. In the case of marine invertebrates, such solutions can become necessary for successful preparation and sometimes are recommended as the basic fluid for fixatives. Among the physical properties of sea water, the salt content must be considered; it varies and is complex, but an arbitrary definition of salt content (salinity) has been calculated. Salinity (S 0/00) equals the weight in grams (in vacuo) of solids in 1 kg of sea water. The major constituent, which is easily determined, is a silver-precipitating halide. Chlorinity (Cl 0/00) is defined as the weight in grams (in vacuo) of chlorides in 1 kg of sea water. Standard sea water is about 34.3243 0/00 salinity and 19.4 0/00 chlorinity (Barnes 1954).

Artificial Sea Water (Hale 1958)

Chlorinity, 19.0/00; salinity, 34.33 0/00. The weights in the right-hand column are for anhydrous form of the salt; those in parentheses include crystallization water as indicated on the reagent bottle.

$NaCl$	23.991 g
KCl	0.742 g
$CaCl_2$	1.135 g
($CaCl_2 \cdot 6H_2O$	2.240 g)
$MgCl_2$	5.102 g
($MgCl_2 \cdot 6H_2O$	10.893 g)
Na_2SO_4	4.012 g
($Na_2NO_4 \cdot 10H_2O$	9.1 g)
$NaHCO_3$	0.197 g
$NaBr$	0.085 g
($NaBr \cdot 2H_2O$	0.115 g)

SrCl$_2$ .... 0.011 g
(SrCl$_2 \cdot$ 6H$_2$O .... 0.018 g)
H$_3$BO$_3$ .... 0.027 g

Dissolve in enough distilled water to make 1 liter.

This solution is to be used only for technical purposes, not for aquaria, for which a proper pH condition is critical and must be maintained (Kelley 1965).

Sea salts (Instant Ocean or other commercially available sea salts) may be purchased and reconstituted by adding water to produce sea water (vendor, Carolina Biological Supplies, Burlington, NC).

Synthetic Spring Water (Kirby 1947)

	(1)	(2)
Na$_2$SiO$_3$	15.0 mg	100.0 mg
NaCl	12.0 mg	12.0 mg
Na$_2$SO$_4$	6.0 mg	6.0 mg
CaCl$_2$	6.5 mg	6.5 mg
MgCl$_2$	3.5 mg	3.5 mg
FeCl$_3$	4.0 mg	4.0 mg
distilled water	1000.0 ml	1000.0 ml

Adjust with HCl to pH 6.8 to 7.0.

Use formula 1 for first transfer from nature; for succeeding transfers use formula 2.

Adhesive Solutions

Not all of the adhesive or buffer solutions will be found or required in every laboratory; however, it is practical to keep formulas of the more widely used solutions at hand, so that the information is available if it is needed.

Mayer Albumen Fixative

Beat white of an egg with an egg beater until well broken up, but not stiff, and pour into a tall cylinder. Let stand until the air brings suspended material to the top (overnight). Pour off liquid from bottom and to it add an equal volume of glycerin. A bit of sodium salicylate, thymol, Merthiolate, or formalin (1:100) will prevent growth of molds. Filter the solution through gauze.

Faulkner and Lillie (1945b) substitute dried egg white. A 5% solution of dried egg white in 0.5% NaCl is shaken at intervals for one day. Do not allow it to froth. Filter in a small Büchner funnel with vacuum. Add an equal amount of glycerin and 0.5 ml of 1:10,000 Merthiolate to each 100 ml of solution.

Masson Gelatin Fixative

gelatin .... 50.0 mg
distilled water .... 25.0 ml

This is recommended by many for alkaline silver techniques, when sections tend to float off during or after impregnation. Float sections on solution on slide, and place on warm plate. When sections have spread, drain excess gelatin and blot dry with filter paper. Place in formalin vapor, 40–50°C: overnight.

Haupt Gelatin Fixative (1930)

gelatin .... 1.0 g
distilled water .... 100.0 ml

Dissolve at 30°C (not above) in water bath or oven. Add:

phenol (carbolic acid) crystals .... 2.0 g
glycerol .... 15.0 ml

Stir well and filter.

Use 2% formalin when mounting sections. This hardens the gelatin; water is not adequate. Some may find the formalin fumes irritating to the eyes and nostrils.

Haupt suggests that if sections tend to loosen, a small uncovered dish of concentrated formalin be placed in an oven with the slides while they are being dried. The formalin tends to make the gelatin insoluble and helps to hold the sections in place.

Bissing's (1974) method reduces the formalin fumes. Combine 1 part of above adhesive with 140 parts of 3% formalin (prepare in distilled water). Flood clean slides with the mixture and float the sections on it. Spread on a warm plate, drain off fluid, and dry for 1 hour.

Weaver Gelatin Fixative (1955)

Solution A
gelatin .... 1.0 g
calcium proprionate .... 1.0 g
Roccal (1% benzalkonium chloride) .... 1.0 ml
distilled water .... 100.0 ml

Solution B
chrome alum $[Cr_2K_2(SO_4)_4 \cdot 24H_2O]$ .... 1.0 g
distilled water .... 90.0 ml
formaldehyde (37–40%) .... 10.0 ml

Solution Preparation

Mix 1 part of A with 9 parts of B. Flood slide with adhesive mixture, add paraffin ribbons, and allow to stretch as usual. Drain off excess adhesive and blot. Wipe edges close to paraffin. Deposits of adhesive should be removed because they pick up stain. Good for sections that are difficult to affix.

"Subbed" Slides (Boyd 1955)

Dissolve 1 g of gelatin in 1 liter of hot distilled water on magnetic stirrer-hot plate combination. Cool and add 0.1 g of chromium potassium sulfate. Store in refrigerator. Dip slides several times in the solution. Drain and dry in a vertical position. Store in dust-free box.

Papas (1971) uses 5 g gelatin and 0.5 g chrome alum, but it seems that the added strength is not necessary.

Land Adhesive (Gray 1954)

Solution A

gum arabic	0.5 g
distilled water	50.0 ml

Solution B

potassium dichromate	0.5 g
distilled water	100.0 ml

Smear a small amount of A on a slide. Flood slide with solution B, add sections, and warm to flatten. Drain off fluid against cleansing tissue and dry. Bilstad (personal communication) dries slides in bright sunlight for several hours; she found it more effective than artificial light.

Buffer Solutions

The following are a few of the chemicals most commonly used for buffer solutions. For others consult a chemical handbook.

Table 30-2. Molecular Weights of Buffer Ingredients

acetic acid, CH_3COOH	60.05
borax, $Na_2B_4O_7 \cdot 10H_2O$ (sodium tetraborate)	381.43
boric acid, $B(OH)_3$	61.84
citric acid (anhydrous), $C_3H_4(OH)(COOH)_3$	192.12
citric acid crystals, $C_3H_4(OH)(COOH)_3 \cdot H_2O$	210.14
formic acid, HCOOH	46.03
hydrochloric acid, HCl	36.465
maleic acid, HOOCCH-CHOOH	116.07
potassium acid phosphate, KH_2PO_4	136.09
	(continued)

Table 30-2. (continued)

sodium acetate, CH_3COONa	82.04
sodium acetate, crystals, $CH_3COONa \cdot 3H_2O$	136.09
sodium barbital, $C_8H_{11}O_3N_2Na$	206.18
sodium bicarbonate, $NaHCO_3$	84.02
sodium carbonate, $NaCO_3$	106.00
sodium citrate, crystals, $C_3H_4OH(COONa)_3$ $\begin{cases} \cdot 5H_2O \\ 5\frac{1}{2}H_2O \end{cases}$	348.17 / 357.18
sodium citrate, granular, $C_3H_4OH(COONa)_3 \cdot 2H_2O$	294.12
sodium hydroxide, $NaOH$	40.005
sodium phosphate, monobasic, $NaH_2PO_4 \cdot H_2O$	138.01
sodium phosphate, dibasic, Na_2HPO_4 $\begin{cases} 7H_2O \\ \text{anhydrous} \end{cases}$	268.14 / 141.98
sulfuric acid, H_2SO_4	98.082
tris (hydroxymethyl) aminomethane $C_4H_{11}NO_3$	121.14

0.2 M Acetate Buffer (Gomori)

STOCK SOLUTIONS

Acetic acid
12.0 ml made up to 1000 ml with distilled water.

Sodium acetate
27.2 g made up to 1000 ml with distilled water.

Add a few crystals of camphor to both solutions. For desired pH mix correct amounts as indicated below:

Table 30-3.

pH	Acetic acid (ml)	Sodium acetate (ml)
3.8	87.0	13.0
4.0	80.0	20.0
4.2	73.0	27.0
4.4	62.0	38.0
4.6	51.0	49.0
4.8	40.0	60.0
5.0	30.0	70.0
5.2	21.0	79.0
5.4	14.5	85.5
5.6	11.0	89.0

Acetate-Acetic Acid Buffer (Walpole)

STOCK SOLUTIONS

M/5 *acetic acid*
12.0 ml (99% assay) made up to 1000 ml with distilled water.

M/5 *sodium acetate*
27.2 g made up to 1000 ml with distilled water.
For desired pH mix correct amounts as indicated below.

Table 30-4.

pH	M/5 acetic acid (ml)	M/5 sodium acetate (ml)
2.70	200.0	0.0
2.80	199.0	1.0
2.91	198.0	2.0
3.08	196.0	4.0
3.14	195.0	5.0
3.20	194.0	6.0
3.31	192.0	8.0
3.41	190.0	10.0
3.59	185.0	15.0
3.72	180.0	20.0
3.90	170.0	30.0
4.04	160.0	40.0
4.16	150.0	50.0
4.27	140.0	60.0
4.36	130.0	70.0
4.45	120.0	80.0
4.53	110.0	90.0
4.62	100.0	100.0
4.71	90.0	110.0
4.80	80.0	120.0
4.90	70.0	130.0
4.99	60.0	140.0
5.11	50.0	150.0
5.22	40.0	160.0
5.38	30.0	170.0
5.57	20.0	180.0
5.89	10.0	190.0
6.21	5.0	195.0
6.50	0.0	200.0

0.05 M Barbital Buffer (Gomori)

STOCK SOLUTION

Sodium barbital

1.03 g in 50 ml distilled water.

To 50 ml sodium barbital add 0.1 N HCl according to the table below to obtain desired pH; dilute to total of 100 ml.

Table 30-5.

pH	0.1 N HCl (ml)
8.7	5.0
8.5	7.5
8.3	11.0
8.1	15.0
7.9	19.0
7.65	26.0
7.45	31.0
7.3	36.0
7.15	41.0
6.9	43.5

Boric Acid-Borax Buffer (Holmes)

STOCK SOLUTIONS

M/5 *boric acid*

12.368 g made up to 1000 ml with distilled water.

M/20 *borax*

19.071 g made up to 1000 ml with distilled water.

For desired pH mix correct amounts as indicated below.

Table 30-6.

pH	M/5 boric acid (ml)	M/20 borax (ml)
7.4	90.0	10.0
7.6	85.0	15.0
7.8	80.0	20.0
8.0	70.0	30.0
8.2	65.0	35.0
8.4	55.0	45.0
8.7	40.0	60.0
9.0	20.0	80.0

0.2 M Phosphate Buffer (Gomori)

STOCK SOLUTIONS

Monobasic sodium phosphate
27.6 g made up to 1000 ml with distilled water.

Dibasic sodium phosphate
53.6 g made up to 1000 ml with distilled water.
For desired pH mix correct amounts as indicated below.

Table 30-7.

pH	Monobasic sodium phosphate (ml)	Dibasic sodium phosphate (ml)
5.9	90.0	10.0
6.1	85.0	15.0
6.3	77.0	23.0
6.5	68.0	32.0
6.7	57.0	43.0
6.9	45.0	55.0
7.1	33.0	67.0
7.3	23.0	77.0
7.4	19.0	81.0
7.5	16.0	84.0
7.7	10.0	90.0

Phosphate Buffer (Sörensen)

STOCK SOLUTIONS

M/15 dibasic sodium phosphate
9.465 g made up to 1000 ml with distilled water.

M/15 potassium acid phosphate
9.07 g made up to 1000 ml with distilled water.

For desired pH mix correct amounts as indicated on the facing page: some techniques will require dilution with up to 1000 ml of water.

Table 30-8.

pH	M/15 dibasic sodium phosphate (ml)	M/15 potassium acid phosphate (ml)
5.29	2.5	97.5
5.59	5.0	95.0
5.91	10.0	90.0
6.24	20.0	80.0
6.47	30.0	70.0

(*continued*)

Table 30-8. (continued)

pH	M/15 dibasic sodium phosphate (ml)	M/15 potassium acid phosphate (ml)
6.64	40.0	60.0
6.81	50.0	50.0
6.98	60.0	40.0
7.17	70.0	30.0
7.38	80.0	20.0
7.73	90.0	10.0
8.04	95.0	5.0

Standard Buffer (McIlvaine)

STOCK SOLUTIONS

0.1 M citric acid (anhydrous)
19.212 g made up to 1000 ml with distilled water.

0.2 M sodium phosphate (anhydrous)
28.396 g (· 7H_2O, 52.628 g) made up to 1000 ml with distilled water.
For desired pH mix correct amounts as indicated below.

Table 30-9.

pH	Citric acid (ml)	Disodium phosphate (ml)
2.2	19.6	0.4
2.4	17.76	1.24
2.6	17.82	2.18
2.8	16.83	3.17
3.0	15.89	4.11
3.2	15.06	4.94
3.4	14.3	5.7
3.6	13.56	6.44
3.8	12.9	7.1
4.0	12.29	7.71
4.2	11.72	8.28
4.4	11.18	8.82
4.6	10.65	9.35
4.8	10.14	9.86
5.0	9.7	10.3
5.2	9.28	10.72

(continued)

Table 30-9. (continued)

5.4	8.85	11.15
5.6	8.4	11.6
5.8	7.91	12.09
6.0	7.37	12.63
6.2	6.78	13.22
6.4	6.15	13.85
6.6	5.45	14.55
6.8	4.55	15.45
7.0	3.53	16.47
7.2	2.61	17.39
7.4	1.83	18.17
7.6	1.27	18.73
7.8	0.85	19.15
8.0	0.55	19.45

0.2 M Tris Buffer (Hale)

STOCK SOLUTIONS

0.2 M tris (hydroxymethyl) aminomethane
24.228 g made up to 1000 ml with distilled water.

0.1 N HCl (38% assay)
8.08 ml made up to 1000 ml with distilled water.

To 25 ml 0.2 M tris add 0.1 N HCl as indicated in the table below and dilute to 100 ml.

Table 30-10.

pH	0.1 N HCl (ml)
7.19	45.0
7.36	42.5
7.54	40.0
7.66	37.5
7.77	35.0
7.87	32.5
7.96	30.0
8.05	27.5
8.14	25.0
8.23	22.5

(continued)

Table 30-10. *(continued)*

pH	0.1 N HCl (ml)
8.32	20.0
8.41	17.5
8.51	15.0
8.62	12.5
8.74	10.0
8.92	7.5
9.1	5.0

Tris Maleate Buffer (Gomori)

STOCK SOLUTION
- maleic acid .... 29.0 g
- tris (hydroxymethyl) aminomethane .... 30.3 g
- distilled water .... 500.0 ml

Add 2 g charcoal, shake, let stand 10 minutes, and filter.

To 40 ml of stock solution add N NaOH (4%) as indicated below and dilute to 100 ml.

Table 30-11.

pH	Sodium hydroxide (ml)
5.8	9.0
6.0	10.5
6.2	13.0
6.4	15.0
6.6	16.5
6.8	18.0
7.0	19.0
7.2	20.0
7.6	22.5
7.8	24.2
8.0	26.0
8.2	29.0

BUFFERED SALINE SOLUTIONS

Phosphate buffered saline (PBS), pH 7.4
Add 13.40 g of sodium phosphate dibasic, heptahydrate and 8.0 g of sodium chloride to 900 ml of distilled water. Stir until disolved. Adjust to pH 7.4 by slowly adding 2 N HCl to mixture. Bring to final volume of 1 L with distilled water. Refrigerate.

Tris buffered saline (TBS), pH 7.5–7.6
Add 7.45 g Trizma (trihydroxymethyl amino methane and tris hydrochloride) and 8.75 g NaCl to 1 L of distilled water. Stir. Refrigerate.

STAIN SOLUBILITIES

Stain solubilities are useful only occasionally, but they are definitely needed when a saturated solution must be made. Solubilities of different batches of dyes can vary, but the table on the next three pages can be consulted when preparing saturated solutions. Certain dyes whose solubilities were not available are not included, such as pinacyanole (b), Darrow red (b), Sirius supra blue (a), garnet GB (b), and the azures (b). Gallocyanin is considered an amphoteric dye. For complete details concerning these dyes, see Conn, *Biological Stains* (1969) and Lillie's edition (1977).

Table 30-12. Stain Solubilites

Stain Solubilities

Stain	In water (percent)		In absolute ethyl alcohol (percent)		Basic or acidic	C.I. number
	26°C*	15°C†	26°C*	15°C†		
acid alizarine blue	—	1.0	—	0.75	a	63015
acid fuchsin	—	45.0	—	3.0	a	42685
acridine orange	—	5.0	—	0.75	b	46005
acridine yellow	—	0.5	—	0.75	b	46025
acriflavine	—	15.0	—	1.0	b	46000
Alcian blue 8GX	—	9.5	—	6.0	b	74240
alizarine red S	7.69	6.5	0.15	0.15	a	58005
amido black 10B	—	3.0	—	3.15	a	20470
aniline blue WS	—	50.0	—	0.0	a	—
auramin O	0.74	1.0	4.49	4.0	b	41000

(*continued*)

Table 30-12. *(continued)*

Stain	In water (percent) 26°C*	In water (percent) 15°C†	In absolute ethyl alcohol (percent) 26°C*	In absolute ethyl alcohol (percent) 15°C†	Basic or acidic	C.I. number
aurantia	0.0	0.1	0.33	0.55	a	10360
azocarmine G	—	1.0	—	0.1	a	50085
basic fuchsin	0.26–0.39	1.0	5.93–8.16	8.0	b	42500
Biebrich scarlet	—	5.0	0.5	0.25	a	26905
Bismarck brown Y	1.36	1.5	1.08	3.0	b	21000
brilliant cresyl blue	—	3.0	—	2.0	b	51010
celestin blue B	—	2.0	—	1.5	b	51050
chlorantine fast red	—	1.0	—	0.45	a	28160
chlorazol black E	—	6.0	—	0.1	a	30235
chromotrope 2R	19.3	19.0	0.17	0.15	a	16570
chrysoidin Y	0.86	5.5	2.21	4.75	b	11270
Congo red	—	5.0	0.19	0.75	a	22120
coriphosphine O	—	4.75	—	0.6	b	46020
cresyl violet (cresylecht violet)	0.38	9.5	0.25	6.0	b	—
crystal violet	1.68	9.0	13.87	8.75	b	42555
eosin B	39.11	10.0	0.75	3.0	a	45400
eosin (S) alc. sol. (ethyl eosin)	0.03	0.0	1.13	1.0	a	45386
eosin Y, WS	44.20	44.0	2.18	2.0	a	45380
erythrosin B	11.10	10.0	1.87	5.0	a	45430
erythrosin Y	—	8.5	—	4.5	a	45425
fast green FCF	16.04	4.0	0.35	9.0	a	42053
fluorescein (uranin)	50.02	50.0	7.19	7.0	a	45350
gallocyanin (bisulfite)	—	3.0	—	0.5	b	51030
gallocyanin (hypochlorite)	—	0.5	—	1.25	b	51030
hematein	—	1.5	—	7.5	b	75290
hematoxylin	1.75	10.0	60.0	10.0	b	75290
Janus green B	5.18	5.0	1.12	1.0	b	11050
light green SF yellowish	20.35	20.0	0.82	4.0	a	42095
Luxol fast blue MBS	—	0.0	—	3.0	a	—
malachite green	—	10.0	—	8.5	b	42000
martius yellow	4.57	1.0	0.16	0.0	a	10315
metanil yellow	5.36	5.0	1.45	1.5	a	13065
methyl blue	—	50.0	—	0.0	a	42780
methyl violet 2B	2.93	—	15.21	—	b	42535
methyl violet 6B	—	4.7	—	9.5	b	42535

(continued)

Table 30-12. (continued)

Stain	In water (percent) 26°C*	In water (percent) 15°C†	In absolute ethyl alcohol (percent) 26°C*	In absolute ethyl alcohol (percent) 15°C†	Basic or acidic	C.I. number
methylene blue	3.55	9.5	1.48	6.0	b	52015
naphthol yellow S	—	12.5	—	0.65	a	10316
neutral red	5.64	4.0	2.45	1.8	b	50040
new methylene blue	13.32	—	1.65	—	b	52030
new fuchsin	1.13	1.0	3.2	8.0	b	42520
nigrosin, alc. sol.	—	0.0	—	2.5	b	50415
nigrosin WS	—	10.0	—	0.0	a	50420
Nile blue sulfate	—	6.0	—	5.0	b	51180
oil red O	—	0.0	—	0.5	a	26125
orange G	10.86	8.0	0.22	0.22	a	16230
orcein	—	2.0	—	4.2	b	—
phloxine B	—	10.5	—	5.0	a	45410
picric acid	1.18	1.2	8.96	9.0	a	10305
Poirrier blue C4B (cotton blue)	soluble	—	soluble	—	a	42780
ponceau 2R	—	5.0	—	0.1	a	16150
ponceau S	—	1.35	—	1.2	a	27195
ponceau de xylidine	—	5.0	—	0.1	a	16150
primilin	—	0.25	—	0.03	a	49000
pyronin B	—	10.0	—	0.5	b	45010
pyronin Y	8.96	9.0	0.6	0.5	b	45005
quinoline yellow	—	0.5	—	0.0	a	47005
rhodamin B	0.78	2.0	1.47	1.75	b	45170
rhodamin 6G	—	1.5	—	6.5	b	45160
safranin O	5.45	4.5	3.41	3.5	b	50240
Sudan black B	—	0.0	—	0.25	a	26150
Sudan III	0.0	0.0	0.15	0.25	a	26100
Sudan IV	0.0	0.0	0.09	0.5	a	26105
thioflavine S	—	1.0	—	0.4	a	49010
thioflavine T	—	2.0	—	1.0	b	49005
thionin	0.25	1.0	0.25	1.0	b	52000
Titian yellow	—	1.0	—	0.02	a	19540
toluidine blue O	3.82	3.25	0.57	1.75	b	52040
trypan blue	—	1.0	—	0.02	a	23850
Victoria blue B	—	4.3	—	8.25	b	44045
Victoria blue 4R	3.23	3.0	20.49	20.0	b	42563
water blue I	—	50.0	—	0.0	a	42755
wool green S	—	4.0	—	—	a	44090

* From Conn (1969).   † From Gurr (1960).

Rosin Stock Solution (Colophonium)

> white wood rosin .... 10.0 g
> absolute alcohol .... 100.0 ml

Rosin Working Solution

> rosin stock solution .... 5.0 ml
> 95% alcohol .... 40.0 ml

31
General Laboratory Aids

LABELING AND CLEANING SLIDES

All slide labels should contain complete information: the name or number of the tissue, the stain and date. The fixative used and the thickness of the section are optional but useful information that can also be added if necessary. All additional pertinent notes related to the origin and treatment of the tissues must be impeccably recorded in a laboratory notebook.

Most laboratories use a label that is preglued and is applied by pressure. If ordinary slide labels are glued on, processed slides must be cleaned to prevent loosening of the labels. Dip the slides (when coverslip is thoroughly hardened) in water to which has been added a small amount of ammonia (or glass cleaner). Wipe dry and apply a label. Labels applied by pressure are easier to use and do not require slide cleaning.

Mounting can be untidy, and some of the resin may ooze over the cover glass. If this happens, scrape off excess resin with a razor blade if the slide is dried. Do not clean away the resin too close to the glass; in fact, leave a small band of it to protect the cover from chipping or catching against an object.

RESTAINING FADED SLIDES

Slides that have been standing for a long time or that have been exposed to bright light frequently fade. The slides can be recovered, McCormick offers the following method of bleaching them for restaining.

McCormick Method (1959a)

1. Remove cover glass by soaking slides in xylene until cover glass slips off unaided. Do not force it off; the sections might get torn.
2. Soak longer in xylene to make certain that all of the resin is removed.
3. Hydrate to water.
4. Treat with 0.5% potassium permanganate (0.5 g/100 ml water): 5 minutes.
5. Wash in running water: 5 minutes.
6. Bleach with 0.5% oxalic acid (0.5 g/100 ml water) until colorless: 1 to 2 minutes. If old stain still remains, repeat steps 4, 5, and 6.
7. Wash thoroughly in running water: 5 or more minutes.
8. Restain, using more dilute stains or staining for a shorter time. Potas-

sium permanganate and oxalic acid make tissues especially sensitive to hematoxylin and aniline nuclear stains.

RECOVERING BROKEN SLIDES

Usually it is practical to remove the bits of broken cover glass by soaking the slide in xylene overnight or longer, until the cover glass pieces slide off unaided. Then mount the broken pieces of slide on another slide (as thin a slide as obtainable) with mounting medium, just as if mounting a cover glass. Do not allow the tissue on the slide to dry out during this process. Cover with a new cover glass and mountant. Allow to dry on a warm plate overnight or several days. With a razor blade carefully clean away any mountant exuding from between the slides.

See Geil (1961) for salvaging histological sections from broken slides.

RESTORING BASOPHILIC PROPERTIES

If tissues have been improperly fixed (unbuffered formalin, overtime in Zenker fixative), stored for long periods in a fixative, or excessively decalcified or dried, the nuclei will lose some of their basophilic properties and stain poorly. After unbuffered formalin and over decalcification, extended time in hematoxylin sometimes is sufficient. But more satisfactory results will follow treatment of the deparaffinized sections in 10% aqueous sodium bicarbonate: 6 to 8 hours. This treatment definitely improves the staining of Zenker-fixed tissue. Small tissue blocks can be treated before embedding in a 5–10% aqueous solution of sodium bicarbonate overnight. Wash thoroughly in running water before staining slides or processing blocks. Luna (1968) also suggests using 5% aqueous ammonium sulfide or 5% aqueous periodic acid in the same manner. Humason preferred the bicarbonate treatment.

TWO DIFFERENT STAINS ON ONE SLIDE

Feder and Sidman Method (1957)

This method is good for a quick checking of a tissue against two stains, side by side.

1. Deparaffinize and hydrate sections to water.
2. Blot carefully, and while they are still moist, coat alternate sections with silicone grease. A soft brush may be used, or more efficiently use a 1 ml syringe and a 15 gauge needle that has been clipped off and flattened.

3. Apply first staining procedure.
4. Dehydrate and leave in xylene 15 to 30 minutes to remove coating. (After sections have been in xylene 5 minutes, blotting with filter paper helps to remove grease.)
5. Rehydrate; apply coating to stained sections.
6. Apply second staining procedure.
7. Repeat step 4.
8. Rinse briefly in absolute alcohol and clear in fresh xylene. Mount.

RECLAIMING AND STORING SPECIMENS

Reclaiming Biopsy Material

Graves Method (1943)

If biopsy material has dried, do not try to process it before first softening and rehydrating it. Place it in physiological saline solution for 1 hour, then fix, dehydrate, clear, and embed as usual.

See Haviland (1963) for recovery of rust-stained formalin-preserved gross tissues.

Reclaiming Dried Gross Specimens

(Robert Ingersoll Howes, Jr., is acknowledged for hours of effort in developing a recovery process for the Los Alamos Medical Center in 1959.)

Tissues that have been stored in alcohol or formalin often become completely or partially desiccated and shrunken. If a major catastrophe arises and tissue must be reclaimed, partial recovery can be made with fair returns for microscope identification. Van Cleave and Ross (1947) restored desiccated helminths and invertebrates to normal size by soaking them in a 0.5% aqueous solution of trisodium phosphate. Trial runs of this nature were made on dried formalin-fixed specimens for 4 hours to 30 days, depending on the hardness and size of tissue. Occasionally changing of the fluid seemed to help if the pieces were exceptionally dry; also, warming in a $\pm 40°C$ oven was helpful. Increasing the concentration of the phosphate did not speed recovery. Finally the tissues were washed for 30 minutes to 2 to 3 hours, again depending on size.

If the specimen was fixed in a mercuric chloride fixative, the trisodium phosphate was found to be effective until the tissue had been pretreated as follows: (1) soaked in water with enough Lugol's solution added to color the solution a deep brown: 1 to 2 days (if the solution became colorless, it was renewed); (2) washed in running water: 2 hours; (3) treated with 5% sodium thiosulfate: 1/2 to 1 hour; and (4) washed: 2 hours.

Only fair results were obtained by the above methods. Considerable shrinkage remained in the cells, and the nuclei stained only lightly. The latter condition improved somewhat after mordanting in mercuric chloride or potassium dichromate.

Since recovery was based on the detergent action of trisodium phosphate, the inevitable question arose: Why not try one of the modern household detergents? A 1% solution of Trend in water was added to dried tissues and kept overnight in a paraffin oven at approximately 58°C. This was followed by 6 to 8 hours washing and then routinely processed in the processor and embedded. The results were creditable—good enough for tissue identification and some pathological reading. The tissues averaged 10 to 25 cm in size and were fixed in either formalin or a mercuric chloride fixative. A longer stay in detergent might be required for large pieces. No pretense is made that results are exceptional: considerable shrinkage remains in the cells, and staining is not brilliant, but it is better than after trisodium phosphate. Therefore, if it is essential to recover damaged tissue concerning its identity, a malignancy, or some other "matter of life or death," this method of recovery may save the day.

Sandison Method (1955)

Sandison treated mummified material with the following solution and prepared histological sections with good results.

95% alcohol	30 volumes
1% formaldehyde (37–47%)	50 volumes
5% aqueous sodium carbonate	20 volumes

Leave tissues in solution overnight or until they become soft. Replace one-third of the fluid with 95% alcohol. Repeat until the tissues become firm. Dehydrate, clear, infiltrate, and embed.

Zimmerman (1976) described another method that works for mummified tissues. He applied the same technique for dried pathology specimens and the results were remarkable.

Procedure
Ruffer's solution (Ruffer 1921)

sodium carbonate	6.0 g
distilled water	42.0 ml
absolute alcohol	18.0 ml

Dissolve the sodium carbonate in water, and add the alcohol.

1. Allow tissues to hydrate in solution overnight.
2. Place in absolute alcohol for 24 hours.

3. Continue through xylene (or substitute) and embed in parafffin. Section, mount and stain as usual.

(Use vacuum to facilitate process.)

Preserving Gross Specimens in a Pliable Condition
Palkowsky Method (1960)

It is economical to preserve some gross specimens, pathological ones in particular, for future teaching or demonstration purposes. Fix as soon as possible in Kaiserling solution: 3 to 4 hours.

Kaiserling solution

chloral hydrate	300.0 g
potassium sulfate	6.0 g
potassium nitrate	114.0 g
sodium sulfate	54.0 g
sodium chloride	66.0 g
sodium bicarbonate	60.0 g
formaldehyde (37–40%)	300.0 ml
distilled water	10,000.0 ml

Drain off excess Kaiserling and place specimens in 4 times their volume of 80% alcohol: 18 to 24 hours. Deep freeze at about −29°C, sealed in airtight polyethylene bags not much larger than the specimen.

Because of contained air, lungs will float on the fixative; submerge them under cotton soaked with fixative.

The specimens will keep indefinitely this way, retain their color, and become pliable after thawing. After use, they can be returned to deep-freeze with little deterioration.

Color Preservation in Gross Specimens
McCormick Fixation (1961)

formaldehyde (37–40%)	50.0 ml
Sörenson's buffer, pH 6.4	950.0 ml
Prague powder	8.5 g
ascorbic acid	1.7 g

Storing Gross Specimens

Sealing tissues for storage has been a serious problem. Bottled storage has risks; any seal can spring a leak, and the resulting evaporation culminates in desiccated tissue. Storing in plastic bags is a more reliable system than the use of bottles. A heavy quality of polyethylene plastic is recommended and is widely available. Lighter grades can spit or unseal, whereas the heavy grade, once well sealed, almost supports itself without collapsing. A small amount of formalin included in the bag will keep the tissue moist as long as the bag remains sealed. For long term storage of tissues we recommend 70% alcohol. A quantity of small containers can be sealed together in a large bag, affording additional protection against drying. Plastic tags with data can be enclosed or attached to the outside. Storing with bags also saves storage space, preservative and containers.

REMOVING LABORATORY STAINS FROM HANDS AND GLASSWARE

The most frequently used laboratory stains, with suggested treatment for removal, are listed here. With the routine use of protective gloves with all laboratory procedures, skin stains can be reduced.

basic fuchsin—difficult to remove; try strong acetic acid in 95% alcohol, or dilute HCl
carmine—strong ammonia water or weak HCl; if stain resists, use them alternately
chromic acid—dilute sulfurous acid in water, or concentrated sodium thiosulfate and a few drops of sulfuric acid added
fast green and similar acid stains—ammonia water
hematoxylin—weak acid or lemon juice
hemoglobin—fresh stains: lukewarm to cool water (never hot) Older stains: soften with borax solution, dilute ammonia, or tincture of green soap; finally, treat with 2% aqueous oxalic acid
iodine—sodium thiosulfate
iron alum stains on glassware—strong NaOH (sticks) in water, followed by strong HCl
methylene blue—acid alcohol or tincture of green soap
most dyes—tincture of green soap
osmium tetroxide on glassware—3% H_2O_2 (Carr and Bacsich 1958)
PAS—to remove from clothing, soak in weak solution of potassium permanganate, approximately 0.5%; rinse off and bleach in 1% oxalic acid
picric acid—lithium iodide or carbonate, aqueous
potassium permanganate—dilute sulfurous acid, HCl, oxalic acid, or hyposulfite
safranin and gentian violet—difficult to remove; try acid alcohol
silver—Lugol or tincture of iodine, followed by sodium thiosulfate
Verhoeff—4% aqueous citric acid (Hull and Wegner 1952)

Cleaning Solution for Glassware

Strong

potassium dichromate	20.0 g
water	200.0 ml

Dissolve dichromate in water; when cool add very slowly:

sulfuric acid, concentrated	20.0 ml

Weak

potassium dichromate, 2% aqueous	9 parts
sulfuric acid	1 part

This is strong enough for most purposes.

Acid clean glass ware is necessary for many procedures. Another alternative to acid cleaning glassware is to use disposable plastic slide mailers or jars for the procedures that require acid clean glassware. Concentrated nitric acid or commercial acid cleaning solutions are equally satisfactory for acid cleaning. Thoroughly rinse the glassware for 30 to 60 seconds; then wash in tap water and follow with distilled water.

SUPPLIERS OF EQUIPMENT AND CHEMICALS

This is not intended to be a complete list of companies and the location and recent mergers make such a list almost impossible to obtain.

Major Laboratory Supply Companies

Curtin Matheson Scientific, 9999 Veteran's Memorial Dr., Houston, TX 77038; 800-443-1447

Fisher Scientific, 711 Forbes Ave., Pittsburgh, PA 15219-4785; 800-766-7000

VWR Scientific Products, P.O. Box 66, Bridgeport, NJ 08014; 800-932-5000

Microscopes

Carl Zeiss, Inc., Thornwood, NY 10594; 800-233-2343
Leica, Inc., Deerfield, IL 60015; 800-248-0123
Nikon, Melville, NY 11747; 800-52-NIKON
Olympus America, Melville, NY 11747; 800-844-5000

Electron Microscopy Supplies

Electron Microscopy Sciences, Ft. Washington, PA 19034; 800-523-5874
Polysciences, Inc., Warrington, PA 18976; 800-523-2575
Ted Pella Inc., Redding, CA 96049; 800-237-3526

Microtomes and Cryostats

Carl Zeiss, Inc., Thornwood, NY 10549; 800-233-2343
Leica, Inc., Deerfield, IL 60015; 800-248-0123
Research and Manufacturing Co. (RMC), Tucson, AZ 85714; 520-889-7900
Olympus America, Melville, NY 11747; 800-844-5000
Sakura Finctck USA, Inc., Torrance, CA 90504; 800-725-8723
Shandon-Lipshaw, Pittsburgh, PA 15275; 800-245-6212
Surgipath Medical, Richmond, IL 60071; 800-225-8867

Tissue Processors

Hacker Instruments, Fairfield, NJ 07004; 201-226-8450
Leica, Inc., Deerfield, IL 60015; 800-248-0123
Miles (Bayer) Diagnostic, Division, Tarrytown, NY 10591; 800-445-5901
Research and Manufacturing Co., Tucson, AZ 85714; 520-889-7900
Sakura Finetek USA, Inc., Torrance, CA 90504; 800-725-8723
Triangle Biomedical Sciences, Durham, NC 27705; 919-477-9283
Shandon-Lipshaw, Pittsburgh, PA 15275; 800-547-7429

Embedding Centers

Leica, Inc., Deerfield, IL 60015; 800-248-0123
Research and Manufacturing Co. (RMC), Tucson, AZ 85714; 520-889-7900
Sakura Finetek USA, Inc., Torrance, CA 90504; 800-725-8723
Shandon-Lipshaw, Pittsburgh, PA 15275; 800-547-7429
Surgipath Medical, Richmond, IL 60071; 800-225-8867
Triangle Biomedical Sciences, Durham, NC 27705; 919-477-9283

Knife Sharpeners

Leica, Inc., Deerfield, IL 60015; 800-248-0123
Hacker Instruments, Fairfield, NJ 07004; 201-226-8450
Shandon-Lipshaw, Pittsburgh, PA 15275; 800-547-7429

Slide Stainers

Hacker Instruments, Fairfield, NJ 07004; 201-226-8450
Leica, Inc., Deerfield, IL 60015; 800-248-0123

General Laboratory Aids 489

Sakura Finetek USA, Inc., Torrance, CA 90504; 800-725-8723
Shandon-Lipshaw, Pittsburgh, PA 15275; 800-547-7429

Distillers

B/R Instruments, Easton, MD 21601; 410-820-8800
Shandon-Lipshaw, Pittsburgh, PA 15275; 800-547-7429
Vision Medical, Nixon, TN 37343; 423-875-5299

Automatic Coverslippers

Hacker Instruments, Fairfield, NJ 07007; 201-226-8450
Sakura Finetek USA, Inc., Torrance, CA 90504; 800-725-8723
Shandon-Lipshaw, Pittsburgh, PA 15275; 800-245-6212

Disposable Knife Blades

Leica, Inc. Deerfield, IL 60015; 800-248-0123
Sakura Finetek USA, Inc., Torrance, CA 90504; 800-725-8723
Secure Medical Products, Inc., Mundelein, IL 60060; 847-970-4600
Surgipath Medical, Richmond, IL 60071; 800-225-8867
Triangle Biomedical Sciences, Durham, NC 27705; 919-477-9283

Immunohistochemicals

Accurate Chemical & Scientific Corp., Westbury, NY 11590; 800-645-6264
Becton-Dickinson, San Jose, CA 95131; 800-448-BDIS (2347)
Biomeda, Foster City, CA 94404; 800-341-8787
BioGenex, San Ramon, CA 94583; 800-421-4149
Dako Corp., Carpinteria, CA 93013; 800-235-5763
Enzo Diagnostics, Farmingdale, NY 11791; 800-221-7705
Incstar Corp., Stillwater, MN 55082; 612-439-9710; 800-328-1482
Jackson Immuno Research, West Grove, PA 19390; 800-367-5296
Pierce Chemical Company, Rockford, IL 61101; 800-874-3723
Research Genetics, Huntsville, AL 35801; 800-533-4363
Shandon-Lipshaw, Pittsburgh, PA 15275; 800-245-6212
Sigma Diagnostics, St. Louis, MO 63178; 800-325-3010
Signet Laboratories, Dedham, MA 02026; 800-223-0796
Zymed Laboratories, San Francisco, CA 94080; 800-874-4494

Safety Supplies

Aldrich Safety Supply Co., Milkwaukee, WI 53233; 800-558-9160
Evergreen Scientific, Los Angeles, CA 90058; 800-421-6261
Lab Safety Supply, Janesville, WI 53547; 800-357-0722

Marketing Instrumentation, Topeka, KS 66604; 800-447-0173

3-M Occupational Health & Environmental Safety Products Division, St. Paul, MN 55144; 800-243-4630

Dyes and Chemicals

Cell Point Scientific, Rockville, MD 20850; 800-424-2984
EM Science, Gibbstown, NJ 08027; 800-222-0342
Lerner Laboratories, Pittsburg, PA 15275; 800-547-7429
Newcomer Supply, Middleton, WI 53562; 800-383-7799
Polysciences, Inc. Warrington, PA 18976; 800-523-2575
Poly Scientific, Bay Shore , NY 11706; 800-645-5825
Rowley Biochemical Institute, Rowley, MA 01969; 508-948-2067
Sigma Diagnostics, St. Louis, MO 63178; 800-325-3010
Streck Laboratories, Omaha, NE 68144; 800-843-0912

Others

Anatech Ltd., Battle Creek, MI 49015; 800-ANATECH
Ameresco, Solon, OH 44139; 800-995-0635 (tissue fixative)
Apex Engineering Products Co., Plainfield, IL 60544; 800-451-6291 (RDO decalcifier)
Calibrated Instruments, Hawthorne, NY 10532; 914-741-5700 (mounting medium)
Cargille Laboratories, Cedar Grove, NJ 07009; 201-239-6633 (immersion oil)
Cel Tek, Glenview, IL 60025; 847-803-9495 (odd lab supplies)
Earth Safe Industries, Inc., Clarksville, VA 23927; 804-374-BIOS (tissue fixative)
Imeb, Inc., San Marcos, CA 92069; 800-543-8496 (used equipment; microtome repair)
Instrumedics, Hackensack, NJ 07601; 201-343-1313
Isolab, Inc., Akron, OH 44321; 800-321-9632 (lab acrylics)
Oncogene Research Products, Cambridge, MA 02142; 800-662-2616 (molecular biology products)
Richard-Allen Medical, Richland, MI 49083; 888-831-9800 (slides, coverslips, stains, and processing chemicals)
Ridge Microtomes, Knoxville, TN 37922; 423-966-4327 (microtome repair)
Streck Laboratories, Inc., Omaha, NE 68144; 800-843-0912 (tissue fixative)
Sturkey, Lebanon, PA 17042; 800-274-9446 (microtome knives)

References

d'Ablaing, G., Rogers, E. R., Parker, J. W., and Lukes, R. J.
    (1970) A simplified and modified methyl green pyronin stain. *American Journal of Clinical Pathology* 54:667–69.

Ackerman, A. G.
    (1958) A combined alkaline phosphatase-PAS staining method. *Stain Technology* 33:269–71.

Adams, C. W. M.
    (1956) A stricter interpretation of the ferric ferricyanide reaction with particular reference to the demonstration of protein-bound sulphydryl and disulphide groups. *Journal of Histochemistry and Cytochemistry* 4:32–35.
    (1957) A *p*-dimethylaminobenzaldehydenitrate method for the histochemical demonstration of tryptophane and related compounds. *Journal of Clinical Pathology* 10:56–62.

Adams, C. W. M., and Sloper, J. C.
    (1955) Technique for demonstrating neurosecretory material in the human hypothalamus. *Lancet* 268:651–52.
    (1956) The hypothalamic elaboration of posterior pituitary principles in man, the rat and dog; histochemical evidence derived from a performic acid–Alcian blue reaction for cystine. *Journal of Endocrinology* 13:221–28.

Adams, C. W. M., and Tuqan, N. A.
    (1961) The histochemical demonstration of protease by a gelatin-silver film substrate. *Journal of Histochemistry and Cytochemistry* 9:469–72.

Adamstone, F. B., and Taylor, A. B.
    (1948) The rapid preparation of frozen tissue sections. *Stain Technology* 23:109.

Adley, W. R., Rudolph, A. F., Hine, I. F., and Harritt, N. J.
    (1958) Glees staining of monkey hypothalamus: a critical appraisal of normal and experimental material. *Journal of Anatomy* 92:219–35.

Agrell, I. P. S.
    (1958) Whole mounts of small embryos attached directly to glass slides. *Stain Technology* 33:265–67.

Ahlqvist, J., and Andersson, L.
    (1972) Methyl green-pyronin staining: effects of fixation; use in routine pathology. *Stain Technology* 47:17–22.

Alberts, B. D., Bray, J., Lewis, M., Raff, K., Roberts, K., and Watson, J.
    (1994) *The Molecular Biology of the Cell*, 3rd ed. New York: Garland Publishing Inc.

Aldridge, W. G., and Watson, M. L.
    (1963) Perchloric acid extraction as a histochemical technique. *Journal of Histochemistry and Cytochemistry* 11:773–81.

Alfert, M., and Geschwind, I. I.
  (1953) A selective staining method for basic proteins of cell nuclei. *Proceedings of National Academy of Sciences* 39:991–99.
Alpert, M., Jacobowitz, D., and Marks, B. H.
  (1960) A simple method for the demonstration of lipofuchsin pigment. *Journal of Histochemistry and Cytochemistry* 8:153–58.
Alsop, D. W.
  (1974) Rapid single-solution polychrome staining of semithin epoxy sections using polyethylene glycol 200 (PEG 200) as a stain solvent. *Stain Technology* 49:265–72.
Altmann, P. L., and Dittman, D. W., eds.
  (1964) *Biology Data Book*. Washington, DC: Federation of American Societies for Experimental Biology.
Amano, M.
  (1962) Improved techniques for the enzymatic extraction of nucleic acids from tissue sections. *Journal of Histochemistry and Cytochemistry* 10:204–12.
American Public Health Association.
  (1956) *Diagnostic Procedures for Virus and Rickettsial Disease*. New York: The Association.
Anderson, D. L.
  (1961) Selective staining of ribonucleic acid. *Journal of Histochemistry and Cytochemistry* 9:619–20.
Anderson, D., and Grieff, D.
  (1964) Direct fluorochroming of Rickettsiae. *Journal of Histochemistry and Cytochemistry* 12:194–96.
Anderson, F. D.
  (1959) Dichromate-chlorate perfusion prior to staining degeneration in brain and spinal cord. *Stain Technology* 34:65–67.
Anderson, P. J.
  (1967) Purification and quantitation of glutaraldehyde and its effect on several enzyme activities in skeletal muscle. *Journal of Histochemistry and Cytochemistry* 15:652–61.
Anderson, R. A.
  (1971) Decalcification of snail shells to facilitate sectioning of the whole organism. *Stain Technology* 46:267–68.
Andre, G., Wenger, J. B., Rebolloso, D., Arrington, J., and Mehm, W.
  (1994) Evaluation of clearing and infiltration mixtures (CIMs) as xylene substitutes for tissue processing. *Journal of Histotechnology* 2:137–42.
Angulus, V., and Sepinwall, J.
  (1971) Use of gallocyanine as a myelin stain for brain and spinal cord. *Stain Technology*. 46:137–43.
Aoyama, F.
  (1930) Eine Modifikation der Cajakschen Methode zur Darstellung des Golgischen Binnenapparates. *Zeitschrift für Wissenschaftliche Mikroskopie und Mikroskopische Technik* 46:489–91.
Aparicio, S. R., and Marsden, P.
  (1969) Application of standard micro-anatomical staining methods to epoxy resin-embedded sections. *Journal of Clinical Pathology* 22:589–92.

Arensburger, K. E., and Markell, E. K.
  (1960) A simple combination direct smear and fecal concentrate for permanent stained preparations. *American Journal of Clinical Pathology* 34:50–51.
Armstrong, J. A.
  (1956) Histochemical differentiation of nucleic acids by means of induced fluorescence. *Experimental Cell Research* 11:640–43.
Armstrong, J. A., Niven, J., and Anderson, S. S.
  (1957) Fluorescence microscopy in the study of nucleic acids. *Nature* 80:1335–36.
Arnold, J. S., and Jee, W.
  (1954a) Preparing bone sections for radioautography. *Stain Technology* 29:49–54.
  (1954b) Embedding and sectioning undercalcified bone and its application to radioautography. *Stain Technology* 29:225–39.
Aronson, W., and Pharmakis, T.
  (1962) Enhancement of neotetrazolium staining for succinic dehydrogenase activity with cyanide. *Stain Technology* 37:321.
Atkinson, William B.
  (1952) Studies on the preparation and recoloration of fuchsin sulfurous acid. *Stain Technology* 27:153–60.
Austin, A. P.
  (1959) Iron alum aceto-carmine staining for chromosomes and other anatomical features of Rhodophyceae. *Stain Technology* 34:69–75.
Avers, C. J.
  (1963) An evaluation of various fixings and staining procedures for mitochondria in plant root tissues. *Stain Technology* 38:29–35.
Baker, J. R.
  (1944) The structure and chemical composition of the Golgi element. *Quarterly Journal of Microscopical Science* 85:1–72.
  (1945) *Cytological Technique*. London: Methuen.
  (1946) The histochemical recognition of lipine. *Quarterly Journal of Microscopical Science* 87:441–70.
  (1947) The histochemical recognition of certain guanidine derivatives. *Quarterly Journal of Microscopical Science* 88:115–21.
  (1958) *Principles of Biological Technique*. London: Methuen; New York: John Wiley.
Bancroft, J. D., and Stevens, A.
  (1982) *Theory and Practice of Histological Techniques,* 2nd ed. Edinburgh: Churchill Livingston.
  (1995) *Theory and Practice of Histological Techniques*, 4th ed. Churchill Livingston.
Bancroft, J. D.
  (1990) *Theory and Practice of Histological Techniques*, Edinburgh: Churchill Livingston.
Barley, D. A.
  (1964) Salivary gland chromosomes of black flies. *Turtox News* 44:298–300.
Barnard, E. A.
  (1961) Acylation and diazonium coupling in protein cytochemistry with special

reference to the benzoylation-tetrazonium method. *General Cytochemical Methods*, II. Ed. by J. F. Danielli. New York: Academic Press.

Barnes, H.
(1954) Some tables for the ionic composition of sea water. *Journal of Experimental Biology* 31:582–88.

Barnett, R. J., and Seligman, A. M.
(1952) Demonstration of protein-bound sulfydryl and disulfide groups by two new histochemical methods. *Journal of National Cancer Institute* 13:215–16.

Barr, M. L., and Bertram, E. G.
(1949) A morphological distinction between neurons of the male and female, and the behavior of the nucleolar satellite during accelerated nuceloprotein synthesis. *Nature* 163:676.

Barr, M. L., Bertram, F., and Lindsay, H. A.
(1950) The morphology of the nerve cell nucleus, according to sex. *Anatomical Record* 107:283–92.

Barrett, A. M.
(1944) On the removal of formaldehyde-produced precipitate from sections. *Journal of Pathology and Bacteriology* 56:135–36.

Barrós-Pita, J. C.
(1971) Protective paraffin infiltration of soft tissues in insects to facilitate softening of hard exoskeleton. *Stain Technology* 46:171–75.

Barszcz, C. A., and Yevich, P.
(1976) Preparation of codepods for histological examination. *Transactions of the Microscopic Society* 1:104–5.

Bartholomew, J. W., and Mittwer, T.
(1950) The mechanism of the Gram reaction. 1. The specificity of the primary dye. *Stain Technology* 25:103–10.
(1951) The mechanism of the Gram reaction. 3. Solubilities and some negative evidence on the causative role of sulfhydryl groups. *Stain Technology* 34:147–54.

Bartholomew, J. W., Mittwer, T., and Finklestein, H.
(1959) The phenomenon of Gram-positivity: its definition and some negative evidence on the causative role of sulfhydryl groups. *Stain Technology* 34:147–54.

Baserga, R.
(1961) Two-emulsion autoradiography for the simultaneous demonstration of precursors of deoxyribonucleic and ribonucleic acids. *Journal of Histochemistry and Cytochemistry* 9:586.

Baserga, R., and Malamud, D.
(1969) *Autoradiography. Techniques and Application*. New York: Harper and Row.

Baserga, R., and Nemeroff, K.
(1962) Two-emulsion radioautography. *Journal of Histochemistry and Cytochemistry* 10:628–35.

Batty, I., and Walker, P. D.
(1963) Differentiation of *Clostridium septicum* and *Clostridium chauvoei* by the use of fluorescent labelled antibodies. *Journal of Pathology and Bacteriology* 85:517–20.

Beamer, P. R., and Firminger, H. I.
> (1955) Improved methods for demonstrating acid-fast and spirochaetal organisms in histological sections. *Laboratory Investigation* 4:9–17.

Beçak. M. L., and Nazareth, H. R. S.
> (1962) Karytypic studies of two species of South American snakes. *Cytogenetics* 1:305–13.
>
> (1964) Chromosomes of cold blood animals from whole blood short-term cultures. *Mammalian Chromosomes Newsletter* 14:55–56.

Beckel, W. E.
> (1959) Sectioning large heavily sclerotized whole insects. *Nature* 184:1584–85.

Becker, E. R., and Roudabush, R. L.
> (1945) *Brief Directions in Histological Technique*. Ames, IA: Collegiate Press.

Bedi, K. S., and Horobin, R. W.
> (1976) An alcohol-soluble Schiff's reagent: a histochemical application of the complex between Schiff's reagent and phosphotungstic acid. *Histochemistry* 48:153–59.

Beech, R. H., and Davenport, H. A.
> (1933) The Bielschowsky staining technic: a study of the factors influencing its specificity for nerve fibers. *Stain Technology* 8:11–30.

Bélanger, L. F.
> (1961) Staining processed radioautographs. *Stain Technology* 36:313–17.

Bélanger, L. F., and Bois, P.
> (1964) A histochemical and histophysiological survey of the effects of different fixatives on the thyroid gland of rats. *Anatomical Record* 148:573–80.

Belling, J.
> (1926) The iron-aceto-carmine method of fixing and staining chromosomes. *Biological Bulletin, Woods Hole* 50:160–62.

Bender, M. A., and Eide, P. E.
> (1962) Separation of monkey (*Ateles*) leukocytes for culture. *Mammalian Chromosomes Newsletter* 8.

Benés, K.
> (1960) The chemical specificity of staining mitochondria in frozen dried and frozen substituted tissue with buffered Amidoblack 10B. *Acta Histochemica* 10:255–64.

Bennett, H. S.
> (1951) The demonstration of thiol groups in certain tissues by means of a new colored sulfhydryl reagent. *Anatomical Record* 110:231–47.

Bennett, H. S., and Watts, R. M.
> (1958) The cytochemical demonstration and measurement of sulfhydryl groups by azo-aryl mercaptide coupling with special reference to mercury orange. *General Cytochemical Methods*, I. Ed. by J. F. Danielli. New York: Academic Press.

Bennett, H. S., Wyrick, A. D., Lee, S. W., and McNeill, J. H.
> (1976) Science and art in preparing tissues embedded in plastic for light microscopy, with special reference to glycol methacrylate, glass knives and simple stains. *Stain Technology* 51:71–97.

Bensley, R. R., and Bensley, S. H.
> (1938) *Handbook of Histological and Cytological Technique.* Chicago: University of Chicago Press.

Bensley, S. H.
> (1959) The scope and limitations of histochemistry. *American Journal of Medical Technology* 25:15–32.

Bergeron, J. A., and Singer, M.
> (1958) Metachromasy: an experimental and theoretical re-evaluation. *Journal of Biophysical and Biochemical Cytology* 4:433–57.

Berkowitz, L. R., Riorello, O., Kruger, L., and Maxwell, D. S.
> (1968) Selective staining of nervous tissue for light microscopy following preparation for electron microscopy. *Journal of Histochemistry and Cytochemistry* 16:808–14.

Berrios, M.
> (1994) Device to prepare extruded nuclei and chromosome squashes. *Biotechnic and Histochemistry* 2:78–80.

Berthrong, M., and Barhite, M.
> (1964) A technic for preparation of tissue sections of needle-aspirated material from bone marrow. *American Journal of Clinical Pathology* 42:207–11.

Berube, G. R., Powers, M. M., Kerkay, J., and Clark, G.
> (1966) The gallocyanin-chrome alum stain: influence of methods of preparation on its activity and separation of active staining compound. *Stain Technology* 41:73–81.

Betts, A.
> (1961) The substitution of acridine orange in the periodic acid-Schiff stain. *American Journal of Clinical Pathology* 36:240–43.

Betzig, E., and Trautman, J. K.
> (1992) Near-field optics: microscopy, spectroscopy, and surface modification beyond the diffraction limit. *Science* 257:189–95.

Bielschowsky, M.
> (1902) Die Silber impregnation der Axencylinder. *Neurologisches Centralblatt* 21:579–84.

Binnig, G., Rohrer, H., Gerber, C., and Weibel, E.
> (1982) *Physiological Review Letters* 49:57.

Birge, W. J., and Tibbits, D. F.
> (1961) The use of sodium chloride containing fixatives in minimizing cellular distortion in histological and cytochemical preparations. *Journal of Histochemistry and Cytochemistry* 9:409–14.

Bissing, D. R.
> (1974) Haupt's gelatine adhesive mixed with formalin for affixing paraffin sections to slides. *Stain Technology* 49:116–17.

Bitensky, L.
> (1963) Modification to the Gomori phosphatase technique for controlled temperature frozen sections. *Quarterly Journal of Microscopical Science* 104:193–96.

Black, M. M., and Ansley, H. R.
> (1964) Histone staining with ammoniacal silver. *Science* 43:693.

Black, M. M., Ansley, H. R., and Mandl, R.
  (1964) On cell specificity of histones. *Archives of Pathology* 78:350–68.
Black, M. M., and Jones, E. W.
  (1971) Macular amyloidosis: a study of 21 cases with special reference to the role of the epidermis in its histogenesis. *British Journal of Dermatology* 84:199.
Blank, H., McCarthy, P. L., and Delamater, E. D.
  (1951) A non-vacuum freezing-dehydrating technic for histology, autoradiography, and microbial cytology. *Stain Technology* 26:193–97.
Blaxhall, P. C.
  (1983) Factors affecting lymphocyte culture for chromosome studies. *Journal of Fish Biology* 22:61–76.
Block, D. P., and Godman, G. C.
  (1955) A microphotometric study of the synthesis of deoxyribonucleic acid and nuclear histone. *Journal of Biophysical and Biochemical Cytology* 1:17–28.
Block, M., Smaller, V., and Brown, J.
  (1953) An adaptation of the Maximow technique for preparation of sections of hematopoietic tissue. *Journal of Laboratory and Clinical Medicine* 42:145–51.
Bodian, D.
  (1937) The staining of paraffin sections of nervous tissue with activated Protargol. *Anatomical Record* 69:153–62.
Bogen, E.
  (1941) Detection of tubercle bacilli by fluorescence microscopy. *American Review of Tuberculosis* 44:267–71.
Bogoroch, R.
  (1951) Detection of radio-elements in histological slides by coating with stripping emulsion—the strip-coating technic. *Stain Technology* 26:43–50.
Bohorfoush, J. G.
  (1963) A thiosulfate diluent for Wright's stain. *Stain Technology* 38:292–93.
Bokdawala, F. D., and George, J. C.
  (1964) Histochemical demonstration of muscle lipase. *Journal of Histochemistry and Cytochemistry* 12:768–71.
Borror, A. C.
  (1968) Nigrosin-$HgCl_2$-formalin; a stain-fixative for ciliates (Protozoa, Ciliophora). *Stain Technology* 43:293–95.
Bostwick D. G., Alannouf, N., and Choi, C.
  (1994) Establishment of the formalin free surgical laboratory: Utility of an alcohol-based fixative. *Archives of Pathology and Laboratory Medicine* 118:298–302.
Boyd, G. A.
  (1955) *Autoradiography in Biology and Medicine*. New York: Academic Press.
Boyd, I. A.
  (1962) Uniform staining of nerve endings in skeletal muscle with gold chloride. *Stain Technology* 37:225–30.
Bozzola, J.
  (1992) *Electron Microscopy: Principles and Techniques for Biologists*. St. Louis: Sigma.

Bracegirdle, B.
(1987) *History of Microtechnique.* Chicago: Science Heritage.
Branton, D., and Jacobson, L. J.
(1962) Dry, high resolution autoradiography. *Stain Technology* 37:239–42.
Braunstein, H., and Adriano, S. M.
(1961) Fluorescent stain for tubercle bacilli in histologic sections. *American Journal of Clinical Pathology* 36:37–40.
Brecher, G.
(1949) New methylene blue as a reticulocyte stain. *American Journal of Clinical Pathology* 19:895–96.
Brenner, R. M.
(1962) Controlled oxidation of background grains in radioautographs. *Journal of Histochemistry and Cytochemistry* 10:678.
Bridges, C. H., and Luna, L.
(1957) Kerr's improved Warthin-Starry technic: study of permissible variations. *Laboratory Investigation* 6:357–67.
Brinn, N.
(1983) Rapid metallic histological staining using the microwave oven. *Journal of Histotechnology* 6:125–29.
Brown, C. D., and Fleming, L.
(1965) A modified micromethod for the culturing of lymphocytes from peripheral blood for the assessment of chromosome morphology. *Mammalian Chromosomes Newsletter* 15:110–13.
Bruemmer, N. C., Carver, M. J., and Thomas, L. E.
(1957) A tryptophan histochemical method. *Journal of Histochemistry and Cytochemistry* 5:140–44.
Brunk, U. T., and Ericsson, J. L. E.
(1972) The demonstration of acid phosphatase in *in vitro* cultured tissue cells. Studies on the significance of fixation, tonicity and permeability. *Histochemical Journal* 4:349–63.
Buijs, R., and Dogterom A. A.
(1983) An improved method for embedding hard tissue in polymethyl methacrylate. *Stain Technology* 58:135–41.
Bullivant, S.
(1965) Freeze substitution and supporting techniques. *Laboratory Investigation* 14:1178–95.
Burdette, W. J.
(1962) *Methodology in Human Genetics.* San Francisco: Holden-Day.
Burkholder, P. M., Littlel, A. H., and Klein, P. G.
(1961) Sectioning at room temperature of unfixed tissues frozen in a gelatin matrix, for immunohistologic procedures. *Stain Technology* 36:89–91.
Burrows, R. B.
(1967) Improved preparation of polyvinyl alcohol-$HgCl_2$ fixative used for fecal smears. *Stain Technology* 42:93.
Burrows, W.
(1954) *Textbook of Microbiology.* Philadelphia: W. B. Saunders.
Burstone, M. S.
(1958) The relationship between fixation and techniques for the histochemical

localization of hydrolytic enzymes. *Journal of Histochemistry and Cytochemistry* 6:322–39.

(1959) Acid phosphatase activity of calcifying bone and dentin matrices. *Journal of Histochemistry and Cytochemistry* 7:147–48.

(1960) Postcoupling, noncoupling, and fluorescence techniques for the demonstration of alkaline phosphatase. *Journal of the National Cancer Institute* 24:1199–1218.

(1962) *Enzyme Histochemistry and Its Application in the Study of Neoplasms.* New York: Academic Press.

Burstone, M. S., and Folk, J. E.

(1956) Histochemical demonstration of aminopeptidase. *Journal of Histochemistry and Cytochemistry* 4:217–26.

Burton, G. J.

(1958) Preparation of thick malarial blood smears by the tapping method. *Mosquito News* 18:228–29.

Bussolati, G., and Bassa, T.

(1974) Thiosulfation aldehyde fuchsin (TAF) procedure for staining of pancreatic B cells. *Stain Technology* 49:313–15.

Cameron, M. L., and Steele, J. E.

(1959) Simplified aldehyde-fuchsin staining of neurosecretory cells. *Stain Technology* 34:265–66.

Capko, D. G., Krochmalnic, L., and Penman, S.

(1984) A new method of preparing embedment-free sections for transmission electron microscopy: applications to the cytoskeletal framework and other three-dimensional networks. *Journal of Cell Biology* 98:1878–85.

Cares, A.

(1945) A note on stored formaldehyde and its easy reconditioning. *Journal of Technical Methods* 25:67–70.

Carey, E. J.

(1941) Experimental pleomorphism of motor nerve plates as a mode of functional protoplasmic movement. *Anatomical Record* 81:393–413.

Carleton, H. M., and Leach, E. H.

(1947) *Histological Technique.* New York: Oxford University Press.

Carlo, R.

(1964) Alcian blue-Alcian yellow: a new method for the identification of different acidic groups. *Journal of Histochemistry and Cytochemistry* 12:44–45.

Carmichael, G. G.

(1963) A tetrazolium salt reduction method for demonstrating lipo-proteins in tissue sections. *Journal of Histochemistry and Cytochemistry* 11:738–40.

Caro, L. G.

(1964) High resolution autoradiography. *Methods in Cell Physiology,* I. Ed. by David M. Prescott. New York: Academic Press.

Carr, D. H., and Walker, J. E.

(1961) Carbol fuchsin as a stain for human chromosomes. *Stain Technology* 36:233–36.

Carr, L. A., and Bacsich, P.

(1958) Removal of osmic acid stains. *Nature* 182:1108.

Carr, L. B., Rambo, O. N., and Feichtmeier, T. V.
> (1961) A method of demonstrating calcium in tissue sections using chloranilic acid. *Journal of Histochemistry and Cytochemistry* 4:415–17.

Carson, F.
> (1990) *Histotechnology: A Self Instructional Text.* Chicago: ASCP Press.

Carter, C. H., and Leise, J. M.
> (1958) Specific staining of various bacteria with a single fluorescent antiglobulin. *Journal of Bacteriology* 76:152–54.

Carver, M. J., Brown, F. C., and Thomas, L. E.
> (1953) An arginine histochemical method using Sakaguchi's new reagent. *Stain Technology* 28:89–91.

Casselman, W. G. B.
> (1959) *Histochemical Technique.* London: Methuen; New York: John Wiley.

Castañeda, M. R.
> (1939) Experimental pneumonia produced by typhus Rickettsia. *American Journal of Pathology* 15:467–75.

Caulfield, J. B.
> (1957) Effects of varying the vehicle for osmic acid in tissue fixation. *Journal of Biophysical and Biochemical Cytology* 3:827–29.

Cavanagh, J. B., Passingham, R. J., and Vogt, J. A.
> (1964) Staining of sensory and motor nerves in muscles with Sudan black B. *Journal of Pathology and Bacteriology* 88:89–92.

Cejková, J., Boklová, A., and Lojda, Z.
> (1973) A study of acid mucopolysaccharides in cold microtome sections of normal and experimentally hydrated bovine corneas. *Histochemie* 36:167–72.

Chandley, A. C.
> (1988) Meiosis in man. *Trends in Genetics* 4:79–84.

Chang, J. P.
> (1956) Staining mitochondria in frozen dried tissues. *Experimental Cell Research* 11:643–46.

Chang, J. P., and Hori, S. H.
> (1961a) The section-freeze-substitution technique. I. Method. *Journal of Histochemistry and Cytochemistry* 9:292–300.
> (1961b) The section-freeze-substitution technique. II. Application to localization of enzymes and other chemicals. *Annales d'Histochimie* 6:419–32.

Chang, J. P., Hori, S. H., and Yokoyama, M.
> (1970) A modified section freeze-substitution technique. *Journal of Histochemistry and Cytochemistry* 18:683–84.

Chang, S. C.
> (1972) Hematoxylin-eosin staining of plastic embedded tissue sections. *Archives of Pathology* 93:344–51.

Chaplin, A. J., and Grace, S. R.
> (1976) An evaluation of some complexing methods for the histochemistry of calcium. *Histochemistry* 47:263–69.

Chapman, D. M.
> (1975) Dichromatism of bromphenol blue, with an improvement in the mercuric bromphenol blue technic for protein. *Stain Technology* 50:25–30.

(1977) Eriochrome cyanin as a substitute for hematoxylin and eosin. *Canadian Journal of Medical Technology* 39:65–66.

Chatton, E., and Lwoff, A.

(1930) Imprégnation, par diffusion argentique, de l'infraciliature des ciliés marins et d'eau douce, après fixation cytologique et sans dessiccation. *Comptes Rendus Hebdomadaires des Séances et Mémoires de la Société de Biologie* 104: 834–36.

(1935) Les ciliés apostomes—première partie. Aperçu historique et général étude monographique des genres et des espèces. *Archives de Zoologie Expérimentale et Général* 77:1–453.

(1936) Technique pour d'étude des protozaires, spécialement de leurs structures superficielles (cenétome et argyrome). *Bulletin de la Société Française de Microscopie* 5:25–39.

Chen, T.-T.

(1944a) Staining nuclei and chromosomes in protozoa. *Stain Technology* 19: 83–90.

(1944b) The nuclei in avian parasites. I. The structure of nuclei in *Plasmodium elongatum* with some considerations on technique. *American Journal of Hygiene* 40:26–34.

Chessick, R. D.

(1953) Histochemical study of the distribution of esterases. *Journal of Histochemistry and Cytochemistry* 1:471–85.

Chèvremont, M., and Fréderic, J.

(1943) Une nouvelle méthode histochimique de mise en évidence des substances à fonction sulfydrile: application à l'épiderme, au poil et à la levure. *Archives de Biologie* 54:589–605.

Chiffelle, T. L., and Putt, F. A.

(1951) Propylene and ethylene glycol as solvents for Sudan IV and Sudan black B. *Stain Technology* 26:51–56.

Childs, G. V., ed.

(1986) *Immunocytochemical Technology*. New York: Alan R. Liss.

Christenson, L. P.

(1965) *Human Chromosome Methodology*. Ed. by Jorge J. Yunis. New York: Academic Press.

Christophers, A. J.

(1956) The differential leukocyte count: observations on the error due to method of spreading. *Medical Journal of Australia* 1:533–36.

Chubb, J. C.

(1963) Acetic acid as a diluent and dehydrant in the preparation of whole, stained helminths. *Stain Technology* 37:179–82.

Chung, C. F., and Chen, C. M. C.

(1970) Restoring exhausted Schiff's reagent. *Stain Technology* 45:91–92.

Churukian, C. J.

(1993a) Zinc facts. *Histo-Logic* 1:13.

(1993b) *Manual of the Special Stains Laboratory of the University of Rochester Medical Center*. Rochester, NY.

Churukian, C. J., and Schenk, E. A.
: (1976) Iron gallein elastic model—a substitute for Verhoeff's elastic tissue stain. *Stain Technology* 51:213–17.
: (1988) Staining pneumocystis carinii and fungi in unfixed specimens with ammoniacal silver using a microwave oven. *Journal of Histotechnology* 11:19–21.

Clark, G.
: (1945) A simplified Nissl stain with thionine. *Stain Technology* 20:23–24.

Clark, G., and Allard, L.
: (1983) *History of Staining*. Baltimore: Williams & Wilkins.

Clark, G., Reed, C. S., and Brown, F. M.
: (1973) An evaluation and modification of Cole's hematoxylin. *Stain Technology* 48:189–91.

Clark, R. F., and Hench, M. E.
: (1962) Practical application of acridine orange stain in the demonstration of fungi. *American Journal of Clinical Pathology* 37:237–38.

Clendenin, T. M.
: (1969) Intraperitoneal colchicine and hypotonic KCl for enhancement of abundance and quality of meiotic chromosome spreads from hamster testes. *Stain Technology* 44:63–69.

Cocke, E. C.
: (1938) A method for fixing and staining earthworms. *Science* 87:443–44.

Cohen, I.
: (1949) Sudan black B—a new stain for chromosome smear preparation. *Stain Technology* 24:117–84.

Cole, E. C.
: (1933) Ferric chloride as a mordant for phosphate ripened hematoxylin. Mimeo from author.
: (1943) Studies on hematoxylin stains. *Stain Technology* 18:125–42.

Cole, W. V.
: (1946) A gold chloride method for motor-end plates. *Stain Technology* 21:23–24.

Cole, W. V., and Mielcarek, J. E.
: (1962) Fluorochroming nuclei of gold chloride-stained motor endings. *Stain Technology* 37:35–39.

Coleman, E. J.
: (1965) A simplified autoradiographic dipping procedure—slides handled in groups of five. *Stain Technology* 40:240–41.

Conger, Alan D.
: (1960) Dentist's sticky wax: a cover sealing compound for temporary slides. *Stain Technology* 35:225.

Conger, A. D., and Fairchild, L. M.
: (1953) A quick-freeze method for making smear slides permanent. *Stain Technology* 28:281–83.

Conklin, J. S.
: (1963) Staining reactions of mucopolysaccharides after formalin containing fixatives. *Stain Technology* 38:56–59.

Conn, H. J.
  (1946) The development of histological staining. *Ciba Symposia* 7.
  (1948) *History of Staining.* Geneva, NY: Biotech Publications.
  (1969) *Biological Stains,* 8th ed. Rev. by R. D. Lillie. Baltimore: Williams and Wilkins.
Conn, H. J., and Emmel, V. M.
  (1960) *Staining Procedures.* Baltimore: Williams and Wilkins.
Conrad, M. E., Jr., and Crosby, W. H.
  (1961) Bone marrow biopsy: modification of the Vim-Silverman needle. *Journal of Laboratory and Clinical Medicine* 57:642–45.
Conroy, J. D., and Toledo, A. B.
  (1976) Metachromasia and improved histologic detail with toluidine blue-hematoxylin and eosin. *Veterinary Pathology* 13:78–80.
Controls for Radiation, Inc.
  (1981) *Con-Rad / Joftes Fluid Emulsion Radioautography Instruction Manual.* Cambridge, MA: Controls for Radiation.
Coons, A. H.
  (1958) Fluorescent antibody methods. *General Cytochemical Methods,* I. Ed. by J. F. Danielli. New York: Academic Press, pp. 399–422.
Corliss, J. O.
  (1953) Silver impregnation of ciliated protozoa by the Chatton-Lwoff technic. *Stain Technology* 28:97–100.
Courtright, R. C.
  (1966) Use of polyester resins as a mounting medium for parasites. *Transactions American Microscopical Society* 85:319–20.
Cowdry, E. V.
  (1952) *Laboratory Technique in Biology and Medicine,* 3rd ed. Baltimore: Williams and Wilkins.
Cowell, R. L., and Tyler, R. D.
  (1992) *Cytology and Hematology of the Horse.* Coleta, CA: American Veterinary Publications.
Cramer, A. D., Rogers, E. R., Parker, J. W., and Lukes, R. J.
  (1973) The Giemsa stain for tissue sections: an improved method. *American Journal of Clinical Pathology* 60:148–56.
Crang, R. F. E., and Klomparens, K. L.
  (1988) *Artifacts in Biological Electron Microscopy.* New York: Plenum Publishing.
Crary, D. D.
  (1962) Modified benzyl alcohol clearing on alizarin-stained specimens without loss of flexibility. *Stain Technology* 37:124–25.
Crookham, J. N., and Dapson, R. W.
  (1991) *Hazardous Chemicals in the Histopathology Laboratory: Regulations, Risks, Handling, and Disposal.* Battle Creek, MI: Anatech.
Crozier, R. H.
  (1968) An acetic acid dissociation air-drying technique for insect chromosomes, with aceto-lactic orcein staining. *Stain Technology* 43:171–73.
Culling, C. E. A.
  (1957) *Handbook of Histopathological Technique.* London: Butterworth.

Culling, C., and Vassar, P.
- (1961) Deoxyribose nucleic acid. A fluorescent histochemical technique. *Archives of Pathology* 71:88/76–92/80.

Cumley, R. W., Crow, J. F., and Griffin, A. B.
- (1939) Clearing specimens for demonstration of bone. *Stain Technology* 14:7–11.

Curtis, C. G.
- (1981) *Whole Body Autoradiography*. New York: Academic Press.

Danielli, J. F.
- (1953) *Cytochemistry*. New York: John Wiley.

Daoust, R.
- (1957) Localization of deoxyribonuclease in tissue sections. A new approach to the histochemistry of enzymes. *Experimental Cell Research* 12:203–11.
- (1961) Localization of deoxyribonuclease activity by the substrate film method. *General Cytochemical Methods*, II. Ed. by J. F. Danielli. New York: Academic Press.
- (1964) In vitro binding of nucleic acids to tissue sections after removal of tissue nucleic acids. *Journal of Histochemistry and Cytochemistry* 12:640–45.
- (1968) The localization of enzyme activities by substrate film methods—evaluation and perspectives. *Journal of Histochemistry and Cytochemistry* 16:540–45.

Daoust, R., and Amano, H.
- (1960) The localization of ribonuclease activity in tissue secretions. *Journal of Histochemistry and Cytochemistry* 8:131–34.

Darnell, J., Lodish, H., and Baltimore, D.
- (1994) *Molecular Cell Biology*, 3rd ed. New York: W. H. Freeman and Co.

Darrow, M. A.
- (1952) Synthetic orcein as an elastic tissue stain. *Stain Technology* 27:329–32.

Davenport, H. A.
- (1960) *Histological and Histochemical Technics*. Philadelphia: W. B. Saunders.

Davenport, H. A., and Combs, C. M.
- (1954) Golgi's dichromate silver method. 3. Chromating fluids. *Stain Technology* 29:165–73.

Davenport, H. A., Windle, W. F., and Buch, R. H.
- (1934) Block staining of nervous tissue. IV. Embryos. *Stain Technology* 9:5–10.

Davies, H., and Harmon, P. J.
- (1949) A suggestion for prevention of loose sections in the bodian Protargol method. *Stain Technology* 24:249.

Dawar, B. L.
- (1973) A combined relaxing agent and fixative for Triclads (Planarians). *Stain Technology* 48:93.

deBruyn, P. P. H., Farr, R. S., Banks, H., and Morthland, F. W.
- (1953) In vivo and in vitro affinity of diaminoacridine for nucleoproteins. *Experimental Cell Research* 4:174–80.

Deck, J. D., and Desouza, G.
- (1959) A disrupting factor in silver staining techniques. *Stain Technology* 34:287.

Decosse, J. J., and Aiello, N.
- (1966) Feulgen hydrolysis: effect of acid and temperature. *Journal of Histochemistry and Cytochemistry* 14:601–4.

de Duve, C.
> (1959) Lysosomes, a new group of cytoplasmic particles. *Subcellular Particles.* Ed. by T. Hayashi. New York: Ronald Press.
>
> (1963a) The lysosome. *Scientific American* 208 (May):64.
>
> (1963b) General properties of lysosomes: the lysosome concept. *Lysosomes.* Ed. by A. V. S. deReuck and M. P. Cameron. Boston: Little, Brown.

De Guisti, D., and Ezman, L.
> (1955) Two methods for serial sectioning of arthropods and insects. *Transactions of American Microscopical Society* 74:197–201.

Deitch, A. D.
> (1955) Microspectrophotometric study of the binding of the anionic dye naphthol yellow S, by tissue sections and by purified protein. *Laboratory Investigation* 4:324–51.
>
> (1961) An improved Sakaguchi reaction for microspectrophotometric use. *Journal of Histochemistry and Cytochemistry* 9:477–83.
>
> (1964) A method for the cytophotometric estimation of nucleic acids using methylene blue. *Journal of Histochemistry and Cytochemistry* 12:451–61.

Deitch, A. D., Wagner, D., and Richart, R. M.
> (1968) Conditions influencing the intensity of the Feulgen reaction. *Journal of Histochemistry and Cytochemistry* 16:371–79.

Delameter, E. D.
> (1951) A staining and dehydrating procedure for handling of micro-organisms. *Stain Technology* 26:199–204.

De La Torre, L., and Salisbury, G. W.
> (1962) Fading of Feulgen-stained bovine spermatozoa. *Journal of Histochemistry and Cytochemistry* 10:39–41.

Delez, A. L., and Davis, O. S.
> (1950) The use of oxalic acid in staining with phloxine and hematoxylin. *Stain Technology* 25:111–12.

Delly, J. G.
> (1988) *Photography through the Microscope,* 9th ed. Rochester, NY: Eastman Kodak.

Del Vecchio, P. R., Dewitt, S. H., Borelli, J. I., Ward, J. B., Wood, T. A., Jr., and Malmgren, R. A.
> (1959) Application of millipore fixation technique to cytologic material. *Journal of the National Cancer Institute* 22:427–32.

Demke, D. D.
> (1952) Staining and mounting helminths. *Stain Technology* 27:135–39.

Dempsey, E. W., and Lansing, A. I.
> (1954) Elastic tissue. *International Review of Cytology,* III. Ed. by G. H. Bourne and J. B. Danielli. New York: Academic Press.

Denver, R.
> (1960) A proposed standard system of nomenclature of human mitotic chromosomes. *Lancet* 1:1063–65.

DePalma, P. A., and Young, G. G.
> (1963) Rapid staining of *Candida albicans* in tissue by periodic acid oxidation, basic fuchsin, and light green. *Stain Technology* 38:257–59.

DeRenzis, F. A., and Schechtman, A.
(1973) Staining by neutral red and trypan blue in sequence for assaying vital and nonvital cultured cells. *Stain Technology* 48:135–36.

Deuchar, E. M.
(1962) Staining sections before autoradiographic exposure: excessive background graining caused by celestine blue. *Stain Technology* 37:324.

de Witt, S. H., Del Vecchio, P. R., Borelli, J. I., and Hilberg, A. W.
(1957) A method for preparing wound washings and bloody fluids for cytologic evaluation. *Journal of the National Cancer Institute* 19:115–22.

Diamond, L. S.
(1945) A new rapid stain technic for intestinal protozoa, using Tergitol-hematoxylin. *American Journal of Clinical Pathology* 15:68–69.

Donaldson, P. T., Lillie, R. D., and Pizzolato, P.
(1973) Staining mast cells in sublimate-fixed guinea pig tissue. *Stain Technology* 48:47–48.

Dowding, G. L.
(1959) Plastic embedding of undecalcified bone. *American Journal of Clinical Pathology* 32:245–49.

Drets, M. E., and Shaw, M. W.
(1971) Specific banding patterns of human chromosomes. *Proceedings of the National Academy of Sciences* 68:2073–77.

Dunn, R. C.
(1946) A hemoglobin stain for histologic use based on the cyanol-hemoglobin reaction. *Archives of Pathology* 41:676–77.

Durie, B., and Salmon, S.
(1975) High speed scintillation autoradiography. *Science* 190:1093–95.

Dutt, M. K.
(1974) Cytochemical localization of nucleic acids with gallocyanin. *Acta Histochemica* 48:149–51.

Dykstra, M. J.
(1992) *Biological Electron Microscopy: Theory, Techniques, and Troubleshooting.* New York: Plenum Publishing.
(1993) *A Manual of Applied Techniques for Biological Electron Microscopy.* New York: Plenum Publishing.

Dziabis, M. D.
(1958) Luxol fast blue MBS, a stain for gross brain sections. *Stain Technology* 33:96–97.

Eapen, J.
(1960) The effect of alcohol-acetic formalin, Zenker's fluid, and gelatin on the activity of lipase. *Stain Technology* 35:227–28.

Eastman Chemical
(1992) *Autoradiographic Detection Principles.* Rochester, NY.

Eayres, J. T.
(1950) An apparatus for fixation and supravital staining of tissues by perfusion method. *Stain Technology* 25:137–42.

Edwards, D. A.
(1988) The technique of whole body autoradiography. *Journal of Histotechnology* 4:261–63.

Edwards, J. H., and Young, R. B.
  (1961) Chromosome analysis from small volumes of blood. *Lancet* 2:48–49.
Egozcue, J., and Vilarasau de Egozcue, M.
  (1966) Simplified culture and chromosome preparations of primate leukocytes. *Stain Technology* 41:173–78.
Ehrenrich, T., and Kerpe, S.
  (1959) A new rapid method of obtaining dry fixed cytological smears. *Journal of American Medical Association* 170:94–95.
Einarson, L.
  (1951) On the theory of gallocyanin-chromalum staining and its application for quantitative estimation of basophilia. A selective staining of exquisite progressivity. *Acta Pathologia et Microgiologica Scandinavica* 28:82–102.
Elftman, H.
  (1952) A direct silver method for the Golgi apparatus. *Stain Technology* 27:42–52.
  (1954) Controlled chromation. *Journal of Histochemistry and Cytochemistry* 2:1–8.
  (1957) Phospholipid fixation by dichromate-sublimate. *Stain Technology* 32:29–31.
  (1958) Effects of fixation in lipoid histochemistry. *Journal of Histochemistry and Cytochemistry* 6:317–21.
  (1959a) Aldehyde-fuchsin for pituitary cytochemistry. *Journal of Histochemistry and Cytochemistry* 7:98–100.
  (1959b) A Schiff reagent of calibrated sensitivity. *Journal of Histochemistry and Cytochemistry* 7:93–97.
  (1963) Combined Schiff procedures. *Stain Technology* 38:127–30.
Elias, J. M.
  (1969) Effects of temperature, prostaining rinses and ethanol-butanol dehydrating mixtures on methyl green-pyronin staining. *Stain Technology* 44:201–4.
  (1990) *Immunohistopathology: A Practical Approach to Diagnosis.* Chicago: ASCP Press.
Ellender, M., and Lojda, Z.
  (1973) Studies in lipid histochemistry. XI. New, rapid, simple and selective method for the demonstration of phospholipids. *Histochemie* 36:149–66.
Emmel, V. M., and Cowdry, E. V.
  (1964) *Laboratory Technique in Biology and Medicine.* Baltimore: Williams and Wilkins.
Endicott, K. M.
  (1945) Plasma or serum as a diluting fluid for thin smears of bone marrow. *Stain Technology* 29:25–26.
Enerbäck, L.
  (1969) Detection of histamine in mast cells by O-phthalaldehyde reaction after liquid fixation. *Journal of Histochemistry and Cytochemistry* 17:757–59.
Enlow, D. H.
  (1954) A plastic seal method for mounting sections of ground bones. *Stain Technology* 29:21–22.
  (1961) Decalcification and staining of ground thin sections of bone. *Stain Technology* 36:250–51.

Epple, A.
(1967) A staining sequence for A, B, and D cells of pancreatic islets. *Stain Technology* 42:53–61.

Epple, A., and Brinn, J. E.
(1986) Pancreatic islets. *Vertebrate Endocrinology: Fundamentals and Biomedical Implications.* Ed. by P. K. T. Pang and M. P. Schreibman. New York: Academic Press, vol 1.

Ericsson, J. L. E., and Biberfeld, P.
(1967) Studies on aldehyde fixation. Fixation rates and their relation to fine structure and some histochemical reactions in liver. *Laboratory Investigation* 17:281–98.

Essner, E.
(1970) Observations on hepatic and renal peroximes (microbodies) in the developing chick. *Journal of Histochemistry and Cytochemistry* 18:80–92.

Evans, E. P., Breckon, G., and Ford, C. E.
(1964) An air-drying method for meiotic preparations from mammalian testes. *Cytogenetics* 3:289–95.

Everett, M. M., and Miller, W. A.
(1974) The role of phosphotungstic and phosphomolybdic acids in connective tissue staining. I. Histochemical studies. *Histochemical Journal* 6:25–34.

Ewen, A. B.
(1962) An improved aldehyde-fuchsin staining technique for neurosecretory products in insects. *Transactions of the American Microscopical Society* 81:94–96.

Fahimi, H. D.
(1967) Perfusion and immersion fixation of rat liver with glutaraldehyde. *Laboratory Investigation* 16:737–50.

Farber, E., and Bueling, E.
(1956) Histochemical localization of specific oxidative enzymes. V. The dissociation of succinic dehydrogenase from carriers by lipase and the specific histochemical localization of the dehydrogenase with phenazine methosulfate and tetrazolium salts. *Journal of Histochemistry and Cytochemistry* 4:357–62.

Farber, E., and Louvriere, C. D.
(1956) Histochemical localization of specific oxidative enzymes. IV. Soluble oxidation-reaction dyes as aids in histochemical localization of oxidative enzymes with tetrazolium salts. *Journal of Histochemistry and Cytochemistry* 4:347–56.

Farnsworth, M.
(1963) Handling insect eggs during fixation. *Stain Technology* 38:300–301.

Faulkner, R. R., and Lillie, R. D.
(1945a) A buffer modification of the Warthin-Starry silver method for spirochetes in single paraffin sections. *Stain Technology* 20:81–82.
(1945b) Dried egg white for Mayer's albumin fixative. *Stain Technology* 20:99–100.

Faust, E. C., Russell, P. F., and Jung, R. C.
(1970) *Clinical Parasitology*, 4th ed. Philadelphia: Lea and Febiger.

Favorsky, B. A.
  (1930) Eine Modifikation des silber Impregnations-verfahrens Rámon y Cajal für das periphere Nervensystem. *Anatomischer Anzeiger* 70:376–79.
Feder, N.
  (1962) Polyvinyl alcohol as an embedding medium for lipid and enzyme histochemistry. *Journal of Histochemistry and Cytochemistry* 10:341–47.
Feder, N., and Sidman, R.
  (1957) A method for applying different stains to alternate serial sections on a single microscope slide. *Stain Technology* 32:271–73.
  (1958) Methods and principles of fixation by freeze-substitution. *Journal of Biophysical and Biochemical Cytology* 4:593–600.
Fenton, J. C. B., and Innes, J.
  (1945) A staining method for malaria parasites in thick blood films. *Transactions of the Royal Society of Tropical Medicine and Hygiene* 39:87–90.
Feulgen, R.
  (1914) Über die Kohlenwassenstoffgruppe der echten Nukleinsüre. *Zeitschrift für Physiologische Chemie* 92:154–58.
Feulgen, R., and Rossenbeck, H.
  (1924) Mikroskopisch-Chemischer Nachweis einter Nukleinsaüer von Typus Thymusnucleinsüre und die darauf beruhende elektive Färbung von Zellkernen in mikroskopischen Präparaten. *Zeitschrift für Physiologische Chemie* 135:203–48.
Field, J. W.
  (1941) Further notes on a method of staining malarial parasites in thick blood films. *Transactions of the Royal Society of Tropical Medicine and Hygiene* 35:35–42.
Fitzgerald, P. J.
  (1961) Dry-mounting autoradiographic technic for intracellular localization of water-soluble compounds in tissue sections. *Laboratory Investigation* 10:846–56.
Fitzgerald, P. J., and Pohlmann, M.
  (1966) Use of the silver-hyroquinone sequence for the display of reticular fibers. *Stain Technology* 41:267–72.
Fitzgerald, P. J., Simmel, E., Winston, J., and Martin, C.
  (1953) Radioautography: theory, technic and applications. *Laboratory Investigation* 2:181–222.
Fitz-Williams, W. G., Jones, G. S., and Goldberg, B.
  (1960) Cyrostat techniques: methods for improving conservation and sectioning of tissues. *Stain Technology* 35:195–204.
Flax, M., and Caulfield, J.
  (1962) Use of methacrylate embedding in light microscopy. *Archives of Pathology* 74:387–95.
Flax, M., and Pollister, A. W.
  (1949) Staining of nucleic acids and azure A. *Anatomical Record* 105:536–37.
Foley, J. O.
  (1943) A Protargol method for staining nerve fibers in frozen or celloidin sections. *Stain Technology* 18:27–33.

Foot, N. C.
> (1929) Comments on the impregnation of neurologia with ammoniacal silver salts. *American Journal of Pathology* 51:223–38.

Ford, L.
> (1965) Leukocyte culture and chromosome preparations from adult dog blood. *Stain Technology* 40:317–20.

Frankel, H. H., and Peters, R. L.
> (1964) A modified calcium-cobalt method for the demonstration of alkaline phosphatase. *American Journal of Clinical Pathology* 42:324–27.

Frankel, J., and Heckmann, K.
> (1968) A simplified Chatton-Lwoff silver impregnation procedure for use in experimental studies with ciliates. *Transactions of the American Microscopical Society* 87:317–21.

Freed, J. J.
> (1955) Freeze-drying technics in cytology and cytochemistry. *Laboratory Investigation* 4:106–21.

Freeman, B. L., Moyer, E. K., and Lassek, A. M.
> (1955) The pH of fixing fluids during fixation of tissues. *Anatomical Record* 121:593–600.

Friend, W. G.
> (1963) A microstainer for handling small ova. *Stain Technology* 38:205–6.

Frøland, A.
> (1965) Photographic recording and dye staining of chromosomes for autoradiography and morphology. *Stain Technology* 40:41–43.

Frost, H. M.
> (1959) Staining of fresh, undecalcified thin bone sections. *Stain Technology* 34:135–46.

Gairdner, B. M.
> (1969) Deteriorated paraformaldehyde: an insidious cause of failure in aldehyde fuchsin staining. *Stain Technology* 44:52–54.

Galigher, A. E.
> (1934) *The Essentials of Practical Microtechnique*. Privately published.

Galigher, A. E., and Kozloff, E. N.
> (1964) *Essentials of Practical Microtechnique*. Philadelphia: Lea and Febiger.

Gallimore, J. C., Bauer, E. C., and Boyd, G. A.
> (1954) A non-leaching technic for autoradiography. *Stain Technology* 29:95–98.

Gardner, D. L.
> (1958) Preparation of bone marrow sections. *Stain Technology* 33:295–97.

Gardner, H. H., and Punnett, H. H.
> (1964) An improved squash technique for human male meiotic chromosomes; softening and concentration of cells; mounting in Hoyer's medium. *Stain Technology* 39:245–48.

Garvey, W., Fathi, A., Bigelow, F., Jimenez, C., and Carpenter, B.
> (1987) A new reliable silver impregnation technique for the nervous system. *Journal of Histotechnology* 10:245–47.
> (1990) A new method for myelin at acid pH. *Journal of Histotechnology* 13:279–82.

(1991) Rapid, reliable and economical silver stain for neurofibrillary tangles and senile plaques. *Journal of Histotechnology* 14:39–42.

(1992) Combined modified periodic acid-Schiff and batch staining method. *Journal of Histotechnology* 2:117–20.

Gatenby, J. B., and Beams, H. W.
  (1950) *The Microtomist's Vale-Mecum.* London: J. and A. Churchill.

Geil, R. G.
  (1961) A simple technic for salvage of histologic sections from broken microslides. *Technical Bulletin of the Registry of Medical Technologists* 31:195–96.

Gelei, J. V.
  (1932) Eine neue Goldmethode zur Ciliatenforschung und eine neue Ciliate: *Colpidum pannonicum. Archiv für Protistenkunde* 77:219–30.
  (1935) Eine neue Abänderung der Klein's schen trockenen Silbermethode und das Silberliniensystem von *Glaucoma scintillans. Archiv für Protistenkunde* 84:446–55.

George, J. C., and Ambadkar, P. M.
  (1963) Histochemical demonstration of lipids and lipase activity in rat testis. *Journal of Histochemistry and Cytochemistry* 11:420–25.

George, J. C., and Ipye, P. T.
  (1960) Improved histochemical demonstration of lipase activity. *Stain Technology* 35:151–52.

Geyer, G.
  (1962) Histochemical-methylation with methanol and thionychloride. *Acta Histochemica* 14:284–96.

Gill, G. W., Frost, J. K., and Miller, K. A.
  (1974) A new formula for a half-oxidized hematoxylin solution that neither overstains nor requires differentiation. *Acta Cytologica* 18:300–11.

Giménez, D. F.
  (1964) Staining rickettsiae in yolk sac cultures. *Stain Technology* 39:135–40.

Giolli, R. A.
  (1965) A note on the chemical mechanism of the Nauta-Gygax technique. *Journal of Histochemistry and Cytochemistry* 13:206–10.

Giorno, R.
  (1984) A comparison of two immunoperiodase staining methods based on the Avidin-biotin interaction. *Diagnostic Immunology* 2:161–66.

Gladden, M. H.
  (1970) A modified pyridine-silver stain for teased preparation of motor and sensory nerve endings in skeletal muscle. *Stain Technology* 45:161–64.

Glauert, A. M.
  (1965) The fixation and embedding of biological specimens. *Techniques for Electron Microscopy.* Ed. by D. H. Kay. Philadelphia: F. A. Davis.
  (1975) *Fixation, Dehydration and Embedding of Biological Specimens.* St. Louis: Sigma.

Glauert, A. M., ed.
  (1974) *Practical Methods in Electron Microscopy,* II and III. New York: Elsevier.

Glauert, A. M., and Glauert, R. H.
- (1958) Araldite as an embedding medium for electron microscopy. *Journal of Biophysical and Biochemical Cytology* 4:191–94.

Glegg, R. E., Clermont, Y., and Leblond, C. P.
- (1952) The use of lead tetra-acetate benzidine, O-dianisidine, and a 'film test' in investigating the periodic-acid-Schiff technic. *Stain Technology* 27:277–305.

Glenner, G. C.
- (1957) Simultaneous demonstration of bilirubin, hemosiderin and lipofuscin pigments in tissue sections. *American Journal of Clinical Pathology* 27:1–5.
- (1963) A re-evaluation of the ninhydrin-Schiff reaction. *Journal of Histochemistry and Cytochemistry* 11:285–86.

Glenner, G. G., and Lillie, R. D.
- (1957a) A rhodocyan technic for staining anterior pituitary. *Stain Technology* 32:187–90.
- (1957b) The histochemical demonstration of indole derivatives by the post-coupled p-dimethylaminobenzylidene reaction. *Journal of Histochemistry and Cytochemistry* 5:279–96.
- (1959) Observations on the diazonium coupling reaction for the histochemical demonstration of tyrosine: metal chelation and formazon variants. *Journal of Histochemistry and Cytochemistry* 7:416–21.

Glick, D.
- (1949) *Techniques of Histo- and Cytochemistry.* New York: Interscience Publishers.

Glick, D., and Malstrom, B. G.
- (1952) Studies in histochemistry. XXIII. A simple and efficient freezing-drying apparatus for the preparation of embedded tissue. *Experimental Cell Research* 3:125–235.

Goh, K.
- (1965) Human cytogenetics. *Disease-a-Month*, April.

Goldfischer, S., Essner, E., and Novikoff, A. B.
- (1964) The localization of phosphatase activities at the level of ultrastructure. *Journal of Histochemistry and Cytochemistry* 12:72–95.

Goldman, M.
- (1968) *Fluorescent Antibody Methods.* New York: Academic Press.

Goldstein, D. J.
- (1962) Ionic and non-ionic bonds in staining, with special reference to the action of urea and sodium chloride on the staining of elastic fibers and glycogen. *Quarterly Journal of Microscopical Science* 103:477–92.

Gowali, F. M.
- (1995) Picro hibiscin stain for degenerated muscle fibers. *Laboratory Medicine* 26:470–73.

Gomori, G.
- (1941a) The distribution of phosphatase in normal organs and tissues. *Journal of Cellular and Comparative Physiology* 17:71–84.
- (1941b) Observations with differential stains on human islets of Langerhans. *American Journal of Pathology* 17:395–406.
- (1946) The study of enzymes in tissue sections. *American Journal of Clinical Pathology* 16:347–52.

(1948) Chemical character of enterochromaffin cells. *Archives of Pathology* 45: 48–55.

(1950a) An improved histochemical technic for acid phosphatase. *Stain Technology* 25:81–85.

(1950b) A rapid one-step trichrome stain. *American Journal of Clinical Pathology* 20:662–64.

(1950c) Aldehyde-fuchsin; a new stain for elastic tissue. *American Journal of Clinical Pathology* 20:665–66.

(1950d) Sources of error in enzymatic histochemistry. *Journal of Laboratory and Clinical Medicine* 35:802–9.

(1951) Alkaline phosphatase of cell nuclei. *Journal of Laboratory and Clinical Medicine* 37:526–31.

(1952) *Microscopic Histochemistry.* Chicago: University of Chicago Press.

(1953) Human esterases. *Journal of Laboratory and Clinical Medicine* 42: 445–53.

(1954a) The histochemistry of mucopolysaccharides. *British Journal of Experimental Pathology* 35:377–80.

(1954b) Histochemistry of the enterochromaffin substance. *Journal of Histochemistry and Cytochemistry* 2:50–53.

(1956) Histochemical methods for acid phosphatase. *Journal of Histochemistry and Cytochemistry* 4:453–61.

Gower, W. C.
(1939) A modified stain and procedure for trematods. *Stain Technology* 14:31–32.

Gradwohl, R. B. H.
(1963) *Clinical Laboratory Methods and Diagnosis*, II. St. Louis: C. V. Mosby.

Graupner, H., and Weissberger, A.
(1931) Über der Verwendung des Dioxanes beim Einbetten mikroskopischer Objekte Mitteilungen zur mikroskopischen Technik I. *Zoologischer Anzeiger* 92:204–6.

Graves, K. D.
(1943) Restoration of dried biopsy tissue. *American Journal of Clinical Pathology Technical Section* 7:111.

Gray, P.
(1954) *The Microtomist's Formulary and Guide.* New York: Blakiston.
(1964a) Making dry mounts of Foraminifera and Radiolaria. *Ward's Bulletin* 6.
(1964b) *Handbook of Basic Microtechnique*, 3rd ed. New York: Blakiston.

Green, F.
(1990) *Sigma-Aldrich Handbook of Stains.* St Louis: Sigma.

Greenstein, J. S.
(1957) A rapid phloxine-methylene blue overnight stain for formalin-fixed material. *Stain Technology* 32:75–77.

Gridley, M. F.
(1951) A modification of the silver impregnation method of staining reticular fibers. *American Journal of Clinical Pathology* 21:897–99.
(1953) A stain for fungi in tissue sections. *American Journal of Clinical Pathology* 23:303–7.
(1957) *Manual of Histologic and Special Staining Technics.* Washington, DC: Armed Forces Institute of Pathology.

Grimley, P. M.
- (1964) A tribasic stain for thin sections of plastic-embedded, osmic acid fixed tissues. *Stain Technology* 39:229–33.

Grimley, P. M., Albrecht, J. M., and Michelitch, H. J.
- (1965) Preparation of large epoxy sections for light microscopy as an adjunct to fine-structure studies. *Stain Technology* 40:357–66.

Grizzle, W.
- (1996) Theory and practice of silver staining in histopathology. *Journal of Histotechnology* 19:183–95.

Groat, R. A.
- (1949) Initial and persisting staining power of solutions of iron hematoxylin lake. *Stain Technology* 24:157–63.

Grocott, R. G.
- (1955) Stain for fungi in tissue sections and smears using Gomori's methanamine-silver nitrate method. *American Journal of Clinical Pathology* 25:975–79.

Guard, H. R.
- (1959) A new technic for differential staining of the sex chromatin and the determination of its incidence in exfoliated vaginal epithelial cells. *American Journal of Clinical Pathology* 32:14–51.

Gude, W. D.
- (1968) *Autoradiographic Techniques.* Englewood Cliffs, NJ: Prentice-Hall.

Gude, W. D., and Odell, T. T.
- (1955) Vinisil as a diluent in making bone marrow smears. *Stain Technology* 30:27–28.

Gude, W. D., Upton, A. C., and Odell, T. T.
- (1955) Giemsa staining of autoradiograms prepared with stripping film. *Stain Technology* 30:161–62.

Guillery, R. W., Shirra, B., and Webster, K. E.
- (1961) Differential impregnation of degenerating nerve fibers in paraffin-embedded material. *Stain Technology* 36:9–13.

Gurr, E.
- (1956) *A Practical Manual of Medical and Biological Staining Techniques.* New York: Interscience.
- (1958) *Methods of Analytical Histology and Histochemistry.* London: Leonard Hill.
- (1969) *Encyclopedia of Microscopic Stains.* London: Edward Gurr; Baltimore: Williams and Wilkins.

Gurr, G. T.
- (1953) *Biological Staining Methods.* London: G. T. Gurr.

Guyer, M. F.
- (1953) *Animal Micrology.* Chicago: University of Chicago Press.

Hack, M. H.
- (1952) A new histochemical technique for lipids applied to plasma cells. *Anatomical Record* 112:275–301.

Hafiz, S., Spencer, R. C., Lee, M., Gooch, H., and Duerden, B.
- (1985) Use of Microwaves for Acid and Alcohol Fast Staining. *Journal of Clinical Pathology* 38:1073–84.

Hajian, A.
>   (1961) Note on trichrome stain. *Technical Bulletin of the Registry of Medical Technologists* 31:92.

Hale, A. J.
>   (1957) The histochemistry of polysaccharides. *International Review of Cytology*, VI. Ed. by G. H. Bourne and F. J. Danielli. New York: Academic Press.

Hale, D. M., Cromartie, W. J., and Dobson, R. L.
>   (1960) Luxol fast blue as a selective stain for dermal collagen. *Journal of Investigative Dermatology* 35:293–94.

Hale, L. J.
>   (1958) *Biological Laboratory Data*. London: Methuen; New York: John Wiley.

Halmi, N. S.
>   (1950) Two types of basophils in the anterior pituitary of the rat and their respective cytophysiological significance. *Endocrinology* 47:289–99.

Halnan, C.
>   (1989) *Cytogentics of Animals*. Wallington, United Kingdom: CAB International.

Ham, A. W.
>   (1957) *Histology*. Philadelphia: Lippincott.

Hamlyn, J. H.
>   (1957) Application of the Nauta-Gygax technic for degenerating axons to mounted sections. *Stain Technology* 32:123–26.

Hance, R. T., and Green, F. J.
>   (1961) Behavior of rapidly oxidized hematoxylin. *Stain Technology* 36:253.

Hancox, N. M.
>   (1957) Experiments on the fundamental effects of freeze substitution. *Experimental Cell Research* 13:263–75.

Hanker, J. S., Yates, P. E., Clapp, D. H., and Anderson, W. A.
>   (1972) New methods for the demonstration of lysosomal hydrolyses by the formation of osmium blacks. *Histochemie* 30:201.

Hansen, D. W., Hunter, D. T., Richards, D. F., and Allred, L.
>   (1970) Acridine orange in the staining of blood parasites. *Journal of Parasitology* 56:386–87.

Hanzon, V., and Hermodsson, L. H.
>   (1960) Freeze-drying of tissues for light and electron microscopy. *Journal of Ultrastructure Research* 4:332–48.

Harada, K.
>   (1973) Effect of prior oxidation on the acid-fastness of mycobacteria. *Stain Technology* 48:269–73.
>   (1976) The nature of mycobacterial acid-fastness. *Stain Technology* 51:255–60.
>   (1976) Periodic acid-methenamine silver stain for mycobacteria in tissue sections. *Stain Technology* 51:278–80.

Hardonk, M. J., and van Duijan, P.
>   (1964a) The mechanism of the Schiff reaction as studied with histochemical model systems. *Journal of Histochemistry and Cytochemistry* 12:748–51.
>   (1964b) A quantitative study of the Feulgen reaction with the aid of histochemical model systems. *Journal of Histochemistry and Cytochemistry* 12:752–57.
>   (1964c) Studies on the Fuelgen reaction with histochemical model systems. *Journal of Histochemistry and Cytochemistry* 12:758–67.

Hartman, B. K.
(1973) Immunofluorescence of dopamine-beta-hydroxylase. Application of improved methodology to the localization of the peripheral and central noradrenergic nervous system. *Journal of Histochemistry and Cytochemistry* 21:312.

Hartley, S.E., and Horne, M. T.
(1983) Chromosome polymorphism in the rainbow trout. *Chromosome* 87:461–68.

Hartz, P. H.
(1945) Frozen sections from Bouin-fixed material in histopathology. *Stain Technology* 20:113–14.
(1947) Simultaneous histologic fixation and gross demonstration of calcification. *American Journal of Clinical Pathology* 17:750.

Haupt, A. W.
(1930) A gelatin fixative for paraffin sections. *Stain Technology* 5:97.

Hause, W. A.
(1959) Saw for preparation of blocks of bone. *Technical Bulletin of the Registry of Medical Technicians* 29:101.

Haust, M. D.
(1958) Tetrahydrofuran (THF) for dehydration and infiltration. *Laboratory Investigation* 7:58–67.
(1959) Tetrahydrofuran (THF) for routine dehydration clearing and infiltration. *Technical Bulletin of the Registry of Medical Technicians* 29:33–37.

Haviland, T. N.
(1963) Restoration of rust-stained formalin-preserved gross specimens. *American Journal of Clinical Pathology* 39:364.

Hayat, M. A., ed.
(1986) *Principles and Techniques of Electron Microscopy*. Orlando, FL: Academic Press.
(1993) *Stains and Cytochemical Methods*. New York: Plenum Publishing.

Heady, J., and Rogers, T. E.
(1962) Turtle blood cell morphology. *Proceedings of the Iowa Academy of Sciences* 69:587–90.

Hegner, R. W., Cort, W. W., and Root, F. M.
(1927) *Outlines of Medical Zoology*. New York: Macmillan.

Hendrickson, A., Kunz, S., and Kelly, D. E.
(1968) NaOH-HIO_4 treatment of osmium-collidine fixed epoxy sections to facilitate staining after autoradiography. *Stain Technology* 43:175–76.

Herlant, M.
(1960) Étude critique de deux techniques nouvelles detinées a mettre en évidence les différentes catégories cellulaires présentes dans la glande pituitaire. *Bulletin de Microscopie Appliquée* 10:37–44.

Herman, G. E., Chilpala, E., Sabin, L., and Elfont, E.
(1988) Zinc formalin fixative for automated tissue processing. *Journal of Histotechnology* 11:85–89.

Herr, B. E., Coleman, P. D., and Griggs, R. C.
(1976) A Bodian method for mounted frozen sections. *Stain Technology* 51:261–65.

Hicks, J. D., and Matthaei, E.
: (1955) Fluorescence in histology. *Journal of Pathology and Bacteriology* 70:1–12.
: (1958) A selective fluorescent stain for mucin. *Journal of Pathology and Bacteriology* 75:473–76.

Hilleman, H. H., and Lee, C. H.
: (1953) Organic chelating agents for decalcification of bones and teeth. *Stain Technology* 28:285–86.

Himes, M., and Moriber, L.
: (1956) A triple stain for deoxyribonucleic acid, polysaccharides and proteins. *Stain Technology* 31:67–70.

Hoefert, L. L.
: (1968) Polychromatic stains for thin sections of *Beta* embedded in epoxy resin. *Stain Technology* 43:145–51.

Hoffman, E. O., and Miller, M. J.
: (1975) Immunofluorescent staining of amoebae in routine paraffin-embedded tissues. *Journal of Parasitology* 61:1104.

Hokfelt, T., Fuxe, K., and Goldstein, M.
: (1975). Applications of immunohistochemistry to studies on monoamine cell systems with special references to nervous tissue. *Annals of the New York Academy of Sciences* 254:407.

Holczinger, L., and Bálint, Z.
: (1961) The staining properties of 'masked' lipids. *Acta Histochemica* 11:284–88.

Holgate, C. S., Jackson, P., Cowen, P., and Bird, C.
: (1983) Immunogold-silver staining: new method of immunostaining with enhanced sensitivity. *Journal of Histochemistry and Cytochemistry* 31:938–44.

Hollander, D. H.
: (1963) An oil-soluble anti-oxidant in resinous mounting media to inhibit fading of Romanowsky stains. *Stain Technology* 38:288–89.

Hollister, G.
: (1934) Clearing and dying fish for bone study. *Zoologica* 12:89–101.

Holmes, W. C.
: (1929) The mechanism of staining: the case for the physical theories. *Stain Technology* 4:75–80.

Holt, M., Cowing, R. F., and Warren, S.
: (1949) Preparation of radioautographs of tissues without loss of water soluble P32. *Science* 110:328–29.

Holt, M., Sommers, S. C., and Warren, S.
: (1952) Preparation of tissue sections for quantitative histochemical studies. *Anatomical Record* 112:177–86.

Holt, M., and Warren, S.
: (1950) A radioautograph method for detailed localization of radioactive isotopes in tissues without isotope loss. *Proceedings of the Society for Experimental Biology and Medicine* 73:545.
: (1953) Freeze-drying tissues for autoradiography. *Laboratory Investigation* 2:1–14.

Holt, S. J.
: (1956) The value of fundamental studies of staining reactions in enzyme histo-

chemistry, with reference to indoxyl methods of esterases. *Journal of Histochemistry and Cytochemistry* 4:541–54.

Holt, S. J., Hobbiger, E. E., and Pawan, G. L. S.
(1960) Preservation of integrity of rat tissues for cytochemical staining purposes. *Journal of Biophysical and Biochemical Cytology* 7:383–86.

Holzer, W.
(1921) Über eine neue Methode der gliafraser Farbung. *Zeitschrift für die Gesamte Neurologie and Psychiatrie* 69:354–57.

Hood, R. C. W. S., and Neill, W. M.
(1948) A modification of alizarine red S technic for demonstrating bone formation. *Stain Technology* 23:209–18.

Hori, S. H.
(1963) A simplified acid hematein test for phospholipids. *Stain Technology* 38:221–25.

Hori, S. II., and Chang, J. P.
(1963) Demonstration of lipids in mitochondria and other cellular elements. *Journal of Histochemistry and Cytochemistry* 11:115–16.

Horikawa, M., and Kuroda, Y.
(1959) In vitro cultivation of blood cells of *Drosophila melanogaster* in a synthetic medium. *Nature* 184:2017–18.

Hörmann, H., Grassman, W., and Fries, G.
(1958) Über den Mechanismus der Schiffschen Reaktion. *Justus Liebigs Annalen der Chemie* 616:125–47.

Horn, R. G., and Spicer, S. S.
(1964a) Sulfated mucopolysaccharide and basic protein in certain granules of rabbit leukocytes. *Laboratory Investigation* 13:1–15.
(1964b) Sulfated mucopolysaccharides in azurophile granules of immature granulocytes of the rabbit. *Journal of Histochemistry and Cytochemistry* 12:33.

Horobin, R. W., Flemming, L., and Kevill-Davies, I. M.
(1974) Basic fuchsin-ferric chloride: a simplification of Weigert's resorcin-fuchsin stain for elastic fibers. *Stain Technology* 49:207–10.

Hotchkiss, R. D.
(1948) A microchemical reaction resulting in the staining of polysaccharide structures in fixed tissues. *Archives of Biochemistry* 16:131–41.

Hsu, S.-M., Raine, L., and Fanger, H.
(1981a) Use of the avidin-biotin peroxidase complex (ABC) in immunoperoxidase techniques: a comparison between ABC and unlabelled antibody (PAP) procedure. *Journal of Histochemistry and Cytochemistry* 29:255.
(1981b) Use of Avidin-biotin peroxidase complex in immunoperoxidase techniques: A comparison between Avidin-biotin-peroxidase and unlabeled antibody procedures. *Journal of Histochemistry and Cytochemistry* 29:577–80.

Hull, S., and Wegner, S.
(1952) Removal of stains. *Stain Technology* 27:224.

Humason, G. L., and Lushbaugh, C. C.
(1960) Selective demonstration of elastin, reticulum, and collagen by silver, orcein, and aniline blue. *Stain Technology* 35:209–14.
(1961) A quick pinacyanole stain for frozen sections. *Stain Technology* 36:257–58.

(1969) Sirius supra blue FGL-CF; superior to aniline blue in the combined elastin reticulum and collagen stain. *Stain Technology* 44:105–6.

Humphrey, A. A.
(1936) A new rapid method for frozen section diagnosis. *Journal of Laboratory and Clinical Medicine* 22:198–99.

Hungerford, D. A.
(1965) Leukocytes cultures from small inocula of whole blood and the preparation of metaphase chromosomes by treatment with hypotonic KCl. *Stain Technology* 40:333–38.

Hunter, E.
(1993) *Practical Electron Microscopy*. New York: Cambridge University Press.

Hutchinson, H. E.
(1953) The significance of stainable iron in sternal marrow sections. *Blood* 8:236–48.

Hutner, S. H.
(1934) Destaining agents for iron alum hematoxylin. *Stain Technology* 9:57–59.

Ibanez, M. L., Russell, W. O., Chang, J. P., and Speece, A. J.
(1960) Cold chamber frozen sections for operating room diagnosis of routine surgical stains. *Laboratory Investigation* 9:275–78.

Incstar Corp.
(1989) Instructions for immunocytochemistry. Stillwater, MN: Incstar Corp.

Ingram, R. L., Otken, L. B., and Jumper, J. R.
(1961) Staining of malarial parasites by the fluorescent antibody technic. *Proceedings of the Society for Experimental Biology and Medicine* 106:52–54.

Jagatic, J., and Weiskopf, R.
(1966) A fluorescent method for staining mast cells. *Archives of Pathology* 82:430–33.

Janigan, D. T.
(1965) The effects of aldehyde fixation on acid phosphatase activity in tissue block. *Journal of Histochemistry and Cytochemistry* 13:476–83.

Jaspers, B.
(1987) Practical advice to the PAS reaction. *Journal of Histotechnology* 10:263–66.

Jennings, R. B.
(1951) A simple apparatus for dehydration of frozen tissues. *Archives of Pathology* 52:195–97.

Jha, R. K.
(1976) An improved polychrome staining method for thick epoxy sections. *Stain Technology* 51:159–62.

Joftes, D. L.
(1959) Liquid emulsion autoradiography with tritium. *Laboratory Investigation* 8:131–48.

Joftes, D. L., and Warren, S.
(1955) Simplified liquefied emulsion radioautography. *Journal of the Biological Photographic Association* 23:145.

Jones, R. T., Morton, A. W., and Moghissi, A. A.
(1993) Comparison of deparaffinization agents for automatic immunostainers. *Journal of Histotechnology* 4:367–69.

Kaback, M. M., Saksela, E., and Mellman, W. J.
  (1964) The effect of 5-bromide-oxyuridine on human chromosomes. *Experimental Cell Research* 34:182–86.
Kabat, E. A., and Furth, J.
  (1941) A histochemical study of the distribution of alkaline phosphatase in various normal and neoplastic tissues. *American Journal of Pathology* 17:303–18.
Kallman, J.
  (1971) Aldehyde-fuchsin followed by toluidine blue O for pancreatic islet cells. *Stain Technology* 46:210–11.
Kaniwar, K. C.
  (1960) Note on the specificity of mercuric bromophenol blue for the cytochemical detection of proteins. *Experimentia* 16:355.
Kaplow, L. S.
  (1963) Cytochemistry of leukocyte alkaline phosphatase. *American Journal of Clinical Pathology* 39:439–49.
Kaplow, L. S., and Burstone, M. S.
  (1963) Acid-buffered acetone as a fixative for enzyme cytochemistry. *Nature* 200:690–91.
  (1964) Cytochemical demonstration of acid phosphatase in hematopoietic cells in health and in various hematological disorders using azo dye techniques. *Journal of Histochemistry and Cytochemistry* 12:805–11.
Kasten, F. H.
  (1960) The chemistry of Schiff's reagent. *International Review of Cytology*, X. Ed. by G. H. Bourne and J. F. Danielli. New York: Academic Press.
  (1962) Some comments on a recent criticism of the ninhydrin-Schiff reaction. *Journal of Histochemistry and Cytochemistry* 10:769–70.
  (1965) Loss of DNA and protein, and changes in DNA during a 30-hour cold perchloric acid extraction of cultured cells. *Stain Technology* 40:127–35.
  (1985) Additional Schiff-type reagents for use in cytochemistry. *Stains Technology* 33:39–45.
Kasten, F. H., and Burton, V.
  (1959) A modified Schiff's solution. *Stain Technology* 34:289.
Kasten, F. H., Burton, V., and Glover, P.
  (1959) Fluorescent Schiff-type reagents for cytochemical detection of polyaldehyde moieties in sections and smears. *Nature* 184:1797–98.
Kasten, F. H., Burton, V., and Lofland, S.
  (1962) Schiff-type reagents in cytochemistry. Detection of primary amine dye impurities in pyronin B and pyronin Y (G). *Stain Technology* 37:277–91.
Kasten, F. H., and Lala, R.
  (1975) The Feulgen reaction after glutaraldehyde fixation. *Stain Technology* 50:197–201.
Kasten, F. H., and Sandritter, W.
  (1962) Crystal violet contamination of methyl green and purification of methyl green—a historical note. *Stain Technology* 37:253–55.
Katline, V. C.
  (1962) Retention of nuclear staining by phosphomolybdic-phosphotungstic mordanting. *Stain Technology* 37:193–95.
Kay, D.
  (1965) *Techniques for Electron Microscopy*. Philadelphia: F. A. Davies.

Keeble, S. A., and Jay, R. F.
(1962) Fluorescent staining for the differentiation of intracellular ribonucleic acid and deoxyribonucleic acid. *Nature* 193:695–96.

Kelley, W. E.
(1965) New aquaria and synthetic sea salts provide a notable advance in marine biology. *Ward's Bulletin* 5.

Kellogg, D. S., and Deacon, W. E.
(1964) A new rapid immunofluorescent staining technic for identification of *Treponema pallidium* and *Neisseria gonorrhoeae*. *Proceedings for the Society for Experimental Biology and Medicine* 115:963–65.

Kenney, M., Dyckman, J., and Aronson, S. M.
(1971) Acid fast staining for hooklets of *Echinococcus*. *Stain Technology* 46:160–61.

Kent, S. P.
(1961) A study of mucins in tissue sections using the fluorescent antibody technique. I. The preparation and specificity of bovine submaxillary gland mucin antibody. *Journal of Histochemistry and Cytochemistry* 9:491–97.

Kerr, D. A.
(1938) Improved Warthin-Starry method of staining spirochetes in tissue sections. *American Journal of Clinical Pathology, Technical Supplement* 2:63–67.

Kessel, J. F.
(1925) The distinguishing characteristics of the intestinal protozoa of man. *China Medical Journal* February: 1–57.

Kidder, G. W.
(1933) Studies on *Conchophthirius mytili* De Morgan. I. Morphology and division. *Archiv für Protistenkunde* 79:1–24.

Kimball, R. F., and Perdue, S. W.
(1962) Quantitative cytochemical studies on *Paramecium*. *Experimental Cell Research* 27:405–15.

Kimmel, D., and Jee, W. S. S.
(1975) A rapid plastic embedding technique for preparation of three-micron thick sections of decalcified hard tissue. *Stain Technology* 50:83–86.

King, R.
(1983) *Plastic (GMA) Microtomy: A Practical Approach*. Marietta, GA: Olio.

Kiossoglou, K. A., Wolman, I. J., and Garrison, M. J.
(1963) Fetal hemoglobin-containing erythrocytes. I. Counts of cells stained by the acid solution method compared with alkali denaturation methods. *Blood* 21:553–60.

Kirby, H.
(1947) *Methods in the Study of Protozoa*. University of California Syllabus Series. Berkeley: University of California Press.

Klein, B. M.
(1926) Über eine Eugentumlichkeit der Pellicula von Chilodon uncinatus Ehrbg. *Zoologischer Anzeiger* 67:160–62.

Klessen, C.
(1972) Histochemical staining of zymogen granules of pancreatic acinar cells using a permanganate-HID-technique. *Histochemie* 30:365.

Klinger, H. B.
   (1958) The fine structure of the sex chromatin body. *Experimental Cell Research* 14:207–11.

Klinger, H. B., and Hammond, D. O.
   (1971) Rapid chromosome and sex-chromatin staining with pinacyanol. *Stain Technology* 46:43–47.

Klionsky, B., and Marcoux, L.
   (1960) Frozen storage of incubation media for enzyme histochemistry. *Journal of Histochemistry and Cytochemistry* 8:329.

Klüver, H., and Barrera, E.
   (1953) A method for the combined staining of cells and fibers in the nervous system. *Journal of Neuropathology and Experimental Neurology* 12:400–3.

Knoeff, A. A.
   (1938) Adaptation of the Mallory-Azan staining method to the anterior pituitary of the rat. *Stain Technology* 13:49–52.

Koenig, H.
   (1963) Intravital staining of lysosomes by basic dyes and metallic ions. *Journal of Histochemistry and Cytochemistry* 11:120–21.

Koenig, H., Groat, R. A., and Windle, W. F.
   (1945) A physical approach to perfusion-fixation of tissues with formalin. *Stain Technology* 20:13–22.

Köhler, A.
   (1893) Ein neues Beleuchtungsverfahren für mikrophotographische Zwecke. *Zeitschrift für Wissenschaft Mikroskopie* 9:443–40.

Kok, L. P., and Boon, M. E.
   (1992) *Microwave Cookbook for Microscopists: Art and Science of Visualization*. Leiden, Netherlands: Coulomb Press.

Konno, T., and Takahashi, H.
   (1985) A method for preparing thin ground sections of undecalcified bone for histomorphometry. *Journal of Histotechnology* 8:97–100.

Kopriwa, B. M., and Leblond, C. P.
   (1962) Improvements in the coating technique of radioautography. *Journal of Histochemistry and Cytochemistry* 10:269–84.

Kornhauser, S. I.
   (1930) Hematein: its advantages for general laboratory usage. *Stain Technology* 5:13–15.
   (1943) A quadruple stain for strong color contrasts. *Stain Technology* 18:95–97.
   (1945) A revised method for the 'Quad' stain. *Stain Technology* 20:33–35.

Korson, R.
   (1964) A silver stain for deoxyribonucleic acid. *Journal of Histochemistry and Cytochemistry* 12:875–79.

Koski, J. P., and Reyes, P. F.
   (1986) Silver impregnation techniques with a review of the theory and methods. *Journal of Histotechnology* 9:265–72.

Koss, L.
   (1992) *Diagnostic Cytology and Its Pathology*. Philadelphia: J. B. Lippincott.

Krajian, A. A., and Gradwohl, R. B. H.
   (1952) *Histopathological Technic*. St. Louis: C. V. Mosby.

Krichesky, B.
   (1931) A modification of Mallory's triple stain. *Stain Technology* 6:97–98.
Krishan, A.
   (1962) Avian microchronosomes: as shown by prefixation treatment with colchicine, squashing and hematoxylin staining. *Stain Technology* 37:335–37.
Kristensen, Harold K.
   (1948) An improved method for decalcification. *Stain Technology* 23:151–54.
Kropp, B.
   (1954) Grinding thin sections of plastic embedded bone. *Stain Technology* 29: 77–80.
Krus, S., Andrade, Z. A., and Barka, T.
   (1961) Histochemical demonstration of specific phosphatases of the liver preserved in hypertonic sucrose solution. *Journal of Histochemistry and Cytochemistry* 9:487–90.
Krutsay, M.
   (1960) A versatile Resorcin-fuchsin formula. I. Combined with formaldehyde. II. Used after periodic acid. III. Used after HCl hydrolysis. *Stain Technology* 35:283–84.
   (1962a) Permanganate-resorcin-fuchsin: a selective stain for elastic tissue. *Stain Technology* 37:250–51.
   (1962b) The preparation of iron-hematoxylin from alum hematoxylin. *Stain Technology* 37:249.
Kubie, L. S., and Davidson, D.
   (1928) The ammoniacal silver solution as used in neuropathology. *Archives of Neurology and Psychiatry* 19:888–903.
Kurnick, N. B.
   (1952) Histological staining with methyl green-pyronin. *Stain Technology* 27: 233–42.
   (1955) Histochemistry of nucleic acids. *International Review of Cytology*, IV. Ed. by G. H. Bourne and J. F. Danielli. New York: Academic Press.
Kutlík, V. I. E.
   (1968) Über die Argentaffinität des geformten Bilirubins in den Geweben. *Acta Histochemica* 5:213–24.
   (1970) Nachweis von Eisen in den Geweben mittels Chlorathämatoxylinfärbung. *Acta Histochemica* 37:259–67.
Kuyper, C. M. A.
   (1957) Identification of mucopolysaccharides by means of fluorescent basic dyes. *Experimental Cell Research* 13:198–200.
Kwan, S. K.
   (1970) Sticky wax infiltration in the preparation of sawed undecalcified bone sections. *Stain Technology* 45:177–81.
LaCour, L.
   (1941) Aceto-orcein: a new stain-fixative for chromosome. *Stain Technology* 16: 169–74.
Lagunoff, D., Phillips, J., and Benditt, E. P.
   (1961) The histochemical demonstration of histamine in mast cells. *Journal of Histochemistry and Cytochemistry* 9:534–41.

Lan, H. Y., Mu, W., Nikolic-Patterson, D. J., and Atkins, R. C.
  (1995) A novel, simple, reliable, and sensitive method for multiple immunoenzyme staining: use of microwave oven heating to block antibody cross-reactivity and retrieve antigens. *Journal of Histochemistry and Cytochemistry* 43:97.

Landing, B. H., and Hall, H. E.
  (1955) Differentiation of human anterior pituitary cells by combined metalmordant and mucoprotein stains. *Laboratory Investigation* 4:275–78.
  (1956) Selective demonstration of histidine. *Stain Technology* 31:197–200.

Lane, B. P., and Europa, D. L.
  (1965) Differential staining of ultrathin sections of Epon-embedded tissues for light microscopy. *Journal of Histochemistry and Cytochemistry* 13:579–82.

Langman, J. M.
  (1995) d-Limonene: a safe, effective alternate to xylene? *Journal for Histotechnology* 2:131–37.

Lartique, D. J., and Fite, G. L.
  (1962) The chemistry of the acid-fast reaction. *Journal of Histochemistry and Cytochemistry* 10:611–18.

Lasky, A., and Greco, J.
  (1948) Argentaffin cells of the human appendix. *Archives of Pathology* 46:83–84.

Laufer, I.
  (1949) The differential staining of amphibian yolk granules. *Stain Technology* 24:249–50.

Lawless, D. K.
  (1953) A rapid permanent mount stain technic for the diagnosis of the intestinal protozoa. *American Journal of Tropical Medicine and Hygiene* 2:1137.

Lazarus, S. S.
  (1958) A combined periodic acid-Schiff trichrome stain. *Archives of Pathology* 66:767–72.

Leach, W. B.
  (1960) A method for the histological examination of bone marrow granules. *Canadian Medical Association Journal* 83:717–19.

Leaver, R. W., Evans, B. J., and Corrin, B.
  (1977) Identification of Gram-negative bacteria in histological sections using Sandiford's counterstain. *Journal of Clinical Pathology* 30:290–91.

Leblond, C. P., Messier, B., and Koriwa, B.
  (1959) Thymidine-H^3 as a tool for the investigation of the renewal of cell populations. *Laboratory Investigation* 8:296–308.

Lendrum, A. C.
  (1944) On the cutting of tough and hard tissues embedded in paraffin. *Stain Technology* 19:143–44.

Lendrum, A. C., and McFarlane, D.
  (1940) A controllable modification of Mallory's trichrome staining method. *Journal of Pathology and Bacteriology* 50:38–40.

Leong, A. S-Y., and Milos, J.
  (1986) Rapid immunoperoxidase staining of lymphocyte antigens using microwave irradiation. *Journal of Pathology* 148:183.

Leske, R., and Von Mayersbach, H.
: (1969) The role of histochemical and biochemical preparation methods for the detection of glycogen. *Journal of Histochemistry and Cytochemistry* 17:527–38.

Lev, R., and Stoward, P. J.
: (1969) On the use of eosin as a fluorescent dye to demonstrate mucous cells and other structures in tissue sections. *Histochemie* 20:363–77.

L'Hoste, R., and Torres, M. A.
: (1995) Using zinc formalin as a routine fixative in the histology laboratory. *Laboratory Medicine* 26:210–14.

Lhotka, J. F., and Davenport, H. A.
: (1947) Differential staining of tissue in the block with picric acid and the Fuelgen reaction. *Stain Technology* 22:139–44.
: (1949) Deterioration of Schiff's reagent. *Stain Technology* 24:237–39.
: (1951) Aldehyde reactions in tissues in relation to the Fuelgen technic. *Stain Technology* 26:35–41.

Lieb, E.
: (1947) Permanent stain for amyloid. *American Journal of Clinical Pathology* 17:413–14.
: (1959) The plastic (Mylar) sack as an aid in the teaching of pathology. *American Journal of Clinical Pathology* 32:385–92.

Liebman, E.
: (1951) Permanent preparations with the Thomas arginine histochemical test. *Stain Technology* 26:261–63.

Lillie, R. D.
: (1929) A brief method for the demonstration of mucin. *Journal of Technical Methods and Bulletin of International Association of Medical Museums* 12:120–21.
: (1940) Further experiments with the Masson trichrome modifications of the Mallory connective tissue stain. *Stain Technology* 15:17–22.
: (1944) Acetic methylene blue counterstain in staining tissues for acid fast bacilli. *Stain Technology* 18:45.
: (1945) Studies on selective staining of collagen with acid aniline dyes. *Journal of Technical Methods and Bulletin of International Association of Medical Museums* 25:1.
: (1946) A simplified method of preparation of di-amine-silver hydroxide for reticulum impregnation; comments on the nature of the so-called sensitization before impregnation. *Stain Technology* 21:69–72.
: (1951a) Histochemical comparison of the Casella, Bauer, and periodic acid oxidation Schiff leucofuchsin technics. *Stain Technology* 26:123–26.
: (1951b) Simplification of the manufacture of Schiff reagent for use in histochemical procedures. *Stain Technology* 26:163–65.
: (1951c) The allochrome procedure. A differential method segregating the connective tissues, collagen, reticulum, and basement membranes into two groups. *American Journal of Clinical Pathology* 21:484–88.
: (1952) Staining of connective tissue. *Archives of Pathology* 54:220–33.
: (1953) Factors influencing periodic acid-Schiff reaction of collagen fibers. *Journal of Histochemistry and Cytochemistry* 1:353–61.

(1954a) Argentaffin and Schiff reactions after periodic and oxidation and aldehyde blocking reactions. *Journal of Histochemistry and Cytochemistry* 2:127–36.
(1954b) *Histopathologic Technic and Practical Histochemistry*. 2nd ed. New York: Blakiston.
(1955) The basophilia of melanins. *Journal of Histochemistry and Cytochemistry* 3:453–54.
(1956a) Nile blue staining technic of the differentiation of melanin and lipofuchcins. *Stain Technology* 31:151–53.
(1956b) The mechanism of Nile blue staining of lipofuchsins. *Journal of Histochemistry and Cytochemistry* 4:377–81.
(1956c) The p-dimethylaminobenzaldehyde reaction for pyrroles in histochemistry: melanins, enterochromaffin, zymogen granules, lens. *Journal of Histochemistry and Cytochemistry* 4:118–29.
(1957a) Ferrous iron uptake. *Archives of Pathology* 64:100–3.
(1957b) The xanthydrol reaction for pyrroles and indoles in histochemistry: zymogen granules, lens, neterchromaffin, and melanins. *Journal of Histochemistry and Cytochemistry* 5:188–95.
(1957c) Metal reduction reactions in melanins. *Journal of Histochemistry and Cytochemistry* 5:325–33.
(1957d) Adaption of the Morel Sisely protein diazotation procedure to the histochemical demonstration of protein bound tyrosine. *Journal of Histochemistry and Cytochemistry* 5:528–32.
(1958) Acetylation and nitrosation of tissue amines in histochemistry. *Journal of Histochemistry and Cytochemistry* 6:352–62.
(1960) Metal chelate reaction of enterochromaffin. *Journal of Histochemistry and Cytochemistry* 9:44–48.
(1961) Investigation on the structure of the enterochromaffin substance. *Journal of Histochemistry and Cytochemistry* 9:184–89.
(1962) Glycogen in decalcified tissue. *Journal of Histochemistry and Cytochemistry* 10:763–65.
(1964a) Histochemical acylation of hydroxyl and amino groups. Effect on the periodic acid-Schiff reaction, anionic and cationic dye and van Gieson collagen stains. *Journal of Histochemistry and Cytochemistry* 12:821–41.
(1964b) Studies on histochemical and acylation procedures. I. Phenols. *Journal of Histochemistry and Cytochemistry* 12:522–29.
(1965) *Histopathologic Technic and Practical Histochemistry*, 3rd ed. New York: McGraw-Hill.
(1977) *Conn's Biological Stains*, 9th ed. St. Louis: Sigma.

Lillie, R. D., and Burtner, H. J.
(1953a) The ferric ferricyanide reduction test in histochemistry. *Journal of Histochemistry and Cytochemistry* 1:87–92.
(1953b) Stable sudanophilia of human neutrophil leucocytes in relation to peroxidase and oxidase. *Journal of Histochemistry and Cytochemistry* 1:8–26.

Lillie, R. D., Burtner, H. J., and Henson, J. P.
(1953) Diazosafranin for staining enterochromaffin. *Journal of Histochemistry and Cytochemistry* 1:154–59.

Lillie, R. D., and Earle, W. R.
    (1939) Iron hematoxylins containing ferric and ferrous iron. *American Journal of Pathology* 15:765–70.
Lillie, R. D., and Fulmer, M. H.
    (1976) *Histopathologic Technique and Practical Histochemistry*. New York: McGraw-Hill.
Lillie, R. D., Gilmern, P. R., Jr., and Welsh, R. A.
    (1961) Black periodic and black Bauer methods for tissue polysaccharides. *Stain Technology* 36:361–63.
Lillie, R. D., and Glenner, G. G.
    (1957) Histochemical aldehyde blockade by aniline in glacial acetic acid. *Journal of Histochemistry and Cytochemistry* 5:167–69.
Lillie, R. D., Gutiérrez, A., Madden, D., and Henderson, R.
    (1968) Acid orcein-iron and acid orcein-copper stains for elastin. *Stain Technology* 43:203–6.
Lillie, R. D., and Henderson, R.
    (1968) A short chromic acid-hematoxylin stain for frozen sections of formol fixed brain and spinal cord. *Stain Technology* 43:121–22.
Lillie, R. D., Henson, J. P., Greco, J., and Burtner, H. C. J.
    (1957) Metal reduction reactions of melanins: silver and ferric ferricyanide reduction by various reagents in vitro. *Journal of Histochemistry and Cytochemistry* 5:311–24.
Lillie, R. D., Henson, J. P., Greco, J., and Carson, J. C.
    (1960) Azo-coupling rate of enterochromaffin with various diazonium salts. *Journal of Histochemistry and Cytochemistry* 9:11–21.
Lillie, R. D., Lasky, A., Greco, J., Burtner, H., Jacquelin, P., and Jones, P.
    (1951) Decalcification of bone in relation to staining and phosphatase technics. *American Journal of Clinical Pathology* 21:711–22.
Lillie, R. D., and Pizzolato, P.
    (1968) Histochemical studies of oxidation and reduction reactions of the bile pigments in obstructive icterus, with some notes on hematoidin. *Journal of Histochemistry and Cytochemistry* 16:17–28.
Lillie, R. D., Pizzolato, P., and Donaldson, P. T.
    (1973) Iron Gallein in van Gieson technics, replacing iron hematoxylin. *Stain Technology* 48:348–49.
Lillie, R. D., Zirkle, C., Dempsey, E., and Greco, J. F.
    (1953) Final report of the committee on histological mounting media. *Stain Technology* 28:57–80.
Lison, L.
    (1954) Alcian blue 8G with chlorantine fast red 5B. A technic for selective staining of mucopolysaccharides. *Stain Technology* 29:131–38.
Litwin, J. A., and Kasprzyk, J. M.
    (1976) PAS reaction performed on semithin Epon sections following removal of the resin by NaOH in absolute ethanol. *Acta Histochemica* 55:98–103.
Litwin, J. A., Kasprzyk, J. M., and Cichocki, T.
    (1975) Light microscopic differential staining of epoxy-embedded adenohypophysis. *Acta Histochemica* 52:17–22.

Ljungberg, O.
(1970) Cresyl fast violet—a selective stain for human C-cells. *Acta Pathologica et Microbiologica Scandinavica* 78:618–20.

Lloyd, B., Brinn, N., and Burger, P. C.
(1985) Silver staining of senile plaques and neurofibrillary change in paraffin embedded tissues. *Journal of Histotechnology* 8:155–56.

Lockard, I., and Reers, B. L.
(1962) Staining tissue of the central nervous system with Luxol fast blue and neutral red. *Stain Technology* 37:13–16.

Lodin, Z., Faltin, J., and Sharma, K.
(1967) Attempts at standardization of a highly sensitive Schiff reagent. *Acta Histochemica* 26:244–54.

Lodin, Z., Marés, V., Karásek, J., and Skŕivanová, P.
(1967) Studies on nervous tissue. II. Changes of sizes of nuclei of nervous cells after fixation and after further histological treatment of the nervous tissue. *Acta Histochemica* 28:297–312.

Login, G. R., and Dvorak, A. M.
(1994) *Microwave Tool Book*. Boston: Beth Israel Hospital, Department of Pathology.

Login, G. R., Schnitt, S. J., and Dvorak, A. M.
(1987) Rapid microwave fixation of human tissues for light microscopic immunoperoxidase identification of diagnostically useful antigens. *Laboratory Investigation* 57:585.

Lojda, Z., and Havránková, E.
(1975) The histochemical demonstration of aminopeptidase with bromindolyl leucinamide. *Histochemistry* 43:355–66.

London Conference on the Normal Karyotype.
(1964) Report. *Annals of Human Genetics* 27:295–96.

Long, M. E.
(1948) Differentiation of myofibrillae, reticular and collagenous fibrillae in vertebrates. *Stain Technology* 23:69–75.

Loots, J. M., Loots, G. P., and Joubert, W. S.
(1977) A silver impregnation method for nervous tissue suitable for routine use with mounted sections. *Stain Technology* 52:85–87.

Love, A. M., and Vickers, T. H.
(1972) Durable staining of cartilage in foetal rat skeleton by methylene blue. *Stain Technology* 47:7–11.

Love, R.
(1957) Distribution of ribonucleic acid in tumor cells during mitosis. *Nature* 180:1338–39.

(1962) Improved staining of nucleoproteins of the nucleolus. *Journal of Histochemistry and Cytochemistry* 10:227.

Love, R., and Liles, R. H.
(1959) Differentiation of nucleoproteins by inactivation of protein-bound amino groups and staining with toluidine blue and ammonium molybdate. *Journal of Histochemistry and Cytochemistry* 7:164–81.

Lowry, R. J.
- (1963) Aceto-iron-hematoxylin for mushroom chromosomes. *Stain Technology* 38:149–55.

Lucas, A. M., and Jamroz, C.
- (1961) *Atlas of Avian Hematology*. Agriculture Monograph 25. Washington, DC: U.S. Department of Agriculture.

Luft, J. H.
- (1956) Permanganet—a new fixative for electron microscopy. *Journal of Biophysical and Biochemical Cytology* 2:799–801.
- (1959) The use of acrolein as a fixative for light and electron microscopy. *Anatomical Record* 133:305.

Luna, L. G.
- (1964) Further studies of Bodian's technique. *American Journal of Medical Technology* 30:355.
- (1960) *Manual of Histologic Staining Methods of the Armed Forces Institute of Pathology*, 2nd ed. New York: McGraw-Hill.
- (1968) *Manual of Histologic Staining Methods of the Armed Forces Institute of Pathology*, 3rd ed. New York: McGraw-Hill.

Lycette, R. M., Danforth, W. F., Koppel, J. L., and Olwin, J. H.
- (1970) The binding of Luxol fast blue ARN by various biological lipids. *Stain Technology* 45:155–60.

Lynch, M. J., Raphael, S. S., Mellor, L. D., Spare, P. D., and Inwood, M. J. H.
- (1969) *Medical Laboratory Technology and Clinical Pathology*. Philadelphia: W. B. Saunders.

McCann, J. A.
- (1971) Methyl green as a cartilage stain; human embryos. *Stain Technology* 46:263–65.

McClung, C. E.
- (1939) *Handbook of Microscopical Technique*. New York: Hoeber.
- (1950) *Handbook of Microscopical Technique* for Workers in Animal and Plant Tissues. New York: Hoeber.

McCormick, J. B.
- (1959a) Technic for restaining faded histopathologic slides. *Technical Bulletin of the Registry of Medical Technologists* 29:13–14.
- (1959b) One hour paraffin processing technic for biopsy of bone marrow and other tissues, and specimens of fluid sediment. *American Journal of Clinical Pathology* 31:278–79.
- (1961) Color preservation of gross museum specimens. *Archives of Pathology* 72:82–85.
- (1987) *History of Microscopes and Microscopic Technique*. Chicago: Science Heritage.

McDonald, D. M.
- (1964) Silver impregnation of Golgi apparatus with subsequent nitrocellulose embedding. *Stain Technology* 39:345–49.

McGee-Russell, S. M.
- (1958) Histochemical methods for calcium. *Journal of Histochemistry and Cytochemistry* 6:22–42.

McGuire, S. R., and Opel, H.
(1969) Resorcin-fuchsin staining of neurosecretory cells. *Stain Technology* 44: 235–37.
McManus, J. F. A.
(1948) Histological and histochemical uses of periodic acid. *Stain Technology* 23:99–108.
(1961) Periodate oxidation techniques. *General Cytochemical Methods*, II. Ed. by J. F. Danielli. New York: Academic Press.
McManus, J. F. A., and Mowry, R. W.
(1958) Effects of fixation on carbohydrate histochemistry. *Journal of Histochemistry and Cytochemistry* 6:309–16.
(1960) *Staining Methods: Histologic and Histochemical*. New York: Hoeber.
Maggi, V., and Riddle, P. N.
(1965) Histochemistry of tissue culture cells: a case study of the effects of freezing and of some fixatives. *Journal of Histochemistry and Cytochemistry* 13: 310–17.
Mahoney, R.
(1968) *Laboratory Techniques in Zoology*. London: Butterworth.
Malhotra, S. K.
(1961) Coloration of the Golgi-Nissl network in a vertebrate neuron by Sudan black. *Quarterly Journal of Microscopical Science* 102:387–89.
Malinin, T. I.
(1961) Feulgen-oxidized tannin-azo (FOTA) technique for concomitant staining of DNA and protein. *Stain Technology* 36:198–200.
Mallory, F. B.
(1944) *Pathological Technique*. Philadelphia: W. B. Saunders.
Manheimier, L. H., and Seligman, A.
(1949) Improvement in the method for histochemical demonstration of alkaline phosphatase and its use in a study of normal and neoplastic tissues. *Journal of the National Cancer Institute* 9:181–99.
Manns, E.
(1960) Combined myelin-Nissl stain. *Stain Technology* 35:349–51.
Manwell, R. D.
(1945) The JSB stain for blood parasites. *Journal of Laboratory and Clinical Medicine* 30:1078–82.
Margolis, G., and Pickett, J. P.
(1956) New applications of the Luxol fast blue myelin stain. *Laboratory Investigation* 5:459–74.
Martin, J. H.
(1966) A different organism for the demonstration of giant salivary chromosomes. *Turtox News* 4:178–80.
Martin, J. H., Lynn, J. A., and Nickey, W. M.
(1966) A rapid polychrome stain for epoxy-embedded tissue. *American Journal of Clinical Pathology* 46:250–51.
Masek, B., and Birns, M.
(1961) Advantages of the polyvinyl alcohol-glycerol embedding method for enzyme histochemistry. *Journal of Histochemistry and Cytochemistry* 9: 634–35.

Massignani, A., and Malferrari, R.
(1961) Phosphotungstic acid-eosin combined with hematoxylin as a stain for Negri bodies in paraffin sections. *Stain Technology* 36:5–8.

Massignani, A., and Refinetti, E. M.
(1958) The Papanicolaou stain for Negri bodies in paraffin sections. *Stain Technology* 33:197–99.

Masson, P.
(1928) Carcinoids (argentaffin cell tumors) and nerve hyperplasia of the appendicular mucosa. *American Journal of Pathology* 4:181–211.

Mayer, D. M., Hampton, J. C., and Rosario, B.
(1961) A simple method for removing the resin from epoxy-embedded tissue. *Journal of Biophysical and Biochemical Cytology* 9:909–10.

Maynard, J. H.
(1986) A trichrome stain in glycol methacrylate that works. *Laboratory Medicine* 8:471–73.

Mayner, D. A., and Ackerman, G. A.
(1962) Histochemical demonstration of tissue ribonuclease activity. *Journal of Histochemistry and Cytochemistry* 10:687.
(1963) Tissue localization of ribonuclease activity by the substrate film technique. *Journal of Histochemistry and Cytochemistry* 11:573–77.

Mellors, R. C.
(1959) *Analytical Cytology.* New York: McGraw-Hill.

Meloan, S. N., and Puchtler, H.
(1974) Iron alizarin blue S stain for nuclei. *Stain Technology* 49:301–4.
(1986a) On the structure and chemistry of leucofuchsin and Schiff's reagent. *Journal of Histotechnology* 9:119–22.
(1986b) Sulfonic versus sulfinic acid formula of aldehyde-Schiff's compounds. *Journal of Histotechnology* 9:31–33.

Menzies, D. W.
(1963) Red-blue staining of hydrolyzed nucleic acids in paraffin sections. *Stain Technology* 38:157–60.

Mercer, E. H., and Birbeck, C.
(1961) *Electron Microscopy: A Handbook for Biologists.* Springfield, IL: Charles C Thomas.

Merton, H.
(1932) Gestalterhaltende Fixierungversuche an besonders kontractilen Infusorien. *Archiv für Protistenkunde* 77:449–521.

Meryman, H. T.
(1959) Sublimation freeze-drying without vacuum. *Science* 130:628–29.
(1960) Freezing and drying of biological materials. Part III. Principles of freeze-drying. *Annals of the New York Academy of Sciences* 85:501–734.

Metcalf, R. L., and Paton, R. L.
(1944) Fluorescence microscopy applied to entomology and allied fields. *Stain Technology* 19:11–27.

Mettler, F. A.
(1932) The Marchi method for demonstrating degenerated fiber connections within the central nervous system. *Stain Technology* 7:95–106.

Mettler, F. A., and Hanada, R. E.
>(1942) The Marchi method. *Stain Technology* 17:111–16.

Mettler, S., and Bartha, A. S.
>(1948) Brilliant cresyl blue as a stain for chromosome smear preparations. *Stain Technology* 23:27–28.

Metz, G.
>(1976) Mahon's myelin stain for celloidin-embedded nervous tissue. *Stain Technology* 51:59–61.

Meyer, J. R.
>(1983) The preparation of immunoglobulin gold conjugates (IGS reagents) and their use as markers for light and electron microscopic immunocytochemistry. *Immunohistochemistry*. Ed. by A. C. Cuello. Chichester, England: John Wiley and Sons, pp. 347–72.

Mikat, K., and Mikat, D. M.
>(1973) Fixation of tissue by formaldehyde vapor during centrifugation; a means of enhancing morphological visibility by cellular flattening. *Stain Technology* 48:33–37.

Miller, O. L., Stone, G. E., and Prescott, D. M.
>(1964) Autoradiography of water-soluble materials. *Methods in Cell Physiology*, I. Ed. by D. M. Prescott. New York: Academic Press.

Miller, P. J.
>(1971) An elastin stain. *Medical Laboratory Technology* 28:148–49.

Milligan, M.
>(1946) Trichrome stain for formalin-fixed tissue. *American Journal of Clinical Pathology, Technical Section* 10:184–85.

Millonig, G.
>(1961) Advantages of a phosphate buffer for OsO_4 solutions in fixation. *Journal of Applied Physiology* 32:1637.

Mitchell, B. S.
>(1975) A rapid, reliable modification of Mallory's phosphotungstic acid hematoxylin (PTAH) method for astrocytes using Susa fixative as a mordant. *Medical Laboratory Technology* 32:331–33.

Mittwer, T., Bartholomew, T. W., and Kallman, B. J.
>(1950) The mechanism of the Gram reaction. 2. The function of iodine in the Gram stain. *Stain Technology* 25:169–79.

Moffat, D. B.
>(1958) Demonstration of alkaline phosphate and periodic acid-Schiff positive material in the same section. *Stain Technology* 33:225–28.

Mohr, J. L.
>(1950) On the natural coloring matter, grazilin, and its use in microscopical technique. *Micro-notes* 5:4–16.

Mohr, J. L., and Wehrle, W.
>(1942) Notes on mounting media. *Stain Technology* 17:157–60.

Moldovanu, G.
>(1961) L'identification de la chromatine de sexe chez les chimères hématologiques canines. *Revue Française d'Études Cliniques et Biologiques* 6:165–67.

Moline, S. W., and Glenner, G. G.
  (1964) Ultra-rapid tissue freezing in liquid nitrogen. *Journal of Histochemistry and Cytochemistry* 12:777–83.
Moliner, E. R.
  (1957) A chlorate-formaldehyde modification of the Golgi method. *Stain Technology* 32:105–16.
Mollenhauer, H. H.
  (1964) Plastic embedding mixtures for use in electron microscopy. *Stain Technology* 39:111–14.
Molnar, L. M.
  (1974) Double embedding with nitrocellulose and paraffin. *Stain Technology* 49:311.
Moloney, W. C., McPherson, K., and Fliegelman, L.
  (1960) Esterase activity in leukocytes demonstrated by use of naphthol AS-D chloracetate substrate. *Journal of Histochemistry and Cytochemistry* 8:200–207.
Monroe, C. W., and Spector, B.
  (1963) Tannic acid, iron hematoxylin, Alcian blue and basic fuchsin for staining islets and reticular fibers of the pancreas. *Stain Technology* 38:187–92.
Moody, M. D., Ellis, E. C., and Updyke, L.
  (1958) Staining bacterial smears with fluorescent antibody. IV. Grouping streptococci with antibody. *Journal of Bacteriology* 75:553–60.
Moore, A. R.
  (1962) Collecting and preserving the developmental stages of the Pacific coast bat starfish, *Patiria miniata*. *Ward's Bulletin* 1.
Moore, K. L., and Barr, M. L.
  (1955) Smears from the oral mucosa in the detection of chromosomal sex. *Lancet* 269:57–58.
Moore, K. L., Graham, M. A., and Barr, M. L.
  (1953) The detection of chromosomal sex in hermaphrodites from a skin biopsy. *Surgery, Gynecology and Obstetrics* 96:641–48.
Moree, R.
  (1944) Control of the ferric ion concentration in iron-acelocarmine staining. *Stain Technology* 19:103–8.
Morrison, J. H., and Kronheim, S.
  (1962) The cytochemical demonstration of succinic dehydrogenase in mouse leukocytes. *Journal of Histochemistry and Cytochemistry* 10:402–11.
Morrison, M., and Samwick, A. A.
  (1940) Restoration of overstained Wright films and a new method of staining blood smears. *American Journal of Clinical Pathology Technical Supplement* 4:92–93.
Mortreuil-Langlois, M.
  (1962) Staining sections coated with radiographic emulsion; a nuclear fast red, indigo-carmine sequence. *Stain Technology* 37:175–77.
Mote, R. F., Muhm, R. L., and Gigstad, D. C.
  (1975) A staining method using acridine orange and auramine O for fungi and mycobacteria in bovine tissue. *Stain Technology* 50:5–9.

Mowry, R. W.
- (1956) Alcian blue technique for histochemical study of acidic carbohydrates. *Journal of Histochemistry and Cytochemistry* 4:407.
- (1958) Improved procedure for the staining of acid polysaccharides by Mueller's colloidal (hydrous) ferric oxide and its combination with the Fuelgen and the periodic Schiff reactions. *Laboratory Investigations* 7:566.
- (1959) Effect of periodic acid used prior to chromic acid on the staining of polysaccharides by Gomori's methenamien silver. *Journal of Histochemistry and Cytochemistry* 7:288.
- (1960) Revised method producing improved coloration of acidic mucopolysaccharides with Alcian blue 8GX supplied currently. *Journal of Histochemistry and Cytochemistry* 8:323.

Mundkur, B., and Brauer, B.
- (1966) Selective localization of nucleolar protein with Amidoblack 10B. *Journal of Histochemistry and Cytochemistry* 14:94–103.

Murdock, T. F., and Fratkin, J. D.
- (1988) The Jamarri silver technique: a new staining technique modification for senile plaques and neurofibrillary tangles. *Laboratory Medicine* 19:109–10.

Murgatroyd, L. B.
- (1971) Chemical and spectrometric evaluation of glycogen after routine histological fixatives. *Stain Technology* 46:111–19.

Murgatroyd, L. B., and Horobin, R. W.
- (1969) Specific staining of glycogen and hematoxylin and certain anthraquinone dyes. *Stain Technology* 44:59–62.

Nachlas, M., Tsou, K. C., Desouza, E., Cheng, C. S., and Seligman, A. M.
- (1957) Cytochemical demonstration of succinic dehydrogenase by the use of a new p-nitrophenyl substituted ditetrazole. *Journal of Histochemistry and Cytochemistry* 5:420–36.

Naish, S. J., ed.
- (1989) Handbook for immunocytochemical staining methods. Carpenteria, CA: Dak Corporation.

Naoumenko, J., and Feigin, I.
- (1974) A simple silver solution for staining reticulum. *Stain Technology* 49:153–55.

Nash, D., and Plaut, W.
- (1964) On the denaturation of chromosomal DNA in situ. *Proceedings of the National Academy of Sciences* 51:731–35.

Nassar, T. K., and Shanklin, W. M.
- (1961) Simplified procedure for staining reticulum. *Archives of Pathology* 71:611/21–614/24.

Nassonov, D. N.
- (1923) Das Golgische Binnennetz und seiene Beziehungen zu der Sekretion. Untersuchungen über einige Amphibiendrüsen. *Archiv für Mikroskopische Anatomie und Entwicklungsmechnak* 97:136–86.
- (1924) Das Golgische Binnennetz und seine Beziehungen zu der Sekretion. Mophologische und experimentelle Untersuchungen an einigen Säugetier-

drüsen. *Archiv für Mikroskopische Anatomie und Entwicklungsmechanik* 100: 433–72.

Nauman, R. V., West, P. W., Trou, F., and Geake, G. C.
(1960) A spectroscopic study of the Schiff reaction as applied to the quantitative determination of sulfur dioxide. *Annals of Chemistry* 32:1307–11.

Nauta, W. J. H., and Gygax, P. A.
(1951) Silver impregnation of degenerating axon terminals in the central nervous system. I. Technic. Chemical notes. *Stain Technology* 26:5–11.

Nauta, W. J. H., and Ryan, L. F.
(1952) Selective silver impregnation of degenerating axons in the central nervous system. *Stain Technology* 27:175–79.

Navagiri, S. S., and Dubey, P. N.
(1976) Simple method of staining amyloid deposits. *Journal of the Anatomical Society of India* 25:45.

Nayebi, M.
(1971) Immunofluorescent technique for diagnosis of *Entamoeba histolytica* strains. *Medical Laboratory Technology* 28:413.

Nelson, E. V.
(1974) A simple technique for sectioning honey bee abdomens. *Stain Technology* 49:117–18.

Newcomer, E. A.
(1940) An osmic impregnation method for mitochondria in plant cells. *Stain Technology* 15:89–90.

Newcomer, E. A., and Donnelly, G. M.
(1963) Leukocyte culture for chromosome studies in the domestic fowl. *Stain Technology* 38:54–56.

Nolte, D. J.
(1948) A modified technique for salivary gland chromosomes. *Stain Technology* 23:21–25.

Norenburg, J. L., and Barrett, M.
(1987) Steedman's polyester wax embedment and de-embedment for combined light and scanning electron microscopy. *Journal of Electronmicroscopy Technique* 6:35–41.

Norris, W. P., and Jenkins, P.
(1960) Epoxy resin embedding in contrast radioautography of bones and teeth. *Stain Technology* 35:253–60.

Norton, W. T., Korey, S. R., and Brotz, M.
(1962) Histochemical demonstration of unsaturated lipids by a bromine-silver method. *Journal of Histochemistry and Cytochemistry* 10:83–88.

Notenboom, C. D., van de Veerdonk, F., and de Kramer, J. C.
(1967) A fluorescent modification of the Sakaguchi reaction on arginine. *Histochemie* 9:117–21.

Novelli, A.
(1962) A short method for chondriome. *Journal of Histochemistry and Cytochemistry* 10:102–3.

Novikoff, A. B.
(1960) Biochemical and staining reactions of cytoplasmic constituents. *Develop-*

ing Cell Systems and Their Control. Ed. by D. Rudnick. New York: Ronald Press.

(1961a) Mitochondria (chondriosomes). *The Cell,* II. Ed. by J. Brachet and A. E. Mirsky. New York: Academic Press.

(1961b) Lysosomes and related particles. *The Cell,* II. Ed. by J. Brachet and A. E. Mirsky. New York: Academic Press.

Novikoff, A. B., Goldfisher, S., and Essner, E.

(1961) The importance of fixation in a cytochemical method for the Golgi apparatus. *Journal of Histochemistry and Cytochemistry* 9:459–60.

Novikoff, A. B., Shin, W.-Y., and Drucker, J.

(1960) Cold acetone fixation for enzyme localization in frozen sections. *Journal of Histochemistry and Cytochemistry* 8:37–40.

Novotney, G. E. K., and Novotney, E.

(1974) Glees silver for degenerating nerve tissue. *Stain Technology* 49:273.

(1977) Triple staining of normal and degenerating nervous tissue. *Stain Technology* 52:97–99.

Nowell, P. C.

(1960) Phytohemagglutin: an initiator of mitosis in cultures of normal human leucocytes. *Cancer Research* 20:462–66.

Ogawa, K., Mizuno, N., and Okamoto, M.

(1961) Cytochemistry of cultured neural tissue. III. Heterogeneity of lysosomes. *Journal of Histochemistry and Cytochemistry* 9:625.

Ohno, S.

(1965) Direct handling of germ cells. *Human Chromosome Methodology.* Ed. by J. J. Yunis. New York: Academic Press.

Ojeda, J. L., Barbosa, E., and Bosque, P. G.

(1970) Selective skeletal staining in whole chicken embryos: a rapid Alcian blue technique. *Stain Technology* 45:137–38.

Owen, G.

(1955) Use of propylene phenoxetol as a relaxing agent. *Nature* 174:434.

Owen, G., and Steedman, H. F.

(1956) Preservation of animal tissues with a note on staining solutions. *Quarterly Journal of Microscopical Science* 97:319–21.

(1958) Preservation of mollusks. *Proceedings of the Malacological Society of London* 33:101–3.

Paddy, J. F.

(1970) Metachromasy of dyes in solution. NATO Advanced Study Institute, St. Margherita [Report]. Ed. by E. A. Balasz. London: Academic Press.

Palkowsky, W.

(1960) A new method of deep freezing pathologic specimens in a pliable state. *Technical Bulletin of the Registry of Medical Technologists* 30:187.

Pantin, C. F. A.

(1946) *Notes on Microscopical Technique for Zoologists.* Cambridge, England: Cambridge University Press.

Papincolaou, G. N.

(1942) A new procedure for staining vaginal smears. *Science* 95:438–39.

(1947) The cytology of the gastric fluid of carcinoma of the stomach. *Journal of the National Cancer Institute* 7:357–60.

(1954) *Atlas of Exfoliative Cytology.* Cambridge: Harvard University Press.

(1957) The cancer diagnostic potential of uterine exfoliative cytology. *CA Bulletin of Cancer Progress* 7:125-35.

Pappas, P. W.

(1971) The use of a chrome alum-gelatin (subbing) solution as a general adhesive for paraffin sections. *Stain Technology* 46:121-24.

Parsons, D. F.

(1970) *Some Biological Techniques in Electron Microscopy.* New York: Academic Press.

Past, W. L.

(1961) The histologic demonstration of iron in osseous tissue. *American Journal of Pathology* 39:443-49.

Patau, K.

(1960) The identification of individual chromosomes, especially in man. *American Journal of Human Genetics* 12:250-76.

(1961) Chromosome identification and the Denver report. *Lancet* 1:933-34.

(1965) Identification of chromosomes. *Human Chromosome Methodology.* Ed. by J. J. Yunis. New York: Academic Press.

Patten, B. M.

(1952) *Early Embryology of the Chick*, 4th ed. New York: Blakiston.

Patten, S. F., and Brown, K. A.

(1958) Freeze-solvent substitution technic. A review with application to fluorescence microscopy. *Laboratory Investigation* 7:209-23.

Pauly, H.

(1964) Über die Konstitution des Histidins. *Hoppe-Seylers Zeitshcrift für Physiologische Chemie* 42:508-18.

Pawley, J., and Smallcomb, A.

(1992) An introduction to practical confocal microscopy: the ultimate form of biological light microscopy? *Acta Microscopica* 1:58-73.

Pearse, A. G. E.

(1968) *Histochemistry, Theoretical and Applied*, I. 3rd ed. Baltimore: Williams and Wilkins.

(1972) *Histochemistry, Theoretical and Applied*, II. 3rd ed. London: Churchill Livingston.

(1985) *Histochemistry—Theoretical and Applied.* Secaucus, NJ: Churchhill-Livingston.

Pearson, B.

(1958) Improvement in the histochemical localization of succinic dehydrogenase by the use of nitroneotetrazolium chloride. *Journal of Histochemistry and Cytochemistry* 6:112-21.

Pearson, B., Wolf, P., and Andrews, M.

(1963) The histochemical demonstration of leucine aminopeptidase by means of a new indoyl compound. *Laboratory Investigation* 12:712-20.

Pease, D. C.

(1964) *Histological Techniques for Electron Microscopy.* New York: Academic Press.

Pelc, S. R.

(1956) A stripping film technique for autoradiography. *International Journal of Applied Radiation and Isotopes* 1:172-77.

Peltier, L. F.
: (1954) The demonstration of fat emboli in tissue sections using phosphin 3R, a water-soluble fluorochrome. *Journal of Laboratory and Clinical Medicine* 43:321–23.

Perry, R. P.
: (1964) Quantitative autoradiography. *Methods in Cell Physiology*, I. Ed. by D. M. Prescott. New York: Academic Press.

Persidsky, M. D.
: (1954) Restoration of deteriorated temporary aceto-carmine preparations. *Stain Technology* 29:278.

Petkó, M.
: (1974) The use of 1:9 dimethyl-methylene blue for the demonstration of thyroid cells. *Stain Technology* 49:65–67.

Petrusz, P., Sar, M., Ordronneau, P., and DiMeo, P.
: (1976) Specificity in immunocytochemical staining. *Journal of Histochemistry and Cytochemistry* 24:407.

Picciano, D. J., and McKinnell, R. G.
: (1977) A short term culture method for the preparation of chromosome spreads from *Rana pipiens. Stain Technology* 52:101–3.

Pickett, J. P., Bisoph, C. M., Chick, E. W., and Baker, R. D.
: (1960) A simple fluorescent stain for fungi. *American Journal of Clinical Pathology* 34:197–202.

Pickett, J. P., and Klavins, J. V.
: (1961) Demonstration of elastic tissue and iron or elastic tissue and calcium simultaneously. *Stain Technology* 36:371–74.

Pickworth, J. W., Cotton, K., and Skyring, A. P.
: (1963) Double emulsion autoradiography of identifying tritium-labelled cells in sections. *Stain Technology* 38:237–44.

Pienaar, U. V.
: (1962) *Hematology of Some South African Reptiles.* Johannesburg: Witwaters and University Press.

Pitelka, D. R.
: (1945) Morphology and taxonomy of flagellates of the genus *Peranema dujardin. Journal of Morphology* 76:179–90.

Pizzolato, P., and McCrory, P.
: (1962) Light influence on von Kóssa's silver calcium reaction in the myocardium. *Journal of Histochemistry and Cytochemistry* 10:102.

Poirier, L. J., Ayotte, R. A., and Gauthier, C.
: (1954) Modification of the Marchi technic. *Stain Technology* 29:71–75.

Poley, R. W., and Forbes, C. D.
: (1964) Fuchsinophilia in early myocardial infarction. *Archives of Pathology* 77:325–29.

Pollak, O. J.
: (1944) A rapid trichrome stain. *Archives of Pathology* 37:294.

Pollister, A. W., and Pollister, P. F.
: (1957) The structure of the Golgi apparatus. *International Review of Cytology*, VI. Ed. by G. H. Bourne and J. F. Danielli. New York: Academic Press.

Pool, C. R.
  (1973) Prestaining oxidation by acidified H_2O_2 for revealing Schiff-positive sites in Epon-embedded sections. *Stain Technology* 48:123–26.
Popp, R. A., Gude, W. D., and Popp, D. M.
  (1962) Peroxidase staining combined with autoradiography for study of eosinophilic granules. *Stain Technology* 37:243–47.
Pottz, G., Rampey, J. H., and Furmandean, B.
  (1964) A method for rapid staining of acid-fast bacteria in smears and sections of tissue. *American Journal of Clinical Pathology* 42:552–54.
Powell, E. W., and Brown, G.
  (1975) A critique of silver impregnation methods. *Mikroscopie* 31:77–84.
Powers, M., and Clark, G.
  (1955) An evaluation of cresyl echt violet as a Nissl stain. *Stain Technology* 30:82–92.
  (1963) A note on Darrow red. *Stain Technology* 38:289–90.
Powers, M., Clark, G., Darrow, M., and Emmel, V. M.
  (1960) Darrow red, a new basic dye. *Stain Technology* 35:19–22.
Preece, A.
  (1972) *A Manual for Histologic Technicians*. Boston: Little, Brown.
Prescott, D. M.
  (1964) Autoradiography with liquid emulsion. *Methods in Cell Physiology*, I. Ed. by D. M. Prescott. New York: Academic Press.
Prescott, D. M., and Bender, M. A.
  (1964) Preparation of mammalian metaphase chromosomes for autoradiography. *Methods in Cell Physiology*, I. Ed. by D. M. Prescott. New York: Academic Press.
Prescott, D. M., and Carrier, R. F.
  (1964) Experimental procedures and cultural methods for *Eupoltes eurystomus* and *Amoeba proteus*. *Methods in Cell Physiology*, I. Ed. by D. M. Prescott. New York: Academic Press.
Priestly, J. V., and Cuello, A. C.
  (1983) Electron microscopic immunocytochemistry for CNS transmitters and transmitter markers. *Immunohistochemistry*. Ed. by A. C. Cuello. Chichester, England: John Wiley and Sons, pp. 273–322.
Procknow, J. J., Connelly, A. P., Jr., and Ray, C. G.
  (1962) Fluorescent antibody technique in histoplasmosis. *Archives of Pathology* 73:313–24.
Proescher, F.
  (1933) Pinacyanol as a histological stain. *Proceedings of the Society for Experimental Biology and Medicine* 31:79–81.
  (1934) Contribution to the staining of neuroglia. *Stain Technology* 9:33–38.
Proescher, F., and Arkush, A. S.
  (1928) Metallic lakes of the oxazines (gallamin blue, gallocyanin, and celestin blue) as nuclear stain substitutes for hematoxylin. *Stain Technology* 2:28–38.
Proffitt, E., Mills, B., Arrington, J., and Sobin, L. H.
  (1992) *Laboratory Methods in Histotechnology*. Washington, DC: American Registry of Pathology.

Puchtler, H.
- (1958) Significance of the iron hematoxylin method of Heidenhain. *Journal of Histochemistry and Cytochemistry* 6:401–2.

Puchtler, H., Chandler, A. B., and Sweat, F.
- (1961) Demonstration of fibrin in tissue sections by the Rosindole method. *Journal of Histochemistry and Cytochemistry* 9:340.

Puchtler, H., Meloan, S. N., and Terry, M. S.
- (1969) On the history and mechanism of alizarin and alizarin red S stains for calcium. *Journal of Histochemistry and Cytochemistry* 17:110–24.

Puchtler, H., Rosenthal, S. I., and Sweat, F.
- (1964) Revision of the amidoblack stain for hemoglobin. *Archives of Pathology* 78:76–78.

Puchtler, H., and Sweat, F.
- (1960) Commercial resorcin-fuchsin as a stain for elastic fibers. *Stain Technology* 35:347–48.
- (1962) Some comments on the ninhydrin-Schiff reaction. *Journal of Histochemistry and Cytochemistry* 10:365.
- (1963a) A combined hemoglobin-hemosiderin stain. *Archives of Pathology* 75:588–90.
- (1963b) Influence of various pretreatments on the staining properties of connective tissue fibers. *Annales d'Histochemie* 8:189–98.
- (1964a) On the mechanism of sequence iron-hematein stains. *Histochemie* 4:197–208.
- (1964b) Histochemical specificity of stain-methods for connective tissue fibers: resorcin-fuchsin and van Gieson's picro-fuchsin. *Histochemie* 4:24–34.
- (1965) Congo red as a stain for fluorescence microscopy of amyloid. *Journal of Histochemistry and Cytochemistry* 13:693–94.
- (1966) A review of early concepts of amyloid in context with contemporary chemical literature of 1839 to 1859. *Journal of Histochemistry and Cytochemistry* 14:123–34.

Puchtler, H., Sweat, F., and Doss, N. O.
- (1963) A one-hour phosphotungstic acid-hematoxylin stain. *American Journal of Clinical Pathology* 40:334.

Puchtler, H., Sweat, F., and Levine, M.
- (1962) On the binding of Congo red by amyloid. *Journal of Histochemistry and Cytochemistry* 10:355–64.

Puchtler, H., Waldrop, F. S., Meloan, S. N., Terry, M. S., and Conner, H. M.
- (1970) Methacarn (methanol-Carnoy) fixation. Practical and theoretical considerations. *Histochemie* 21:97–116.

Puchtler, H., Waldrop, F. S., Sweat, F., Terry, M., and Conner, H. M.
- (1969) A combined myofibril stain for demonstration of early lesions of striated muscle. *Journal of Microscopy* 89:329–38.

Puchtler, H., Waldrop, F. S., Terry, M. S., and Conner, H. M.
- (1973) Fluorescent microscopic distinction between elastin and collagen. *Histochemie* 35:17–30.

Putt, F. A.
- (1948) Modified eosin counterstain for formaldehyde-fixed tissues. *Archives of Pathology* 45:72.

(1971) Alcian dyes in calcium chloride: a routine selective method to demonstrate mucins. *Yale Journal of Biology and Medicine* 43:279–82.

Putt, F. A., and Huskill, P. B.
(1962) Alcian green, a routine stain for mucins. *Archives of Pathology* 74:169–70.

Quintarelli, G., Cifonelli, J. A., and Zito, R.
(1971) On phosphotungstic acid staining, II. *Journal of Histochemistry and Cytochemistry* 19:648–53.

Raman, K.
(1955) A method of sectioning aspirated bone-marrow. *Journal of Clinical Pathology* 8:265–66.

Ramón y Cajal, S.
(1903) Un sencillo metodo de coloracion selectiva del reticilo proto-plasmico y sus effectos en los diversos organos nerviosos. *Trabajos del Laborartorio Investigaciones Biologica* 2:129–21.
(1910) Las formulas del proceder del nitrato de plata reducido. *Trabajos del Laboratorio Investigaciones Biologica* 8:1–26.

Randolph, L. F.
(1935) A new fixing fluid and a revised schedule for the paraffin method in plant cytology. *Stain Technology* 10:95–96.

Ray, H. N., and Hajra, B.
(1962) Hydrosulfilte-Schiff: limitations in use as shown by periodic-Schiff reactions in protozoa. *Stain Technology* 37:75–77.

Reaven, E. P., and Cox, A. J.
(1963) The histochemical localization of histidine in the human epidermis and its relationship to zinc binding. *Journal of Histochemistry and Cytochemistry* 11:782–90.

Renaud, S.
(1959) Superiority of alcoholic over aqueous fixation in the histochemical detection of calcium. *Stain Technology* 34:267–71.

Repak, A. J., and Levine, A. B.
(1967) A new use for an old stain. *Turtox News* 45:227.

Richards, O. W.
(1941) An efficient method for the identification of *M. tuberculosis* with a simple fluorescence microscope. *American Journal of Clinical Pathology Technical Section* 5:1–8.
(1949) *The Effective Use and Proper Care of the Microtome*. Buffalo, NY: American Optical Company.
(1954) *The Effective Use and Proper Care of the Microscope*. Buffalo, NY: American Optical Company.

Richards, O. W., Kline, E. K., and Leach, R. E.
(1941) Demonstration of tubercle bacilli by fluorescence microscopy. *American Review of Tuberculosis* 44:255–66.

Richardson, K. C., Jarret, L., and Finke, E. H.
(1960) Embedding in epoxy resins for ultrathin sectioning in electron microscopy. *Stain Technology* 35:313–23.

Richmond, G. W., and Bennett, L.
(1938) Clearing and staining of embryos for demonstration of ossification. *Stain Technology* 13:77–79.

Rinderknecht, H.
   (1960) A new technique for the fluorescent labelling of proteins. *Experimentia* 16:430–31.

Rinehart, J. F., and Abul-Haj, S. K.
   (1951a) An improved method for histologic demonstration of acid mucopolysaccharides in tissues. *Archives of Pathology* 52:189–94.
   (1951b) Histological demonstration of lipids in tissue after dehydration and embedding in polyethylene glycol. *Archives of Pathology* 51:666–69.

Ritter, C., Di Stephano, H. S., and Farah, A.
   (1961) A method for the cytophotometric estimation of ribonucleic acid. *Journal of Histochemistry and Cytochemistry* 9:97–102.

Roden, D. B.
   (1975) Nitrocellulose sectioning of heads of larval Cerambycidae (Celopter). *Stain Technology* 50:207–8.

Rodriquez, J., and Deinhardt, F.
   (1960) Preparation of a semipermanent medium for fluorescent antibody studies. *Virology* 12:316–17.

Roels, F., Schiller, B., and Goldfisher, S.
   (1970) Peroxisomes and lysosomes in the toad, *Bufo marinus*. *Journal of Histochemistry and Cytochemistry* 18:681–82.

Rogers, A. W.
   (1967) *Techniques of Autoradiography*. New York: Elsevier.

Roman, N., Perkinds, S. F., Perkinds, E. M., Jr., and Dolnick, E. H.
   (1967) Orcein-hematoxylin in iodized ferric chloride as a stain for elastic fibers, with methanil yellow counterstaining. *Stain Technology* 42:199–202.

Romeis, B.
   (1948) *Mikroskopische Technik*. München: Leibniz Verlag.

Rønne, M., Shibaski, Y., Poulsen, P. S., and Anderson, O.
   (1987) The high resolution R-banded karyotype of Rattus noruegicus. *Cytogenetics-Cell Genetics* 45:113–17.

Roozemond, R. C.
   (1969) The effect of calcium and formaldehyde on the release and composition of phospholipids from cryostat sections of rat hypothalamus. *Journal of Histochemistry and Cytochemistry* 17:273–79.

Roque, A. L., Jafarey, N. A., and Coulter, P.
   (1965) A stain for the histochemical demonstration of nucleic acids. *Experimental and Molecular Pathology* 4:266–74.

Rosenthal, S. I., Puchtler, H., and Sweat, F.
   (1965) Paper chromatography of dyes. *Archives of Pathology* 80:190–96.

Rosewater, J.
   (1963) An effective anesthetic for giant clams and other molluscs. *Turtox News* 41:300–2.

Roth, L. J., and Stump, W. E.
   (1969) *Autoradiography of Diffusible Substances*. New York: Academic Press.

Rothenbacher, H. J., and Hitchcock, D. J.
   (1962) Heat fixation and Giemsa staining for flagella and other cellular structures of trichomonads. *Stain Technology* 37:111–13.

Rowson, L. R. A.
: (1974) The role of research in animal production. *Veterinary Record* 95:276–80.

Rubin, R.
: (1951) A rapid method for making permanent mounts of nematodes. *Stain Technology* 26:257–60.

Ruffer, M. A.
: (1921) *Studies in the Paleopathology of Egypt*. Chicago: University of Chicago Press.

Rumpf, P.
: (1935) Recherches physico-chimiques sur la reaction colorie des aldehydes, dite 'Reaction de Schiff.' *Annales de Chimie* 3:327.

Rungby, J., Kassem, M., Ericksen, E. F., Danscher, G., and Aarhus Bone and Research Group.
: (1993) The von Kossa reaction for calcium deposits: silver lactate staining increases sensitivity and reduces background. *Histochemistry* 25:446.

Russell, R.
: (1973) The Bodian stain for nerve fibers and nerve endings. *Laboratory Medicine* 4:40.

Russell, W. O.
: (1941) Histologic technique: the substitution of zinc chloride for mercuric chloride in Zenker's fluid. *Journal of Technical Methodology* 21.

Rutenberg, A. M., Gofstein, R., and Seligman, A. M.
: (1950) Preparation of a new tetrazolium salt which yields a blue pigment on reduction and its use in the demonstration of enzymes in normal and neoplastic tissues. *Cancer Research* 10:113–21.

Rutenberg, A. M., Wolman, M., and Seligman, A. M.
: (1953) Comparative distribution of succinic dehydrogenase in six mammals and modification in the histochemical technic. *Journal of Histochemistry and Cytochemistry* 1:66–81.

Sabatini, D. D., Bensch, K. G., and Barrnett, R. J.
: (1963) Cytochemistry and electron microscopy. The preservation of cellular ultrastructure and enzymatic activity by aldehyde fixation. *Journal of Cell Biology* 17:19–58.

Sabatini, D. D., Miller, F., and Barrnett, J.
: (1964) Aldehyde fixation for morphological and enzyme histochemical studies with the electron microscopy. *Journal of Histochemistry and Cytochemistry* 12:57–71.

Sacks, J.
: (1965) Tracer techniques: stable and radioactive isotopes. *Physical Techniques in Biological Research*, II. Ed. by G. Oster and A. W. Pollister. New York: Academic Press.

St. Amand, G. S., and St. Amand, W.
: (1951) Shortening maceration time for alizarine red S preparations. *Stain Technology* 26:271.

Saksela, E., and Moorehead, P. S.
: (1962) Enhancement of secondary constrictions and the heterochromatic X in human cells. *Cytogenetics* 1:225–44.

Salthouse, T. N.
- (1958) Tetrahydrofuran and its use in insect histology. *Canadian Entomologist* 90:555–57.
- (1962) Luxol-fast blue ARN: a new solvent dye with improved staining qualities for myelin and phospholipids. *Stain Technology* 37:313–16.
- (1964) Luxol-fast blue G as a myelin stain. *Stain Technology* 39:129.

Sams, A., and Davies, F. M. R.
- (1967) Commercial varieties of nuclear fast red; their behavior in staining after autoradiography. *Stain Technology* 42:269–76.

Sanders, P. C., and Humason, G. L.
- (1964) Culture and slide preparation of leukocytes from blood of *Macaca*. *Stain Technology* 39:209–13.

Sanderson, C., and Bloebaum, R. D.
- (1993) Advances in the staining of ground section histology. *Histo-Logic* 1:1–2.

Sandison, A. T.
- (1955) The histological examination of mummified material. *Stain Technology* 30:277–83.

Santamarina, E.
- (1964) A formalin-Wright staining technique for avian blood cells. *Stain Technology* 39:267–74.

Sasaki, H., and Hogan, B. L.
- (1993) Differential expression of multiple for need-related genes during gastrulation and axial pattern formation in the mouse embryo. *Development* 118:47–52.

Sasaki, M.
- (1961) Observations on the modification in size and shape of chromosomes due to technical procedures. *Chromosoma* 11:514–22.

Sasaki, M., and Makino, S.
- (1963) The demonstration of secondary constrictions in human chromosomes by means of a new technique. *American Journal of Human Genetics* 15:24–33.

Sato, T., and Shamoto, M.
- (1973) A simple rapid polychrome stain for epoxy-embedded tissue. *Stain Technology* 48:223–27.

Saunders, A. M.
- (1962) Acridine orange staining for the identification of acid mucopolysaccharides. *Journal of Histochemistry and Cytochemistry* 10:683.

Sawicki, W., and Pawinska, M.
- (1965) Effect of drying on unexposed autoradiographic emulsion in relation to background. *Stain Technology* 40:67–68.

Saxena, P. N.
- (1957) Formalin-chloride fixation to improve silver impregnation of the Golgi apparatus. *Stain Technology* 32:203–7.

Schajowicz, F., and Cabrini, R. L.
- (1955) The effect of acids (decalcifying solutions) and enzymes on the histochemical behavior of bone and cartilage. *Journal of Histochemistry and Cytochemistry* 3:122–29.
- (1959) Histochemical demonstration of acid phosphatase in hard tissues. *Stain Technology* 34:59–63.

Schantz, A., and Schecter, A.
: (1965) Iron-hematoxylin and safranin O as a polychrome stain for epon sections. *Stain Technology* 40:479–82.

Schenk, R. K., Olah, A. J., and Hermann, W.
: (1984) Preparation of calcified tissues for light microscopy. In *Methods of Calcified Tissue Preparation*. Ed. by G. R. Dickson. New York: Elsevier, pp. 1–56.

Schiff, R., Quinn, L. Y., and Bryan, J. H. D.
: (1967) A safranin fast green stain for the differentiation of the nuclei of rumen protozoa. *Stain Technology* 42:75–80.

Schiffer, L. M., and Vaharu, T.
: (1962) Acridine orange as a useful chromosome stain. *Technical Bulletin of the Registry of Medical Technologists* 32:91–92.

Schleifstein, J.
: (1937) A rapid method for demonstrating Negri bodies in tissue sections. *American Journal of Public Health* 27:1283–85.

Schmid, W.
: (1965) Autoradiography of human chromosomes. *Human Chromosome Methodology*. Ed. by J. J. Yunis. New York: Academic Press.

Schmidt, R. W.
: (1956) Simultaneous fixation and decalcification of tissue. *Laboratory Investigation* 5:306–7.

Schneider, J. D.
: (1963) A simplified Gram stain for demonstrating fungi in tissues. *Technical Bulletin of the Registry of Medical Technologists* 33:195–97.

Schreck, C. B., and Moyle, P. B.
: (1990) *Methods for Fish Biology*. Bethesda, MD: American Fisheries Society.

Schreibman, M. P.
: (1986) The pituitary gland. *Vertebrate Endocrinology: Fundamentals and Biomedical Implications*. Ed. by P. K. T. Pang and M. P. Schreibman. New York: Academic Press, 1:11–56.

Schubert, M., and Hamerman, D.
: (1956) Metachromasia: chemical theory and histochemical use. *Journal of Histochemistry and Cytochemistry* 4:159–89.

Schultz, J., and St. Lawrence, P.
: (1949) A cytological basis for a map of the nucleolar chromosome in man. *Journal of Heredity* 40:31–38.

Schultze, B.
: (1969) Autoradiography at the cellular level. *Physical Techniques in Biological Research*, III. 2nd ed. Ed. by A. W. Polliser. New York: Academic Press.

Scott, H. R., and Clayton, B. P.
: (1953) A comparison of the staining affinities of aldehyde-fuchsin and the Schiff reagent. *Journal of Histochemistry and Cytochemistry* 1:336–52.

Scott, J. E.
: (1976) Phosphotungstic acid 'Schiff-reactive' but not a 'glycol reagent.' *Histochemistry* 48:1084–85.

Scott, J. E., and Mowry, R. W.
: (1970) Alcian blue—a consumer's guide. *Journal of Histochemistry and Cytochemistry* 18:842.

Seal, S. H.
(1956) A method for concentrating cancer cells suspended in large quantities of fluid. *Cancer* 9:866–68.

Seligman, A. M., Wasserkrug, H. L., Chandicharan, D., and Hanker, J. S.
(1968) Osmium-containing compounds with multiple basic or acidic groups as stains for ultrastructure. *Journal of Histochemistry and Cytochemistry* 16:87–101.

Serra, J. A.
(1946) Histochemical tests for proteins and amino acids; the characterization of basic proteins. *Stain Technology* 21:5–18.
(1958) Cytochemical demonstration of masked lipids. *Science* 128:28–29.

Shanklin, D. R., and Laite, M. B.
(1963) Pickett-Sommer film strip technique. *Archives of Pathology* 75:91–93.

Shanklin, W. M., and Nassar, T. K.
(1959) Luxol fast blue combined with periodic acid Schiff procedure for cytological staining of kidney. *Stain Technology* 34:257–60.

Shanklin, W. M., Nasser, T. K., and Issidorides, M.
(1959) Luxol fast blue as a selective stain for alpha cells in the human pituitary. *Stain Technology* 34:55–58.

Shapiro, S., and Sohns, L.
(1994) Rapid microwave phosphotungstic acid hematoxylin stain for paraffin and glycomethacrylate sections. *Journal of Histotechnology* 17:127–30.

Shaver, E. L.
(1962) The chromosomes of the opossum, *Didelphis virginiana*. *Canadian Journal of Genetics and Cytology* 4:62–68.

Sheehan, D.
(1960) A comparative study of the histologic techniques for demonstrating chromaffin cells. *American Journal of Medical Technology* 26:237–40.

Sheehan, D., and Hrapchak, B.
(1980) *Theory and Practice of Histotechnology*, 2nd ed. Columbus, OH: Battelle Press.

Shelley, W. B.
(1963) Rapid technic for obtaining leukocytes from small samples of blood. *American Journal of Clinical Pathology* 39:433–35.
(1969) Fluorescent staining of elastic tissue with rhodamine B and related xanthene dyes. *Histochemie* 20:244–49.

Shires, T. K., Johnson, M., and Richter, K. M.
(1969) Hematoxylin staining of tissues embedded in epoxy resins. *Stain Technology* 44:21–25.

Shuman, H., Murray, J. M., and DiLullo, C.
(1989) Confocal microscopy. *Biotechniques* 7:154–62.

Shute, P. G., and Maryon, M. E.
(1966) *Laboratory Technique for the Study of Malaria*, 2nd ed. London: J. and A. Churchill.

Shyamasundari, K., and Rao, K. H.
(1975) A procedure for the simultaneous demonstration of neurosecretory and mucosubstances in tissue sections. *Acta Histochemica* 54:272–74.

Sidman, R. I., Mottla, P. A., and Feder, N.
  (1961) Improved polyester wax embedding for histology. *Stain Technology* 36: 279–84.
Sieracki, J., Michael, J. E., and Clark, D. A.
  (1960) The demonstration of beta cells in pancreatic islets and their tumors. *Stain Technology* 35:67–69.
Sills, B., and Marsh, W. H.
  (1959) A simple technic for staining fat with oil-soluble azo dyes. *Laboratory Investigation* 8:1006–9.
Silver, M. L.
  (1942) Colloidal factors controlling silver staining. *Anatomical Record* 82: 507–29.
Silverstein, A. M.
  (1957) Contrasting fluorescent labels for two antibodies. *Journal of Histochemistry and Cytochemistry* 5:94–95.
Simmel, E. B.
  (1957) The use of a fast, coarse grain stripping film for radioautography. *Stain Technology* 32:299–300.
Simmel, E. B., Fitzgerald, P. J., and Godwin, J. T.
  (1951) Staining of radioautography with metanil yellow and iron hematoxylin. *Stain Technology* 26:25–28.
Simpson, C. F., Carlisle, J. W., and Mallard, L.
  (1970) Rhodanile blue: a rapid and selective stain for Heinz bodies. *Stain Technology* 45:221–23.
Singer, M.
  (1952) Factors which control the staining of tissues with acid and basic dyes. *International Review of Cytology*, I. Ed. by G. H. Bourne and J. F. Danielli. New York: Academic Press.
Sinha, R. N.
  (1953) Sectioning insects with sclerotized cutile. *Stain Technology* 28:249–53.
Sisken, J. E.
  (1964) Methods for measuring the length of the mitotic cycle and the timing of DNA synthesis for mammalian cells in culture. *Methods in Cell Physiology*, I. Ed. by D. M. Prescott. New York: Academic Press.
Slidders, W., Fraser, D. S., Smith, R., and Lendrum, A. C.
  (1958) On staining the nucleus red. *Journal of Pathology and Bacteriology* 75: 466–68.
  (1969) A stable iron-hematoxylin solution for staining chromatin of cell nuclei. *Journal of Microscopy* 90:61–65.
Sloper, J. C., and Adams, C. W. M.
  (1956) The hypothalamic elaboration of posterior pituitary principles in man. Evidence derived from hypophysectomy. *Journal of Pathology and Bacteriology* 72:587–602.
Smith, E. J., Puchtler, H., and Sweat, F.
  (1966) Investigation of the chemical mechanism of trichrome stain. *Laboratory Investigation* 15:1141–42.

Smith, G. S.
>(1943) A danger attending the use of ammoniacal solutions of silver. *Journal of Pathology and Bacteriology* 55:227–28.

Smith, L.
>(1947) The acetocarmine technique. *Stain Technology* 22:17–31.

Smith, R.
>(1994) *Microscopy and Photomicrography: A Working Manual*, 2nd ed. Boca Raton, FL: CRC Press.

Smyth, J. D.
>(1944) A technic for mounting free-living protozoa. *Science* 100:62.

Snodgrass, A. B., and Lacey, L. B.
>(1961) Luxol fast blue staining of degenerated myelinated fibers. *Anatomical Record* 140:83–89.

Snodgrass, M. A., Dorsey, C. H., Bailey, G. W. H., and Dickson, L. G.
>(1972) Conventional histopathologic staining methods compatible with Epon-embedded osmicated tissue. *Laboratory Investigation* 26:329–37.

Sodeman, W. A., Jr., and Jeffery, G. M.
>(1966) Indirect fluorescent antibody test for malaria antibody. *Public Health Reports* 81:1037–41.

Solcia, E., Capella, C., and Vassallo, G.
>(1969) Lead hematoxylin as a stain for endocrine cells. *Histochemie* 20:116–26.

Solcia, E., Vassallo, G., and Capella, C.
>(1968) Selective staining of endocrine cells by basic dyes after acid hydrolysis. *Stain Technology* 43:257–63.

Spicer, S. S.
>(1961) Differentiation of nucleic acids by staining at controlled pH and by Schiff-methylene blue sequence. *Stain Technology* 36:337–40.
>(1962) Histochemically selective acidophilia of basic nucleoproteins in chromatin and nucleoli at alkaline pH. *Journal of Histochemistry and Cytochemistry* 10:691–703.

Spicer, S. S., Horn, R. G., and Leppi, T. J.
>(1967) The connective tissue. *Histochemistry of Connective Tissue Mucopolysaccharides*. International Academy of Pathology Monograph. Ed. by B. M. Wagner and D. E. Smith. Baltimore: Williams and Wilkins.

Spicer, S. S., and Lillie, R. D.
>(1961) Histochemical identification of basic proteins with Biebrich scarlet at alkaline pH. *Stain Technology* 36:365–70.

Spicer, S. S., and Meyer, D. B.
>(1960) Histochemical differentiation of acid mucopolysaccharides by means of combined aldehyde fuchsin-Alcian blue staining. *American Journal of Clinical Pathology Technical Section* 33:453–56.

Spurlock, B. O., Skinner, M. S., and Kattine, A. A.
>(1966) A simple rapid method for staining epoxy-embedded specimens for light microscopy with the polychromatic stain, Paragon-1301. *American Journal of Clinical Pathology* 46:252–58.

Srivastava, P. K., and Lasley, J. F.
>(1968) Leucocyte culture and chromosome preparation from pig blood. *Stain Technology* 43:187–90.

Staples, T. C.
 (1991) Modified silver impregnation stain for nerve tissue. National Society for Histotechnology Workshop.
Staples, T. C., and Clark, L. P.
 (1990) Dilute ammoniacal silver nitrate solutions for the demonstration of reticulum and argentaffin granules. *Journal for Histotechnology* 13:137–39.
Staples, T. C., and Grizzle, W. E.
 (1986a) The effects of temperature on the argentaffin reaction: development of a rapid high temperature argyrophil procedure. *Stain Technology* 62:41–49.
 (1986b) A methyl green nuclear stain for argyrophil procedures. *Laboratory Medicine* 17:532–34.
Stearn, A. E., and Stern, E.
 (1929) The mechanism of staining explained on a chemical basis. I. The reaction between dyes, protein, and nucleic acid. *Stain Technology* 4:111–19.
 (1930) The mechanism of staining explained on a chemical basis. II. General considerations. *Stain Technology* 5:17–24.
Steedman, H. F.
 (1947) Ester wax: a new embedding medium. *Quarterly Journal of Microscopical Science* 88:123–33.
 (1960) *Section Cutting in Microscopy*. Springfield, IL: Charles C Thomas.
Steil, W.
 (1936) Modified Wright's method for staining blood smears. *Stain Technology* 11:99–100.
Steinman, I. D.
 (1955) A vinyl plastic coating for stained thin smears. *Stain Technology* 30:49–50.
Stenram, U.
 (1962) Loss of silver grains from radioautographs stained by gallocyanin–chrome alum. *Stain Technology* 37:231–34.
Sterchi, D. L., and Eurell, J.
 (1989) A new method for preparation of undecalcified bone sections. *Stain Technology* 64:201–5.
 (1995) An evaluation of methylmethacrylate mixtures for hard tissues. *Journal of Histotechnology* 1:45–49.
Sterling, C., and Chichester, C. O.
 (1956) Autoradiography of water-soluble materials in plant tissues. *Stain Technology* 31:227–30.
Sternberger, L. A.
 (1979) *Immunocytochemistry*. New York: John Wiley and Sons.
Sternberger, L. A., Hardy, P. H., Cuculis, J. J., and Meyer, H. G.
 (1970) The unlabeled antibody-enzyme method histochemistry: preparation and properties of soluble antigen-antibody complex (horseradish peroxidase-antihorseradish peroxidase) and its use in identification of spirochetes. *Journal of Histochemistry and Cytochemistry* 18:315–33.
Stone, G. E., and Cameron, I. L.
 (1964) Methods for using *Tetrahymena* in studies of the normal cell cycle. *Methods in Cell Physiology*, I. Ed. by D. M. Prescott. New York: Academic Press.
Stowell, R. E.
 (1945) Feulgen reaction for thymonucleic acid. *Stain Technology* 20:45–58.

(1946) The specificity of the Feulgen reaction for thymonucleic acid. *Stain Technology* 21:137–48.

Straus, W.
(1964) Factors affecting the cytochemical reaction of peroxidase with benzidine and the stability of the blue reaction product. *Journal of Histochemistry and Cytochemistry* 12:462–69.

(1967) Methods for the study of small phagosomes and their relationship to lysosomes with horseradish peroxidase as a 'marker' protein. *Journal of Histochemistry and Cytochemistry* 15:375–80.

Stumpf, W., and Roth, L. J.
(1964) Vacuum freeze drying of frozen sections for dry-mounting high-resolution autoradiography. *Stain Technology* 39:219–23.

Swaab, D. F., Pool, C. W., and Van Leeuwen, F. W.
(1977) Can specificity ever be proved in immunocytochemical staining? *Journal of Histochemistry and Cytochemistry* 25:338.

Swank, R. L., and Davenport, H. A.
(1934a) Marchi's staining method. I. Studies of some of the underlying mechanisms involved. *Stain Technology* 9:11–19.

(1934b) Marchi's staining method. II. Fixation. *Stain Technology* 9:129–35.

(1935a) Marchi's staining method. III. Artifacts and effects of perfusion. *Stain Technology* 10:45–52.

(1935b) Chlorate-osmic-formalin method for staining degenerating myelin. *Stain Technology* 10:87–90.

Sweat, F., Meloan, S. N., and Puchtler, H.
(1968) A modified one-step trichrome stain for demonstration of fine connective tissue fibers. *Stain Technology* 38:179–85.

Sweat, F., Puchtler, H., and Rosenthal, S. I.
(1964) Sirius red F3BA as a stain for connective tissue. *Archives of Pathology* 78:69–72.

Sweat, F., Puchtler, H., and Sesta, J. I.
(1968) PAS-phosphomolybdic acid-Sirius supra blue FGL-CF. *Archives of Pathology* 86:33–39.

Sweat, F., Puchtler, H., and Woo, P.
(1964) A light-fast modification of Lillie's allochrome stain. *Archives of Pathology* 78:73–75.

Swigart, R. H., Wagner, C. E., and Atkinson, W. B.
(1960) The preservation of glycogen in fixed tissues and tissue sections. *Journal of Histochemistry and Cytochemistry* 8:74–75.

Swisher, B. L.
(1987) Modified Steiner procedure for microwave staining of spirochetes and nonfilamentous bacteria. *Journal for Histotechnology* 10:241–43.

Taft, P. D., and Arizaga-Cruz, J. M.
(1960) A comparison of the cell block, Papanicolaou, and Millipore filter technics for the cytological examination of serous fluids. *Technical Bulletin of the Registry of Medical Technologists* 30:189–92.

Takaya, K.
(1967) Luxol fast blue MBS and phloxine; a stain for mitochondria. *Stain Technology* 42:207–11.

Tan, K. H.
> (1973) Peracetic acid as an oxidizer to replace permanganate in the staining of neurosecretory cells. *Stain Technology* 48:140–41.

Tanaka, R.
> (1961) Aceto-basic fuchsin as a stain for nucleoli and chromosomes of plants. *Stain Technology* 36:325–27.

Tandler, C. J.
> (1955) The reaction of nucleoli with ammoniacal silver in darkness; additional data. *Journal of Histochemistry and Cytochemistry* 3:196–202.
> (1974) A method for the selective removal of deoxyribonucleic acid from tissue sections. *Stain Technology* 49:147–52.

Taylor, C. R.
> (1993) *Immunomicroscopy: A Diagnostic Tool for the Surgical Pathologist*. Philadelphia: Saunders.

Taylor, J. D.
> (1965) Gelatin embedding on the tissue carrier for thin serial sections in the cryostat. *Stain Technology* 40:29–31.

Taylor, W. R.
> (1967) An enzyme method of clearing and staining small vertebrates. *Proceedings of U.S. National Museum* 122:1–17.

Tepper, H. B., and Gifford, E. M.
> (1962) Detection of ribonucleic acid with pyronin. *Stain Technology* 37:52–53.

Terner, J. Y., and Clark, G. R.
> (1960a) Gallocyanin–chrome alum: I. Technique and specificity. *Stain Technology* 35:167–68.
> (1960b) Gallocyanin–chrome alum: II. Histochemistry and specificity. *Stain Technology* 35:305–11.

Thieme, G.
> (1965) Small tissue dryers with high capacity for rapid freeze-drying. *Journal of Histochemistry and Cytochemistry* 13:386–89.

Thomas, J. T.
> (1953) Phloxine-methylene blue staining of formalin fixed tissue. *Stain Technology* 28:311–12.

Thomas, L. E.
> (1950) An improved arginine histochemical method. *Stain Technology* 25:143–48.

Thompson, E. C.
> (1961) Simultaneous staining of reticulocytes and Heinz bodies with new methylene blue N in dogs given iproniazid. *Stain Technology* 36:38–39.

Thompson, S. W.
> (1966) *Selected Histochemical and Histological Methods*. Springfield, IL: Charles C Thomas.

Thorgaard, G. H., and Disney, J. E.
> (1990) Chromosome preparation and analysis. *Methods for Fish Biology*. Ed. by Schreck, C. B. and Moyle, P. B. Bethesda, MD: American Fisheries Society, pp. 171–190.

Thurston, J. M., and Joftes, D. L.
- (1963) Stains compatible with dipping radioautography. *Stain Technology* 38: 231–34.

Titford, M.
- (1993) George Grubler and Karl Hollborn: two founders of the Biological Stain Commission. *Journal of Histotechnology* 2:155–58.
- (1996) Safety considerations in the use of silver stains. *Journal of Histotechnology* 19:197–202.

Tomlinson, W. J., and Grocott, R. G.
- (1944) A simple method of staining malaria protozoa and other parasites in paraffin sections. *American Journal of Clinical Pathology* 14:316–26.

Toren, D. A.
- (1963) A Giemsa-trichrome stain for mast cells in paraffin sections. *Stain Technology* 38:249–50.

Trott, J. R.
- (1961a) The presence of glycogen in the rat liver following in vitro processing in decalcifying agents. *Journal of Histochemistry and Cytochemistry* 9:699–702.
- (1961b) An evaluation of methods commonly used for the fixation and staining of glycogen. *Journal of Histochemistry and Cytochemistry* 9:703–10.

Trott, J. R., Gorenstein, S. L., and Peikoff, M. D.
- (1962) A chemical and histochemical investigation of glycogen in rat liver and palate following treatment with various fixatives and ethylenediamine tetraacetic acid. *Journal of Histochemistry and Cytochemistry* 10:245–49.

Trump, B. F., Smuckler, E. A., and Benditt, E. P.
- (1961) A method for staining epoxy sections for light microscopy. *Journal of Ultrastructure Research* 5:343–48.

Tuan, H.
- (1930) Picric acid as a destaining agent for iron alum hematoxylin. *Stain Technology* 5:135–38.

Turchini, J., and Malet, P.
- (1965) Long conservation of histoenzymatic activities of fresh tissues in glycerol. *Journal of Histochemistry and Cytochemistry* 13:405–6.

Upadhya, M. D.
- (1963) The use of a α-bromo-naphthalene, rapid hot fixation, and distributed pressure squashing for chromosomes of Triticinae. *Stain Technology* 38:293–95.

Vacca, J.
- (1985) *Laboratory Manual of Histochemistry*. Hagerstown, MD: Lippincott-Raven.

Vacek, Z., and Plackova, A.
- (1959) Silver impregnation of nerve fibers in teeth after decalcification with ethylenediaminetetracetic acid. *Stain Technology* 34:1–3.

Van Cleave, H. J., and Ross, J. A.
- (1947) A method of reclaiming dried zoological specimens. *Science* 105:318.

Van den Pol, A. N.
- (1986) Tyrosine hydroxylase immunoreactive neurons throughout the hypothalamus receive glutamate decarboxylase immunoreactive synapses: a double

pre-embedding immunocytochemical study with particulate silver and HRP. *Journal of Neuroscience* 6:877–91.

Van Duijn, P.
    (1961) Acrolein-Schiff, a new staining method for proteins. *Journal of Histochemistry and Cytochemistry* 9:234–41.

Vassar, P. S., and Culling, C. F. A.
    (1959) Fluorescent stains with special reference to amyloid and connective tissue. *Archives of Pathology* 68:487–98.
    (1962) Fluorescent amyloid staining of casts in myeloma nephrosis. *Archives of Pathology* 73:59–63.

Verino, D. M., and Laskin, D. M.
    (1960) Sex chromatin in mammalian bone. *Science* 132:675–76.

Verrling, J. M., and Thompson, D. E.
    (1972) A polychrome stain for use in parasitology. *Stain Technology* 47:164–65.

Vidal, O. R., Aya, T., and Sandberg, A. A.
    (1971) Glutaraldehyde-ammoniacal silver carbonate-formaldehyde staining of histones of mitotic chromosomes. *Stain Technology* 46:89–92.

Villyaneuva, A. R.
    (1979) Decalcifying solution. *Journal of Histotechnology* 2:24.

Villyaneuva, A. R., Haltner, R. S., and Frost, H. M.
    (1964) A tetrachrome stain for fresh, mineralized bone sections useful in the diagnosis of bone disease. *Stain Technology* 39:87–94.

Vlachos, J.
    (1959) Desaturation of staining solutions of lipids, a means of avoiding precipitation on stained sections. *Stain Technology* 34:292.

von Kossa J.
    (1901) Ueber die im organismus kuenstlich erzeugbaren verkakung. *Beiträge für Pathologische Anatomie* 29:62.

Vyas, A. B.
    (1972) Taking the sting out of preserving scorpions. *Turtox News* 49:25.

Wachstein, M., and Meisel, E.
    (1964) Demonstration of peroxidase activity in tissue sections. *Journal of Histochemistry and Cytochemistry* 12:538–44.

Wall, P.
    (1950) Staining and recognition of the fine degenerating nerve fibers. *Stain Technology* 25:125–26.

Walls, G. L.
    (1936) A rapid celloidin method for the rotary microtome. *Stain Technology* 11:89–92.
    (1938) The microtechnique of the eye with suggestions as to material. *Stain Technology* 13:69–72.

Walsh, R. J., and Love, R.
    (1963) Studies of the cytochemistry of nucleoproteins. II. Improved staining methods with toluidine blue and ammonium molybdate. *Journal of Histochemistry and Cytochemistry* 11:188–96.

Wartman, W. B.
    (1943) Notes on the Field's method of staining parasites in thick blood films. *Army Medical Bulletin* 68:173–77.

Wasserung, R. J.
  (1976) A procedure for differential staining of cartilage and bone in whole formalin-fixed vertebrates. *Stain Technology* 51:131–34.
Weaver, H. L.
  (1955) An improved gelatine adhesive for paraffin sections. *Stain Technology* 30:63–64.
Weiss, J.
  (1954) The nature of the reaction between orcein and elastin. *Journal of Histochemistry and Cytochemistry* 2:21–28.
Weissmann, G.
  (1964) Lysosomes. *Blood* 24:594–606.
  (1968) The many-faceted lysosome. *Hospital Practice* 31–39.
Welshons, W. J., Gibson, B. H., and Scandyln, B. J.
  (1962) Slide processing for the examination of male mammalian meiotic chromosomes. *Stain Technology* 37:1–5.
Werth, V. G.
  (1953) Fluoreszez mikroskopische Beobachtungen an menschlichem Knochenmark. *Acta Haematologica* 10:209–22.
White, L. E., Jr.
  (1960) Enhanced reliability in silver impregnation of terminal axonal degeneration—original Nauta method. *Stain Technology* 35:5–9.
Wickramasinghe, H. K.
  (1992) *Scanned Probe Microscopy*. New York: American Institute of Physics.
Wieland, H., and Scheuing, G.
  (1921) Die Farbreaktion mit Aldehyden. *Berichte der Deutschen Chemischen Gesellschaft* 54:2527–55.
Wijffels, C. C. B. M.
  (1971) Freeze-drying by cryosorption; a simple device having a tissue capacity of 500 milligrams. *Stain Technology* 46:33–34.
Wilcox, A.
  (1943) *Manual for the Microscopical Diagnosis of Malaria in Man*. National Institute of Public Health, Bulletin 180. Washington, DC: U.S. Government Printing Office.
Wilder, H. C.
  (1935) An improved technique for silver impregnation of reticular fibers. *American Journal of Pathology* 11:817–19.
Wilhelm, W. E., and Smoot, A.
  (1966) Permanent demonstration of food vacuoles in *Paramecium* stained in relief. *Turtox News* 44:158.
Williams, D. L., Lafferty, D. A., and Webb, S. L.
  (1970) An air drying method for the preparation of dictyotene chromosomes from ovaries of Chinese hamsters. *Stain Technology* 45:133–35.
Williams, G., and Jackson, D. S.
  (1956) Two organic fixatives for acid mucopolysaccharides. *Stain Technology* 31:189–91.
Williams, T.
  (1962) The staining of nervous elements by the Bodian method. I. The influence

of factors preceding impregnation. *Quarterly Journal of Microscopical Science* 103:155–61.

Witten, V. H., and Holmstrom, V.
(1953) New histologic technics for autoradiography. *Laboratory Investigation* 2:368–75.

Wittman, W.
(1962) Aceto-iron-hematoxylin for staining chromosomes in squashes of plant material. *Stain Technology* 37:27.
(1963) Permanent-type mounting of aceto-iron-hematoxylin squashes in corn syrup. *Stain Technology* 40:161–64.

Wolberg, W. H.
(1965) Darrow red-light green as a stain for autoradiographs. *Stain Technology* 40:90.

Wolfe, H. J.
(1964) Techniques for the histochemical localization of extremely water soluble acid mucopolysaccharaides. *Journal of Histochemistry and Cytochemistry* 12:217–18.

Wolman, M.
(1956) A histochemical method for the differential staining of acetic tissue components (particularly ground-substance polysaccharides). *Bulletin of the Research Council of Israel, Section E, Experimental Medicine* 6E:27–35.
(1961) Differential staining of acidic tissue components by the improved bi-col method. *Stain Technology* 36:21–25.

Wolters, W. K., Chrisman, C. L., and Libey, G.
(1981) Induction of triploidy in channel catfish. *Transactions of American Fisherges Society* 110:310–12.

Woodruff, L. A., and Norris, W. P.
(1955) Sectioning of undecalcified bone: with special reference to radioautographic applications. *Stain Technology* 30:179–88.

Woods, P. S., and Pollister, A. W.
(1955) An ice-solvent method of drying frozen tissue for plant cytology. *Stain Technology* 30:123–31.

Wright, J. H.
(1902) A rapid method for the differential staining of blood films and malarial parasites. *Journal of Medical Research* 7:138–44.

Wróblewska, J.
(1969) Chromosome preparations from mouse embryos during early organogenesis: dissociation after fixation, followed by air drying. *Stain Technology* 44:147–50.

Wynnchuk, M.
(1993) Evaluation of xylene substitutes for paraffin tissue processing. *Journal of Histotechnology* 2:143–49.

Yaeger, J. A.
(1958) Methacrylate embedding and sectioning of calcified bone. *Stain Technology* 33:220–39.

Yamada, K.
(1963) Staining of sulfated polysaccharides by means of Alcian blue. *Nature* 197:789.

Yamaguchi, B. T., Jr., and Braunstein, H.
> (1965) Fluorescent stain for tubercle bacilli in histologic sections. II. Diagnostic efficiency in granulomatous lesions of the liver. *Technical Bulletin of the Registry of Medical Technologists* 35:184–87.

Yang, J., and Scholten, T. H.
> (1976) Celestin blue B stain for intestinal protozoa. *American Journal of Clinical Pathology* 65:715–18.

Yasuma, A., and Ichikawa, T.
> (1953) Ninhydrin-Schiff and Alloxan-Schiff staining. A new histochemical staining method for protein. *Journal of Laboratory and Clinical Medicine* 41:296–99.

Yensen, J.
> (1968) Removal of epoxy resin from histological sections following halogenation. *Stain Technology* 43:344–46.

Yerganian, G.
> (1957) Cytologic maps of some isolated human pachytene chromosomes. *American Journal of Human Genetics* 9:42–54.
> (1963) Cytogenetic analysis. *Methodology in Mammalian Genetics*. Ed. by W. J. Burdette. San Francisco: Holden-Day.

Yetwin, I. J.
> (1944) A simple permanent mounting medium for *Necator americansus*. *Journal of Parasitology* 30:201.

Young, I.
> (1962) The use of a sponge base as an aid in preparing frozen sections of small objects. *Technical Bulletin of the Registry of Medical Technologists* 32:76.

Youngpeter, J. M.
> (1964) Spalteholz preparations. *Ward's Bulletin* 3:7.
> (1967) Spalteholz preparations. *Ward's Bulletin* 6:7.

Zambernard, J., Block, M., Vatter, A., and Trenner, L.
> (1969) An adaptation of methacrylate embedding for routine histologic use. *Blood* 33:444–50.

Zeiger, K., Harders, H. I., and Müller, W.
> (1951) Der Strugger-effekt an der Nervenzelle. *Protoplasma* 40:76–84.

Zimmerman, M.
> (1976) Rehydration of accidently desiccated pathology specimens. *Laboratory Medicine* 7:13–17.

Zinn, D. J., and Morin, L. P.
> (1962) The use of commercial citric juices in gold chloride staining of nerve endings. *Stain Technology* 37:380–91.

Zirkle, C.
> (1940) Combined fixing, staining and mounting media. *Stain Technology* 15:139–53.

Zlotnik, I.
> (1960) The initial cooling of tissues in the freeze-drying technique. *Quarterly Journal of Microscopical Science* 101:251–54.

Zugibe, F. T.
> (1970) *Diagnostic Histochemistry*. St. Louis: C. V. Mosby.

Zugibe, F. T., Brown, K. D., and Last, J. H.
> (1958) A new technique for the simultaneous demonstration of lipid and acid

mucopolysaccharides on the same tissue section. *Journal of Histochemistry and Cytochemistry* 7:101–6.

Zugibe, F. T., Kopaczyk, K. C., Cape, W. E., and Last, J. H. (1958) A new carbowax method for routinely performing lipid, hematoxylin and eosin and elastic staining techniques on adjacent freeze dried or formalin fixed tissues. *Journal of Histochemistry and Cytochemistry* 6:133–38.

Index

ABC method, 367–68
Acetate buffer 0.2M (Gomori), 470; acetate-acetic acid buffer (Walpole), 471
Acetic acid, 19
Acetocarmine staining of chromosomes in squash preparations, 411
Aceto-orcein: solution, 412; staining sex chromosomes in squash preparations, 411
Acetone, 19
Acetone fixation and embedding. *See* Gomori
Acetylation, 242
Acid alcohol, 124, 463
Acid-fast staining: fluorescent method, 306; Harada (Ziehl-Neelson type), 305
Acid fuchsin, 126; short acid fuchsin method (Novelli), 297; solution, 133
Acid hematein for phospholipids, 251
Acid-hematein method for staining mitochondria, 298
Acid mucopolysaccharides, 273–74; alcian blue method:—at pH 1, 259; —at pH 2.8, 257; colloidal iron method, 260; iron diamine method, 259; metachromasia, 274
Acid phosphatase, 345–48; azo coupling method, 347; Gomori, 346
Acidified hematein, 113
Acidified water, 125, 127
Acids used in microtechnique, 8
Acridine orange, 279, 316
Adamstone-Taylor cold knife technique for histochemistry, 336
Adhesive (affixing) solutions: Haupt gelatin fixative, 468; Land adhesive, 469; Masson gelatin fixative, 468; Mayer albumen fixative, 467; "subbed" slides, 469; Weaver gelatin fixative, 468
AEC (3-amino-9-ethylcarbazole)/H_2O_2, 367
Agar (Chatton) for blocking small organisms, 464
Alcian blue, 259, 280; for acid mucopolysaccharides: —at pH 1.0, 259; at pH 2.5, 257; —at pH 2.8, 256
Alcohol-chloroform, 155
Alcohols, 20; acid, 463; alkaline, 463; analine, 124

Aldehydes, 20–21; aldehyde-fuchsin, 282–83
Alfert-Geschwind method, 232
Alizarine red S: for calcium, 223; staining whole embryos and small vertebrates (bone formation), 138; stock solution, 139
Alkaline alcohol, 463
Alkaline phosphatase: azo-coupling method, 341; Gomori, 340; Kaplow, 343
Altmann aniline fuchsin, 295
Altmann fixative for cell inclusions, 31
Altmann method for staining Golgi apparatus, 295
Alums used in microtechnique, 8
Aminopeptidase, Burstone and Folk method, 349
Ammoniacal silver method, 232; ammoniacal silver hydroxide, 148; ammoniacal silver nitrate, 150
Ammonium sulfide, 216
Amyloid staining: congo red, 269; crystal violet, 271; fluorescent method, 270; metachromasia, 271
Anesthetizing and narcotizing agents for handling of invertebrates: asphyxiation, 423; chloretone, 423; cocaine, 422; cold, 423; ether and alcohol, 423; magnesium chloride and magnesium sulfate, 422; menthol, 422; propylene phenoxetol, 423
Aniline alcohol, 124
Aniline blue stain, 125
Annelida, 428
Antibody, 362
Antigen, 362
APAAP, 365
Aqueous mounting media: fluorescent mounting medium, 99; Kaiser glycerine jelly, 98; lactophenol, 99; Von Apathy gum syrup, 99; Yetwin mounting medium for nematodes and ova, 99; techniques:—ringing cover glasses, 100; —using two cover glasses, 100
Argentaffine reaction, 264
Arginine staining, Sakaguchi reaction, 233
Arthropoda, 428

559

560   Index

Artificial sea water, 466
Asphyxiation, for anesthetizing invertebrates, 423
Astrocytes, staining with phophotungstic acid hematoxylin, 157
Auramine stain, 307
Automatic tissue processors, 54
Autoradiography, 449–57; chromosome cultures for autoradiography, 457; emulsion application: —liquid emulsion ("dipping") method, 452; —stripping film method, 452; fixation and slide preparation for, 450; use of isotopes in chromosome studies with tritium, 456
Avidin-biotin procedures, 365; avidin-peroxidase immunostaining, 367
Azan stain, for connective tissues, pituitary, pancreas, and others, 124
AZF fixative, 27
Azocarmine, 124
Azo-coupling method for bile pigment, 218; acid phosphatase, 347; alkaline phosphatase, 341; staining enterochromaffin, 268
Azure A-Schiff reagent, 191
Azure-eosinate stock solution, 201

B-5 Fixative for cell inclusions, 37
Bacteria staining, 302–8: acid-fast, 305; capsule, 308; Gram staining, 302
Baker acid hematein, 298
Balanced salt solutions, 464
Barbital buffer 0.05 M (Gomori), 472
Barium hydroxide, 233
Barr bodies, 408, 410
Basic fuchsin stain, 134
Benzidine solution, 358
Benzoylation, 243
Best carmine for glycogen staining, 254
Biebrich scarlet solution, 409; Biebrich scarlet/acid fuchsin, 399
Bielschowsky method for blocks of nerve tissue, 164
Bile pigment (bilirubin) staining: azo-coupling method, 218; bile pigments, 227; Glenner method, 217
Bilirubin, 214
Biliverdin, 214
Bilstat (modified Carnoy's) for cell inclusions, 31
Biological Stain Commission, history of, 5
Biotinylated secondary antibody, 366, 368

Bleaching solution (sulfurous acid), 187
Blocking methods for chemical groups: aldehydes, 243; carboxyl groups or other acid groups, 244; lyosine, 245; reactions of protein end groups for amines, 242; tryptophane and cystine, 245
Blocking serum used in immunocytochemistry, 376
Blood smears: preparation of thick blood films, 198; preparation of thin smears, 193; thick film methods: —Field-Wartman, 199; —Wilcox, 200; Wright stain, 195; —for birds, 197; —for cold-blooded vertebrates, 197
Blood tissue elements and inclusion bodies: Giemsa stain, 201; —for birds, 204; —for cold-blooded vertebrates, 204; Jenner-Giemsa stain for malarial parasites, 203; reticulocyte staining, 205
Bodian method for nerve fibers, 165
Bohorfoush diluent, 196
Bone marrow staining: Maximow eosin-azure, 208; phloxine-methylene blue, 211
Bone staining, 136–40; alizarine red S staining method for whole embryos and small vertebrates (for bone formation), 138; staining decalcified sections, 136; staining hand-ground, undecalcified sections, 137
Borax, 175
Borax-ferricyanide, 177, 298
Bordeaux red, 107
Boric acid, 175; boric acid-borax buffer (Holmes), 472
Borror nigrosin stain for protozoa, 440
Bouin fixative, 27; Bouin chrome alum, 284; Bouin-Duboscq (alcoholic Bouin) fixative, 28
Brachiopoda, 427
Brain fixation, 38–39
Brazilin, 85
Brown-Brenn Gram staining. See Leaver et al.
Bryozoa, 427
BSA, 374
Buffalo black for hemoglobin, 205
Buffer solutions, 469–77; acetate-acetic acid buffer (Walpole), 471; 0.2 M acetate buffer, 470; 0.05 M barbital buffer (Gomori), 472; boric acid-borax buffer (Holmes), 472; 0.2 M phosphate buffer (Gomori), 473; phosphate buffer (Sörensen), PBS, 473, 477; standard buffer

(McIlvaine), 474; TBS, 477; 0.2 M tris buffer (Hale), 475; tris maleate buffer (Gomori), 476
Burstone-Folk method for aminopeptidase, 349

Cacodylate buffer, pH 7.2, 329
Calcium deposits: removal of, 40; staining: —alizarine red S, 223; —Von Kossa method, 222
Calcium formalin fixation for enzyme histochemistry, 328
Cameron-Steele aldehyde fuchsin, 281
Capsule staining of bacteria by Hiss method, 308
Carbohydrate and glycoprotein staining. *See* Periodic acid-Schiff
Carbohydrates (saccharides), 253
Carbol-fuchsin, 305
Carbol-xylol, 463
Carbon, 226
Carmine (and cochineal), 84; best carmine, 254
Carnoy fixative for cell inclusions, 31; Carnoy-Lebrun fixative for cell inclusions, 32
Castañeda method for staining rickettsiae, 318
Caulfield solution, 298
Celestin blue, 106
Cell blocks, 52
Celloidin, 337; celloidin technique (nitrocellulose embedding), 391
Cestodes, 425–26
Champy fixative for cell inclusions, 32
Chang et al. method for histochemistry, 334
Chatton agar for blocking small organisms, 464
Chelating agents for decalcification, 42
Chick embryos, preparation of, 431
Chlorantine fast red, 257
Chloretone for anesthetizing invertebrates, 423
Chloroform, 48, 423, 428
Chromatin, 408
Chromatization postfixation treatment, 40
Chrome-hematoxylin, Bargmann modification for neurosecretory substance (NSS), 284–86
Chromic acid-hematoxylin for nerve tissue, 176
Chromium-hematoxylin-phloxine for pancreatic islet cells, 286

Chromium trioxide (chromic acid), 9, 19
Chromogen, 365
Chromosome analysis, 419–21; preparation of karyotypes, 415, 419, 420
Chromosome cultures for autoradiography, 457
Chromosomes: chromosome squashes, 410; acetocarmine or aceto-orcein staining, 411; pachytene, 414; somatic, culturing and spreading techniques: —culture methods for slide preparation of, 416; —general, 414; —Giemsa method for specific banding, 418
Cleaning solution for glassware, 487
Clearing, infiltrating, and embedding, paraffin method, 48–54
Cocaine for narcotizing invertebrates, 422
Cochineal and carmine staining, 84; cochineal-hematoxylin for staining whole mounts, 437
Coelenterates, 424–25
Cole hematoxylin, 104
Cole method for muscle and nerve, 170
Collagen and elastin staining, 131–36; basic fuchsin, 134; iron gallein elastin stain, 134; orcein, 135; picro-Ponceau with hematoxylin, 131; Verhoeff elastin stain, 133
Collidine buffer, 329
Colloidal iron method for acid mucopolysaccharides, 260–62
Color preservation in gross specimens, McCormick fixation for, 485
Compound microscope, 67
Confocal microscopy, 82
Congo red for staining amyloid, 269
Connective tissue and muscle staining, 121
Corals, 425
Counterstains (plasma stains): eosin, 106; eosin (Putt), 107; eosin-orange G, 107; orange G, 107; other counterstains, 107
Cover glass mounting, 96
Cramer Giemsa stain, 201
Creosote, 436; combined with xylene, aniline, or carbol, 436
Cresyl cochineal-hematoxylin for staining whole mounts, 437
Cresyl violet: in dopa oxidase method, 357; for nissl substance, 158
Cryostat sectioning, 334

Crystal violet, 155, 304; for metachromatic staining of amyloid, 271
Culture methods for somatic chromosome slide preparation, 416
Cyanol reaction, 206
Cytochrome oxidase, 359
Cytology, 404

DAB (diaminobenzidine)/H_2O_2, 367
Dark-field microscopy, 74, 370
Darrow red, 116
Deacetylation (saponification), 243
Deamination, 242
Decalcification, 40: acid reagents, 41; chelating agents for decalcification (EDTA), 42; combined with dehydration using Jenkins fluid, 43; combined with fixation using Lillie fluid, Lillie alternate fluid, and Perenyi fluid, 42; Villaneuva fixative, 41
Decalcified sections, staining of decalcified bone, 136
Deformalization postfixation treatment, 40
Degenerating axons: Marchi method-B, 180; Nauta-Gygax method, 177
Dehydrating and clearing combinations: clearing, infiltrating, and embedding by paraffin method, 50; dioxane method, 49; Haust method, 50; tetrahydrofuran (THF) for dehydration, 49
Dehydrating and embedding: Chang et al., 334; in freeze-drying, 331; in freeze substitution, 332; in freeze substitution of sections, 333
Dehydration: dehydration and clearing combinations, 49; ethanol and butanol, 46; general principles, 45; small colorless tissues, 47; stock solutions, 46
Delafield hematoxylin, 102; Delafield (or Harris) hematoxylin: I (progressive method), 108; II (regressive method), 110
Demethylation (saponification), 244
Deoxyribonucleic acid (DNA) fluorescent technique, 189
Diazo-safranin method for enterochromaffin staining, 266
Dieterle method for staining spirochetes, 309
Dimethylsulfoxide (DMSO), 306
Dioxane method for dehydration and clearing, 49
Direct silver method for staining Golgi apparatus, 293
DNA. *See* Deoxyribonucleic acid (DNA) fluorescent technique; Feulgen reaction; Nucleic acid and nucleoprotein staining; RNA removal
Dopa stock solution, 356
Dried gross specimens, restoration, 482
Drosophila larvae, 412
Dyes, 84

Earle solution, 465
Earthworms, 428
Echinodermata, 430
EDTA: anticoagulant, 210; decalcification, 42
Ehrlich hematoxylin, 102
Elastin. *See* Collagen and elastin staining
Electron microscopy, 79, 371
Embedding (blocking): embedding cellular contents of body fluid "cell blocks," 52; gelatin, 389; short Gomori method, 338; using ethyl alcohol, 52; using isopropyl alcohol, 53; with paraffin, 51; timing schedule for paraffin method, 52
Embryos, staining. *See* Whole mounts, of hydras, embryos, and flukes
Endocrine gland staining, 276
Enterochromaffin (EC) cell staining: azo-coupling method, 268; diazo-safranin, 266; ferric-ferricyanide method (Schmorl), 267; Fontana method, 264
Environmental solutions: artificial sea water, 466; synthetic spring water, 467
Eosin staining, 106; eosin-azure, 208, 406; eosin-orange G, 107
Epon (epoxy resin): processing, 400; staining, 402
Eriochrome cyanine R, 116
Erythrosin B, 107, 279
Ester wax embedding methods, 392: polyester wax method of Norenburg and Barrett, 394; Sidman modification, 394; Steedman method, 393
Esterases (nonspecific) and lipases: esterase (Maloney et al. modification) 351; lipase, 352
Ether and alcohol for anesthesia, 423
Ethers, 9
Ethylene glycol, 9
Ewen modification of Cameron-Steele aldehyde-fuchsin method for pituitary, 281
Exfoliative cytology, 404; gynecological cytology, 404; Papanicolau method, 406
Extraction techniques for control slides, 240

Faded slides, 481
Fast green, 127, 141, 187, 232
Feder-Sidman method, 482
Ferro-ferricyanide method for melanin, 221; modified Schmorl technique for staining enterochromaffin, 267
Ferrous sulfate, 221
Fetal hemoglobin, 207
Feulgen reaction, 187; fluorescent, 189; hydrolysis chart, 188
Fibrin, staining of, ledrum acid picro-Mallory method, 212
Field-Wartman thick film method for blood films, 199
Fish and reptiles, histochemistry and special procedures, 447
Fixation: blocking and sectioning of frozen tissues, 386; general principles, 17–19; by perfusion, 38; properties of an ideal fixative, 17; and slide preparation for autoradiography, 450; for staining (see Rapid staining methods for frozen sections)
Fixation for histochemistry: aldehydes, 328; glutaraldehyde, 328; glycerine, 331; gum sucrose, 330; paraformaldehyde, 329
Fixatives, chemicals commonly used in, 18: acetic acid, acetone, alcohols, chromium trioxide, 19; aldehydes, formaldehyde and formalin, 20; mercuric chloride, osmium tetroxide, picric acid, 22; potassium dichromate, 24; salts (indifferent) in fixatives, 24
Fixatives for cell inclusions and special techniques, 31; Altmann, Carnoy, Bilstat (modified Carnoy), methacarn, 31; B-5 fixative, zinc formalin (ZF), 37; Carnoy-Lebrun, Champy, Flemming, 32; formalin-calcium, 294; formol-alcohol, Gendre, Gilson, 33; Johnson, Kolmer, Lavdowsky, Navashin, 34; Perenyi, Regaud, Rossman, Sanfelice, 35; Schaudinn, 36; Schaudinn PVA, 445
Flemming fixative for cell inclusions, 32
Flukes, staining. See Whole mounts, of hydras, embryos, and flukes
Fluorescein-isothiocyanate, 370
Fluorescence microscopy, 78
Fluorescent method for staining amyloid, 270; acid-fast bacteria, 306; for fungi, 316; lipid, 251; for mucin, 263
Fluorescent mounting medium, 99
Fluorescent Schiff reagent, 189

Fontana method for enterochromaffin staining, 264
Foraminifera and radiolaria, dry mounts, 442
Formaldehyde and formalin, 8, 20; formol-alcohol fixatives for cell inclusions, 33; formalin pigment: —formation, 21; —removal, 226
Formic acid, 41
Freeze-substitution of sections for histochemistry, 332–33
Freezing techniques, 386
Frozen section staining. See Rapid staining methods for frozen sections
Frozen sections fixed in acetone for histochemistry, 338
Fungi staining, 312: fluorescent method, 316; Gomori methenamine-silver nitrate, 314; Gridley, 313

GAG (glucose oxidase-antiglucose oxidase complex), 365
Gallocyanin, substitute for hematoxylin, 105
Garnet GBC, 268, 350
Gelatin embedding for frozen sectioning, Pearse method, 389
Gelatin fixative (Weaver), 468
Gendre fluid fixative for cell inclusions, 33
General laboratory aids, 481
Gentian violet. See Crystal violet
Giemsa stain, 201; for birds and cold-blooded vertebrates, 204; Giemsa stock solution, 203; Gude et al. modification for staining autoradiographs, 455; for staining rickettsiae, 320
Gill's hematoxylin, 103
Gilson fixative, 33
Glassware cleaning solution, 487
Glees method for nerve tissue, 168
Glenner method for bile pigment staining, 217
Glia cells, 155
Globus hydrobromic acid, 156
Glutaraldehyde fixation for histochemistry, 328
Glycerine for storage of tissues for enzyme histochemistry, 331
Glycogen staining, best carmine for, 254
Glycol distearate, 393
Glycol methacrylate processing for thin sections: fixation, 397; staining with hematoxylin and eosin, 398; staining with modified Maynard's trichrome, 398
Gold chloride, 147; stock solution, 463

Golgi apparatus staining, 290: direct silver, 293; Nassonov-Kolatchew, 291; osmium tetroxide (Ludford method), 290; Saxena method, 292; Sudan black, 294
Gomori: 1-2-3 fixative, 28: one-step trichrome, 129; acetone fixation and embedding for histochemistry, 337; acid phosphatase, 346; alkaline phosphatase, 340; methenamine-silver nitrate method for staining enterochromaffine cells, 264; method: —for silver impregnation, 149; —Sweat et al. modification, 130; —for trachoma, 321; —short method, 338; for staining fungi, 314
Gomori buffers: acetate buffer 0.2M, 470; barbital buffer 0.05M, 472; phosphate buffer 0.2M, 473; tris maleate buffer, 476
Gram staining for bacteria, 302; Gram-Weigert method, 303
Grenacher borax carmine for staining whole mounts, 435
Gridley method for staining fungi, 313
Groat variation of Weigert hematoxylin, 103
Grocott's methenamine silver nitrate stain using microwave, 383
Guard's method for staining sex chromatin in squash preparations, 409
Gum mastic, 161, 309
Gum sucrose storage of tissues for enzyme histochemistry, 330
Gynecological cytology, 404

Hale 0.2 M tris buffer, 475
Halmi mixture, 282
Handicapped student safety, 11
Handling small tissue samples, 448
Hank solution, 465
Harada method for acid-fast staining, 305
Harris hematoxylin, 102; Harris (or Delafield) hematoxylin: I (progressive method), 108; II (regressive method), 110
Haupt gelatin fixative, 468
Haust method for embedding using THF/paraffin, 50
Heidenhain iron hematoxylin, 111; Heidenhain azan for pituitary gland staining, 278
Helly fixative (Zenker formol), 29; for special techniques, 37
Hematein (hemalum), Kornhauser, 106; hematein, acid, 252; for staining whole mounts, 437
Hematoidin, 214

Hematoxylin and eosin staining using microwave, 382; for rapid staining of frozen sections, 389
Hematoxylin staining, 84, 101; chromium hematoxylin, 286; Cole hematoxylin, 107; lead hematoxylin, 281; procedures: —acidified hematein, 113; —Delafield (or Harris) progressive method, 108; —Delafield (or Harris) regressive method, 110; —Heidenhain iron hematoxylin, 111; —Mayer hematoxylin, 111; —phosphotungstic acid hematoxylin solution, 114; procedures for hematoxylin substitutes, 115; —eriochrome, cyanine RC, 116; —gallocyanin, 115; —hematein (hemalum), 115; —red nuclear staining, 116; —scarba red, 117; Slidder hematoxylin, 104, 257; substitutes for, 105; —celestin blue B, 106; —gallocyanin, 105; —hematein (hemalum), 106; —iron alizarine blue S, 106; testing of staining solutions, 105; using double solutions: —Groat variation of Weigert hematoxylin, 103; —Heidenhain iron hematoxylin, 105; Krutsay iron hematoxylin, 104; —Weigert iron hematoxylin, 103; using single solutions: —Delafield method, 101; —Ehrlich method, 102; —Gill method, 103; —Harris method, 102; —Mayer method, 102
Hemoglobin staining: buffalo black for hemoglobin, 205; cyanol reaction, 206; fetal hemoglobin, 207
Hemosiderin, 214, 227
Herlant pituitary stain I and II, 279, 280
High iron diamine method for mucopolysaccharides, 260
Hiss method for bacterial capsule staining, 308
Histochemistry, 327; control slides, 339; preservation of incubation media, 339; stock solutions, 338; storage of slides and tissues, 339
Histone protein staining: Alfert-Geschwind, 233; ammoniacal silver, 233
History of microtechnique. See Microtechnique, history of
Hollande Bouin fixative, 29
Holmes boric acid-borax buffer, 472
Holzer method for glia fibers, 154
Hotchkiss-McManus method for staining fungi, 312

Humphrey method for rapid staining of frozen sections, 390
Hydra, 424, staining. *See* Whole mounts of hydras, embryos, and flukes
Hydroquinone, 161
Hypochlorite solution, 234

Immunoglobulins, 364
Immunohistochemistry, 361; general immunohistochemical staining, directions and hints, 375; high background, 373; immunostaining: —absence of staining, 374; —direct method, 362; —indirect methods, 362 (*see also* Soluble enzyme immune complex methods); —immunofluorescence, indirect immunohistochemistry for immunostaining, 370; —troubleshooting, dealing with unexpected results of immunostaining, 373
In situ hybridization, 277
Inclusion bodies, 320
Indifferent salts, 24
Indirect methods of immunostaining, 362. *See* Soluble enzyme immune complex methods
Indophenol method for lipofuscin, 221
Infiltrating with paraffin, 50
Insects and arachnids, 428
Interference microscopy, 76
Intestinal protozoa, 433
Invertebrates: anesthetizing with menthol, 422; preparation of, for whole mounts and sections, 422
Invertebrates, tips on special handling for histochemistry: annelida, 428; arthropoda, 428, brachiopoda, 427; bryozoa, 427; coelenterates, 424; cestodes, 425; corals with extended polyps, 425; echinodermata, 430; freshwater mollusks, 428; hydrae, 424; insects and arachnids, 428; jellyfish, 424; medusae, 424; microfilaria, 427; mollusca, 427; mussels, 427; nemathelminths, 426; nemertines, 426; nudibranchs, 428; planaria, 425; platyhelminths, 425; porifera, 424; rotiferae, 426; sea anemones, 424; snails, 427; trematodes, 425
Iodine, 9; Gram iodine solution, 464; Langeron, 114; Lugol, 463; Weigert, 463
Iron alizarine blue S (substitute for hematoxylin), 106
Iron alum, 126
Iron hematoxylin, 103–5, 111

Iron methods: iron diamine methods for acid mucosubstances, 259; iron gallein elastin stain, 134
Iron, staining of, 214: Prussian blue reaction, 214; iron method (Hutchinson), 215; Turnbull blue for ferrous iron, 216
Isopropyl alcohol, 20, 46; processing schedule, 115
Isotope use in autoradiographic studies of chromosomes, 456

Janus green B stock solution, 300
Jellyfish, 424
Jenkins fluid, decalcification combined with dehydration, 43
Jenner solution, 318; Jenner-Giemsa stain for malarial parasites, 203
Johnson fixative for decalcification, 34

Kaiser glycerine jelly, 98
Kaiserling solution, 484
Kaplow method, modified, for alkaline phosphatase, 343
Karyotypes, 415–20
Kernechtrot (nuclear fast red), 215; Kernechtrot (or Mayer's hemalum) staining of autoradiographs, 455
Klinger-Hammond method for staining sex chromatin in squash preparations, 408
Kohn stain, combined fixation and stain for animal parasites, 443
Kollmer fixative for cell inclusions, 34
Kristensen fluid, 41
Krutsay iron hematoxylin, 104

LAB method, 367
Labeling and cleaning slides, 481
Laboratory safety, 7; chemicals, 8; material safety data sheets (MSDS), 8; practical aspects of, 7; regulatory agencies, 7; specific chemicals and safety, 8; stain removal, 486
Lactophenol mounting medium, 99
Laidlaw dopa oxidase method, 356
Land adhesive, 469
Langeron iodine solution, 114
Lavdowsky fixative, 34
Lawless rapid method for animal parasites, 445
Lead hematoxylin method for pituitary, 283
Leaver et al. Gram staining, 304
Ledrum acid picro-Mallory method for fibrin, 212

Leviditi method for block staining of spirochetes, 311
Lewitsky saline modification, 33
Light green, 107, 313
Lillie alternate fluid for decalcification and fixation, 42
Lillie fluid for decalcification, 42
Lillie nile blue method for staining melanin and lipofuscin, 219; alternate Nile blue method for melanin and lipofuscin, 220
Lipases and esterase staining: esterase (Maloney et al. modification), 351; lipase, 352
Lipid staining: acid hematein for phospholipids, 252; fluorescent method, 251; general principles for, 247; naphthalene yellow, 249; oil red O, 248; osmium tetroxide, 250; Sudan black B for phospholipids (Elftman), 250; Sudan IV or Sudan Black B (Chifelle and Putt), 248
Lipofuscin staining. *See* Melanin and lipofuscin staining
Liquid emulsion ("dipping") method for autoradiography, 452
Lithium carbonate, 159
Locke solution, 465
Lugol solution, 463
Luxol fast blue: Luxol fast blue ARN, 172; Luxol fast blue G, 173; Luxol fast blue-Holmes silver nitrate: —for mitochondria, 299; —for nerve fibers, 174; Luxol fast blue MBSN, 172; for nerve tissue and for muscle, 172

Maceration of tissues using maceration fluids, 24
Magnesium chloride and magnesium sulfate, 422
Malachite green, 304
Malarial pigment, Gridley method of removal, 226
Mallory staining: azan stain, 124; Mallory I, 122; Mallory II, 123; triple connective tissue stain, 121
Mann osmic sublimate, 290
Marchi method-B for degenerating axons, 180
Massignani-Malferrari method for negri bodies, 322
Masson: gelatin fixative, 468; trichrome stain (Gurr modification), 126
Mast cell staining for metachromasia: quick toluidine blue, 275; thionin, 274; toluidine blue, 274
Material safety data sheets (MSDS), 10

Maximow eosin-azure, 208
Mayer: albumen fixative, 467; carmalum for staining whole mounts, 436; hemalum or kernechtrot for audioradiograph staining, 455; hematoxylin, 111
Maynard's trichrome (modified) for glycol methacrylate processing, 397
McCormick: fixation for color preservation in gross specimens, 485; method for restaining faded slides, 481
McIlvaine, standard buffer, 474
Measuring objects with ocular micrometer, 73
Mechanical aids, 95
Medium for nematodes and ova, 99
Medusae, 424
Melanin and lipofuscin staining: ferro-ferricyanide, 221; indophenol, 221; Lillie alternate Nile blue method, 220; Lillie Nile blue method, 219
Mercuric bromphenol blue method, 230
Mercuric chloride (corrosive sublimate): removal from fixed tissues, 22; safe disposal, 22
Mercury, 9; and safety, 10
Merton method for making slides of protozoa, 439
Metachromasia, 271; acid mucopolysaccharides, 273; amyloid, 271; mast cells, 274; thionine, 273
Metanil yellow, 314
Methacarn fixative for cell inclusions, 31
Methenamine silver for enterochromaffin staining, 264; acetone fixation and embedding for histochemistry, 337; acid phosphatase, 346; alkaline phosphatase, 340; methenamine-silver nitrate stock solution, 315; method for trachoma, 321; short method, 338; for staining fungi, 314
Methyl alcohol (methanol). *See* Alcohols
Methyl green for staining nucleic acids: methyl green-pyronine, 239; methyl green-thionine, 240
Methylation, 244
Methylene blue solution, 319; methylene blue-azure solution, 211
Microfilaria, 427
Micrometers, ocular, 73
Microscopes: bright-field, 67; compound, 67; dark-field, 74; efficient microscopy, hints for, 72; electron, 79; fluorescence, 78; history of, 6; operation of a microscope, 69; near-field optical scanning, 81; ocu-

lar micrometer, 73; research microscopes, 70; student microscopes, 69; use of oil immersion objective, 71; vernier, reading, 73
Microtechnique: Biological Stain Commission, 5; development of the microscope, 1; differential staining, 4; fixatives, use of, 3; fundamental steps in microtechnique, 2; history of, 1; microtome development, 2
Microtomes, 55; knives, 56; knife sharpening, 57
Microwave histology, 380: drying slides, 381; fixation, 381: —methods for microwave stains, 382; —routine tissue processing, 382; Grocott's methenamine silver nitrate, 383; hematoxylin and eosin staining, 382; Perl's iron stain (Kok-Boon modification), 383; rapid Papanicolaou method, 384
Milligan trichrome stain, 140
Millon reaction: Gomori, 228; Romeis, 229
Mitochondria staining, 295; enzymes, 353–56; methods: —acid hematein, 298; —Altmann, 295; —osmic, 296; —short acid fuchsin, 297; —luxol fast blue, 299
Molar solutions, 461
Mollusca, 427, 428
Mordants used in staining, 85
Mounting and staining procedures, 95
Mounting media (mountants): aqueous, 98; —Kaiser glycerin jelly, 98; —Von Apathy gum syrup, 99; —lactophenol, 99; —Yetwin for nematodes and Ova, 99; —mounting medium for fluorescence, 99; natural resins, 98; synthetic resins, 98
Mounting sections: mounting frozen sections, 388; ringing cover glasses or mounting two cover glasses when using aqueous mounting media, 100; serial sections, 66; slides and cover glasses, 63; techniques, 64
Mucin metachromasia: thionin, 273; toluidine blue, 272
Muscle staining, 140
Mussels, 427
Myelin staining: chromic acid hematoxylin, 176; Luxol fast blue, 172; Luxol fast blue-Holmes silver nitrate, 174

Naoumenko-Feigin method for silver impregnation of reticulum, 146

Naphthalene yellow for lipid staining, 250
Naphthol solution, 234
Naphthol yellow S reagent, 191
Nassonov-Kolatchew method for staining Golgi apparatus, 291
Natural dyes used in staining: cochineal and carmine, 84; hematoxylin, 84; use of mordants, 85
Natural resins as mountants, 98
Nauta-Gygax method: for axons, 167; for degenerating axons, 177
Navashin fixative (for cell inclusions), 34
Near-field scanning optical microscopy (NSOM), 81
Negri bodies, 322
Nemathelminths, 426
Nematodes, 426
Nemertines, 426
Nerve cells, processes, and fibrils: Bielschowsky method for blocks, 174; Bodian method, 165; Cole method, 170; fluorescent method, 164; Glees method, 168; Nauta-Gygax method, 167; pyridine-silver method, 163, 171; Ramón y Cajal method for blocks, 160; silver impregnation for nerve tissue, 161
Neurosecretory substance (NSS). See Pituitary cells
Neutral buffered formalin (NBF), 28
Neutral dyes, 89
Neutral red stock solution, 300; neutral red-janus green method for supravital staining, 299
Nile blue method, 219
Ninhydrin reaction, 231; ninhydrin-Schiff reaction, 232
Nissl substance staining: cresyl violet, 158; gallocyanin, 160; thionin, 159
Nitro BT method for succinic dehydrogenase, 355
Nitrocellulose, 9, 337; embedding method (celloidin technique), 391
Nitroprusside solution, 358
Nolte method for making permanent mounts of chromosome squashes, 413
Norenburg-Barrett, polyester wax method of embedding, 394
Normal solutions, 461
Novelli short acid fuchsin method, 297
NSS (neurosecretory substance). See Pituitary cells
Nuclear fast red (Kernechtrot), 215

Nucleic acid and nucleoprotein staining, 236; pyronine-methyl green, 238, 239; thionin-methyl green, 240; nucleoprotein staining: —blocking protein end groups and reactions for amines, 243; —control slide techniques, 241; —extraction techniques, 241
Nudibranchs, 428

Occult iron, 214
Ocular micrometer, 73
Oil immersion objective, 71
Oil red O for lipids, 248
Orange G, 107, 141
Orcein, 135
Ordway-Machiavello method for staining rickettsiae, 320
OSHA rules and regulations, 9
Osmium black methods for enzyme histochemistry, 360
Osmium tetroxide (osmic acid), 9, 22, 32; lipid staining, 250; staining Golgi apparatus (Ludford method), 290; staining mitochondria, 296
Ova, worm, 426
Oxalic acid, 147
Oxidase, Laidlaw dopa oxidase method, 336

Pachytene chromosomes, squashes, 414
Palkowsky method for preserving specimens in a pliable condition, 484
Pancreatic islet cells, chromium-hematoxylin-phloxine for, 286
Pantin method for Mallory staining, 122
PAP immunostaining. *See* Peroxidase-antiperoxidase [PAP] immunostaining method
Papanicolaou method for exfoliative cytology, 406
Pappenheim stain (modified) for rickettsia, 318
Paraffin: embedding, 51; infiltration, 50; sectioning and mounting, 58
Paraformaldehyde fixation for histochemistry, 329
Paramecia, 439
Parasites, animal: Kohn stain, combined fixation and stain, 443; Lawless rapid method, 445; preparing concentrate smears, 442; smear techniques for intestinal protozoa, 442
Parloidin, 391

PAS. *See* Periodic acid-Schiff technique
PBS (phosphate buffer saline used in immunocytochemistry, 365, 477
Pearse method of gelatin embedding for frozen sectioning, 389
Penfield modification of del Río-Hortega silver carbonate method, for glia, 155
Peracetic acid, 225
Perenyi fixative for cell inclusions, 35
Perenyi used to fix and decalcify, 42
Performic acid, 225
Periodic acid-Schiff (PAS), feulgen techniques, and related reactions, 182; PAS techniques (Schiff reactions) aqueous, 183; alcoholic, 186; for carbohydrate and carbohydrate-protein, 185; periodic acid, 9. *See* Schiff reactions
Perl's iron stain (Kok-Boon modification) with microwave, 383
Permanent mounts: Smith method, 413; Nolte method, 413; other permanent methods, 413; restoration of deteriorated slides, 414
Permanganate solution, 280
Peroxidase anti-peroxidase (PAP) immunostaining method, 365; PAP complex, 363, 365; PAP pre-embedding immunostaining for use with electron microscope, 371
Peroxides, 9
Phase microscopy, 75
Phenylhydrazine, 244, 374
Phloxine-Methylene blue, for bone marrow staining, 211; phloxine solution, 211
Phosphate buffer $0.2M$ (Gomori), 473; phosphate buffer (Sörensen), 473
Phosphate-buffered fixative, 329
Phosphate buffered saline, 477
Phospholipid staining: acid hematein, 251; Luxol fast blue, 172; Sudan stains, 250
Phosphomolybdic acid, 279; phosphomolybdic/phosphotungstic acid solution, 399
Phosphotungstic acid-hematoxylin, 114; for astrocytes, 157; phosphotungstic acid–eosin stain, 323
Physiological buffer, 329
Physiological solutions (balanced salt solutions [BSS]), 464; Hank solution, 465; Earle solution, 465; physiological saline, 465; Ringer solution, 466
Picric acid, 9, 23
Picro-Ponceau with hematoxylin, 131–32

Pigment removal, 224; formalin pigment, malarial, and carbon, 226; hemosiderin and bile, 227; melanin, 224
Pigments and minerals, staining, 214
Pinacyanol method for rapid staining of frozen sections, 390
Pituitary cells, 276: Cameron-Steele (Ewen modification) for NSS, 281; chrome hematoxylin for NSS (Bargmann modification), 284; Heidenhain Azan, 278; Herlant pituitary stains I and II, 279, 280; lead hematoxylin method, 283
Planaria, 425
Platyhelminths, 425
Polarizing microscopy, 77
Pollak rapid method trichrome for connective tissue staining, 128
Polychroming, 89
Polyester wax, 395; sectioning, 393
Polystyrene, 335
Ponceau de xylidine, 126
Porifera, 424
Postfixation treatments, chromatization, deformalization, 40
Potassium cyanide, 428
Potassium dichromate, 24
Potassium ferricyanide, 216, 221
Potassium iodide, 155
Potassium permanganate, 282
Preparation for thin blood smears, 193
Preparation of invertebrates for whole mounts and sections, 422
Prescott-Carrier method for making slides of protozoa, 439
Preserving gross specimens in a pliable condition, Palkowsky method, 484
Procedures for autoradiographs, 450; staining: —Giemsa stain (Gude et al. modification), 455; —Mayer hemalum or Kernechrot, 455
Progressive and regressive methods: method I, excess mordant, 87; method II, acids, 87; method III, oxidizers, 87; use of accentuators and accelerators, 88
Propanediol stock solution, 343
Propylene oxide, 372
Propylene phenoxetol for anesthetizing invertebrates, 423
Protargol solution, 165
Protein staining: histone protein, 233; mercuric bromphenol glue, 230; Millon reaction: —Gomori, 228; —Romeis, 229; ninhydrin reaction, 231; ninhydrin-Schiff reaction, 232
Proteins and nucleic acids, staining of, 228
Protozoa, methods for handling, 438; Borror Nigrosin stain, 440; Chen cover glass method, 438; Merton method, 439; Prescott-Carrier method, 439; sectioning, Stone-Cameron agar method, 441; whole mounts, 440
Prussian blue reaction for staining iron, 214
Pyridine-silver method for muscle, 171
Pyronine-methyl green for nucleic acids: Elias, 239; Kurnick, 238

Quick toluidine blue metachromatic staining of mast cells, 275

Radiolaria and forminifera, dry mounts, 442
Ramón y Cajal method for blocks, 160; pyridine-silver method for blocks, 163; reducing solution, 292
Rapid Papanicolaou staining method with microwave, 384
Rapid staining methods for frozen sections: hematoxylin and eosin, 389; Humphrey, 390; Pinacyanol, 390; thionin or toluidine blue, 390
Reclaiming and storing specimens, 483; biopsy material, 483; dried gross specimens, discussion of methods, 483; Graves method, 483; Sandison method, 484
Recovering broken slides, 482
Regaud fixative for cell inclusions, 35
Removal of pigments: bile pigments, 227; carbon, 224, 226; hemosiderin, 227; melanin pigments: —bromine, 225; —chlorate, 225; —chromic acid, 225; —performic or peracetic acid, 225; —permanganate, 225; —peroxide, 226
Removing laboratory stains from hands and glassware, 486; cleaning solution for glassware, 487; suggested treatments for removing frequently used laboratory stains, etc., 486
Reptiles and fish, histochemistry and special procedures, 447
Research microscope operation, 70
Resins, mounting media, natural, synthetic, 98
Restaining faded slides, McCormick method, 481

Restoration of deteriorated permanent slides made by squash method, 414
Restoring basophilic properties, 482
Rhodamine-isothiocyanate, 370
Rickettsiae and inclusions staining: Castañeda method, 318; Giemsa method, 320; Massignani and Malferrari for negri bodies, 322; modified Gomori for trachoma, 321; modified Pappenheim stain: —Ordway-Machiavello method, 320; —Schleifstein method for negri bodies, 322
Ringer solution, 466
RNA removal, 241
Romanovsky stain, 89
Rosin stock solution (colophonium), 480; working solution, 480
Rossman fixative for cell inclusions, 35
Rotiferae, 426
Routine fixatives and fixing procedures: AZF fixative, Bouin, 27; Bouin-Duboscq formalin, Gomori, neutral buffered formalin, 28; glutaraldehyde, 328; Helly (Zenker formol), Hollande Bouin, Orth, 29; Stieve, Susa, Zenker, 30
Ruffer's solution, 484

Saccharides, 263
Safety in the laboratory, 7; handicapped students, 11; resources, 12; safety regulations, OSHA, 9
Safranin, 21
Sakaguchi reaction, 233
Salt (saline) solutions, 464; buffered saline, 477
Sandiford stain, 304
Sandison method for reclaiming gross dried specimens, 484
Sanfelice fixative for cell inclusions, 35
Saxena method for staining Golgi apparatus, 292
Scarba red for red nuclear staining, 117
Schaudinn fixative for special techniques, 36; Schaudinn-PVA fixative, 445
Schiff reactions: deoxyribonucleic acid (DNA) fluorescent technique, 189; feulgen rection, 187; PAS aqueous technique, 183; —and alcoholic, 186; triple stain for DNA, polysaccharides, and proteins, 190
Schiff reagent, 182; solution, 184
Schleifstein method for negri bodies, 322
Schmidt fluid fixes and decalcifies, 43

Schmorl's technique for enterochromaffine, 267
Scientific supplies, 487
Scott solution, 464
Sea anemones, 424
Sea water, artificial, 466
Sectioning, 58; difficulties and remedies, 60; sectioning for histochemistry: —Adamston-Taylor cold knife technique, 306; —sectioning polyester waxes, 393; serial sections, 66
Sex chromatin slide preparation: Guard's method, 409; Klinger-Hammond method, 408
Sharpening microtome knives, 57
Short acid-fuchsin method (Novelli) for staining mitochondria, 297
Sidman modification for ester wax embedding, 394
Silver compounds, 9; silver nitrate solution, 146
Silver impregnating reticulum, 143; Gomori method, 149; Gridley method, 148; Naomenko-Feigin method, 146; silver impregnating and staining neurological elements, 153; Wilder method, 150
Silver-intensified immunogold immunofluorescence staining, 373
Silver staining of glia: Holzer method, 155; Penfield modification of del Rio-Hortega silver carbonate method, 155
Sinha fixative for cell inclusions, 36
Slidder iron hematoxylin, 104
Slides and cover glasses, mounting sections, 63
Small tissue samples, handling, 448
Smith fixative for special techniques, 36; method for making permanent mounts of chromosome squashes, 413
Smith sodium thiosulfate solution (hypo) 5%, 464
Snails, 427
Sodium metabisulfite, 280
Solubility of stains, table, 477
Soluble enzyme immune complex (unlabeled antibody) methods, 365; avidin/peroxidase immunostaining, 367; indirect immunofluorescence for immunostaining, 370; PAP pre-embedding immunostaining for use with electron microscope, 371; peroxidase anti-peroxidase (PAP) immunostaining method, 365;

silver-intensified immunogold staining, 373; and using free-floating section method, 369
Solution preparation; buffer solutions, 469; environmental solutions, 466; general purpose solutions, 463; molecular solutions, 461; normal solutions, 461; percentage solutions, 461
Solutions, general purpose: acid alcohol, 463; alkaline alcohol, 463; carbol-xylol, 463; Chatton agar for blocking small organisms, 464; gold chloride stock solution, 463; Lugol solution, 463; physiological solutions (balanced salt solutions [BSS]), 464; Scott solution, 464; sodium thiosulfate solution (hypo) 5%, 464
Sörensen phosphate buffer, 473
Special techniques, fixatives for, 31
Specialized microscopy: confocal, 82; electron, 79; fluorescence, 78; interference, 76; near-field scanning optical microscopy (NSOM), 81; phase, 75; polarizing, 77
Spirochete staining: Dieterlie method, 308; Leviditi method for block staining, 311; Warthin starry silver method, 310
Spring water, synthetic, 467
Spurr's resin, 373
Stain solubility table, 477
Staining: of cellular elements, 264; procedures (*see specific names*); by progressive and regressive methods, 87, 88; with two different stains on one slide (Feder and Sidman method), 482
Stains and staining action, 84; nature of, 92; standardization of stains, 93
Standard buffer (McIlvaine), 474
Steedman method for ester wax embedding, 393
Sterling gentian violet, 303
Stieve fixative, 30
Stone-Cameron agar method for agar sectioning, 441; invertebrates, histochemistry of, 430 (*see* Invertebrates, tips on special handling for histochemistry)
Storing gross specimens, 485
Straus method for peroxidase, 358
Strepavidin, 367
Stripping film method for autoradiography, 452
Structure of synthetic dyes, 91
Student microscope operation, 69

"Subbed" slides, 469
Substrate film methods for enzyme histochemistry, 359
Succinic dehydrogenase, nitro BT method, 353, 355
Sudan black B method for phospholipids (Elftman), 250; for staining Golgi apparatus (Baker), 294; Sudan black B or Sudan IV for phospholipids (Chifelle and Putt), 248
Sulfurous acid, 173
Suppliers: automatic cover slippers, 489; disposable knife blades, 489; distillers, 489; dyes and chemicals, 490; electron microscopy supplies, 488; embedding centers, 488; immunohistochemicals, 489; knife sharpeners, 488; major laboratory supply companies, 487; microscopes, 487; safety supplies, 489; slide stainers, 488; tissue processors, 488
Supravital staining, 299; neutral red-janus green method, 299; vital-nonvital stain, 301
Susa fixative, 30
Synthetic spring water, 467

Tandler extraction method for nucleic acids, 242
Tetrahydrofuran (THF), 9; for dehydration, 49
Tetrazolium, 354
Thin smear method, Wright stain, 195
Thionin for Nissl substance, 159; metachromatic staining, 271; acid mucopolysaccharides, 273; mast cells, 274; mucin, 273; thionin method for rapid staining of frozen sections, 390; thionin-methyl green for nucleic acids, 240
Thyroid cells, fixing and staining methods, 288
Tissue preparation, other methods, 43
Toluidine blue method for rapid staining of frozen sections, 390; metachromatic staining of mast cells, 275; mucin, 272
Trematodes, 425
Trichloracetic acid, 232
Trichrome staining, 126; Gomori method, 126; Gomori one-step trichrome, 129; Masson trichrome staining, 126; trichrome staining solutions, 128, 130
Triple stain for DNA, polysaccharides, and proteins, 190

Tris buffer, pH 7.19, 349; 0.2M (Hale), 475; tris buffered saline (TBS), 477
Triton X-100, 369
Trypan blue, 301
Turnbull blue method for ferrous iron, 216
Two different stains on one slide, Feder-Sidman method, 482

Uranium nitrate, 150

Van Gieson substitute, nonfading, 131
Verhoeff elastin stain, 133
Vernier, reading, 73
Vibratome, 371
Villaneuva fixative for decalcification, 41
Viruses, 317
Vital-nonvital stain, 301
Von Apathy gum syrup, 99
Von Kossa method for staining calcium, 222

Walpole acetate-acetic acid buffer, 471
Warthin-Starry silver method for staining spirochetes, 310
Weaver gelatin fixative, 468
Weigert: iodine, 463; iron hematoxylin, 103
Whole mounts: glycerine jelly mounts, 433; —of hydras, embryos, and flukes, 435, staining: —cochineal-hematoxylin, 437; —grenacher borax carmine, 435; —hematein, 437; —Mayer carmalum, 436
Wilcox thick film method for blood films, 200
Wilder method for silver impregnation, 150
Wright stain for blood: staining solution, 195; —for birds, 197; —for cold-blooded vertebrates, 197; Wright-Giemsa stock solution, 200

Xylene (xylol): for clearing, 48; safety, 8

Yetwin mounting medium for nematodes and ova, 99

Zenker formol (Helly fixative), 29; Zenker fixative, 30; Zenker's fluid used for special techniques, 37
Ziehl-Neelsen type for acid-fast staining, 305
Zinc formalin (ZF) used for special techniques, 37
Zirkle: for dehydration, 46; fixative for osmic acid staining of mitochondria, 296

About the Authors

Gretchen Lyon Humason was a histologist at the Associated Universities in Oak Ridge, Tennessee, from 1963 until her retirement in 1989. Her career in tissue research spanned sixty years. She died in 1991 at the age of 84.

Janice Keller Presnell, B.S., H.T. (ASCP), is a histology technician with over twenty years of experience.

Martin Paul Schreibman, Ph.D., is Distinguished Professor in the Department of Biology, Brooklyn College, City University of New York. He has taught histology and microtechnique for more than thirty years. He uses histology and immunohistochemistry routinely in his research programs.

Library of Congress Cataloging-in-Publication Data

Presnell, Janice K.
    Humason's animal tissue techniques. —5th ed. / Janice K. Presnell and Martin P. Schreibman.
       p. cm.
    Includes bibliographical references and index.
    ISBN 0-8018-5401-6 (alk. paper)
    1. Histology—Technique.   2. Stains and staining (Microscopy) 3. Histochemistry—Technique.   I. Schreibman, Martin P.
II. Humason, Gretchen L. Animal tissue techniques.   III. Title.
QM556.P74   1997
578'.9—dc20   96-26473
                                                                 CIP
                                                                 r96